ENVIRONMENTAL TECHNOLOGY AND BIOSPHERE MANAGEMENT

ENVIRONMENTAL TECHNOLOGY AND BIOSPHERE MANAGEMENT

Har Darshan Kumar

Formerly Professor of Botany and Co-ordinator,
Biotechnology Programme, Banaras Hindu University
Varanasi, India

Science Publishers, Inc.
Enfield (NH), USA   Plymouth, UK

CIP data will be provided on request.

SCIENCE PUBLISHERS, Inc.
Post Office Box 699
Enfield, New Hampshire 03784
United States of America

Internet site: *http://www.scipub.net*

sales@scipub.net (marketing department)
editor@scipub.net (editorial department)
info@scipub.net (for all other enquiries)

ISBN 1-57808-194-7

© 2002, Copyright reserved

The chapter on Urbanisation (chapter 9) draws on the material presented in the book Hardoy, Jorge E, Diana Mitlin and David Satterthwaite (1992), *Environmental Problems in Third World Cities*, Earthscan Publications, London. The author gratefully acknowledges the kind consideration shown by Hardoy et al. for allowing the author to use the material from their book.

Published by Science Publishers, Inc., Enfield, NH, USA
Printed in India.

Preface

In the scheme of things, the Earth is just an inherently changeable and small planet in space. Life on Earth has played an important role in shaping its physical and chemical nature. The new global environment requires new approaches, novel ideas and innovative technologies to address the emerging challenges in areas as varied as climate change and genetic engineering.

The twentieth century has brought about a deluge of technical inventions, innovations and developments, possibly more than humankind has ever experienced over the millennia. The fundamental changes of civilization and society involve almost every human being on the planet and impact diverse aspects in daily life, technically as also environmentally.

The previous decade witnessed a growing concern over the extent of soil degradation in some parts of the world. This poses a threat to agricultural production. However, while the health and productive capacity of soils in some countries presents negative nutrient balances, the picture is more varied at field and farm level where farmers have developed a variety of management strategies. Appropriate measures to support better soil and farm husbandry are needed to assess the happenings at the farm and field level. The management of soil fertility varies amongst different fields, farmers and locations, and this diversity has implications for making effective interventions with the objective of improving the management of soil fertility.

Conflict over environmental resources such as land, water and forests has for long been widespread. The dimensions and intensity of conflict can vary greatly. So too can the opportunities for resolving the conflicts. Today, the information revolution taking place in the industrialized

North is transforming the developing nations of the south. Countries that only a decade ago seemed rather remote, now have acquired the potential to become active players in the process of global development. Information, communication and environmental technologies have generated hopes for many less-developed countries, particularly for participating in the emerging era of precision (site- or field-specific) agriculture. Technologies are now available for laying a strong interdisciplinary foundation for conserving and perpetuating biological resources of crop plants within the systems in which they have evolved.

The turbulence the world is currently experiencing is not just because of the beginning of the Information Revolution, fundamentally, it signals the end of the Baconian Age, which began four centuries ago when Sir Francis Bacon postulated his concepts of progress and development—concepts that have shaped human endeavours ever since. The end of that era means much to the future of technological development everywhere and also to the ways in which we manage the Earth's environment in the coming few decades has slipped out of the ecological balance and the environment, becoming a wasteful killer of life. Population growth is the single most important factor involved in increasing environmental stress. Our use of fossil fuels continues to grow. If ever declines in the coming years, it will be more than offset by an increase in coal burning, thereby further increasing the levels of greenhouse gases and particulate matter in the atmosphere. Anthropogenic deeds (and misdeeds) committed largely, but not entirely, in several industrially-advanced countries in the twentieth century have outcompeted nature in changing the global environment.

But the situation is not beyond hope. Improved management of natural resources in the light of proper engineering, technological and business practices combined with both the older, indigenous knowledge and the modern agroindustrial technologies, biotechnology, informatics, and micro-electronics can go a long way in preventing further damage and restoring the earth system by introducing feedbacks on climate which can help stabilize and regulate its environment.

While discussing the above issues, I have attempted to introduce appropriate technologies that are compatible with the environment and which assure an optimal utilization of resources, thereby contributing to socioeconomic development. In the past, 'appropriate' technology used to be considered under the 'small is beautiful' frame. It has now advanced to a level expressing what technology should really be about anyway—a means of enhancing the livelihood of humankind in tune with a healthy and clean environment.

One of the major challenges facing mankind is to ensure sustainability in the field of energy consumption. This requires improved management of planetary ecosystems and a substantial reduction of the noxious emissions. In order to feed the growing number of people in the coming years, the current model of large-scale, industrialized monoculture will have to give way to, or be complemented by, diversified, precision agriculture that can meet the needs of marginal and small farmers. Besides these aspects, some other issues related to the environment, e.g. transportation, climate, energy, water, air, industry, technology and waste management have been reviewed in a similar context.

This book could not have been completed without help and cooperation from a large number of authors, editors, societies, agencies and other professionals who have supplied monographs, journals, reprints, newsletters and other literature.

I wish to thank the Council of Scientific and Industrial Research, New Delhi (Dr R.A. Mashelkar), the Indian National Science Academy, New Delhi (Drs S. Varadarajan and G. Mehta), Dr J.C. van der Leun, Chair, UNEP Ozone Effects Panel, Utrecht (The Netherlands), and the Korean Society for Applied Microbiology, Seoul, for their help, cooperation and encouragement in various ways. Several drafts of the manuscript were patiently and ably composed by Mangala Prasad Dubey.

14 February, 2001   *HD Kumar*

Contents

Preface — v

1. People–Environment Interfaces — 1

Introduction *1* Science for Basic Human Needs *6* Climate Changes *7* Agroecology *10* Ethnoecology *11* Human Activities and the Environment *12* Human Demography *15* Enlightened Use of Energy and Materials *22* Economic Instruments and Environmental Problems *24* Global Change *28* People, Tech-nology and the Environment *32* Climate, Weather and Health *35* Health Impacts of Haze-Related Air Pollution *37* Animal Farming and Greenhouse Gases *38* Vegetation Fires and Global Climate *39* References *43*

2. Water — 45

Introduction *45* The Hydrologic Cycle *47* Water as a Life-Sustaining Substance *49* Groundwater *53* Sustainable Exploitation of Groundwater Resources *57* Water Quality *58* Water Management *59* In-house Water Conservation *61* Water for Cities *62* Water Resources and Environ-mental Planning *64* Water and Climate *66* Water and Agriculture *68* Water and Energy *71* Water and Industry *72* Water and Health *74* Global Issues Related to Continental Aquatic Systems *76* Sustainable Management of Water Use *79* Aquatic Recovery From Acidification *84* Protection and Management of Tropical Lakes *85* Conclusions *96* References *96*

3. Atmosphere and Climate Change — 99

Introduction *99* Water Vapour *107* Water Vapour and Climate *108* Holistic Approach and Biospheric Processes *109* Role of the Stratosphere in the Climate System *113* Man-Made

Climate Change *115* Global Climate Change *118* The Active Carbon Cycle *120* Greenhouse Gases and Climate Change *122* Recent Assessment Reports of Inter-governmental Panel on Climate Change *124* Climate Change and Carbon Dioxide *128* El Niño *134* Sustainable Forestry for Carbon Mitigation *138* Biological Impacts of Global Climate Change *140* Global Warming *147* Vulnerabilities and Potential Consequences *148* Clouds and Atmospheric Radiation *150* Clouds and Aerosols *155* World's Webs *159* Prevention of Climate Changes *160* References *161*

4. Energy and the Environment   165

Introduction *165* Energy Resources, Reserves and Availability *168* Renewable Energy *175* Energy and Development *180* Future of Renewable Energy *182* Energy Events of the 1970s *186* Bioenergy *195* Rational Use of Energy in Developing Countries *201* Co-generation Power Plants and Efficient Energy Utilization *203* Energy Wastage in the Steam Boiler *204* Photovoltaic Systems *204* Local and Regional Energy and Environmental Issues *207* Coal Industry *208* Energy Demands *210* A Flame-free Future *214* Energy and Environmental Recovery *216* Energy Use and Pollution in Asia *217* Air Pollution Control *218* Transportation *219* Energy Transition *220* Energy Modelling, Economics and Environment *221* References *226*

5. Industry, Energy and Technology   228

Introduction *228* Clean Production *231* Leather Industry *237* Technological Dynamism in Asia *238* Energy *239* Fossil Fuels and the Environment *241* Energy Choices and Technological Change *244* Energy Myths *245* Conversion Technologies *247* Ecologically-sustainable Industrial Development for Asia and the Pacific Region *248* Solar Energy and Photovoltaic Pumps *250* Plant Oil for Use in Engines *253* Technology Transfer *255* Non-energy Benefits of Industrial Energy Efficiency *258* References *260*

6. Rural Development   262

Introduction *262* Rural Transformation Through Appropriate Technology *264* Appropriate Energy Technology and Photovoltaics *269* Rural Labour *271* Biogas Production Technology *272* Solar Cooking *277* Plants as Energy Source *277* Culture and Biodiversity *279* Agrobiodiversity *283* Rural Development in Third World *284* Integrated Rural Develop-ment *287* References 288

7. Wastes and Pollution   289

Introduction *289* Environmental Engineering, Waste Disposal and Resource Recovery *291* Wastewater Technology *294*

Activated Sludge Plants for Enhanced Suspended Solids Removal *296* Optimization in Wastewater Treatment Technology *299* Urban Wastewater Use for Plant Biomass Production *302* Assessment and Management of Hazardous Wastes *302* Hazardous Waste Treatment *304* Non-point Sources of Pollution *304* Pollution, Sanitation and Human Health *306* Major Sources of Pollution *307* Combating Pollution *308* Recycling Human Waste *311* Treatment of Sanitary Landfill Leachates *314* Bioremediation *319* Phyto-remediation *325* Plastics Recycling and Waste Management *328* References *329*

8. Animal Husbandry and Wildlife Management 333

Introduction *333* Livestock Diversity *337* Animal Husbandry and Sustainable Tropical Farming Systems *339* Animal Husbandry and Global Warming *341* Animal Husbandry and the Environment *345* Animal Nutrition *349* Impact of Animal Breeding on Conservation of Natural Resources *355* People-Wildlife Interactions and Wildlife Management *357* Threatened Mammals *360* Nature Conservation, Hunting Tourism and Sustainable Development *361* References *363*

9. Urbanization 366

Introduction *366* Urbanization as a Global Process *367* Urban Ecodevelopment *369* Environmentally sustainable Transportation System *376* Emissions from Sea Transport *377* Urban Agriculture *379* Consequences of Urbanization *383* Humane Urban Development *390* Urbanization and Sustainable Development in Developing Countries *391* Environmental Problems *392* Indoor Environment *395* Neighbourhood Environment *396* City Environment *396* Regional Impacts and Rural-Urban Interactions *399* Global Impacts *400* Sustainable Development as a Transformation Process *401* Rural-Urban Interfaces *403* References *406*

10. Biodiversity, Biotechnology and Food Security 408

Introduction *408* Agriculture, Ethics and Equity *409* Beyond Paradox *411* Protecting the Environment *412* Systems of Biodiversity Management *413* Ethics of Humanity's Common Future *416* Land Degradation *416* Food Security *417* Future Food Supply Prospects *422* Global Food Prospects to 2025 *425* Biodiversity, Food Security and Indigenous People *427* From Seeds to Genes (1970s to 1990s) *430* Biodiversity and Agroforestry *431* Forests and Farming Societies *433* Modern Farming and Forests *438* Food, Farming and Biodiversity *439* Agricultural Biotechnology *441* Precision Farming *453* Transgenic Plants and Pest Management *455* Integrated Pest Management *462* Integrated Disease Management *464* References *465*

xii Environmental Technology and Biosphere Management

11. **Sustainable Development** — 469

   Introduction 469 Definitions and Concepts 472 The Trilogy of Sustainability 476 Goals of Sustaining Natural Systems 477 Sustainable Development and Public Policy 478 The Demographic Imperative 481 Perspectives and principles 482 Traditions 483 Resistance to Destruction 485 Transition Towards Sustainability 486 Sustainable Management of Ecosystems and Natural Resources 488 Resource Management Sustainability 490 Reform and Redevelopment of Degraded Regions 492 Geochemistry and Sustainable Development 493 Sustainable Development and Nuclear Energy 495 Sustainable Development at the Local Level 496 Transition to Sustainable Society 499 The Environment-Development Interface 502 Rural Poverty 504 Environmentally-Sound Development 505 Technological Change and Global Society 506 Systems Analysis for Sustainable Development 507 Energy Use and Sustainable Development 510 Ethics of Sustainable Development 514 Tomorrow's Sustainable Society 518 References 520

12. **Biosphere Management** — 525

   Introduction 525 Valuation of Ecosystem Services 527 Mineralogy and Human Welfare 530 Trenchless Technologies 530 Environmental Technology Performance Verification 531 Bottom Sediments and the Secondary Pollution of Aquatic Environments by Heavy Metals 532 Some Passive Systems for the Treatment of Acid Mine Drainage 533 Mechanisms of Contaminants Removal 535 Types of Passive Treatment Systems 535 Water Hyacinth 536 Mountain Resources 538 Traditional Resource Management 539 Conservation and Management of Resources 541 Assessment of Forest Condition 544 Forest Management 544 New Forestry 548 Tropical Forest Synergies 549 Forest Disturbance and Land Cover Impacts in Southeast Asia 551 Managing Excess CO_2 553 Total Assessment Audits 554 Human Health Effects 556 Environmental Health and its Management 558 Indigenous Knowledge for Sustainable Development 561 Pollution Prevention 562 Human Settlements 565 People-centred Development 568 Human Development: The Challenges Ahead 568 Enclosed Ecosystem (Biosphere 2) 573 References 574

Abbreviations and Acronyms — 577

Glossary — 583

Index — 609

Chapter 1

People–Environment Interfaces

INTRODUCTION

In the finite world in which we live, the ever-increasing numbers of human beings are steadily consuming, destroying, or otherwise using up natural resources. Essentially, we are behaving as badly-adapted parasites or uncontrolled cancer cells. We tend to lose sight of the fact that the world is unique. We have no other habitable planet to invade and parasitize—and no other place in which to try again.

The three chief underlying causes of most environmental issues are population, affluence and technology.

Demographic trends are undoubtedly going to have a very marked bearing on future development. The population in many European countries is aging fast, while that in the developing countries will continue to increase unabated for many decades to come in spite of declining fertility rates. This implies not only population growth, but a young, outward looking South–facing an old, inward looking North.

Very often, protecting the environment and promoting economic development seem to be antithetical. The notion persists that a choice must be made between the two. The situation is compounded by a lack of institutions to encourage their integration, the absence of practical models to show the way and the little amount of information to help people see things differently. With very few exceptions, neither governmental agencies nor non-government organizations have succeeded in integrating protection of the environment with development programmes designed to improve the livelihoods of the rural poor.

Although the real test of meshing economic growth with the sustainable use of resources takes place at local and national levels, international activities have become increasingly important during the last decade. This has partly been the result of the threat of climate change and the thinning of the ozone layer, both of which will require innovative international agreements. Growing global interdependence and trade also have an important effect on resource and energy use at local levels. There is mounting concern that resource shortages, particularly of water, could undermine national security in some regions such as the Middle East.

As the world grapples with climate change, acid rain, ozone depletion, and other global environmental problems (see Table 1.1), it is being compelled to confront issues traditionally regarded as part of international relations and security. Tensions between industrialized and

Table 1.1 Evolution of public involvement in the environmental field

Events of the Decade	The Public's Mood
1950—Economic Optimism and Growth	
Post-War era: Rise in industrialization and standard of living	Trust in government and industry. Focus on growth, not environmental impacts
1960—Rise in Environmental Awareness	
Silent Spring by Rachel Carson, 1962	Awareness of interrelationship of people and the environment.
Vietnam War	Protests and demonstrations.
Man lands on the Moon—1969	Recognition of the Fragility of Spaceship Earth
1970—Era of Regulation	
Earth Day, 1970. Enactment of environmental laws: (USA, developed world)	Demand to protect the environment.
Chipko movement	Appreciation of importance of forests and trees
1980—Era of Intervention and Disillusionment	
Ozone hole; CFCs responsible for ozone thinning	Montreal Protocol (agreements)
Environmental disasters: Bhopal, Chernobyl	Distrust in industry
Awareness of potentially harmful effects of byproducts of technology	Demand for voice in policy-making.
Stringent regulations require closure of existing facilities, increase need for siting new ones	Focus on pollution prevention and waste minimization
1990—Era of Enlightenment and Creativity	
Earth Day—1990	Global perspectives on environ-mental crisis
Complexity of environmental issues increases: acid rain, global warming	Willingness to pay more for environmentally-safe products

developing countries and questions of national sovereignty have become intertwined with the new generation of environmental problems.

In the span of a single human generation, the Earth's life-sustaining environment has changed more rapidly than it has over any comparable period of human history. Today, the scale of human interventions in nature is increasing and the physical effects of our decisions are spilling across national borders. The growth in economic interaction between nations amplifies the wider consequences of national decisions. Economics and ecology bind us in ever-tightening networks. Today, many regions face risks of irreversible damage to the human environment that threaten the basis for human progress. Worldwide economic and technological activities are contributing to rapid and potentially-stressful changes in our global environment in ways that we are only now beginning to understand. The effects of these changes may profoundly impact the generations to come. Natural forces have affected and shaped the Earth's environment over the course of its lifetime. The uniqueness and challenge posed by the changes facing us today lie not only in the magnitude and rate at which these changes are occurring, but also in humankind's ability to inadvertently affect such changes. Increasing atmospheric concentrations of greenhouse gases, due partly to the burning of fossil fuels, significantly alter our climate. Agriculture, forestry and other land-use practices, industrial activities, waste disposal and transportation have altered terrestrial and coastal ocean ecosystems, thus affecting, for instance, biological productivity, water resources and the chemistry of the global atmosphere. These fundamental changes, evident in the decline of stratospheric ozone and in acid precipitation too extend beyond the conventional boundaries of scientific disciplines and have potential impacts that reach beyond the domains of individual countries.

There is growing awareness throughout the world of the necessity to preserve the global ecological balance better. This includes serious threats to the atmosphere, which could lead to future climate changes. Air, lakes, rivers, oceans and seas are becoming increasingly polluted resulting in hazards such as acid rain, dangerous waste substances and rapid desertification and deforestation. Such environmental degradation endangers species and undermines the well being of individuals and societies.

Decisive action is urgently needed to understand and protect the Earth's ecological balance. Human societies must work together to achieve the common goals of preserving a healthy and balanced global environment in order to meet shared economic and social objectives as also carry out obligations to future generations.

Several interactions of biological, chemical and physical processes govern changes in the Earth system which are most susceptible to human perturbation. There is need to develop a predictive understanding of the Earth system, especially in relation to changes that affect the biosphere.

Careful consideration of how natural and human forces contribute to global change has to be kept in mind with the following properties:

1. How is the chemistry of the global atmosphere regulated and what is the role of terrestrial processes producing and consuming trace gases?
2. How do the biogeochemical processes of the ocean influence and respond to climate change?
3. How does vegetation interact with physical processes in the hydrological cycle?
4. How will climatic change affect terrestrial ecosystems?

Atmospheric concentrations of several trace gases have been increasing rapidly, mainly due to human activities. Increases in greenhouse and other trace gases result in changes in the composition of the atmosphere, thus affecting physical aspects of the climate system. Global changes are brought about by forces of both natural and human origin, the respective roles of which must be considered in the development of each research priority. In interpreting and understanding past records with regard to linkages among the physical and biogeochemical aspects of the climate system, the roles of solar and orbital variations, solid Earth phenomena such as vulcanism, and relatively recent human-induced disruptions in the major elemental cycles must be included. Human activities are likely to be the dominant force in the near future, although we must allow for yet unknown cycles in solar activity and episodes of volcanic activity. Climate change and increased nutrient deposition from the atmosphere directly affect soils, plant productivity, vegetation structure and species composition. Temporal and spatial patterns for temperature, precipitation and the occurrence of extreme weather events effect not only natural ecosystems, but also exert regional constraints on agriculture and forestry. Indirectly, climate change can also affect biogeochemical cycles with important consequences for ecosystems. Major changes in ecosystem extents can have important feedback effects on physical aspects of the climate system through associated changes in the roughness and reflectivity of the Earth's surface.

A multiplicity of environmental problems weighs on the future of our planet. Beyond the well-known phenomena of population growth and increasing urbanization, industrial, agricultural and transport activities significantly transform the global environment with serious consequences for human health and the productivity of ecosystems. Human activities have also started to affect the functioning of global life support systems such as the climate system. The need to adopt precautionary principles, initiate anticipatory research, take preventive action, and indeed, make sustainability an essential ingredient in any model of development, has become urgent at a time when societies, cultures, economies and environments are becoming more and more interdependent.

One notable feature of our times is the emergence of organized sectors of society demanding participation in democratic debates and decision-making, as well as transparency on all public issues. Alongside traditional actors such as trade unions and political parties, strong new groups are emerging, including the communication media, citizen movements and a variety of non-governmental organizations, such as associations of parliamentarians, industrial professions and entrepreneurs. Many of these are concerned with current environmental issues. Women as a majority of the world population are claiming an increased role in all walks of life.

Over time, many discoveries, applications and know-how that constitute an unprecedented source of knowledge, information and power have accumulated. Never have discoveries and innovations promised a greater increase in material progress than today but, neither has the productive—nor destructive—capability of humankind left un-resolved so many uncertainties. The major challenge of the twenty-first century will lie in the ground between the power which humankind has at its disposal and the wisdom which it can exercise in using it. There is an urgent need for profound changes of attitude and approach to problems of development, especially to their social, human and environmental dimensions. Science must be put to work for sustainable peace and development in a progressively responsive and democratic framework and scientists must correspondingly recognize their ethical, social and political responsibilities. Also traditional barriers between the natural and the social sciences need to be demolished and interdisciplinarity adopted as a common practice. Moreover, since the processes underlying present global problems and challenges need the concurrence of all scientific disciplines, it is imperative to attain a proper balance in their support.

The new communication and information technologies have already become an important element of change, giving rise to new directions, methodologies, scenarios and new ways of producing, accessing and using information. The growing impact and potential of the new technologies make it necessary for scientists and institutions to adapt themselves in order to fully benefit from the advantages they can bring about. Computing and information systems reflective of the diverse cultures, languages, technical resources, habits and wants of people around the world are the current need.

SCIENCE FOR BASIC HUMAN NEEDS

Food, water, shelter, access to health care, social security and education are cornerstones of human well-being. Poverty and dependence that affect several countries can only be escaped through social and economic transformation, political determination, a comprehensive and upgraded educational system, and the appropriate development and use of science and technology. Scientific knowledge needs to be applied to find ways and means of reducing the imbalance, injustice and lack of resources that particularly affect the marginalized sectors of society and the poorer countries in the world.

Science, Environment and Sustainable Development

One major challenge facing the world community in the twenty-first century is the attainment of sustainable development, calling for balanced inter-related policies aimed at economic growth, poverty reduction, human well-being, social equity and the protection of the Earth's resources, commons and life-support systems. There is no doubt that sustainable management as well as use of resources and sustainable production and consumption patterns, in general, are the only routes to meeting developmental and environmental needs of present and future generations. Scientific capabilities need to be enhanced to develop sustainably, with special reference to the following objectives: (i) to strengthen capacity and capability in science for sustainable development, with particular emphasis on the needs of developing countries; (ii) to reduce scientific uncertainty and improve the long-term prediction capacity for the prudent management of environment-development interactions; and (iii) to bridge the gap between science, the productive sectors, decision makers and major groups in order to broaden and strengthen the application of science.

Science and Technology

Science, technology and engineering are some of the chief drivers of industrial and economic development. Innovation in all sectors is increasingly characterized by bi-directional feedback between the basic research system, and technology development and diffusion. This is changing the requirements for successful technology transfer and upgrading of innovation capabilities in the developing countries, with implications for domestic policies and international cooperation. One of their main priorities must now be to promote the development of national scientific and technological infrastructures and corresponding human resources.

Science and Society

Science should serve humanity as a whole and contribute to improving the quality of life for every member of present as well as future generations. Those fields that promise to address issues of social interest should be placed high on the agenda. Different individuals, sectors or groups can have widely varying needs and requirements, according to parameters such as age, education, health, professional training, working place, living place, economic status, gender and cultural background. Identifying these diverse needs and finding possible ways to address and fulfil them, require the concerted effort of scientists from different disciplines. The new reciprocal commitment between science and society will require that the scientific community take account of these challenges and also that the cooperation mechanisms be resolute in promoting a strategy to meet them.

Modern science by no means is the only form of knowledge, and closer links need to be established between this and other forms, systems and approaches to knowledge, for their mutual enrichment and benefit. Traditional societies, many of them with strong cultural roots, have nurtured and refined systems of knowledge of their own, relating to such diverse domains as meteorology, geology, ecology, botany, agriculture, physiology, psychology and health. These knowledge systems represent an enormous wealth. Not only do they harbour information as yet unknown to modern science, but they are also expressions of alternate ways of living in the world, other relationships between society and nature, and other approaches to the acquisition and construction of knowledge. Special action is called for to conserve and cultivate this fragile and diverse world heritage, in the face of globalization and the growing dominance of a single view of the natural world as espoused by science. A closer linkage between science and other knowledge systems is expected to bring important advantages to both sides (see *The Declaration on Science and the Use of Scientific Knowledge. The Science Agenda — Framework for Action*. World Conference on Science, Budapest, 26 June to 1 July 1999).

CLIMATE CHANGES

The growing world population, expanding human activities and production processes since the Industrial Revolution, especially since the middle of the twentieth century, have resulted in the additional emission of certain naturally-occurring gases into the atmosphere (Table 1.2). These gases reduce the amount of heat radiation from the earth and cause the so-called greenhouse effect which may lead to global warming.

8 Environmental Technology and Biosphere Management

Table 1.2 Some more important greenhouse gases (after Müssig, 1994)

GAS	CO_2	CH_4	FREON 11	FREON 12	N_2O
Pre-industrial concentration (ppm)	280	0.8	0	0	280
Concentration today	360	1.7	280	480	310
Approx. residence time in the atmosphere (years)	50-200	10	65	130	150

The concentration of carbon dioxide, an important greenhouse gas, in the atmosphere has increased *inter alia* by the combustion of fossil energy carriers. Besides carbon dioxide, chlorofluorocarbons (CFCs), methane and dinitrogen monoxide also contribute to the greenhouse effect in the following proportions: Carbon dioxide (CO_2) 55%; CFCs 24%; methane (CH_4) 15%; and N_2O and others 6%.

Fossil energy carriers contribute to the greenhouse effect through their emission of CO_2 and CH_4, but they do not emit chlorofluorocarbons or N_2O.

Depending on the specific carbon content of fossil energy carriers, and in relation to the energy content utilized through combustion, different amounts of CO_2 are emitted (Table 1.3). The following are the values for different types of fuel in relation to their gross calorific values.

Table 1.3 CO_2 emissions of various fossil energy carriers (after Müssig, 1994)

	$kgCO_2/kwh$	%
Brown coal (lignite)	0.40	121
Coal	0.33	100
Crude oil (petroleum)	0.27	82
Natural gas	0.19	58

The climatic effect of the various greenhouse gases (GHG) is expressed in terms of a numerical factor, the Global Warming Potential (GWP), which indicates how much stronger the effect of a molecule of GHG is by comparison to that of a CO_2 molecule emitted at the same time. (The climatic effect of CO_2 is taken as 1 for this purpose.)

Since the rate of breakdown in the atmosphere varies from one GHG to another, the time span is an important factor for calculating the GWP.

Methane has a retention time of approximately 10 years in the atmosphere, whereas CO_2 has a mean span of 120 years. Therefore compared to CO_2, the GWP for methane decreases with time.

The following illustrates this decrease in the GWP of methane as compared to that of carbon dioxide in relation to the time elapsed:

Period (years) →	20	100	500
GWP →	23	7.6	3.3

If methane emission is added to the specific CO_2 emission, expressed as an equivalent of CO_2 emission, in terms of the GWP and related to a given time span, we arrive at the following values (see Müssig, 1994).

Year	1	20	50	100
		($kgCO_2$/kwh)		
Brown coal (lignite)	0.40	0.40	0.40	0.40
Coal	0.52	0.48	0.44	0.40
Crude oil (petroleum)	0.30	0.29	0.28	0.27
Natural gas	0.24	0.23	0.22	0.21

It can be seen that of all fossil energy carriers, natural gas has the lowest value in any time span examined. It follows that increased use of natural gas is one of the measures that could be taken to reduce CO_2 emissions greatly. Besides, natural gas also allows for a significant increase in the efficiency of energy conversion by using some modern technologies.

CO_2 emissions can be reduced by adopting certain primary and secondary measures. Some primary measures include avoidance of unnecessary use, reduction of specific energy consumption, improvement of specific energy conversion, energy recuperation and renewable energy sources.

Secondary measures are those operations which prevent the emission of CO_2 into the atmosphere but do not result in a saving of primary energy. These include: replacement of gas turbines or motors by electric motors (local improvement of CO_2 balance) and removal of CO_2 from natural gases and/or combustion gases, followed by depositing CO_2 on the sea floor or in depleted gas or crude oil reservoirs (Müssig, 1994).

The greater part of methane emissions usually occurs in local distribution and through end users. In this area, the current measures for emission reduction have proved effective. Some measures, applicable to all aspects of the natural gas economy, include replacement of gas by air in control systems used in gas production and processing; using flash gases in one's own heat generators; re-compression of flash gases and re-introduction into the process; intermediate storage of gas released during the loading and unloading of liquids; use of compressed air start-up devices for gas turbines; optimization of seals in compressors; and replacement of old supply lines made of steel and ductile casts. By adopting these measures, methane emission can be significantly reduced (Müssig, 1994).

AGROECOLOGY

Agroecology, the application of ecological science to the study of sustainable agricultural management, includes an analysis of socio-economic, ecological and cultural factors that affect the sustainability of rural communities. A sustainable agroecosystem is one that maintains the rsource base, upon which it depends or relies on a minimum of artificial inputs from outside the farm system; manages pests and diseases through internal regulating mechanisms; and is capable of recovering from the disturbances caused by cultivation and harvest.

In practice, however, it is quite difficult to point to an actual agroecosystem and identify it as sustainable or not, or to specify exactly how to build a sustainable system. Ultimately, sustainability is a function of time. An agroecosystem is sustainable if it has continued to be productive over a fairly long period of time without degrading the resource base on which it depends, not all of which is local.

The process of identifying indicators of sustainability begins with two kinds of existing systems, viz., natural ecosystems and traditional agroecosystems. Both have stood the test of time in terms of maintaining the productivity over long periods. Each of these systems offers a different kind of knowledge base. Natural ecosystems are an important reference point for understanding the ecological basis of sustainability, while traditional agroecosystems exemplify agricultural practices that are actually sustainable and also offer insights into how social systems whether cultural, political, or economic—fit into the sustainability equation. Sustainable food systems, for instance, are based on a system that has cultural, social, economic and political components. These components shape the way human actors design and manage agroecosystems, and also place constraints on it. The natural ecosystem provides the raw materials and physical context for the agro-ecosystem. It has both local and global components, such as solar radiation and climatic patterns. Table 1.4 shows some important aspects of social and ecological systems that interact at each level in sustainable food systems. Good understanding of these systems can allow devising principles, practices, and designs with which to convert conventional, unsustainable agroecosystems into sustainable ones.

In the context of sustainability and the industrialization of agriculture, it appears that the shift towards large-scale agriculture accelerates environmental degradation and reduces the quality of rural life. Some

Table 1.4 Some important aspects of social and ecological systems that interact at each level in sustainable food systems

Condition/parameters	The social system	The ecological system
Sustainability	Equitability Satisfaction Efficiency Cultural stability	Stability Resilience Health Permanence
Agroecosystem function	Dependence on external forces Land tenure relationship Role in food product economy Share of return to workers	Biotic diversity Soil fertility and structure Rates of erosion Rates of nutrient recycling
Agroecosystem structure and function	Farmers and farm workers Landowners Consumers of food products Knowledge and know-how	Crop plants and their genomes Non-crop organisms Soil quality Nutrient cycling

Source: Gliessman, S.R. *Agroecology: Ecological Processes in Sustainable Agriculture*. Ann Arbor Press, Michigan (1998).

authors have predicted that farms in the future may need to be smaller rather than larger if they are to remain productive and competitive in the post-industrial, knowledge-based era of economic and social development (see D'Souza and Gebremedhin, 1998). Family, consensus, community pressure and the land-tenure system are crucial factors that determine whether or not a farmer wishes to adopt sustainable farming practices. It is a social issue rather than a purely technical one. Change from conventional to sustainable agriculture requires a paradigm shift regarding beliefs about nature and the environment.

ETHNOECOLOGY

The field of ethnoecology is concerned with the study of human perceptions of the natural world. Human-ecological relations are examined across the globe, and due consideration is given to indigenous knowledge. As a discipline, however, ethnoecology has remained largely isolated within the realms of cognitive studies, linguistics and anthropology (see Nazarea, 1999).

The central feature of ethnoecology is its concentration on local communities' perceptions of the environment and their interactions with it. While sharing this common perspective, ethnoecologists are now shifting their focus. Rather than remaining a largely academic discipline studying percetions of nature, they wish to make practical contributions to global concerns, mainly in areas relating to sustainable development, for instance, in areas such as plant genetic resources, traditional resource rights, nutrition, medicinal plants and intellectual property rights.

HUMAN ACTIVITIES AND THE ENVIRONMENT

Modern society depends strongly on petroleum for energy. The term 'fossil fuel' conveys that the oil comes from the bodies of dead plants and animals, altered initially by microbial decay and then 'cooked' under pressure at relatively high temperatures during burial to several kilometres' depth.

Humans have exploited this resource extremely rapidly, and they may eventually run out of this precious commodity. But now, Gold (1999) has postulated that oil and gas are of primordial, abiological origin, part of the solid material that accreted to form the planet some 4.5 billion years ago. As the planet later heated up, only partial melting occurred, producing the different layers of the Earth (crust, mantle and core), and enabling the primordial hydrocarbons and/or their precursors to survive. These are concentrated at depth of 100 to 300 km and are constantly moving towards the surface as fluids and gases via conduits within rocks and collecting in oil reservoirs (Parkes, 1999).

This controversial theory conflicts with the basis for current petroleum exploration and has been broadly rejected by scientists. An additional aspect of the theory is that oil and gas should not be restricted to sedimentary deposits, as dictated by the biogenic theory of oil formation, but should be widespread in all rock types, including rocks formed from magma.

One serious problem is that petroleum contains 'molecular fossils', clear fingerprints of a biological origin, and hence, cannot be of primordial, abiological origin. Gold, however, explains this problem by hypothesizing that, as the hydrocarbons migrate upwards, they are intercepted by a deep sub-surface bacterial biosphere, which uses some of the hydrocarbons as an energy source. Hence, it is these bacteria that provide the biological fingerprint to oil and not organisms that originally lived at the surface and gained their energy from sunlight via photosynthesis. According to Gold, this deep biosphere probably extends down to 10 km, which would mean high temperatures (between 150°C and 300°C), and so he termed it 'the deep, hot biosphere'. The current temperature maximum for bacterial life is 113°C; hence, bacteria living at even higher temperatures would have to exist to occupy this zone, and 300°C is improbably high. Pressures at this depth would also be limiting. Further, hydrocarbons are relatively resistant to bacterial degradation, especially under the anaerobic conditions that must prevail in the deep subsurface. Therefore, it would be difficult for bacteria to live under the conditions outlined by Gold (Parkes, 1999).

Surprisingly, large bacterial populations have been found at depths up to a few km in a range of different sub-surface environments, including terrestrial aquifers, granites and basalts, shales, marine seediments and the rocks beneath. In some locations, bacterial activity and populations have even increased in deeper zones where geospheric processes may be involved in fuelling a bacterial biosphere to much greater depths than was previously believed possible.

Burning of coal, oil and natural gas to heat our homes, energize our cars, and light up cities produces carbon dioxide (CO_2) and other greenhouse gases as by-products. Burning of fossil fuels is estimated to add 6 billion metric tons of caron each year to the atmosphere. Burning and logging of forests contributes another 1-2 billion tons annually by reducing the storage of caron by trees. Deforestation and clearing of land for agriculture also release significant quantities of such gases. In the twentieth century, humans have emitted greenhouse gases to the atmosphere faster than natural processes could remove them. Therefore, atmospheric levels of these gases climbed steadily and will continue their steep ascent as global economies grow. There is a marked human influence on global climatic change. The greenhouse effect naturally warms the Earth's surface. Without it, Earth would be 60°F cooler than it is today and the planet would be uninhabitable for modern forms of life.

Consequently, the atmospheric level of CO_2, the most important human-derived greenhouse gas, has increased 30%, from 280 to 360 parts per million (ppm) since 1860. Over the same time period, agricultural and industrial practices also greatly increased the levels of other potent greenhouse gases—methane concenrations doubled and nitrous oxide levels rose by about 15%. As these gases have atmospheric lifetimes ranging from decades to centuries, it seems certain that today's emissions will be affecting the climate well into the twenty-first century.

The overall emissions of greenhouse gases are growing at about 1% per year. For millennia, there has been a clear correlation between CO_2 levels and the global temperature record. The current level of CO_2 is already much higher than it has been at any point during this period. If current emissions trends continue over the next century, concentrations will rise to levels not seen on the planet for 50 million years!

In 1995, 73% of the total CO_2 emissions from human activities came from the developed countries. The United States was the largest single source, accounting for 22% of the total, with caron emissions per person now exceeding 5 tons per year. Over the next few decades, 90% of the world's population growth will take place in the developing countries, some of which are also undergoing rapid economic development. Per capita energy use in the developing countries, which is currently only 1/10 to 1/20 of the US level, will also increase. If the current trends continue, the developing countries are likely to account for more than half of total global CO_2 emissions by 2035. China, which is currently the second largest source, will most probably displace the United States as the largest emitter by 2015.

The Effects of Warming

A warmer Earth speeds up the global water cycle involving the exchange of water among the oceans, atmosphere and land. Higher temperatures cause more evaporation and soils dry out faster. Increased amounts of water in the atmosphere mean more rain or snow overall. Ecosystems also react to warming.

Global mean sea level has risen up to about 22 cm over the last century because water expands when heated. The melting of glaciers, which has occurred worldwide over the last century, also contributes to the rise. Melting and tundra warming also leads to decay of organic matter and the release of trapped caron and methane, creating an additional source of greenhouse gases.

Although *Homo sapiens* is just one species among the millions of living organisms, it has a particularly dominant impact upon nature. In this *biological*-oriented perspective, this human effect is very commonly destructive and *Homo sapiens* must, therefore, be treated as a disturbing factor. Consequently, conservation of nature often involves attempts to defend nature (i.e. ecosystems and biodiversity) against the activities and onslaught of mankind in a slightly different (*geographical*) perspective. The landscape may be viewed as the typical ecological scene and as being a product of nature and man. Humans have simultaneously as been regarded both creators and destroyers. Unspoiled nature has no value as such. Habitat and species biodiversity must be arranged in such a way that they get along well — which means surviving only if mankind has enough space for itself.

Unfortunately, both the biological and the geographical viewpoints have some flaws. The purely *biological* view can underestimate the human impact, whereas the purely *geographical* view tends to overlook the millions of other species and their long evolutionary path before we humans came into the picture. Since we are an intrinsic part of the earth's biodiversity, the latter danger is a greater threat because our existence is impossible without the earth's biodiversity (Nentwig, 1999).

The environment with which people interact is mostly man-made, comprising cultural landscapes, villages and urban zones. Throughout history, humans have converted virgin ecosystems into intensively managed areas, with natural biodiversity either disappearing or adapting. Agroecology and urban ecology deal only with those species that are able to survive under heavily disturbed conditions and human-regulated material and energy flows. Culturally, humans have created artificial ecosystems, heavily influenced landscapes and have distanced themselves greatly from mother nature. Advanced human culture and civilization are, in a way, the antithesis of nature and can be maintained only with regular inputs of material and energy. In the modern materialistic society, only useful things have a value, and non-usable ecosystems and species are immaterial.

The ever-growing human influence leads to a considerable reduction in biodiversity. The future existence of human kind critically depends on the quality of its environment, which includes the richness of our global biodiversity. The human future and the extent of biodiversity being closely linked, they are two sides of the same coin. Human ecology concerns not only the ecology of humans but also virtually all aspects of global ecology.

People influence their environment strongly. Each individual person contributes to and/or impacts on such issues as food and energy, nitrogen and carbon dioxide; availability of oil and minerals; crop varieties and urban growth; pollution of soil, water and atmosphere. Although each human being contributes to these effects, the contribution in industrialized countries is much greater than that in developing countries. Since population may be expected to go on increasing and also since also developing nations will continue to develop, it may be expected that the destructive human influence on the environment will increase, rather than decrease, in the coming decades (Nentwig, 1999).

HUMAN DEMOGRAPHY

Human demography is chiefly determined by two factors: birth rate (fertility) and death rate (mortality). These parameters are not constant and, in most countries, the basic pattern of these changes shares common characteristics. During the past two centuries, the mortality rate has decreased in some industrialized European countries as the result, essentially, of improvements in nutrition, hygiene and medical treatment.

Generally, a few decades after the decrease in mortality, the birth rate also tends to decline.

The change from high levels of birth rate and death rate to low levels is termed as demographic transition (Fig. 1.1). Population growth is highest in the second one-third of this transition, when mortality rate is already low, but birth rate is high. At the end of the demographic transition, population growth approaches zero or even falls below zero.

Reproductive patterns and fertility rates vary greatly from country to country. Industrialized countries have made much progress in their demographic transition, and some of them have an annual growth rate of

16 Environmental Technology and Biosphere Management

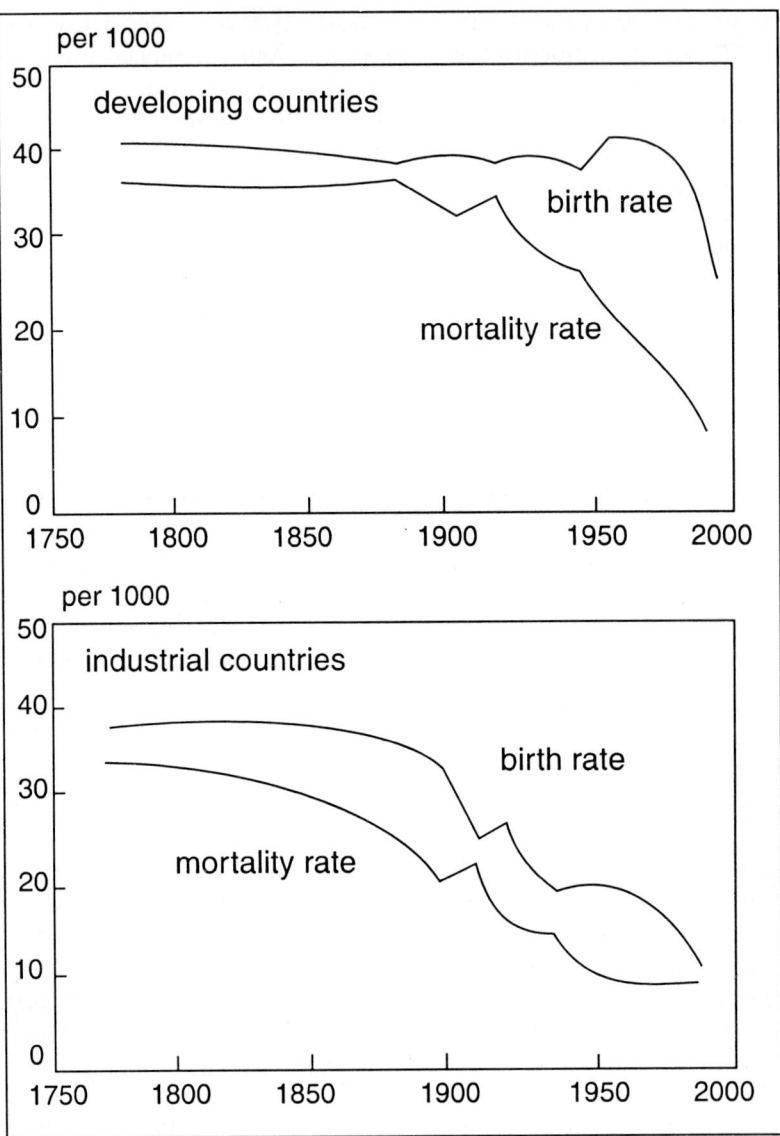

Fig. 1.1 Changes in birth rate and mortality rate in industrialized and developing countries (after Nentwig 1999).

approximately 0.0% or even lower. Nations with such a shrinking population are restricted to Europe and have growth rates between –0.1% and –0.7%. Six nations now have zero growth (Sweden, Greece, Italy, Spain, Portugal, Slovenia), and five have 0.1% growth (Belgium, Austria,

Table 1.5 Population growth rate and fertility rate (1997) for all nations in the world, by region (after Nentwig, 1999)

Population growth	Africa	America	Asia	Europe	Oceania
A. High					
Number of nations	41	18	19	1	6
Population growth (%)	2.6 (1.4-3.6)	2.2 (0.9-3.3)	2.3 (1.1-3.6)	0.8	2.7 (1.8-4.0)
Births/woman	5.6 (2.1-7.4)	3.5 (2.0-5.2)	4.3 (1.9-7.2)	2.0	5.0 (3.0-7.2)
B. Satisfactory					
Number of nations	11	16	21	24	3
Population growth (%)	2.8 (1.2-3.6)	1.5 (0.3-2.6)	2.0 (0.6-3.4)	0.3 (-0.4-1.7)	1.5 (0.7-2.9)
Births/woman	5.9 (2.1-7.0)	2.6 (1.5-4.8)	3.6 (1.6-6.4)	1.6 (1.1-2.8)	3.0 (1.8-5.3)
C. Low					
Number of nations	1	1	6	15	0
Population growth (%)	2.0	0.8	1.1 (0.2-2.8)	-0.2 (0.7-0.4)	–
Births/woman	5.0	2.3	2.5 (1.3-5.7)	1.4 (1.2-1.7)	–

Moldavia, Croatia, Poland). Japan and the United Kingdom have quite large population, now growing at 0.2%. The average annual growth rate in industrialized countries is only 0.1%, but Europe as a whole shows negative growth (-0.1%).

The combined population of China and India is 2.231 billion, i.e. 37.7% of the world's population. The Chinese constitution mandates one-child families, and China has succeeded in halving its growth rate to 0.9%. Indian policies much less successful, and so India managed to reduce its growth rate by only one-quarter, to 1.6%.

Out of some 140 developing countries, about 100 conform to the demographic transition model and have reduced fertility rates. In most countries, however, population size is still increasing substantially. There are still 85 countries with annual growth rates of 2% or more. In many cases, however, growth rate is on the decline.

Some developing countries do not conform to the demographic transition model and show either constant or even increasing fertility rates, growing by an annual average of 2-3%.

In general, it appears that the demographic process in the industrialized world conforms largely to the demographic transition model. About two-third of all developing countries also appear to be at some stage of demographic transition. More than one-half of developing countries still have annual growth rates higher than 2%, and their demographic success is far from clear. One-third of developing countries still have increasing growth rates (Nentwig, 1999).

Table 1.5 compares the population growth sites and fertility rates in different parts of the world by region.

The Earth's Carrying Capacity for Humans

Ecologically, there is always a sustainable level for a given species in any given area. When any species invades a new area, its density increases until this threshold is reached. The final population curve tends to oscillate around the sustainability capacity of the habitat depending on the availability of resources. For humans, the most limiting resources are food and energy. In the late 1970s, nearly one billion persons in the world were undernourished, which is about 20% of the total population. Hunger occurs mostly in underdeveloped areas. Chronic malnutrition aggravates the effect of diseases, so hunger is involved directly or indirectly in virtually 50% of all causes of mortality. FAO (1998) has listed 40 nations with inadequate food supply, and estimated a total of 840 million persons suffering from hunger, the causes being poverty, unfavourable climatic conditions, lack of technology and increasing population growth (see Bread for the World Institute 1994).

Because of differing population densities, industrialized countries have more arable land (0.52 ha. per capita in 1990) than the developing countries (0.17 ha.). By 2025, these figures will probably have further declined, respectively, to 0.46 ha. and 0.13 ha. If the world population doubles, an average of only 0.1 ha. will be available per capita (Oerke and Dehne, 1997). This means that new areas must be cultivated to compensate for at least some of the loss. Sadly, this is commonly being achieved by deforestation, mainly of tropical forests, and irrigation of large areas. This reduces noncrop areas and decreases the biodiversity of ecosystems, habitats and species. The earth is being transformed into a gigantic food bowl for humans.

For the most important crops, the average area under cultivation has increased in recent times by roughly 1% yearly. Agriculture has been intensified, with a focus on fewer crop species. Increases in yield can be achieved by increasing agrochemical inputs. Industrialized countries today use fertilizers and pesticides in such high amounts that they pollute the environment and reduce biodiversity. Developing countries have so far not reached this level, and increased environmental contamination is sure to occur when they catch up (Nentwig, 1999).

Food production is correlated with energy use, since the increasing number of humanity needs more and more energy. More than 80% of all primary energy is derived from fossil fuels, so using this source of energy increases the level of atmospheric CO_2. This changes the chemical and physical properties of the atmosphere as well as the climate. The greenhouse effect will raise the average earth temperature and also make hurricanes, floods and other catastrophes more frequent.

There is no doubt that the earth's capacity to sustain mankind has already been reached. A population of six billion is much more than the earth can sustain (see Fig. 1.2).

To give economic aid to a developing country so as to enable it to develop through the demographic transition to a life-style resembling that in industrialized countries is self-destructive, when it leads to overpopulation, extensive environmental damage, and finally, population decline. Rather, a much more appropriate developmental aid programme should comprise the following elements (see Nentwig, 1999): 1. Controlling population growth; 2. Encouraging small family norms by giving tax benefits, etc.; 3. The global goal should be the one-child family until the current increase in population growth stops; and 4. The affluent should sacrifice some of their luxurious life styles, adopt ecologically efficient technologies, and reduce their resource and energy consumption (the 'sufficiency revolution') to enable other parts of the world to undertake necessary development.

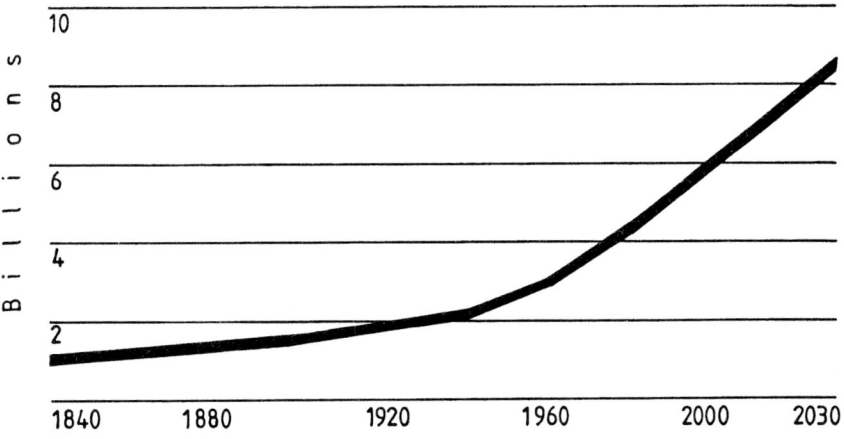

Fig. 1.2 World population, 1840-2030 A.D. (Source : Report of the Brundtland Commission. An annual rise of 1.2% has been used in extrapolating the curve to 2030 as in the report).

Global markets, global technology, global ideas and global solidarity can enrich the lives of people everywhere. But the challenge is to ensure that the benefits are shared equitably and that this increasing interdependence works for people — not just for profits. Globalization is not a new phenomenon but the present era of globalization, driven by competitive global markets, is outpacing the governance of markets and the repercussions on people.

Characterized by shrinking space and time and disappearing borders, globalization has opened up the door to breakthroughs in communications technologies and biotechnology, which if directed for the needs of people, have the potential to usher in many advances for humankind. But markets can go too far and squeeze the non-market activities so vital for human development. Whereas financial squeezes can constrain the provision of social services, a time squeeze reduces the supply and quality of caring labour, and an incentive squeeze harms the environment. Globalization also increases human insecurity as the spread of global crime, disease and financial volatility outpaces actions to tackle them (see UNDP, 1999). The agenda for action recommended by UNDP (1999) consists of reforms of global governance to ensure greater equity, new regional approaches to collective action and negotiation and national and local policies to catch opportunities in the global marketplace and translate them more equitably into human advancement.

UNDP's analyses of the trends in human development from 1975 to 1997 for 79 countries reveal that overall, countries have made substantial progress in human development, but that the speed and extent of progress have been quite uneven.

For the first time in history, the whole world has virtually become capitalist. Even the few remaining command economies have started a development programme through their links to global capitalist markets. This version of capitalism is both very old and fundamentally new — old because it involves relentless competition in the pursuit of profit, new because it is driven by information and communication technologies that lie at the root of new productivity sources, new organizational forms, and the building of a global economy (see Castells, 1999).

For almost a century, the role of international institutions in the communication field has primarily been to co-ordinate national policies, independently shaped by sovereign governments. Today, however, the scope for independent national policy making is shrinking as the international policy context increasingly takes precedence over all others.

Castells examined the profile of a world centred around multinational corporations, global financial markets and a highly concentrated system of technological research and development. He stressed the extreme flexibility of the system, which allows it to connect up everything that is valuable according to dominant interests, while delinking everything that is not valuable, or becomes devalued. This leads to social underdevelopment, at the threshold of the potentially most promising era of human fulfilment.

According to Castells, the reintegration of social development and economic growth in the information age will warrant massive technological upgrading of countries, firms and households around the world — a strategy of the highest interest for everyone, including businesses. It also requires the establishment of a worldwide network of science and technology, so as to reverse the downward spiral of exclusion and make use of the human potential that is now being wasted.

Living conditions in large cities in developing countries have recently been worsening for the poor and middle classes. In several Indian states (e.g., Uttar Pradesh and Bihar), there is no electric supply for 5-10 hours everyday and water supply is available for only 3-4 hours daily. The condition of roads has deteriorated to such an extent that many of them have virtually become death traps. At the same time, liberalization of national economies, privatization of services and state industries and the increasing integration of local economies into the international trading system have stimulated rapid modernization of the physical and social infrastructure serving areas where elites reside.

The lop-sided ('inverted') progress in India is best illustrated by the fact that whereas high-quality air conditioners, washing machines and microwave ovens are freely available in the Indian market, often one cannot use them because the power supply is totally unreliable. The same is true for many roads. India has acquired the capability to manufacture world class cars but its roads are such that these same cars cannot ply on them smoothly.

There are many reasons for this sorry state of affairs — corruption, red tapism, bureaucracy, lack of accountability and ministers and politicians adopting cheap, populist policies to attract votes for short-term gains.

Historically, most relationships between the private sector and non-governmental organizations (NGOs) have been marred by conflict. Happily, however, this pattern of business-NGO relations has changed in recent years, turning long-standing adversaries into partners for sustainable development (see Murphy and Bendell, 1999). Three notable instances of this have occurred in protest and partnership initiatives in the forest products, oil and sporting goods industries.

Civil society organizations play a strong role in promoting pro-poor policy change at the national and international levels. They are faced with the challenge of moving from protest and opposition to constructive forms of engagement with both the state and the private sector.

The role of NGOs in the promotion of corporate responsibility for sustainable development has expanded. There is great potential for wider replication of NGO-driven corporate environmentalism in developing countries, in the context of global processes including the globalization of business, trade and finance, advances in communication technologies and new governance challenges.

According to Murphy and Bendell, growing corporate responsibility for the environment is neither a rational business response to so-called 'win-win' opportunities, nor is it simply a public relations exercise, as some critics contend. Rather, companies are responding to various forms of pressure from civil society organizations and movements. 'Civil regulation' has become more and more important in motivating corporate environmental and social responsibility. Nevertheless, uncertainty still surrounds many areas relating to corporate environmentalism, and there is, thus, the need for greater international collaboration in this area.

An increasing number of large firms claim that they are adopting policies and practices conducive to sustainable development, but in several cases, change is occurring slowly and in a piecemeal fashion (Utting, 2000). There is need to consider if there may be certain powerful economic, political and structural forces that might permit a scaling-up of initiatives associated with corporate responsibility, thus enabling business to make a more meaningful contribution to sustainable development. Conceivably, corporations may well become more responsive in the coming years to environmental and social concerns. However, the process of change is likely to remain fragmented, spread unevenly in terms of companies, countries and sectors (Utting, 2000).

ENLIGHTENED USE OF ENERGY AND MATERIALS

According to Amory B. Lovins (see Hawken et al., 1999), oil will become uncompetitive even at low prices before it becomes unavailable even at high prices. People need to adopt a radical new approach to the way they live. In the Industrial Revolution, humans mobilized natural resources without restraint. Now that diverse natural materials resources, energy sources, water, topsoil, forests and the climate itself are being degraded and depleted, there is a strong compulsion for humans to move towards a capitalism that safeguards the natural resource base which ultimately underpins all economic activities. The remarkable inefficiency of conventional capitalism is well known. In most industrialized countries of the West, materials used by industry's metabolism every day amount to many times the total weight of a country's population. Yet, only about 1-5% of these materials end up in products that are still in use a year after being sold, the rest being junked. A far better strategy is to design industrial parks where each manufacturer will use the wastes of others until emissions are finally reduced to zero. This manner of industrial ecology can eliminate not only waste, but the concept of waste itself—as has been practised by nature with its closed-loop ecosystems (Myers, 1999).

To achieve sustainable economies, people must reduce their materials and energy intensity (the amount used per unit product) by at least one-half worldwide. Since developing countries will be reluctant as well as unable, to do this for some time, developed countries should aim to cut their consumption by at least 75%, if not 90%. Several manufacturers have found that they can dispense with 95% of 'secondary' packaging around toothpaste tubes, ice-cream cartons and other products.

The strategy of 75% reduction in materials is known as 'factor 4'; and that of 90% reduction as 'factor 10'. Austria, the Netherlands and Norway have committed themselves to pursuing factor 4 efficiencies, which means that, by using existing technologies, they can enjoy twice as much well-being while using half as many materials and causing half as much waste (Hawken et al., 1999; Myers, 1999). The same strategy has been approved by the European Union as the new paradigm for sustainable economies. Sweden and the OECD (the Organization for Economic Cooperation and Development) wish to go in for factor 10, as has the World Business Council for Sustainable Development.

The radical new approach to people-environment interface in the new millennium may be illustrated by the concept of the 'hypercar' as conceived by Amory Lovins. This 'hypercar' would be largely made of advanced polymer composites, especially carbon fibre and use one-third less aluminium, three-fifths less rubber, four-fifths less platinum and nine-tenths less steel. It would weigh only one-third as much as the cars of today. Because of these and other design efficiencies, notably a hybrid-electric drive, it would run over 300 km per imperial gallon, would be 95% less polluting and would be almost entirely recyclable. Hypercars could eventually save as much oil as OPEC sells today. These hypercars would be veritable computers on wheels (Myers, 1999).

Some other 'techno-breakthroughs' being envisaged include: diodes that emit light for 20 years without bulbs; ultrasound washing machines that use no water, heat or soap; deprintable and reprintable paper; reusable as well as compostable plastics; roofs and roads that double up as solar-energy collectors: extra-light materials even stronger than steel; and quantum semi-conductors that store vast amounts of information on chips no bigger than a dot.

People could get far greater benefit from biological resources, which till now have been over exploited and under utilized. Biomimicry may well inspire design from spiders which, using flies as feedstocks, make silk as strong as Kevlar yet much tougher, and without needing high-temperature extruders. The abalone manufactures an inner shell twice as tough as the best ceramics and diatoms make glass, both processes using sea water and no furnaces (Myers, 1999).

The paramount need today is to abandon exploitation of the planetary resources in support of the human cause, and instead, exploit the brain power of ingenious, brilliant people in support of the planetary cause which, no doubt, will increase the prospect of securing the human cause as well.

ECONOMIC INSTRUMENTS AND ENVIRONMENTAL PROBLEMS

Besides global environmental issues, people in developing countries have to contend with and tackle their local environmental problems, some of which, e.g. urban air pollution, have already become serious for these countries.

Broadly, the following two approaches are available for the protection of the environment.

1. The traditional approach relies on indigenous methods, including local community conservation practices to achieve sustainability. This approach is preferred by environmentalist NGOs. It is well known that traditional conservation practices used to be quite effective in the past and failed only due to tampering with the system, mostly through governmental intervention that came in the form of regulations or economic activities (exploitation of local resources for use elsewhere, e.g. logging activities). Increasing pressure on resources, e.g. due to increased population and the increased economic activities of people, have also been responsible for the failure of the system. Today, the traditional approach can be helpful in a limited way but it alone cannot suffice for achieving environmental protection in a world with ever-increasing resource requirements. Resource use is expected to increase substantially, specially in developing countries.
2. The second approach is based on either a *command and control* principle or a *market-based* method. Typically, in the initial stages, the approach has been control through regulation on limits to discharge (a relatively inefficient approach), and slowly moving to efficient market-based approaches (Painuly, 1995).

In direct regulation or command and control approach, regulatory instruments are used along with systems for monitoring the standards, and sanctions are imposed for non-compliance. For example, certain standards for ambient air and water quality are proposed. This is followed by imposition of emission and discharge limits. It could also extend to process and product standards. These are monitored and compliance is mandatory. Often, the polluters are consulted and limits are imposed through bargaining, which improves the chances of compliance by reducing uncertainties for both polluters and regulators.

Direct regulation is the preferred approach with the regulators all over the world. Standard-setting through regulation can be in any of the following manners: limit on maximum discharge from a source; pollution density limit on discharges/emissions; pollution control for specific pollutants to a given degree (say in percentages); need to implement a technology that can reduce pollutants to the chosen best technology level; discharge limits related to inputs or outputs in a production process; and discharge bans for specific pollutants (Painuly, 1995).

Market mechanisms or economic instruments are mechanisms that send appropriate economic signals to the polluters to modify their behaviour. The approach usually involves *financial transfers* between polluters and the community and affects relative prices. But the polluters have freedom to respond and adjust in the manner they want, e.g. by choosing the least cost option to meet the requirements. It is a more efficient approach compared to the approach based on standards and regulations.

The choice of economic instruments implies reduced direct governmental intervention and leads to cost effective pollution control. It also implies that instead of depending on end-of-pipe pollution abatement (as is the case with standards), preventive measures like process control and alternative procsses, etc., are preferable.

The economic instruments can be of the following types: pollution taxes or charges (effluent or emissions charges; user charges; product charges; tax differentials; and administrative changes); subsidies (grants; soft loans; tax allowances); market creation; market creation (pollution rights, e.g. market intervention; liability insurance, etc.).

Benefits of Economic Instruments

Economic development in both market- and non-market economies has led to some degradation of the environment. It is possible to reverse this process in a market economy with least pain (cost), where economic instruments can help in utilizing the power of the market to achieve environmental goals. Through incentives, individuals can be prompted to change their choices in such a way that they match with social choices. Here, societal costs are internalized by the polluter, while at the same time, the polluter enjoys flexibility to select an optimal approach. To achieve a given level of emissions reduction, an individual can use his superior information to select the best means compared to the case where regulating authorities prescribe the technology to be used (Titenberg, 1990).

In the traditional *command and control approach,* the cost increase tends to become steep as standards become stricter. This implies that pollution control becomes inefficient with stricter standards as the costs to society increase. The approach leaves little incentive for innovation in pollution control for the polluter, as standards are based on a given best or practically achievable technology.

According to Vernon (1990), market mechanisms do not suffer from these limitations, and the benefits include cost effectiveness; incentive for reducing pollution; flexibility and resource transfer.

Several countries have adopted economic instruments in one form or other to control pollution. User charges, administrative charges and subsidies have been used in almost all the OECD countries. Major instruments used by various countries comprise air emission taxes, water emission taxes, waste emission taxes, noise taxes, user charges, product taxes, administrative charges, tax differentials, subsidies and market creation (Painuly, 1995).

The cost effectiveness of the economic instruments follows from the assumption that individual participants are cost minimizers. This is true so long as they operate in a competitive market or cannot fully pass on the increased costs resulting from pollution control requirements to their customers. In this case, an emission charge, or emissions trading system introduced by the authorities results in a cost-effective allocation of control responsibilities to reach a target reduction in pollution, even though regulatory authorities may not have information about control possibilities.

However, several issues need to be resolved before economic instruments can be used to address environmental problems. Further, issues differ in application of the instruments at national and international levels. Some common issues are toxic pollutants, side effects, regressive nature and pre-requisite of a well-developed market.

The application of economic instruments in some developing countries is particularly problematic in view of poor infrastructure and also because market and other conditions necessary for application of economic instruments vary from one developing country to other. In India, for instance, a variety of problems have been encountered in application of economic instruments which require that transactions take place in a well developed market that responds to price signals. India has a large non-market sector (*the unorganized sector or informal sector*). In the case of industries, this includes unregistered small and medium scale enterprises. Non-market activities include items like fuel collection, garbage collection, transport services, unregistered establishments like tea stalls, repair services, leather production units, brick making and so on. The activities of this sector often result in adverse environmental impacts due to deforestation, effluent generation or air pollution, etc. Activities like livestock raising on common property resources also degrade such lands. The non-market sector makes up a sizeable segment of the Indian economy. Over 30% of the primary energy requirement is provided by non-commercial fuels like fuelwood, animal dung and crop residues. In villages, over 70% of household energy needs are met by these fuels. These fuels are a major cause of local environmental pollution and soil degradation. Similarly, transport and farm services in rural areas are provided by animal energy. In the industry sector, the unorganized sector accounts for more than 40% of the value added. This includes industries like food processing, textiles, wood and wood products, paper, electrical machinery and transport equipment (Painuly, 1995).

For India, the command and control approach may be a good option for non-point sources of pollution like automobiles and economic instruments should have a role in the improvement of the environment. This will mean that the pollution control system would also mature along with the economy, which is in a phase of liberalization and expansion. Problems such as the lack of a sizeable market and technology transfer are disappearing with the opening up of the economy.

As a first step, pollution charges should be introduced, followed by market creation. The latter is a preferred alternative for the long term. Pollution charges depend on the amount of pollutant emitted which, in turn, depends on technology. A good knowledge of industries, their technologies and outputs is the chief requirement to introduce the tax. Part of the tax revenue may be used as a subsidy for strengthening pollution control and part to provide compensation to victims. The second phase can involve a shift to the permit system for more efficient pollution control.

Painuly (1995) identified the following steps for the introduction of economic instruments:

I. Pilot phase

(i) Review of existing standards and regulations to identify areas of inefficiency and high control costs.
(ii) Identification of some prospective areas suitable for application of economic instruments.
(iii) Fixing environmental targets (like pollution levels) and pollution tax rates or allocating rights to pollute, depending on the type of instrument. Appropriate regulatory measures should also be decided to implement the system.
(iv) Finalizing the tax scheme or pollution rights and monitoring mechanisms in consultation with the polluters.
(v) Testing and evaluation of the framework through a pilot phase.

II. Implementation phase

(i) Rectification or modification of the scheme based on the experience and evaluation of the pilot phase.

(ii) Effective implementation of the revised scheme.
(iii) Evaluation of the programme and its rectification and extension to other areas.

Application of Economic Instruments at International Level

As there exist wide disparities among various countries in income standards, per capita and gross emissions, and overall responsibility for the accumulation of GHGs in the environment, standards offer no solution at the international level to reach an agreement to reduce GHG emissions.

Nevertheless, in view of the theoretical superiority of the economic instruments over other mechanisms, proposals have been made for their application at international level to control global pollution. Since the atmospheric accumulation of GHGs is a major threat to the global environment, efforts are required to limit anthropogenic emissions of the GHGs. CO_2 accounts for the largest share in global GHG emissions. Chloroflurocarbons (CFCs) and methane (CH_2) are other important GHGs, but emissions of CFCs have already been covered by the Montreal Protocol and there is a lot of uncertainty about the role of various sources of CH_4 emissions. Therefore, the focus of most current discussions has been on CO_2. Three notable economic instruments and mechanisms widely discussed in several forums for application at the international level to reduce the GHG emissions have related to: (i) pollution charges as carbon tax; (ii) tradeable permit system; and (iii) joint implementation. Of these, tradeable emissions permit system particularly offers enormous opportunities to resolve global environmental problems in an efficient and equitable manner and joint implementation, a mechanism suitable for emissions reductions through joint efforts of two countries, has good potential to simulate a tradeable emissions permit system (Parikh, 1994; Painuly, 1995).

GLOBAL CHANGE

Significant contributions have been made in the past few years in the study of the global energy system; world agricultural production, trade, and consumption; the global forest products system; comparative national demographics; and the potential impacts of climate change on global agriculture and vegetation. Work on models has been done in transboundary air pollution and ozone for international negotiations, energy management and conservation, forest resources management, water management, technological evolution, and risk analysis and economic transition. Continued concerns about human welfare have also added two other related research areas: the future of social security systems and implications of future catastrophic events on society and its institutions. In much of the research on the above areas, analytical multidisciplinary approaches link natural and social sciences, databases and geographic information systems. The databases contain extensive inventories of carbon mitigation technologies, statistics on soils, forests, and land use in Asia, data on multi-pollutant emissions in Europe, and global as well as national population projections.

Earthquakes score a high place on the agenda of global concerns today. About 10,000 earthquakes with tremors strong enough to be felt occur each year, and about 100 of these tremors cause much damage. Every year or so, some places in the world experience a super-catastrophic event akin to the strong earthquake in Turkey in the summer of 1999 that left 15,000 dead. Many earthquakes often take place in unpopulated areas, but they can be expected, and we had better be ready for them, especially in vulnerable, heavily populated areas. Shoddy construction, poor building codes, limited public awareness and inadequate disaster-response training make developing countries and their megacities particularly vulnerable (Schaeffer, 1999). Half a century ago, about 50% of the world's urban earthquake-threatened population lived in developing countries. Today, it is 85%. Earthquakes can devastate economies, especially those of developing countries.

Geological disasters such as earthquakes, volcanic eruptions and landslides occur in the outer shell of the solid earth—the lithosphere or rock domain—which consists of blocks that move relative to each other, through fluid interactions, friction, buckling and fracturing. These processes create strong instabilities, turning the lithosphere into a 'hierarchical' chaotic system.

Short-term earthquake prediction is virtually impossible. Yet, with sufficient information about a site's geology, it is possible to compute the local ground motion that would result from sizeable earthquakes. Scientists usually focus on two distinct questions: What happens to the ground during a quake, and where and when will such events take place? The first makes use of knowledge of ground and experimental data collected from actual events to create hazard maps used, for example, in devising building codes (Schaeffer, 1999).

Several big cities including Beijing, Cairo, Delhi, Mexico City and Rome face a wide range of seismic risks, which require different strategies for attaining sufficient levels of preparedness. Data needs to be collected from several fields such as soil engineering, geophysics and lithology. Information has been gathered for tectonics, paleoseismology and seismotectonic models. Detailed models are used to predict seismic ground motion and prepare hazard maps. These efforts are made possible through data input from a network of global observation points, together with the application of modern theories and powerful computers (Schaeffer, 1999).

The security of nuclear power reactors in some countries, e.g. Bulgaria, Hungary, Romania and Slovenia has been examined to determine how to retrofit existing plants to make them more secure.

Non-linear dynamics of lithosphere blocks are being studied and algorithms for earthquake prediction are being tested to narrow the forecast and area in which an event will likely occur and to develop predictive measures of its magnitude. The global predictive capability of this approach has exceeded 75%. The 1994 quake in Northridge, California was successfully predicted. This quake caused at least US$ 20 billion in economic damage. The prediction, made well in advance, covered an area of 400 km^2 and time span of 18 months. On the other hand, the recent earthquake in Turkey could not be predicted.

Four unstable systems—earth, life, engineering and society—are tied together in megacities. So, a powerful quake can cause intense havoc. Basic research exploring chaotic systems might help improve our efforts to limit the impact of earthquakes and other types of disasters.

Science and technology play a crucial role in the development process. There has been much concern about the widening gap in science and technology between the developing and developed world. This ominous trend has caused increased hunger and disease, and has led to such serious global problems as deforestation, desertification and climate change. To ensure sustainable development, aid from developed to developing countries should be channelled primarily to projects that foster human resource development.

There is no doubt that science and technology have been the two prime forces that resulted in increased human productivity over the past two centuries and that they would remain prime forces behind efforts to address global problems in the twenty-first century. As science and technology are not worlds unto themselves but form part of society, development and application of science and technology should conform to larger ethical principles. Like economics, science and technology are now global enterprises. Each new scientific discovery and technological invention is a by-product of human ingenuity on a global scale. For this reason, science and technology should serve the global community and not just the needs of a few select nations. Considerable scientific progress has taken place in the south over the past few decades and there is growing awareness among scientists in developing countries of their responsibilities in promoting science-based development within their own lands. Indeed, there is a strong need for each nation to advance science within its borders while promoting international cooperation.

People–Environment Interfaces   31

Some of the challenges that humanity must address in the twenty-first century include sustainable food production, availability of sufficient quantity of freshwater of satisfactory quality and mitigation of the impacts of natural disasters. Humanity must also address such environmental concerns as climate change, ozone layer depletion and pollution control. In view of the increasing demands from fast growing populations, renewed commitments in science and technology will have to be made if all countries are to experience sustained socio-economic development through environmentally-friendly ways. The geosciences that deal with the Earth's basic life systems such as air, water, sea and land, also have significant roles to play in these endeavours. Perhaps the best support that the geosciences can provide in addressing the challenges facing people is to accurately predict the future state of the atmosphere and, in particular, the occurrences of extreme weather and climate events with sufficient lead time to allow actions to save lives, minimize property damage and protect the environment. The prediction and early warning of extreme weather and climate events have become possible because the enhancement of the basic global, regional and national infrastructure that comprises the World Meteorological Organization's Global Observing System. The global challenge is to further improve and strengthen this unique and vital observation network for the continuous monitoring of weather and climate systems, as also for the protection of the atmosphere.

Impact studies today are far from being simple 'driver-impact-consequence' sequences based on linear thinking. Rather, they have

Climate Change Impact Studies

"Linear, Pollution-pipe" Models

[Climate Change Scenario] [Bio-physical Impact] [**Socio-economic Consequences**]

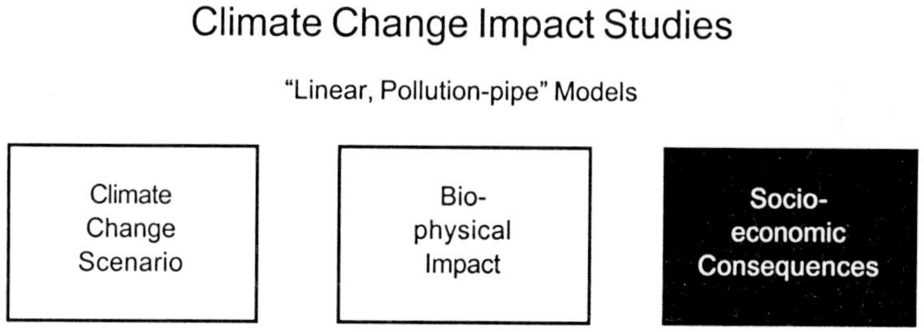

Fig. 1.3 The old 'pollution pipe' approach to climate change impact studies (after Steffen, 1999).

evolved into integrated assessments, in which a complex, linked biophysical-socioeconomic system is influenced by multiple, interacting drivers (Figs. 1.3 and 1.4).

The current range of global environmental problems are unprecedented in their complexity and extent. To tackle them will require all the cleverness and wisdom that humans can muster. We must effectively integrate the social and natural sciences. Too often the social sciences are

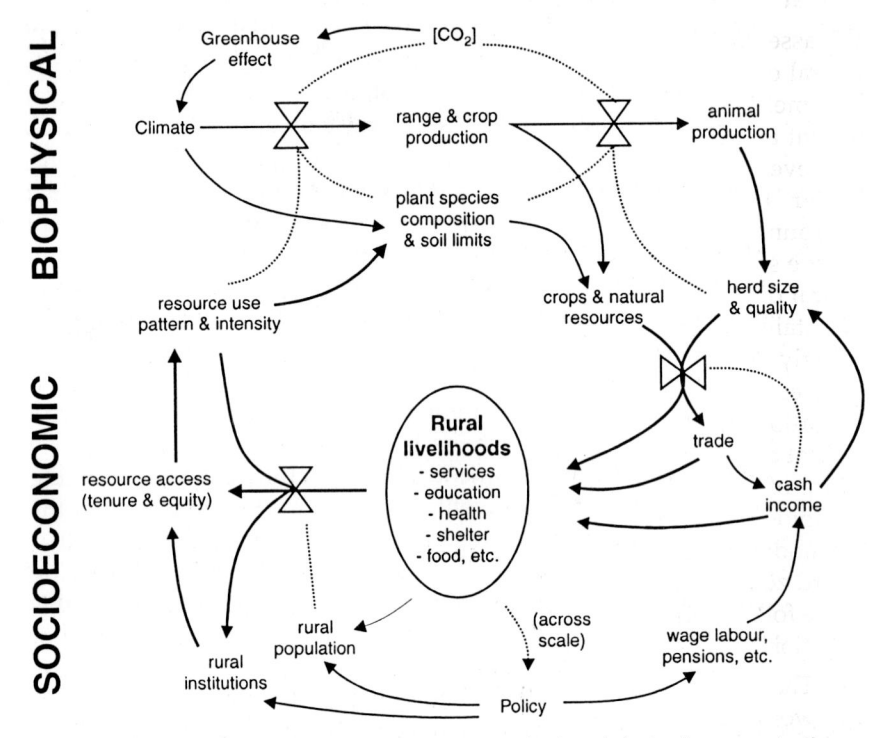

Fig. 1.4 An integrated systems approach to global change studies (after Steffen, 1999).

viewed as an 'add-on' at the end of a project. Effective integration must start at the design stage of the work with some fundamental issues: Whose questions? Whose agenda? Whose paradigms? Whose approaches? (Steffen, 1999).

PEOPLE, TECHNOLOGY AND THE ENVIRONMENT

Humanity is now confronted with global problems that cannot be solved without science. Historically, technology has been generally independent of science but now technology is being given a systematic scientific cover.

Today, destruction of the global environment poses a threat to humankind. But many of the problems of the global environment owe their genesis to human activities. Paradoxically, the stated intent of these human activities has always been to create safer and more comfortable lives for people.

Natural Disaster Reduction

Risk assessment and sustainable development have attracted the focus of natural disaster mitigation. Technological disasters have also come into the limelight. About one-fourth of the world population is at risk at present and the economic losses continue to rise to the extent that they have even tripled since the 1960s. The apprehension is that natural and similar technological disasters having an adverse effect on the environment will continue to cause major disruptions of society and impede social and economic development in the coming decades. Many natural disasters are, in fact, not natural but triggered or aggravated by unsustainable human activities. Population growth and urbanization strongly increase global disaster vulnerability.

Recent progress has taken place in disaster management and mitigation, the best example being the strongly reduced death toll of tropical cyclone disasters in Bangladesh. Fundamental knowledge of the nature of extreme events has increased and the application of modern technology, such as satellite remote sensing, disaster information systems and modelling has advanced. These developments have led to improved hazard zoning and more reliable early warning systems.

The following three major fields of future research relating to disaster reduction have emerged in recent times:

1. The risks in large urban/industrial areas, where over a half of the world's population is concentrated and where enormous property damage by natural and technological disasters is expected;
2. The effects of changing land use and pollution patterns on disaster vulnerability, especially in marginal rural areas; and
3. The disasters related to decadal climatic fluctuations such as *El Niño*, the Southern Oscillation and the North Atlantic Oscillation, that strongly affect droughts, river floods and mountain disasters, causing loss of life and property, social distress and food shortage.

Approximately, one hundred earthquakes of the millions that occur annually throughout the world are potentially disastrous because of their magnitude and proximity to an urban centre. Post-earthquake investigations provide useful case histories and are the ultimate scientific laboratory, because they constitute a reality check on the relative vulnerability of buildings and infrastructure at risk in a community. As in other areas of disaster research, post-earthquake investigations make use of geologic, seismological, engineering, health care and social science studies so as to understand the effects of earthquakes on people, their environment and their administrative structures. Such studies have brought out the vulnerability of a community to an earthquake depends on physical factors such as the frequency of earthquakes having different magnitudes, the earthquake's focal depth and proximity to the urban centre, the direction of the energy release, the geometry and physical properties of the soil and rock underlying the structure as well as societal factors such as the availability of earthquake insurance and the degree of prevention, mitigation and preparedness measures that are adopted as public policy and enforced by community.

Successful preparedness for natural disasters and mitigation work require the full participation of all members of the population. The development of a broad-based natural disasters programme improves the capacity of the general public to make sensible decisions relative to the natural disasters. There are advantages if an entire country puts its efforts into the education and community-based efforts to mitigate effects.

Both private and public bodies should be concerned with disaster management and disaster prevention. Assistance activities focus not only upon emergency response, but also upon technical, scientific, planning and educational research and planning. The different activities for disaster prevention and assistance in pursuance of the national objectives include: elaborating natural hazard maps and determining maximum risk zones; surveying and forecasting natural hazards; including the risk variable as determinant in municipal and regional development plans; vulnerability analysis and relocation of high risk zones housing; reconstructing after disasters and restoring degraded drainage areas. All these tasks are reinforced with education, public information and training programmes.

The gradual onset of certain natural disasters differs sharply from the sudden occurrence of others. For example, whereas earthquakes occur quickly and are of relatively short duration, onset of droughts may take months or years. Droughts have a duration from several to many years, and have far-ranging effects, from famine to mass migrations of the population in the region. Often, drought and famine cycles exemplify the effect of climate change and its consequences in the fostering of natural disasters.

A climatological drought can become a disaster event when rainfall deficit impacts a particular agrarian system and accentuates broader socio-economic forces to inflict hardship on a vulnerable population. Sometimes, just as drought occurs without causing famine, so can famine occur in the absence of drought, for example when political upheaval disrupts agricultural production and food distribution. If not the sole cause, drought is a significant trigger of famine and part of the explanation for poverty.

In Europe, the Bay of Naples coast is an area under constant threat of volcanic action. Here, Mt Vesuvius looms large and dangerously close to the excavated site of Pompeii — a popular tourist spot. However, while the tourists are interested in the history and scenic attractions of the area, the coastline of the Phlegrean Fields near the town of Pozzuoli, about 16 km from Vesuvius, has been uplifting as a platform for the past three decades. In 1985, the area rose by 2 metres in a few days and led to the evacuation of 40,000 people. Volcanologists have discovered that there is a rock plug jamming the 30 metre-wide conduit of Vesuvius. As the magma chamber fills, the increasing pressure on this plug will force it out and a cloud of molten rock, ash and gas will be thrown 2 km into the atmosphere and will be held there for some hours by the heat generated by the volcano. When it cools, it will sink and flow along the ground at speeds of abut 160 k.p.h. and the pyroclastic flow of super-heated ash and poisonous gases at 500°C will reach some nearby towns within only a few mintues, reaching the outskirts of Naples which has a population of about 3 million. It is feared that the next eruption of Vesuvius could destroy everything within 8 km, including one million people, in only a few minutes. Some scientists feel that it might be possible to reduce the impact of a moderately explosive eruption by placing 30 metre high barriers about 3 km and 5 km from the central vent to block the downward surge of pyroclastic flow. With the addition of shelters on the lower slopes of the volcano, the people living around Vesuvius may receive some protection.

CLIMATE, WEATHER AND HEALTH

It has long been realized that weather and climate affect human health and well being (see Fig. 1.5). Folklore everywhere is rich in belief about the effect of the seasons and weather fluctuations on physical and mental health. Fevers vary seasonally; so do mood and various psychological disorders; aches and pains in the joints flare up in winter; and heatwaves can produce a debilitating effect.

The apparent increased instability of weather patterns in many parts of the world in recent years, new insights into cyclical phenomenon such as *El Niño,* and the evidence pointing to the global climate changing in response to greenhouse gas emissions, have directed new attention on health consequences of the climate.

Epidemiologists have traditionally viewed the effects of weather and climate on health as part of the natural backdrop to life, and have shown no interest in the health effects of climate variations in their own right. Happily, however, this attitude is now changing. The health impacts of extreme weather events are now being quantified. Studies of excess deaths associated with heatwaves have been carried out in many countries. We are learning why populations are more vulnerable in some cities than in others; why some populations show clear thresholds for increased daily

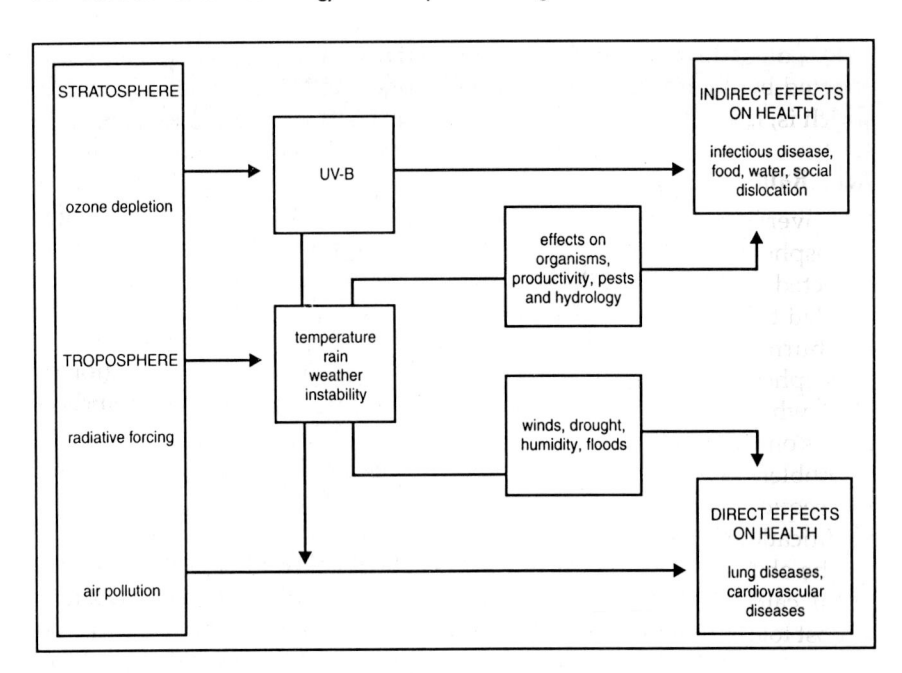

Fig. 1.5 Indirect and direct effects of major constituents of the atmospheric system on human health (after McMichael, 1999).

mortality when critical temperatures are exceeded; why urban populations are typically more vulnerable than rural populations; and the extent to which the excess mortality is due to short forward displacement of the time of death in susceptible persons (McMichael, 1999). It has emerged that certain types of air mass are more hazardous than others, and that this is location-specific.

Changes in weather patterns sometimes exert negative health consequences that are caused by other environmental conditions. During the strong *El Niño* event of 1997–98, the knock-on health effects of regional air pollution resulting from the disastrous forest fires in Indonesia were studied. In neighbouring Papua New Guinea, the same *El Niño*-related drought resulted in one of the worst famines last century.

Studies of the ENSO cycle in relation to outbreaks of malaria and, to a lesser extent, dengue fever, have revealed strong correlations around the world—in Pakistan, north India, East Africa and Brazil. Some understanding of how and why cholera tends to break out in certain coastal populations, such as in Bangladesh, when local sea-surface temperatures rise and algal blooms occur, is slowly emerging.

In temperate zones, the transmission of certain water borne diseases is affected by the quality and distribution mechanisms of drinking water, which is, in turn, affected by rainfall patterns (McMichael, 1999).

Acid Rain

For over a century, the emission of acidifying compounds to the atmosphere has been increasing steadily right up to the 1970s. This trend reflected the increasing industrialization in Europe and North America and led to the release of sulphur and nitrogen compounds, largely from the burning of fossil fuels. These pollutants are transported in the atmosphere throughout the Northern Hemisphere and deposited as 'acid rain', which causes acidification of soils and surface waters, with adverse effects on terrestrial and freshwater biota. In recent decades, the emissions of problematic sulphur dioxide and nitrogen oxides have been reduced. Consequently, the chemistry of surface waters is showing signs that the acidification process is reversing (Jenkins, 1999).

The observed recovery was as per expectation because evidence from small catchments covered with roofs, to remove sulphur deposition almost totally, led to chemical recovery almost immediately.

HEALTH IMPACTS OF HAZE-RELATED AIR POLLUTION

The haze episodes in Southeast Asia have led to increased respiratory-related complaints requiring hospitalization, increased asthma attacks among children and persistent decreases in lung functioning in many school-going children in Malaysia.

The level of particulate matter attained in Indonesia, Malaysia, the Philippines, Singapore and Thailand, when observed in the past in other areas, has been found to be associated with increased daily mortality, hospitalization, emergency treatment, increased respiratory problems, exacerbation of asthma and chronic obstructive pulmonary disease and decreased lung functioning—primarily in the elderly, the very young and in individuals with pre-existing respiratory and/or cardiac illness.

There is no evidence that particles from different combustion sources have different impacts on health. The available data strongly suggest that combustion-source particulates, including those produced during forest fires, are associated with a wide range of adverse health outcomes.

To address these issues, priority should be given to preventing and extinguishing fires, with incentives to reduce the use of fire as a means of clearing land; and haze shelters should be set up in areas that are most at risk.

ANIMAL FARMING AND GREENHOUSE GASES

Animal farming contributes to air pollution and the greenhouse effect. Dairy farms, piggeries and livestock account for about three-quarters of the global emissions of ammonia and also adds some carbon dioxide and around one-fifth of the methane emissions to the global budget of greenhouse gases.

Methane has a 19% share in the greenhouse effect. Although methane is a trace gas, it makes a substantial contribution to the greenhouse effect because of its greater (10-70 times) effect per molecule (Lashof and Ahuja, 1990).

The total emission of methane from all sources amounts to 500 ± 200 Tg/a (Tg/a = 1 terragram/annum = 10^{12} g/a = 10^6 t/a) (Pelchen and Peters, 1994). During the last decade, the atmospheric methane content has risen by roughly one per cent per year.

About 32% methane emissions come from wet rice cultivation, about 21% from animal farming, and the remainder from burning biomass, swamps and anthropogenic sources (Pelchen and Peters, 1994).

Over 95% of the methane emitted from animal farming in the world comes from ruminants, 74% of which is from cattle, eight% from water buffalo and sheep, and the remainder from camels and goats. If we ignore the monogastric animals, then these ratios roughly correspond to the contributions of each species to the total biomass of all ruminants in the world (Table 1.6). The ruminants are so significant because they show a high number of methane-forming bacteria in their rumen, which are rare or absent in monogastric animals. The share of all other agricultural animals (monogastric animals), therefore, amounts to only 4%. These ratios tend to vary slightly from region to region, depending on the significance of each species in the region.

About 75% of the total methane emissions are concentrated in the Northern Hemisphere. A maximum occurs between 20° and 30° North, and two more peaks between 40° and 50° North and 20° and 30° South. The maximum is caused mainly by the large numbers of cattle in India and China, the water buffalo of Southeast Asia, and the goats in India. The second peak is caused mainly by milch cows and sheep, but also by the other cattle, where the higher feeding level over-compensates for the relatively low numbers of animals. The emission maximum in the Southern Hemisphere comes mainly from the cattle in Brazil and Argentina, and the sheep in Australia and South Africa (Pelchen and Peters, 1994).

Table 1.6 Proportions of biomass of individual ruminant species in the total biomass of ruminants worldwide (Source : FAO, 1998 and Pelchen and Peters, 1994).

Species	Number (in 1,000)	Biomass* (t)	Proportion (%)
Cattle	1,263,330	260,246	78.4
Sheep	1,165,320	34,960	10.5
Buffaloes	138,374	35,594	10.4
Goats	459,700	13,791	4.1
Camels	24,605	741	0.2

*Assumed weight (kg) of individual animals: buffalo = 250, cattle = 206, goat = 30, sheep = 30, camel = 307.

Theoretically, with the same additional feed energy input, it is possible to achieve a greater saving on methane emissions in the developing countries than in the industrial countries since, if the maintenance requirement is doubled, the reduction in energy loss through methane would be greater than if the maintenance level were quadrupled rather than trebled, as would be necessary in animal farming in the industrial countries (Pelchen and Peters, 1994).

VEGETATION FIRES AND GLOBAL CLIMATE

Coal seams found in different parts of the world bear witness to the most ancient forest fires that our earth has ever experienced. The charcoal deposits in these seams are remnants of forest fires, which over the millennia, became transformed into bogs where they later formed the coal deposits. These forest fires, which took place over 300 million years ago, were caused by lightnings and volcanic eruptions.

Hominids appear to have been able to use fire since about 1.5 million years, for a variety of purposes in the earliest cultural stages of human development. Besides purely cooking and heating, fire was used for hunting, to clear forested landscape for security reasons and, later, because it was the only effective tool for clearing land (slash-and-burn technique) and keeping it open for grazing (Goldammer, 1994).

In some countries, traditional burning techniques are still in vogue. The vegetation cover of tropical grass savannahs is burnt for maintaining or improving pastoral systems.

In many extensively farmed regions, there are areas in which uncontrolled natural and man-made fires show specific characteristics and effects (fire regime). The most important characteristics of a fire regime are the following:

1. The combustible material influences the properties and effects of a fire. For example, in the open grass, bush and forest landscapes (e.g. savannahs) it is mainly the grass layer that burns. In continuous forests, which contain more combustible phytomass, not only ground-level fires occur but also canopy fires (Goldammer, 1994).
2. The season of fire outbreaks affects their characteristics and the ecological consequence, e.g. at the start, peak or end of the dry season the combustible material shows different moisture contents and the fire can occur at various stages in the course of vegetative development (Crutzen and Goldammer, 1993).
3. The frequency of fires is important. The interval between consecutive fires determines the amount of vegetation or the accumulation of combustible material that has developed between fires. Succession is also influenced by the intensity of the fires.

Deliberate fires resemble uncontrolled wildfires in their appearance and effect, but differ in their specific and deliberate use, for example in slash-and-burn clearing, to obtain pastureland, and for disposing of agricultural and silvicultural wastes which cannot be used for any other purpose. Figure 1.6 shows the interrelationships between the various ecological and socio-economic factors which determine or are determined by fires.

Savannahs are usually the largest, regularly burned areas of vegetation. Savannahs, located between the Tropics of Cancer and Capricorn and in the adjacent subtropical regions, cover worldwide area of about 2.6 billion hectares. They are burned at intervals of 1 to 3 years for managing game and domestic animals. The significance of fires caused by lightning has declined greatly compared to that of man-made fires. The fire intervals are determined largely by the productivity of the savannahs. Humid savannahs which are highly productive usually burn once a year or once in two years.

Dry savannahs, being less productive, burn less frequently. A dry grass layer weighing about two tonnes per hectare constitutes the lower limit at which fires can spread over extensive areas. If the grass layer is degraded by overuse, or if the combustible layer becomes more discontinuous (desertification), the frequency of fires also declines (Goldammer, 1994).

Conversion of evergreen tropical rain forests into other forms of land use is intimately linked with the use of fire. Primary forests are clear-felled after the harvesting of marketable trees. The rest of the biomass is burned or allowed to rot. In non-marketable secondary forests, very often, all the plant biomass is burned after the forests have been clear-cut.

Clearing of the forests creates a net flux of carbon into the atmosphere. When forested land is converted into pasture land, fire is subsequently used for the same purpose and at similar intervals as in the savannahs.

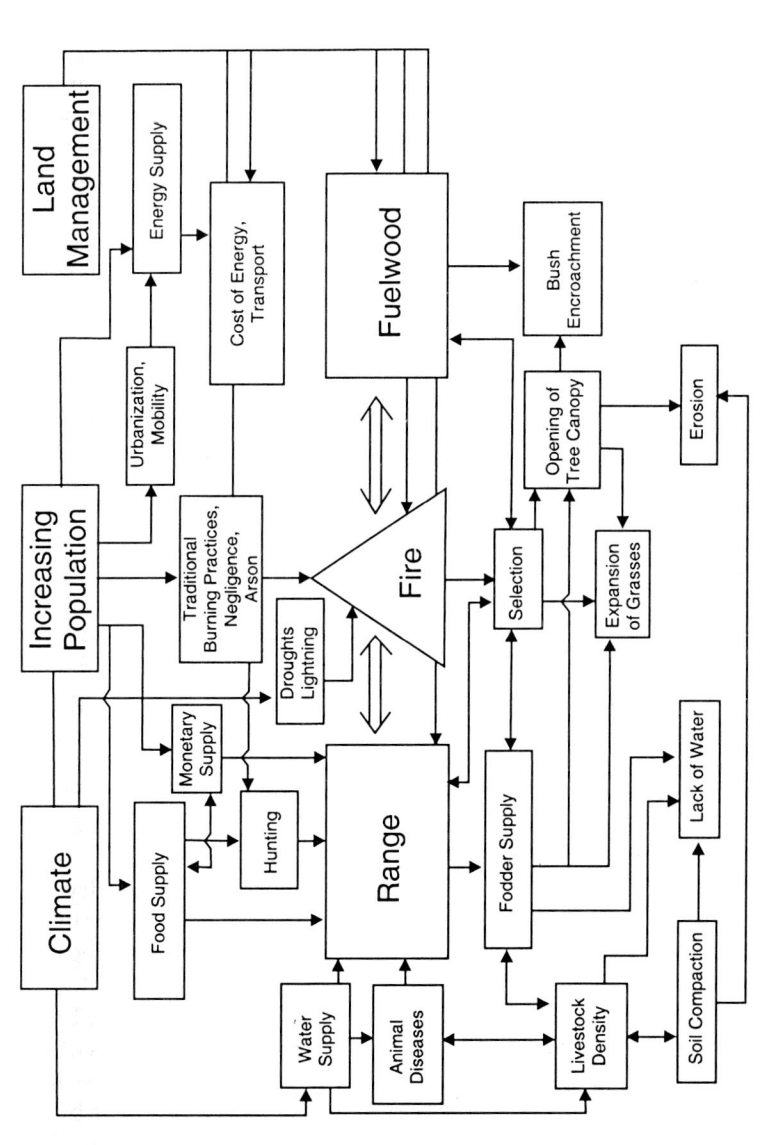

Fig. 1.6 Interrelationships of factors which determine the fire and grazing regimes and the use of fuelwood in some savannahs (after Goldammer, 1994).

The worldwide loss of tropical rain forests is now estimated to involve an annual deforestation rate of about 20 million hectares. Uncontrolled surface fires are also often caused by extreme droughts.

The dry and semi-dry forests of the wet/dry tropics and subtropics, which constitute an ecotone to the tree savannahs, cover some 700-800 million hectares. In the lowlands, the climate has marked dry seasons conducive to the formation of deciduous or semi-deciduous humid and dry forests. Leaves shed after the rainy season dry up quickly, making a highly inflammable litter under the defoliated tree canopies. The composition of the tree species is determined by their ability to resist the surface fires which usually spread every year through these stands. The highly productive, economically important fire-climax forests of southern Asia have abundant teak and other Dipterocarpaceae. Many non-wood forest products are also stimulated and harvested with the aid of fire and provide socio-economic benefits. If fire is kept out of these forests, fire-sensitive species tend to dominate, exerting an adverse effect on the silvicultural and non-silvicultural uses of the forest.

In the sub-montane to montane tropics and subtropics, the seasonal deciduous trees are replaced by fire-adapted pine forests, which also represent a successional stage.

In the densely populated cultivated landscapes of the industrialized countries in the temperate zone of Central Europe, uncontrolled forest fires and other types of vegetation fire play a less important role than in equatorial and boreal regions.

The boreal forest landscapes (taiga) cover about 1.2 billion hectares worldwide, over 70% being located in Eurasia, mostly in the Russian Federation.

Forest firest in Canadian taiga have tended to increase, and 3-5 million hectares are being burned annually. The entire circumpolar belt of coniferous forests is naturally exposed to fire, strongly determining its overall structure and dynamic characteristics (Goldammer, 1994). The low intensity surface fires in which only the litter layer and the understorey vegetation are burned, consume much less of biomass than does a high-intensity crowning fire.

Large-scale clearcuts, the use of heavy machinery on some sensitive soils, and frequent fires prevent the regeneration of the forests and lead to bog formation.

Most of the carbon released during combustion is released as carbon dioxide, followed by carbon monoxide in order of importance, the ratio of the two depending on the efficiency of the combustion. 'Hot' fires fueled by sufficient oxygen, emit very little CO. In smouldering fires (pyrolysis and incomplete oxidation of the combustible material), large volumes of methane, other hydrocarbons, hydrogen (H_2) and organic acids are emitted. Ozone is formed by photochemical processes ('smog') from CO, hydrocarbons and nitrogen oxides. Besides the gaseous compounds, aerosols are also released.

Depending on the type of vegetation and fire, a small amount of carbon is released in elemental form (as soot) and deposited in the soil, in freshwater bodies or in the oceans. Since this carbon is withdrawn from the total C-cycle, it represents a potential climatically effective carbon sink.

Vegetation fires can aid unravelling the complexity of the interactions between vegetation, humans and atmosphere and climate (see Andreae and Goldammer, 1992; Levine, 1991).

REFERENCES

Andreae, M.O., Goldammer, J.G. Tropical wildland fires and other biomass burning: environmental impacts and implications for land-use and fire management. In Cleaver, K. *et al.* (eds) *Conservation of West and Central African Rainforests.* pp. 79-109. The World Bank, Washington, D.C. (1992).

Bread for the World Institute. *Causes of Hunger.* Bread for the World Institute. Silver Spring (1994).

Castells, M. Informational capitalism and social exclusion (DP 114). UNRISD, Geneva (1999).

Crutzen, P.J., Goldammer, J.G. (eds) *Fire in the Environment: The Ecological, Atmospheric, and Climatic Importance of Vegetation Fires.* Dahlem Workshop Reports. Environmental Sciences Research Report 13. Wiley, Chichester (1993).

D'Souza, G.E., Gebremedhin, T.G. *Sustainability in Agricultural and Rural Development.* Ashgate Publishing Co., Aldershot, Hants. (1998).

FAO. Food and Agriculture Organization. *FAO Production Yearbook.* FAO, Rome (1998).

Gliessman, R.R. Agroecology: Ecological Processes in Sustainable Agriculture, Ann Arbor Press, Michigan (1998).

Gold, T. *The Deep Hot Biosphere.* Springer, Berlin (1999).

Goldammer, J.G. Vegetation fires and their effects on global climate. *Natural Resources and Develop.* 40: 99-111 (1994).

Hawken, P., Lovins, A.B., Lovins, L.H. *Natural Capitalism: The Next Industrial Revolution.* Earthscan; London (1999).

Jenkins, A. End of the acid reign? *Nature* 401: 537 (1999).

Lashof, D.A., Ahuja, D.R. Relative contribution of greenhouse gas emissions to global warming. *Nature* 344: 529-530 (1990).

Levine, J.S. (Ed.). *Global Biomass Burning.* MIT Press, Cambridge, MA (1991).

McMichael, A.J. Climate and human health. *World Climate News* No. 14, pp. 2-3. World Meteorol. Organizn., Geneva (Jan., 1999).

Murphy, F., Bendell, J. Partners in time? Business, NGOs and sustainable development (DP 109). UNRISD, Geneva (1999).

Müssig, S. Possibilities for the reduction of CO_2 and CH_4 emissions of natural gas. *Natural Res. Develop.* 40: 42-55 (1994).

Myers, N. Making sense, making money. *Nature* 402: 13-14 (1999).
Nazarea, D. (Ed.) *Ethnoecology: Situated Knowledge, Located Lives.* University of Arizona Press, Tucson AZ (1999).
Nentwig, W. The importance of human ecology at the threshold of the next millennium: How can population growth be stopped? *Naturwissenschaften* 86: 411-421 (1999).
Oerke E.C., Dehne H.W. Global crop production and the efficacy of crop protection — current situations and future trends. *Eur. J. Plant Pathol.* 103: 203-215 (1997).
Painuly, J.P. *Economic Instruments : Application to Environmental Problems.* Indira Gandhi Institute of Development Research (IGIDR), Mumbai and UNEP, Riso, Denmark (1995).
Parikh, J.K. *North-South Cooperation in Climate Change Through Joint Implementation.* Indira Gandhi Institute of Development Research (IGIDR), Mumbai (1994).
Parkes, R.J. Oiling the wheels of controversy. *Nature* 401: 644 (1999).
Pelchen, A., Peters, K.J. Pollutant gases from animal farming — influences on the greenhouse effect. *Natural Resources and Development* 40: 56-68 (1994).
Schaeffer, D. Shook up. *Sci. Internatl.* (Newsletter) No. 71: 20-21 (Dec. 1999).
Steffen, W. Global change science in the next century : a personal perspective. *IGBP Newsletter* 40: 4-5 (1999).
Titenberg, T.H. Economic instruments for environmental regulation. *Oxford Review of Economic Policy* 6 (1): 17-33 (1990).
UNDP. *Human Development Report 1999.* Oxford University Press, Oxford (1999).
Utting, P. Business responsibility for sustainable development (OPG 2). UNRISD, Geneva (2000).
Vernon, J.L. *Market Mechanisms for Pollution Control : Impacts on the Coal Industry.* IEA Coal Research, Vienna (1990).

Chapter 2
Water

INTRODUCTION

Humans have long regarded water as a freely available natural commodity. The notion that water is a finite resource to be conserved and protected is relatively recent and not yet universally shared. Generally, only in regions of chronic shortage is freshwater respected as a giver-of-life. Elsewhere, water tends to be treated as an inexhaustible resource, free to be wasted and even abused.

Some parts of the world worry about water shortages. Other regions are aware of the tremendous potential of the development of water resources for expanding agriculture, energy, industry and transportation. The manner in which this development is carried out can have enormous consequences, both locally as well as globally.

Lack of water is the limiting factor for many household and community-based activities for millions of people living in dryland areas. Rural water supply programmes tend to focus on only two social aspects: improved access to domestic supply and improved sanitation. Less attention has been paid to how communities prefer to use water for their own livelihoods. This is due, in part, to the difficulties of extracting sufficient reliable groundwater in dryland areas, and partly to a misunderstanding of why wells and boreholes fail, which leads to a general belief that abstraction should be limited to domestic supply to conserve the resource.

Implementation of roof and ground rainwater catchment systems can be helpful for meeting either total or supplementary household water requirements. However, the importance of social, economic and environmental considerations when planning and implementing projects needs to be appreciated.

The new awareness of the natural environment has focused attention on the vulnerability of water. A growing number of people now understand that water can no longer be treated as a free commodity to be exploited and polluted at will. Awareness is increasingly growing as to how critical water will be for the attainment of greater self-sufficiency in food (Eckholm, 1976). The availability of pure water for all is both a right as well as a possibility that can be fulfilled (Quigg, 1977).

The severe droughts that have affected every continent in recent years, as powerful reminders of the variability and unpredictability of weather, have come at a time of ever-increasing water consumption, mostly in the developed world. The possibility of serious conflict between nations sharing the same river basins is a matter of genuine concern today.

The world can generally meet much of its need for water merely by using it more efficiently, by careful planning based on an understanding of how the hydrosphere relates to the total environment, and by realizing that the wasteful practices of the developed countries cannot be extended to be whole world—or maintained any longer in the developed world.

Ninety-five per cent of the world's water is contained in the seas. Only 5% is fresh water, out of which 4% is frozen in the polar regions. All the water in lakes and rivers, the soil and vegetation, and all the water underground add up to only 1% of the total. Of this, 1% is liquid freshwater, 0.05% biological, 0.1% in rivers, 0.1% in the atmosphere, 0.2% in soils, 1% in lakes and about 98.5% is groundwater.

Groundwater is not found everywhere. It occurs most commonly in porous sand and gravel beneath the soil or is sandwiched between impermeable layers of clay or rock. Groundwater makes up roughly half the flow of rivers and streams. Such saturated areas form underground reservoirs (aquifers) which commonly become recharged by surface water.

Groundwater has 4 important characteristics: (i) It often is found where surface water is insufficient or negligible, e.g. in the Sahara; (ii) Such reservoirs of so-called fossil water are a non-renewable resource that were deposited millennia ago. They are not being replenished and may soon be exhausted; (iii) Underground water is less vulnerable to contamination than surface water and has some capacity to cleanse itself, especially of pathogenic bacteria; and (iv) If an aquifer does become infected with contaminants such as synthetic chemical compounds and toxic metals, it can remain polluted for generations, posing a continuing hazard to people and the environment.

Extraction of groundwater in arid regions where there is no possibility of recharge is a practice referred to as 'mining'. Where increased water supply is vital for growing populations and there are no alternatives to the mining of groundwater, the decisions as to how much water should be mined and how fast become very difficult indeed.

Water is the lifeline of the environment. Changes in water availability in terms of quantity and quality greatly affect all people. This is best exemplified by the impacts of droughts on the one hand and of flooding, on the other. A wide range of issues related to water affect humans everywhere. These include its physical characteristics; its availability both above and below ground; the uses we make of it; and how we share and manage it. Everyone needs to ponder and consider what one, as an individual can do to help conserve this precious resource for one's own use, and for that of future generations.

THE HYDROLOGIC CYCLE

Ever since prehistoric time when water first appeared, it has remained constant in quantity and continously in motion. Little has been added or lost over the years. The same water molecules have been transferred repeatedly from the oceans into the atmosphere by evaporation, dropped upon the land as precipitation, and transferred back to the sea by rivers and groundwater. This endless circulation is known as the hydrologic cycle. At any instant, about 5 litres out of every 100,000 litres is in motion (Anonymous, 1991).

Over time, major cyclic changes in climatic processes have resulted in the appearance of deserts and ice cover across entire continents. Today, regional short-term fluctuations in the order of days, months or a few years in the hydrologic cycle result in droughts and floods.

Rivers are connected to regional and continental-scale hydrology through interactions among soil, water, evapotranspiration and runoff in terrestrial ecosystems. River systems, indeed the entire global water cycle, control the movement of constituents over long distances from the continental land-masses to the world's oceans and to the atmosphere. Rivers are a central feature of human settlement and development. Rivers begin with soil moisture and runoff, each of which is linked directly to land-use and land-cover. Rivers terminate into, and are key features of, the coastal zone. Rivers deliver much more than just water to the ocean. They contribute nutrients to coastal oceans and hence have a bearing on coastal fisheries. The climate-fishery-human dynamics link is now being well appreciated.

Clearly, any understanding the hydrological cycle and water resources is bound up with properly understanding the human system as well as understanding the biogeochemical and climate systems.

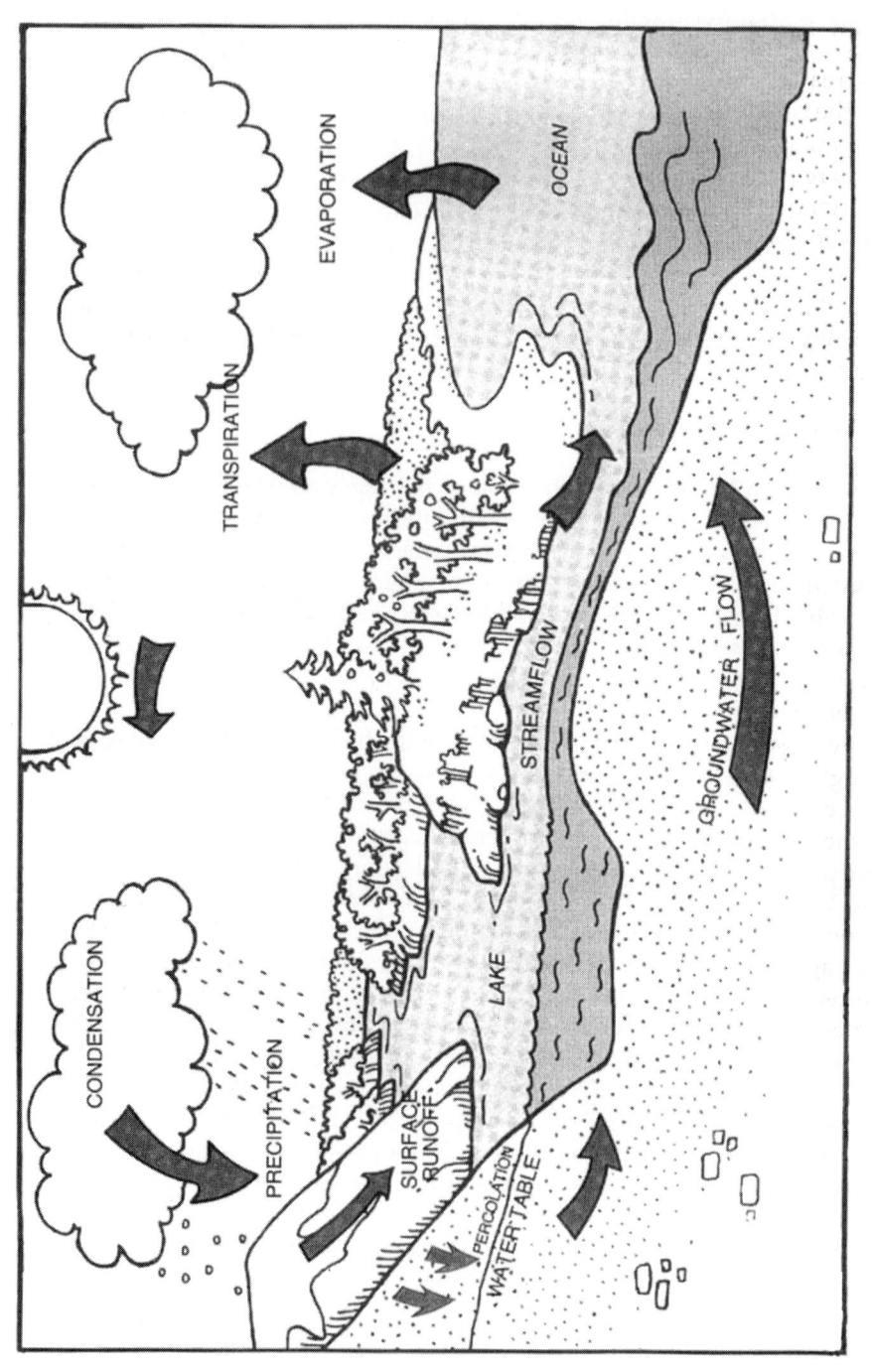

48 Environmental Technology and Biosphere Management

Fig. 2.1 Schematic sketch of the hydrologic cycle.

WATER AS A LIFE-SUSTAINING SUBSTANCE

The great importance of water for developing countries is closely linked with the issues of food and nutritional security. This is because irrigation farming, as far as it is environmentally adapted, is believed to play a crucial role in providing for the growing world population. However, the importance of water for food and nutritional security stems not only from its use for increasing the food supply by irrigation farming, but also from its role as drinking and utility water. Both water quality and access to it affect nutritional security directly. Appreciation of the dual importance of water (as a means of irrigation and drinking and utility water) clarifies the distinction between food security and nutritional security. Food security refers to the supply of a given population with sufficient quantities of food, and can be facilitated by irrigation measures. Nutritional security refers to the individual's dependence on a well-balanced diet, equitable food distribution and comprehensive healthcare for adequate energy and nutrient intake. This presupposes access to clean drinking and utility water (Herbers, 1999).

Health hazards from the consumption of or physical contact with contaminated drinking or utility water are chiefly conferred by pathogens or chemical residues contained in the water. Also, chemical toxins (nitrates, heavy metals, synthetic organochlorocompounds, etc.) in water can lead to diverse symptoms such as allergies, breathing problems, kidney damage, or foetal malformation. The risk of health impairment through chemical contaminants is quite high in industrial countries and large cities in developing countries. Because of insufficient sanitary facilities and the inadequacy or absence of water treatment for drinking or other purposes, many people in the rural as well as urban areas of the Third World are threatened with diverse infectious diseases. As a result of the increasing use of commercial fertilizers, insecticides, pesticides and other agrochemicals, chemical water pollution is also rising in rural regions, causing deleterious health effects. The greatest public health problem in developing countries is that caused by water-borne pathogens (see Fig. 2.2).

Access to proper sanitary facilities linked to effective disposal systems is an important prerequisite for the provision of clean drinking or utility water and for checking faecal-oral transmission of infectious diseases.

Health impairment through pathogens and parasites can also be the consequence of insufficient water availability. This imposes limits on personal and household hygiene (e.g. washing vegetables and fruit, cleaning dishes and household appliances, washing clothes). As a result, pathogens and other contaminants can adhere to peoples' hands, food, clothes, or any other kind of surface and, eventually, gain entry via the

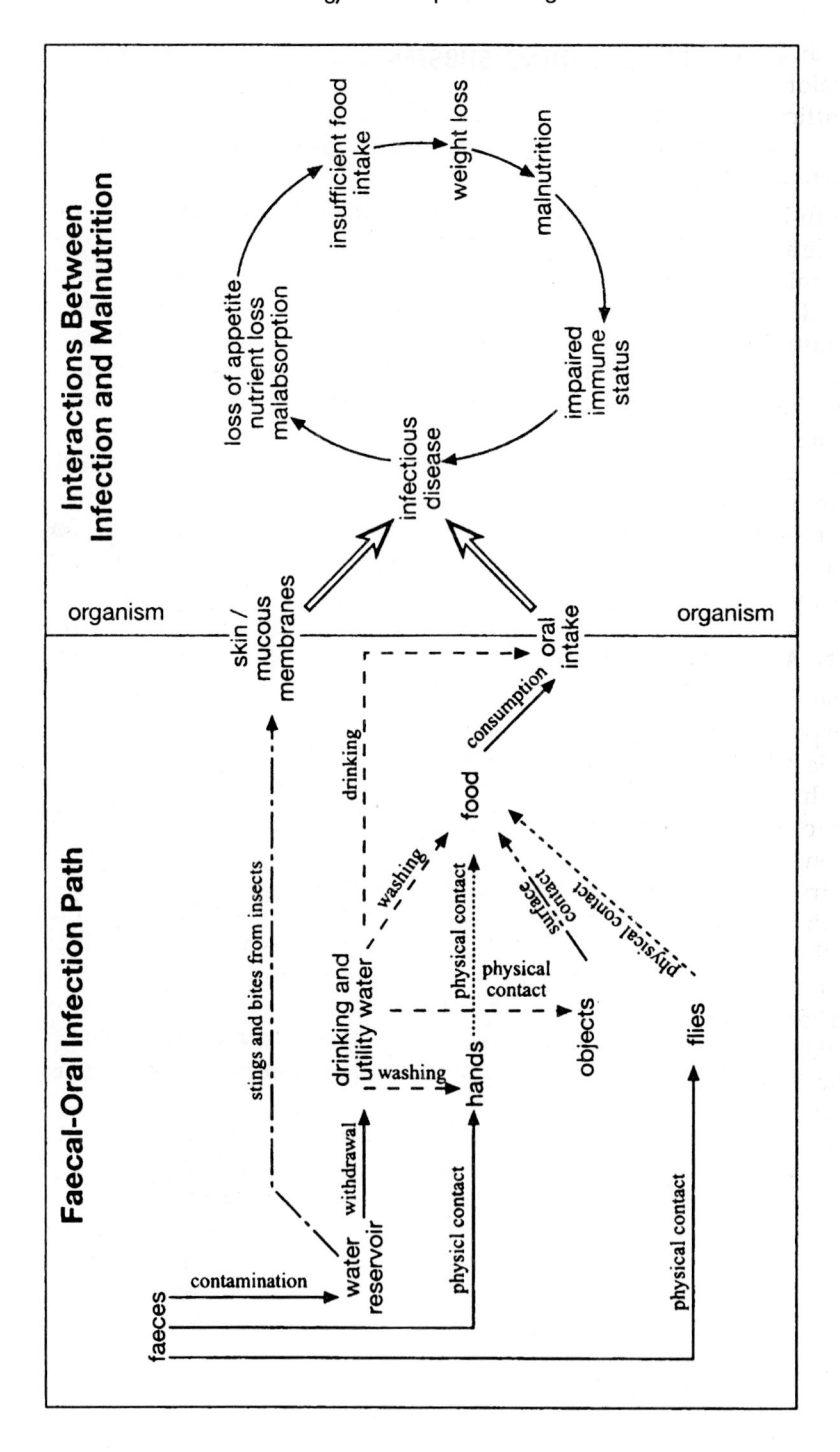

Fig. 2.2 Transmission of infectious diseases and interaction with malnutrition (after Herbers, 1999).

mouth, especially during mealtimes (Fig. 2.2). Clean water alone is, therefore, no safeguard against contamination; it also has to be available in sufficient quantity.

Water, Disease, and Nutrition

Health and nutrition are interrelated. In developing countries, many people are often deficient in both. Malnourished people often have a weak immunity and, hence, they are more vulnerable to infectious diseases that spread through contaminated drinking water or utility water. Malnutrition and infectious disease cause and reinforce one another, often making it impossible to tell which is causing which (Ramalingaswami et al., 1996). This so-called 'malnutrition-infection complex' (World Bank, 1993) is particularly dangerous for infants because their immune system is not yet fully developed. The worldwide predominant classes of infectious diseases are those of acute respiratory infections and diarrhoea—diseases that spread primarily through drinking water and utility water of insufficient quality. Nutritional security is thus in a large measure an issue of water (Herbers, 1999).

Water Availability in South Asia

In South Asia, the paucity of clean drinking water and utility water afflicts large parts of the population in developing countries. In this respect, rural people are disadvantaged vis-á-vis the urban population (Fig. 2.3 A). At least the drinking water situation is a little less critical, offering up to four-fifths of the total population access to potable water under satisfactory hygienic conditions. However, this proportion varies widely between countries, with extremes of 34% in Bhutan and 84% in Bangladesh, and also shows large disparities between rural and urban regions. Nevertheless, there has been much recent progress. From 1980 to 1990, the proportion of the Pakistani population supplied with clean drinking water rose from 25% to 68%, whereas the same proportion in Nepal rose from 8% to 42% (UNDP, 1995). Yet, despite this encouraging development, large parts of the South Asian population still continue to be without a regular supply of clean drinking water and sanitary facilities (see Fig. 2.3 B). To some extent, the progress made in developing elementary health protection facilities has been neutralized by the increased population growth in this region.

Healthcare is a basic link in the cause-and-effect chain from drinking and utility water to infectious disease and on to nutritional security. To check the adverse effects of water-borne and other diseases on people, it is necessary to have access to immediate and effective medical treatment. This is not so in many villages.

52 Environmental Technology and Biosphere Management

In developing countries, over one-third of all deaths can be attributed to infectious and parasitic diseases (WHO, 1996). In keeping with the global trend, respiratory disorders and diarrhoea tend to be the most frequent causes of infant mortality.

Pathogens and parasites taken up by humans with water, whether orally or via the skin, use up the body's nutrients. In malnourished people, they seriously compete for the body's limited energy and food reserves, resulting in weight loss and impairment of the person's nutritional status. Water is, thus, a critical factor of nutritional security in developing countries (Herbers, 1999).

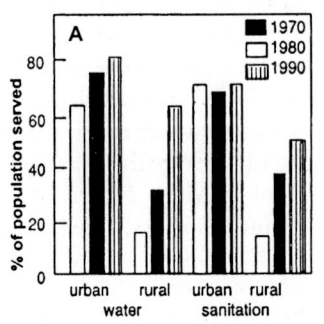

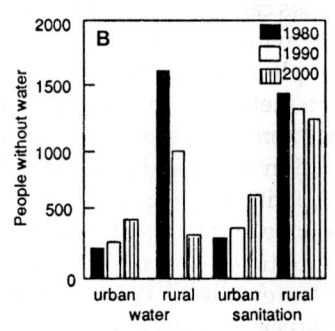

Fig. 2.3 Water supply situation in developing countries. A. Number of people without water supply and sanitation. B. Water supply and sanitation coverage (after UNEP, 1992).

While adequate food intake is essential for the development and maintenance of a human's physical and mental abilities, it is certainly not a sufficient one. Another need is the elimination of noxious environmental agents. For this, it is essential to improve the quality and supply of drinking water and utility water. Water quality can be improved by implementing a multibarrier concept which promotes the installation of sanitary facilities, provides hygiene advisory services, and improves practical healthcare. This concept can materialize if all measures are carried out in parallel. By helping to prevent disease, improved water and sanitary facilities also serve as a safeguard against malnutrition. Conversely, by avoiding nutritional deficiencies, one can reduce the probability of infectious diseases and other ailments.

Nutritional security will continue to be at risk in Third World countries as long as the present adverse situation of the water supply and sanitary facilities persist even when food supply is adequate. Energy and nutrient deficiencies resulting from poor appetite and resorption losses in disease lead to such anthropometric deficiencies as stunting, wasting and underweight which, in turn, prevent the affected person from developing his or her full physical and mental abilities. Contaminated drinking water and utility water is a serious impediment to development in Third World countries, both for the individual as well as society as a whole. Water is not only a critical factor of nutritional security, but also an important development factor that determines the future of countries in the Third World (Herbers, 1999).

Paradox of the Plankton

In aquatic ecosystems, the biodiversity puzzle is particularly troublesome, and known as the paradox of the plankton (Wilson, 1992). Competition theory predicts that, at equilibrium, the number of coexisting species cannot exceed the number of limiting resources (see Grover, 1997). For phytoplankton, only a few resources are potentially limiting: nitrogen, phosphorus, silicon, iron, light, inorganic carbon, and sometimes a few trace metals or vitamins. However, in natural waters, dozens of phytoplankton species coexist. Huisman and Weissing (1994, 1999) offered a solution to the plankton paradox. First, they show how resource competition models can generate oscillations and chaos when species' compete for three or more resources. Second, these oscillations and chaotic fluctuations in species abundances allow the coexistence of many species on a few resources. According to them, the model of planktonic biodiversity may be broadly applicable to the biodiversity of many ecosystems.

The work of Huisman and Weissing has some wider implications beyond the plankton system studied by them. Their results demonstrate that competition, in its broadest sense, is not a simple process. Competitive systems may display highly dynamical phenomena, with continuous shifts and changes in species composition. Also that competition is not necessarily a destructive force. Competitive interactions that generate oscillations and chaos may allow the persistence of a great diversity of competitors on only a few limiting resources.

GROUNDWATER

Two-thirds of the world's fresh water is found underground. This water exists in aquifers and appears at the surface as springs. Quite often, groundwater is interconnected with the lakes and rivers (see Fig. 2.4).

Groundwater occurs in the tiny spaces between soil particles (silt, sand and gravel) or in cracks in bedrock, much like a sponge holds water. The underground aquifers are the source of wells and springs. It is the top of the water in these aquifers that forms the water table.

Groundwater resources are depleted or 'mined' when pumping from an aquifer is not matched by recharge. This can happen in two ways: (1) by overpumping or; (2) by decreasing recharge due to drought, for

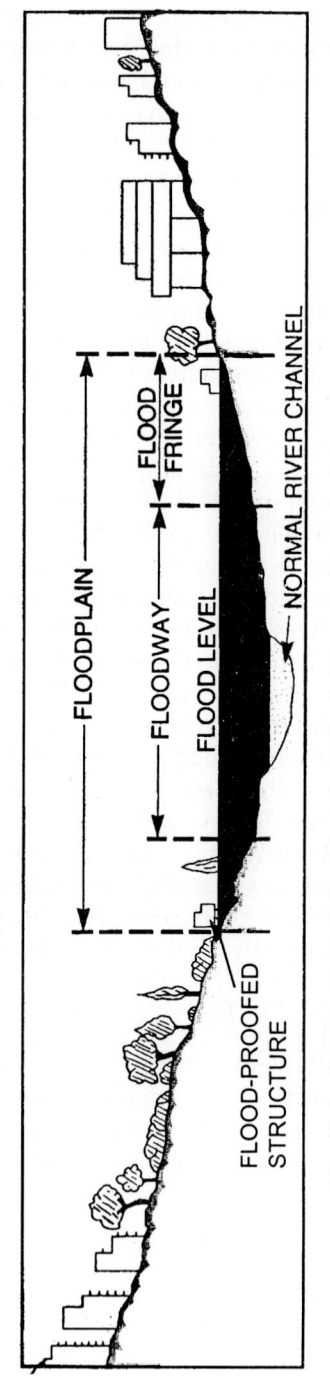

54 Environmental Technology and Biosphere Management

Fig. 2.4 Schematic sketch of flooding, floodplain and floodway (after Anonymous, 1991).

example. The drying up of an aquifer is not to be confused with the failure of individual wells in that aquifer, which occurs more frequently. A well may fail if it is too shallow, so that a temporary decline in water level lowers the water table below the bottom of the well. Or, plugging of the screen at the opening of the pipe at the bottom of the well by mineral and/or bacterial deposits can occur, leading to well failure.

Groundwater recharge is the replenishment of water in an aquifer. Much of the natural recharge of groundwater occurs as the result of the melting snow or from streams in mountainous regions where the water table is usually below the bottom of the stream bed. It can also occur during local heavy rainstorms. Often, groundwater discharges into a river or lake, maintaining its flow in dry seasons.

Groundwater is extremely important in supplying fresh water to meet the needs of people. The interdependency of surface water, groundwater and atmospheric water is very important in the hydrologic cycle. The role of groundwater is critical and its most significant function is its gradual discharge to rivers in order to maintain streamflow during dry weather periods.

Besides supplying human needs, groundwater is also used for livestock watering, aquaculture and mining.

Groundwater becomes contaminated when anthropogenic substances are dissolved in water recharging the groundwater zone. Examples of this are road salt, petroleum products leaking from underground storage tanks, nitrates from the overuse of chemical fertilizers or manure on farmland, excessive applications of chemical pesticides, leaching of fluids from landfills and dumpsites, and accidental spills. Contamination also results from an overabundance of naturally-occurring iron, manganese and substances such as arsenic. Excess iron and manganese are the most common natural contaminants. Another form of contamination results from the radioactive decay of uranium in bedrock, which creates the radioactive gas radon. Methane and other gases sometimes cause problems. Seawater can also seep into groundwater, and is a common but isolated problem in coastal areas. It is referred to as saltwater intrusion (Anonymous, 1991).

Groundwater is generally safer than surface water for drinking because of the filtration and natural purification processes in the ground. These processes can become ineffective by sewage, fertlizers and toxic chemicals which seep into the ground.

Household, commercial and industrial wastes that end up in dumps, waste lagoons, or septic systems, can pollute groundwater. Acid rain can recharge aquifers with contaminated water.

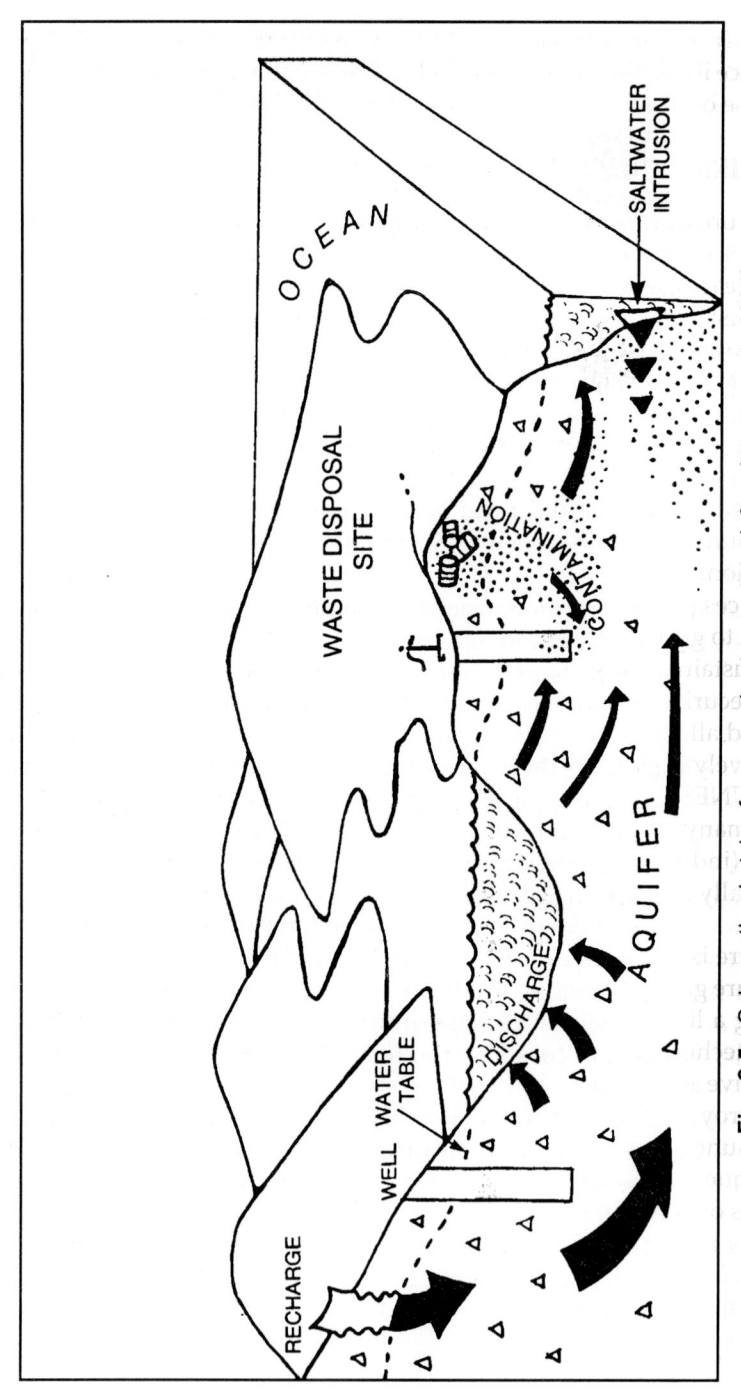

56 Environmental Technology and Biosphere Management

Fig. 2.5 Schematic sketch of a groundwater system (after Anonymous, 1991).

Generally, groundwater is not as easily contaminated as surface water, but once it has become contaminated, it is much more difficult to clean up because of its relative inaccessibility.

SUSTAINABLE EXPLOITATION OF GROUNDWATER RESOURCES

Water undoubtedly is one of the most pervasive substances on Earth, but only 2% of it is freshwater. The current rate of withdrawal is about 3500 km^3/year, some 2100 km^3 for consumptive use, while 1400 km^3 of wastewater is returned to rivers. Groundwater withdrawals (percentage by sectors) are: domestic (8), industry (23) and agriculture (69) (Shiklomanov, 1991). Margat (1991) ranked the top 15 countries, in terms of the amount of groundwater pumpage (in km^3 per annum) as follows: India 150, USA 101, China 74.6, USSR (former) 45, Pakistan 45, Iran 29, Mexico 23, Japan 13.1, Italy 12.1, Germany 9.5, Saudi Arabia 7.4, France 7, Spain 6.3, Turkey 5 and Madagascar 4.9.

Groundwater is mainly threatened from point and diffuse source pollution. Therefore, quality protection is the key issue of groundwater resources policy in the industrialized countries. There are many sources of risk to groundwater. The current practices in many countries lead to a non-sustainable use of groundwater systems. Human health and welfare, food security, industrial development and the ecosystems on which they depend all face risk, unless water and land resources are managed more effectively in the present decade and beyond than they have been in the past (UNESCO-WMO, 1992).

In many countries, sustainable use of groundwater for drinking and other (industrial, ecological) functions is being seriously threatened, especially in the agricultural and industrial core regions of the European Union.

There is need to satisfy present needs without jeopardizing the ability of future generations to satisfy theirs. Overexploitation may be permitted during a limited period to allow better use of other resources, or while other technologies develop. Uncontrolled aquifer development leads to extensive aquifer exploitation and even to severe forms of overexploitation (Villarroya and Aldwell, 1998).

Groundwater overexploitation results in a series of adverse consequences, such as increasing water cost, environmental changes (effects on wetlands, salinity problems), reduction of other water sources already in use, water salinization and impairment of its quality. The net result, however, may be either negative or positive. Positive results sometimes dominate at a regional level, but negative results tend to be long-term.

In order to evaluate groundwater exploitation, not only negative effects but also positive ones have to be considered. Perhaps, the most serious cause of aquifer exploitation may be ignorance of what is happening, and negligence in producing the data needed to evaluate the hydrogeological and economic situation correcly. Another harmful effect is the irresponsible over-reaction of water authorities, especially when they are poorly informed or lack the scientific and technical skills to evaluate emerging problems correctly. Any exploitation of water resources that is not managed in an integrated way, taking into account the needs of the present and future generations, puts these resources at risk. Such development, therefore, is patently unsustainable.

Overexploitation of an aquifer also has a marked influence on the planning and uses of land. Water users' associations should play an important role in regulating the exploitation of natural resources in general, and especially of water resources. The participation of citizens through NGOs such as the water users' associations can be very effective and, on many occasions, has alleviated serious problems concerning the exploitation of the water resources of a region. The associations can be effective as watchdogs to ensure that restrictions and regulations are being complied with. Water authorities should treat them as allies, not opponents.

In the short term, groundwater quality problems are more serious than those of quantity. In the near future, the implementation of measures to prevent pollution of groundwater from diffuse and point sources will probably be the main issue.

Only with solidarity, subsidiarity and involvement can sustained groundwater development be achieved. Practical application demands a parallel effort in education and information for the general public.

WATER QUALITY

Water quality refers to the chemical, physical and biological content of water. The water quality of rivers and lakes changes with the seasons and geographic areas, even when there is no pollution present. There is no single measure that constitutes good water quality. For instance, water suitable for drinking can be used for irrigation, but water used for irrigation often does not meet drinking water guidelines.

Many factors influence water quality. Substances present in the air affect rainfall. Dust, volcanic gases and natural gases in the air such as carbon dioxide, oxygen and nitrogen are all dissolved or entrapped in rain. When other substances such as sulphur dioxide, toxic chemicals or lead are in the air, they are also collected in the rain as it falls to the ground.

Estimates vary, but it is commonly believed that there are up to 100,000 chemicals in commercial use worldwide, with about 1000 new chemicals entering the market every year.

The runoff from urban areas collects debris littering the streets and takes it to the receiving stream or water body. Urban runoff worsens the water quality in rivers and lakes by increasing the concentrations of such substances as nutrients (phosphorus, nitrogen), sediments, animal wastes, petroleum products and road salts.

Industrial, farming, mining and forestry activities also significantly affect the quality of rivers, lakes and groundwater. Farming can increase the concentration of nutrients, pesticides and suspended sediments. Industrial activities increase concentrations of metals and toxic chemicals, add suspended sediment, increase temperature, and lower dissolved oxygen in the water. All these effects have a negative impact on the aquatic ecosystem and/or make water unsuitable for established or potential uses.

Good quality drinking water is free from disease-causing organisms, harmful chemical substances and radioactive matter. It tastes good, is aesthetically appealing, and free from objectionable colour or odour.

Municipalities have the responsibility to provide their citizens with safe drinking water and provide sufficient warning about pollution risks. Samples should be regularly collected and analyzed to check drinking water quality.

There is a difference between 'pure water' and 'safe drinking water.' Pure water, often defined as water containing no minerals or chemicals, does not exist naturally in the environment, but under ideal conditions, it may be made by distilling water. Safe drinking water, on the other hand, may retain naturally-occurring minerals and chemicals such as calcium, potassium, sodium or fluoride which, in proper amounts, are actually beneficial to human health and may also improve the taste of the water. Where the minerals or chemicals occur naturally in high concentrations, certain water treatment processes are used to reduce or remove the substances. In some places where the desirable chemicals are deficient (e.g. fluoride), they are actually added to produce good drinking water. (Fluoride reduces dental cavities).

WATER MANAGEMENT

Increasing use of water with urban, industrial and agricultural expansion leads to increasing competition for the same water supply and necessitates water management. This involves the anticipation and/or resolution of user conflicts in a manner that protects the environment. The objective of any good water resource management is to maintain a balance between growing social and economic demands, and the continued ability of our fresh water resources to support them.

All water development projects affect some part of the environment. Smaller projects such as operating a water intake pipe harm the environment to a lesser degree than, for instance, a large-scale hydroelectric project that diverts and stores large quantities of water. Happily, the impacts of such a megaproject can be reduced. After a site has been chosen, field studies and a literature review will provide an understanding of the existing environmental conditions, permitting prediction of the impacts that the project will have on the environment. Design engineers working with other professionals such as biologists may then mitigate or minimize impacts by adjusting the project design.

Screening determines whether a water resource project proposal, such as filling a reservoir or dredging a harbour, will have adverse environmental effects and whether these can be corrected. If so, the project may be allowed to go ahead without detailed environmental impact studies but only after scientists and water managers have designed adequate protective measures.

For water development projects, these measures might include the construction of fish ladders for migrating fish, replacement of wetland habitat for waterfowl nesting and industrial processes to reclaim pollutants and prevent their entry into the hydrologic cycle. These measures, along with their costs, are then considered as a part of the overall project.

An ecosystem or holistic approach to water management requires an understanding of the interrelationships of the biological, chemical and physical properties of an aquatic ecosystem. Once we understand the ecosystem, we can take certain measures to minimize large-scale and long-term impacts resulting from human uses.

In cases where these impacts cannot be avoided, alternative measures can be taken. For example, the loss of fish habitat caused by dam construction may require the operation of a fish hatchery to replace young fish which can no longer be supplied by the lost habitat (Anonymous, 1991).

Water management also deals with flood protection. The most popular methods of flood protection are engineering works such as dams, dykes and diversion channels, but these are expensive to build and maintain and yet they do not necessarily provide a complete guarantee of safety. During extreme events, floodwaters can sometime overtop dykes and exceed the capacity of reservoirs and diversion channels.

It is now possible to protect existing buildings or new construction by various flood-proofing methods. For example, buildings may be elevated on posts, piers or landfill. Floodwalls or ring dykes can protect groups of

Water   61

buildings. Foundations and basements can be designed to allow some flooding. Consideration needs also to be given for the protection of electrical, sewage and other services.

Water research is necessary for water management. The information/data generated can go a long way towards protecting and conserving the aquatic environment, and to manage it in ways that will continue to make it available for use by us and and our future generations.

Environmental monitoring, resource inventories and field studies all provide a record of past or present water resource conditions. The data describe the state of the resource for different geographic locations and at different times; the physical, chemical, biological characteristics of the water; and the economic, social and institutional makeup of the system of which the water resource is a part.

Other information is obtained by studying changes in the resource over time. For this, an understanding of the cause-effect relationships between different environmental components, the water resource and human activities is essential.

IN-HOUSE WATER CONSERVATION

As the supplies of clean, usable water are limited, people must learn to use them more wisely. Water conservation begins at home, and individuals can do their share by observing the following DOs and DON'Ts in and around the house.

In the Kitchen

- Be sure to always turn taps off tightly so they do not drip.
- Promptly repair any leaks in and around your taps and faucets.
- When hand-washing dishes, do not run water continuously. Wash dishes in a partially-filled sink, and then rinse them using the spray attachment on the faucet.
- When cleaning fruit and vegetables, never do so under a continuously running tap. Wash them in a partially filled sink and then rinse them quickly under the tap.
- When boiling vegetables, save water by using just enough to cover them and use a tightly fitting pot lid.

In the Bathroom

About 75% of indoor home water use occurs in our bathrooms, and toilets are the single greatest water users. Similar rules apply here as they do in the kitchens.

- When washing or shaving, partially fill the sink basin and use that water rather than running the tap continuously.

- When brushing teeth, turn the water off while actually brushing instead of running it continuously. Then use the tap again for rinsing, and use short bursts of water for cleaning the brush.
- Be sure to always turn taps off tightly so they do not drip.
- Promptly repair any leaks in and around taps and faucets.
- Reduce water usage by about 20% by placing a weighted plastic bottle filled with water, in the water tank of your toilet. You can reduce water usage by 40% to 50% by installing low-flush toilets.
- Low-cost 'inserts' for the toilet tank are an alternative to plastic bottles. With a toilet insert, a family of four could save 45,000 litres of water per year. Toilet inserts are now available at hardware and pulmbing supply stores in most industrialized countries.
- Toilet should be flushed only when really necessary. It should not be used as a garbage can to dispose of cigarette butts or paper tissues.

WATER FOR CITIES

One of the most important prerequisites for sustaining human life and achieving sustainable development is freshwater supply. With human populations growing and freshwater demand increasing worldwide, particularly in urban areas, the challenge of supplying sufficient water to meet the needs of society has become a most urgent problem. Over a billion people around the world lack access to satisfactory supplies of freshwater, not only in rural areas but also in cities.

There are two options to ensure adequate water supply: (1) finding additional water resources using conventional centralized approaches; or (2) more efficiently utilizing the limited water resources already available. Much attention has been given to the former, while only limited attention has been given to optimizing water management systems. Better water resource management is the need of the hour to increase the availability of water for urban use.

Health and sanitation need to be considered in the planning of water supply systems, especially since the lack of sanitation continues to be one of the greatest threats to urban populations. There is need for governments to focus on establishing long-term capacity and changing people's attitudes toward water resources.

Harvesting and utilization of rainwater is a suitable option for villages and large cities at the household, community, and institutional/commercial levels. Due attention needs to be given to planning and design of rain water collection and utilization systems; water quality and regulatory aspects; costs and innovative financing options; public acceptance; and, institutional barriers. There should be an integrated approach to improving water resource management, leading to the establishment of a more resilient, autonomous hydrological cycle at the local level.

Reuse of treated human wastes is a major source of water that is often ignored, and on-site decentralized treatment and reuse should be analyzed in terms of cost-effectiveness, health risks, ecological implications and overall benefits. Reliable measures are required to prevent the use of raw sewage in fields of edible crops, while at the same time it is necessary to promote and create incentives to connect flows of treated effluents to agricultural areas.

Augmentation of Groundwater Resources through Aquifer Recharge

Aquifer recharge has several advantages: decentralized treatment with subsequent aquifer recharge in strategically located recharge sites is a cost-effective option for mega cities but it is important that the level of nitrogen in aquifer recharge water is controlled and monitored. Storm water aquifer storage and recovery is a viable process and needs to be further developed for establishing storm-water detention ponds in urban areas.

Leakage Control and the Reduction of Unaccounted Water (UFW)

New approaches are needed for reducing water loss, specifically the types of leaks related to pipe material performance, age, and other technical considerations, as well as labour costs, with a view to more cost-effective management of both public and private water supply systems.

Water Demand Management

The social and economic characteristics of managing water resources also need to be considered carefully. Water conservation is the cheapest and largest available source of water within cities and appropriate water management is an integral element in sustainable development. Complete metering is essential for demand-side management.

Greater focus is needed on water demand management in low-income urban areas in developing countries. Often, the urban poor only have access to expensive water and, thus, these people possess a high level of awareness about water conservation. The key issue is not that consumption should be curbed among the poor, but how to provide water to the vast majority of the urban population. This implies conservation by the more affluent sectors. A significant amount of water is also lost due to leakage. Water losses must be reduced. In many countries, reducing the amount of water used in toilets and eliminating leaks are two of the most efficient approaches to water conservation (Fujita, 1999).

WATER RESOURCES AND ENVIRONMENTAL PLANNING

Environmental planning means the regulation and development of the interaction between natural environment and social development, or the control of interrelationships between nature and society, to optimize the interdependence between the natural basis of society and the structure of social activities and production.

In industrialized countries, environmental planning is chiefly determined by the existing governmental authorities of classical material infrastructure and sector planning such as highway construction, water supply, electric power supply, etc. These institutions have, in principle, been following the classical hierarchic approach of a technical and bureaocratic sector activity. A modern environmental planning system demands a comprehensive approach and effective structures suitable for integrated and holistic planning.

Water resources development or management exemplifies one of the classical sector policies. Water resources development has traditionally dealt with existing parts of nature and with its exploitation and alterations. There is a need to redefine its goal of management as ecologically sustainable development (see Förch, 1992). The chief object-ives of water resources development may be defined as; (1) the protection of the water resources of a region against overexploitation and pollution, and (2) the optimization of economic water use or utilization for different purposes.

It is the aim of water resources planning to maintain or to re-establish the balance between the water cycle (Fig. 2.6) and the demands of the man-made social systems utilizing the natural source of water.

This general hychological cycle is divided into several subcycles, following regional or structural needs.

In respect of water, environmental planning involves the following aspects:

1. It is the protection of the natural environment against any kind of destruction in order to keep nature in good function or state as the basis of society.
2. It is the balancing of human needs with nature's capacities.
3. It is no longer the mere adaptation of nature to satisfy human demands or the exploitation of nature into a handy tool for human needs (like in modern biotechnology).

Water resources planning is now becoming an active measure for nurturing the interrelationships between natural environment and social development to sustain the co-existence of nature and humans.

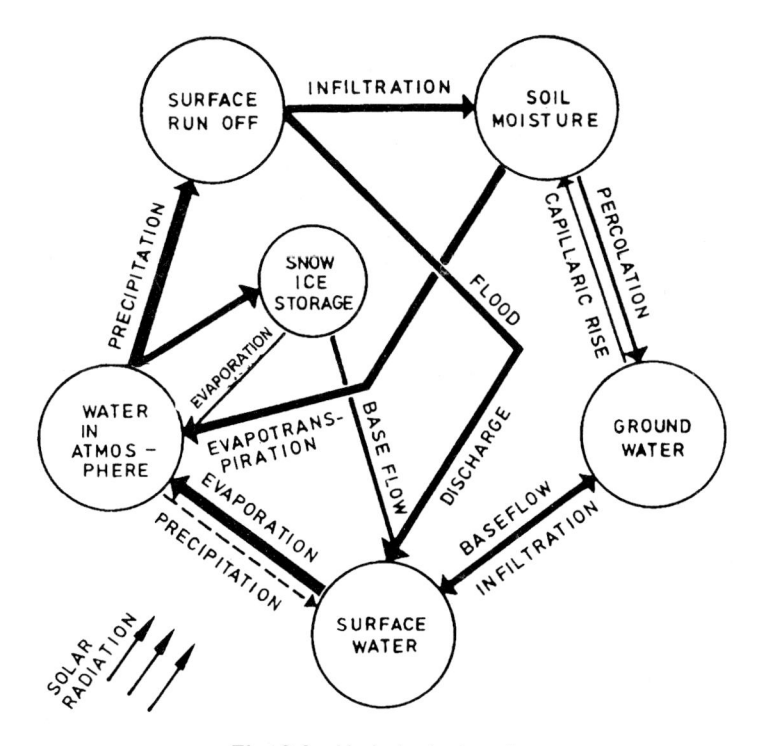

Fig. 2.6 Hydrological cycle.

Listed below are some areas where the 'water demand' directly requires the use or consumption of the source effecting its amount, availability and quality:

1. Water intake (from surface or groundwater) depletes the local source, the temporarily stored volume, with effects on the hydraulic system, aquatic life and vegetation.
2. Water use for irrigation additionally increases the amount of evaporation and infiltration with effects on soil structure, groundwater and water quality.
3. Discharge of used water (into surface or ground water) increases discharge volume with effects on water quality, aquatic life and vegetation.
4. The use of hydropower for generating electricity influences the discharge volumes; if storage in reservoirs is used, it increases evaporation and infiltration, with effects on vegetation and local climate (Förch, 1992).

The following factors influence the natural water cycle indirectly:

1. Human settlements affect the water cycle because of increasing surface runoff due to surface sealing and eradication of natural vegetation and changes of local climate.
2. Traditional forestry and agriculture generally interfere with the natural water cycle as they involve the use of chemical fertilizers, pesticides, herbicides, etc. Monocultures increase surface runoff and interflow, whilst decreasing groundwater recharge.
3. River basin development for improving transport facilities, flood control, etc., tends to change discharge conditions for surface and subsurface water with effects on vegetation and aquatic life.

All human activities related to urbanization and industrialization somehow influence the water cycle. Some of these interferences may be tolerated, but some are to be avoided while some can be minimized. Balancing the demand with the available supply helps solve not only quantitative but also qualitative problems.

WATER AND CLIMATE

There is a strong relationship between water and climate. From a water resource perspective, the climate of a region largely determines the water supply in that region, depending on the precipitation available and on the evaporation loss. Large water bodies such as the oceans and large lakes moderate the local climate by acting as a large source and sink for heat. Regions near these water bodies generally have milder winters and cooler summers than would be the case if the nearby water body did not exist.

The evaporation of water into the atmosphere requires an enormous amount of energy, which ultimately comes from the sun. When water vapour in the atmosphere condenses to precipitation, this energy is released into the atmosphere. Thus water acts as an energy transfer and storage medium for the climate system (Anonymous, 1991).

Because there is an intimate relationship between climate and the hydrologic cycle, changes in the climatic regimes directly affect the average annual water flow, its annual variability and seasonal distribution. Greater climatic variability means changing the frequency of extreme weather events and increasing the incidence of dry and wet year sequences.

Drought

A drought is a sustained and regionally-extensive occurrence of significantly below-average natural water availability, in the form of precipitation, streamflow or groundwater. Droughts are natural events which have occurred throughout history and may be considered as temporary features caused by fluctuations in the climate system. They can occur anywhere. Regions with a semi-arid or arid climate and which have only marginal annual precipitation to meet the water demands are particularly vulnerable to droughts.

Table 2.1 Summary of habitat and biological parameters in SHMAK.

Parameter	How?	Importance in stream ecosystem	Notes on scoring system
Water velocity	Timing a flouting object (an orange).	Different organisms tolerate different velocities; indicates dissolved oxygen (e.g. rapid flows = higher DO).	Score 10 for 0.3–0.7 m/s, ideal for much stream life; lowest scores for very slow and very fast flows, which exclude organisms.
pH	pH papers, range 5 to 9	Many organisms have a preferred range; high pH can increase toxicity (e.g. ammonia).	Around neutral (pH 6.5–7.5, score (0) is ideal. Very high (>9.5) or very low (<5) knocks out stream life (score –5).
Water temperature	Spirit thermometer.	Organisms have preferred range/tolerances. Often maximum temperature reached during the day is important.	Temperatures of 10 to 15°C are ideal (score 10). Over 30 can be lethal to some invertebrates (–5).
Conductivity	Conductivity meter, range 0 to 1000 mS/cm.	Measures amount of dissolved substances (from rocks (natural) and other inputs. Can indicate nutrient content. Influences plant/periphyton growth.	Under 50 suggests clean water (score 20). Over 250 suggests inputs from rock, but could also indicate runoff or point source inputs (score 6). Over 400, scores 1.
Water clarity	SHMAK clarity tube.	Aesthetics – clear water looks cleaner; visibility for invertebrates; light for plants.	Score 10 for "clear to end of tube"; decreasing to 1 as water gets more turbid.
Composition of stream bed	Visual estimate, % cover.	Habitat for stream life; a major determinant of the type of life in a stream.	Large cobbles provide both shelter and stability ans score highest (20). Silt and sand exclude many organisms (score –20).
Deposits	Visual assessment	Indicates recent settling of suspended material; large amounts smother habitat.	No deposits score 10; thick deposits, –10 because of the potential to degrade habitat.
Bank vegetation	Visual estimate, % cover.	May provide shade, bank stability, habitat for adult invertebrates, buffer to absorb nutrient inputs.	Best score (10) for native trees, wetland veg.; short pasture/ bare ground promote runoff (score–10).

Contd.

Table 2.1 *Contd.*

Invertebrates	Identify from charts, count.	Have preferences for (or tolerance of) certain habitats and water quality.	Stoneflies, etc., need clean, oxygenated waters (score 10). Worms, crustaceans tolerate pollution (score 1–2).
Periphyton	Identify from charts, estimate % cover.	Responds to flow regime and increased nutrients. Changes in types good indicator of inputs.	Thin brown films grow in clean, swift flows (score 10), long filaments in low flows, high nutrients (score 1).

Floods

Floods are almost always natural occurrences. Exceptionally, flooding can occur by the collapse of a dam. Many conditions and variables determine whether a lake or river overtops its banks or an ocean rises along its shores. Heavy rainfall can cause floods, so also melting of heavy winter snowcover.

A floodplain is usually divided into two categories; the floodway and the flood fringe. Damage is most extensive in the floodway where the water volumes and velocities are the greatest. It is within these areas that one should not build, since considerable damage can occur not just once in a lifetime but again and again. Building work can be undertaken within the flood fringe providing that some protective measures such as adequate floodproofing are adopted.

WATER AND AGRICULTURE

The development and management of water resources is an essential prerequisite for achieving food security (Eckholm, 1976; Layne, 1976).

The earliest civilizations emerged in areas where soils were rich and water plentiful and vanished when soil became waterlogged and saline through faulty irrigation, when watersheds became deforested, and when soil erosion and silting destroyed the very basis of agriculture.

Today, irrigation may be considered to be as much a problem as a solution to improved agriculture. About 10 million hectares of arable land are being abandoned every year because of saturation, salinity or alkalinization. The problem occurs in the tropics and, indeed, in every country with substantial irrigation.

Salt is a common mineral in earth and water, and creates no problem because rain flushes it through topsoil, or seasonal flooding washes it away. But if these natural processes are interfered with as a result of dams and irrigation works, salt accumulates as the water (but not the salt) evaporates. Salinity can soon reach levels that are toxic for most plants.

In recent decades, several high dams have been built in many places. The multi-purpose uses of the waters they impound—irrigation, power generation, flood control, urban water supply, fisheries, transportation, recreation—are claimed to offer solutions to diverse problems in one majestic construction. Yet experience has often been that high dams can create more problems than they solve. Some negative aspects of high dams are outlined below.

1. **Silting.** Silt, a precious commodity in agriculture but it amounts to a double loss when it accumulates behind a dam. It fails to reach the floodplains where it is needed to renew the soil, and it gradually fills up the reservoir, reducing the dam's storage capacity for power generation.

2. **Salting and waterlogging.** High dams raise waer tables until subsoils are perpetually sodden; controlled irrigation made possible by the dams leads to a fatal build-up of salts; fields that no longer have a chance to dry out become waterlogged.

3. **Evaporation and seepage.** In the tropics, wherever man-made reservoirs have been created, evaporation and seepage have been underestimated. This severe loss not only requires recalculation of a reservoir's potential for power and irrigation, but may also lead to even less foreseen consequences. For example, as a result of seepage from Nagarjunasagar Dam in India, the water table rose, increasing the alkalinity of the soil and changing the chemical balance of sorghum, the staple diet of the region's poor. The effect was to increase their intake of molybdenum which, in turn, caused a high rate of excretion of copper. That resulted in a crippling bone disease (*Genu valgum*) so deforming that some victims are unable to walk (Agarwal, 1975; United Nations, 1976).

4. **Destruction of habitats.** People have to be evicted from their ancestral homes in order to flood a great river valley and rehabilitated elsewhere. This is a difficult task. The destruction of habitats of wildlife and the drowning of creatures isolated by rising waters also deserves consideration.

5. **Waterborne diseases.** Man-made reservoirs and the irrigation works they serve greatly increase the incidence of waterborne diseases.

6. **Earthquakes.** Seismologists believe that the almost incomprehensible weight of large man-made reservoirs increases the risk of earthquakes. In areas around the Koyna Dam in India, where there had been no significant seismic activity before construction began, hundreds of shocks have been recorded, several above 5 on the Richter scale. A magnitude 6 earthquake did extensive damage and killed 200 people.

7. **Fisheries.** The hope that man-made reservoirs can be turned into productive fisheries has often not materialized. The deep waters contained by most high dams tend to be sterile, with no food for fish. High dams have damaged or destroyed remote estuarian and coastal fisheries by cutting off the nutrients that formerly flowed down to the sea. Dams can eliminate the catch of migratory fish by preventing them from reaching their spawning grounds.

8. **Waterweeds.** The damming of watercourses aggravates the proliferation of water hyacinth and other aquatic weeds. In places where waterweeds were once deterred by turbid river flow or were flushed away by annual flooding, with damming, they have free opportunity to take hold and multiply in reservoirs and irrigation channels. In India, over 1,600 kilometers of canals remain unused because weeds have blocked the flow of water (Quigg, 1977).

High dams are built primarily for generating electricity. But when the construction of huge man-made lakes is rationalized on the basis of their value for irrigation, some alternatives are available.

1. Existing water sources can be used more sparingly and efficiently. Irrigation uses up the largest part of the world's available water supply, so even modest increases in efficiency mean large savings. Unlined canals may lose up to 80% of their water before the water reaches the crops, and seepage contributes to salinization and waterlogging. In many instances, it is more profitable to line existing canals than to increase the volume of water.

2. Much can be done to upgrade existing irrigation works. Desilting of existing canals and small reservoirs, and cutting of vegetation from canal embankments, pay large dividends in health as well as agriculture.

3. New small-scale projects can be undertaken to capture surface water that now runs off in feast-or-famine cycles.

4. Wells dug into adequately recharged aquifers are more economical and efficient than large man-made reservoirs. They deliver water at the site of use, avoiding the enormous cost of piping it there from a reservoir. They do not disrupt the environment or suffer loss from evaporation. Well water can be easily protected from contamination. By using more efficient wind and solar pumps, the advantage of wells can be even greater.

5. Old irrigation methods based on maximum economy of water deserve re-appraisal. In pitcher irrigation, seedlings are planted around a buried earthenware pot filled with water. The unglazed pot releases enough moisture for fruits and vegetables to thrive without wasting any water.

6. New methods of irrigation need to be explored. Drip irrigation, for instance, is a more expensive and labour-saving version of pitcher irrigation.

WATER AND ENERGY

Global use of energy has increased greatly during the past few decades. In some places, the availability of water has become more of a constraint on energy production than the availability of primary fuel. The mining of coal, the world's most abundant fossil fuel, poses serious hazards to water. It pollutes both surface water and groundwater with increased sedimentation and soluble minerals from runoff and soil seepage; or, when aquifers are above or below coal seams, underground water may be depleted or destroyed entirely.

Wherever electricity is produced from coal, oil, or nuclear energy, vast amounts of water have to be used as a coolant. In most industrialized countries, 80% of thermal pollution is created in generating electricity, and if nuclear power is further expanded, the extent of thermal pollution as well as the absolute amount will increase, with unknown consequences.

Per capita consumption of energy is also increasing greatly in developing countries.

For some countries, hydroelectric power is an attractive means of meeting a large part of expected energy demand. Hydropower is the least polluting of conventional, large-scale methods of generating electricity. This encourages the tendency to assume that only the financial costs are significant, ignoring the large volumes of water being used. Hydroelectric schemes may involve diversion of water away from agriculture, they may significantly alter the chemistry and biology of lakewater, and other important ecological consequences may result from seasonal drawdown.

Every hydroelectric project involves some consequences for the environment that are common, and others that are unique. They should be clearly understood before any kind of decisions are made. Obviously, some sites and projects will be ecologically safer than others and the environmental impact should be weighed in the light of economic and other considerations.

There exist some good alternatives to large-scale systems of power generation. In terms of economy, resource conservation and environmental protection the merits of small, decentralized systems using renewable resources are compelling. For rural electrification, alternative energy sources generating power where it is consumed have already demonstrated their potential.

Other possibilities include solar energy for cooling, drying crops and pumping water; wind energy for pumping and small-scale production of electricity; production of biogas and fertilizer through fermentation of manure and agricultural wastes; using small turbines to convert the power in natural waterfalls to a few hundred kilowatts of electricity. But all these alternative technologies also have certain limitations. A really efficient, inexpensive solar pump, for example, has yet to be designed. But where these systems have failed, the problem has been not so much technical as social.

Nuclear energy is another option but the threat to surface water and groundwater deters planners from preferring this alternative.

Forestry

A crucial relationship between water and energy is the effect of deforestation on water supply and water flow. The strongest cause of deforestation—and the most devastating—is the need for firewood, primarily for cooking. It is mostly in the places where people are most dependent on wood for fuel that population growth has been the most explosive, exceeding the capacity of forests to provide one of the most basic human needs. The situation is desperate in both human and ecological terms. Much more time is consumed in searching for firewood, or a larger proportion of family income is consumed in purchasing it. Dung now is more widely used for fuel, and crops lose this precious source of nutrients.

When forests are denuded, the soil erodes, washing downstream to silt up reservoirs and irrigation works. Flooding occurs more frequently and is more destructive. Even where rainfall is adequate, wastelands may be created.

Reforestation and watershed management are central to the subject of water. Wherever populations bear heavily on the carrying capacity of the land, subsidies for methane digesters, solar cookers and kerosene may be the best investment a government could make.

WATER AND INDUSTRY

Five industries—primary metals, chemicals, petroleum, pulp and paper, and food processing—account for two-thirds of water's industrial use, yet water represents only about one per cent of their manufacturing costs. For a given unit of output, some industrial plants withdraw upto 20 times as much water as other plants manufacturing the same product. Withdrawal, of course, does not mean consumption in the literal sense, for by far the largest amount is discharged with varying degrees of thermal, biological, or chemical pollution. Industry is highly wasteful and there is little in-plant recirculation of water. The biggest users of water may be industries coming up rapidly in the developing countries. These industries are also among the most polluting and wasteful. Food processing, for example, uses quantities of water to carry away organic wastes which could be used to improve soils and manufacture methane. Instead, the wastes inflict a heavy and insupportable burden on the water's capacity to cleanse itself (Quigg, 1977).

The developing countries should wisely exploit their comparative advantage in having cleaner environments than the industrialized countries. By decentralizing new industry and using the capacity of air and water to cleanse themselves, developing countries could benefit by trading with nations that require expensive antipollution measures. As poverty is perhaps the worst pollutant, anything that lessens poverty should prove to be environmentally beneficial.

Decentralizing industry helps disperse pollutants and lessen the environmental impact. Unfortunately, however, in many developing countries, transportation systems are undeveloped, the generation and distribution of electricity is limited, and urban unemployment very high. Therefore, it is very difficult to start new industries far from major urban centres. The fact is that industry in developing countries is highly concentrated. The rivers running through the cities of the Third World are as polluted as any. Though much of the contamination may be raw sewage, an increasing proportion is industrial effluent, which is generally more toxic, more persistent, and more damaging to the environment.

The pulp and paper industry is one of the most polluting ones. In developed countries, pulp and paper manufacturers are required to make the largest investment in antipollution equipment, relative to total capital, of any industry. Furthermore, to manufacture pulp and paper requires more water than any other industry—up to 417,000 litres for a metric ton of paperboard. Clearly, pulp and paper is an industry which places a particularly heavy burden on the aquatic environment.

Arid and semi-arid lands make up one third of the Earth's land area and support about 20% of its population. The extreme spatial and temporal variability of precipitation and the high potential evapotranspiration in drylands make water resources, and hence plant productivity unpredictable — often resulting in great human suffering in developing economies. Human population growth with its associated changes in land use, and potential climate change are now threatening the integrity of dryland ecosystems and the success of traditional agricultural practices in these regions.

Water availability, vulnerability of natural and socio-economic systems, changes in land use, and a sound science base for developing policies for sustainable use of water resources are some of the major issues of current concern.

The consumption of water containing bacteria and parasites, water-borne diseases, and the consumption or contact with water containing non-biological pollutants all impair human health. By linking the natural and social sciences to the health sciences, the problems of poor water quality can be reduced.

Water resources may be divided into 'blue' water, i.e. surface water, which is in, or bound for, rivers, and 'green' water—water used by plants.

There are much data available on 'blue' water, but they are fragmented, difficult to obtain and the collection networks are deteriorating. There is a need to reappraise these data and to improve their quality and accessibility.

Most people do not live near large rivers, but obtain their water from small catchments or groundwater As water implies food security, there is a need to increase the knowledge of the 'green' water component—how that water can be used more effectively to increase food production. The conventional physical approach to studying water use needs to be integrated with a social science approach that includes the human aspects.

Integrated Land and Water Management can be the essential framework for bringing together the physical, social, economic and legal aspects of water resource management and development. It aims to optimize, simultaneously, social equity, economic efficiency and environmental sustainability. It entails the following aspects :

1. Horizontal integration: integration among adjacent land users and land uses within catchments; between upstream and downstream users; among domestic, industrial, urban and other users; and among governments sharing river systems; and
2. Vertical integration: integration among the range of ortganizations and institutions functioning at different scales and strives to achieve: (a) maintenance of adequate amounts and quality of water to all water users; (b) prevention of soil degradation; (c) food security; and (d) prevention and resolution of conflicts between water users.

WATER AND HEALTH

Of the various environmental ills, contaminated water is perhaps the most devastating in its consequences. Millions of deaths are directly attributable to water-borne intestinal diseases every year. One-third of humanity suffers a perpetual state of illness or debility as a result of impure water; another third is threatened by the release into water of chemical substances of unknown long-term effects.

It is by now apparent that tropical forest land may be entirely unsuited to agriculture. Tropical soils are usually thin and fragile. When exposed to intense heat and rain, they soon become unworkable and useless and their humus and nutrients are destroyed. If the land is exposed to grazing, the grasses are either coarse and unpalatable, or so deficient in nutrients that the number of cattle sustained is very low. Clearcutting of timber proves particularly destructive, for the soil either becomes rocky or is heavily eroded by rain. Land made impermeable to water is the ultimate degradation. Not only does it become unproductive, but the whole hydrological cycle is short-circuited, and the ecosystem may die.

Undoubtedly, water quality is one of mankind's biggest environmental problems. To provide pure water for everyone is a formidable task. Providing clean water will mean nothing without efficient sewage disposal and environmental sanitation. This involves education as much as engineering.

The difficulty of encouraging sanitary practices is illustrated by the problem of providing rural people with potable water. The benefits of deep wells often have evaporated by people's unwillingness to dig and use latrines. Some economists estimated that the cost of providing India with adequate sanitary facilities may exceed the gross national product (Quigg, 1977).

But this gigantic task cannot be accomplished by imitating the water supply and disposal systems of highly industrialized countries. Two issues are pertinent in this context.

1. Dual Water Systems

To use water of the highest quality for all purposes may be inherently wasteful. Only a fraction of the water that people use needs to be of drinking quality, yet with very few exceptions, water to carry away wastes is of the same grade as that flowing from taps. Water, which is upgraded to meet standards of purity for drinking, is often used to wash away excreta and industrial pollutants. No serious effort has been made to use water sequentially for multiple purposes. Water from baths, basins, sinks and washing machines should be filtered and recycled into toilet tanks or used for watering lawns and washing cars. If industrial wastes carrying chemicals and metals are not released into common sewers, household wastes can be sprayed on agricultural land. A few decades ago, water mains and sewer lines used to be laid by uncoordinated authorities, and a dual system was impractical. With new programmes of providing water to large urban areas, it is often possible to go in for some recycling into the system and to reduce the volume of water requiring treatment.

2. Water-borne Sewage

A more serious issue is whether water-borne waste disposal is feasible for the burgeoning urban populations. It involves criminal wastage of water and is beyond the water resources of many countries. Where they exist, flush toilets account for 40% of residential water use, consuming up to 20 litres for every flush. The costs of installation and operation of sewers and water treatment plants are prohibitive for the poor and have to be heavily subsidized.

Organic material carried away by water-borne sanitation systems is not only wasted but becomes a nuisance and a threat to health. Getting rid of sewage sludge is a serious environmental problem.

But alternatives exist. There are composting toilets that produce a clean, rich fertilizer. There are biological toilets that leave no residue, and incinerating toilets that leave a sterile ash. There are oil-flushed toilets that constantly recycle the oil, and vacuum systems using only a litre of water per flush. There are aerobic tanks in which a small air pump speeds up decomposition, and there are digesters (fermentation tanks) that can produce methane as well as fertilizer. However, most have been developed by individuals or small entrepreneurs. Except for the digesters, they are largely untested, and their cost efficiency in large-scale use is not known.

Water-borne diseases represent another serious problem. Some of these are epidemic, like typhoid and cholera. Others are endemic, like dysentery and malaria. Some are caused by ingesting impure water, while others are transmitted by worms or insects dependent on water for some part of their life cycle. Diseases in the latter group—generally the endemic ones—are much more difficult to eradicate. Schistosomiasis and onchocerciasis (river blindness) exemplify this category; they are environmental diseases for which medicine alone has no solutions. They also show that an understanding of the interaction between the natural and human environments is crucial to the issue of water management and development (Quigg, 1977).

There is no doubt that the relationship between water and health is extremely complex and involves the total environment. To bring clean drinking water to those lacking it is a laudable goal but only a partial solution to the acute problem of water-borne diseases.

GLOBAL ISSUES RELATED TO CONTINENTAL AQUATIC SYSTEMS

Continental aquatic systems may be defined from the point where precipitation reaches the Earth's surface until it reaches the sea in full marine conditions (for the exorheic regions) or until it reaches the final base level in closed basins (endorheic regions). The downstream boundary of this realm varies in time and space and includes what is commonly defined as the coastal zone. Some examples of continental aquatic system include rivers, lakes, wetlands, estuaries, groundwater and coastal zones.

The fluvial systems are exposed to two types of major global changes: climate change and direct anthropogenic changes in land use and water use, such as river damming and channelization, agriculture, irrigation and water transfer, industry and mining, and urbanization and population growth (see Table 2.2). They have multiple impacts on physical, hydrological, chemical and biological processes which affect the levels, fluxes, and occurrence of continental water resources, river-borne material and aquatic biota, as well as their spatial and temporal distribution. Most of these impacts are global.

Over long (geological) periods, climate influences river networks and riverine fluxes permanently, yet the acceleration of climate variations due to anthropogenic activities may result in a faster response of all continental waters.

Direct human impacts on aquatic systems are of a more local or regional nature, but they may occur even in desertic and/or remote regions of the world, through river damming, mining and smelting operations, deforestation, and long-range atmospheric transport of SO_2, NO_x and micropollutants.

Impacts on continental aquatic systems result in numerous global issues (see Table 2.2).

Human health through the development of water–related diseases (e.g. malaria), water–borne diseases (e.g. cholera and other pathogens), the increase of chemicals in drinking water and the toxic algal development resulting from eutrophication;

Water availability through enhanced evaporation, development of water use transfer mainly for agriculture, fragmentation of river networks;

Water quality for domestic, agricultural and industrial uses resulting from salinization, eutrophication, increase of chemical inputs;

Global carbon balance through the storage of organic and inorganic dissolved and particulate carbon species, the destorage of organic carbon (permafrost melt), the global change in silicate minerals weathering by atmospheric CO_2;

Fluvial morphology in the form of shift of river courses, erosion or sedimentation in river beds;

Aquatic biodiversity through the general loss of pristine headwaters, the segmentation of river networks, the change of water quality; and

Coastal zone impacts from the modification of sediment transport, the enhanced inputs of nutrients and related eutrophication, the increasing levels of pollutants and of microbial pathogens (Meybeck, 1998).

Table 2.2 Major global threats to continental aquatic systems and related issues (After Jones, 1997).

Environmental state changes	Major impacts	A	B	C	D	E	F	G
Climate change	Development of non-perennial rivers		●	●	●	●	●	●
	Segmentation of river networks					●	●	●
	Development of extreme flow events		●			●	●	●
	Changes in wetland distribution	●	●	●		●	●	●
	Changes in chemical weathering							
	Changes in soil erosion							
	Salt water intrusion in coastal groundwater	●	●	●	●			●
	Salinization through evaporation							
River damming and channelization	Nutrient and carbon retention			●	●	●	●	●
	Retention of particulates							
	Loss of longitudinal and lateral connectivity	●						●
	Creation of new wetlands							
Land-use change	Wetland filling or draining			●	●	●	●	●
	Change in sediment transport						●	
	Alteration of first order streams	● ●	●	●	●	●	●	●
	Nitrate and phosphate increase				●			
	Pesticide increase	●		●			●	
Irrigation and water transfer	Partial to complete decrease of river fluxes	● ● ● ● ●	●	●			●	
	Salinization through evaporation							
Release of industrial and mining wastes	Heavy metals increase			●	●		●	●
	Acidification of surface waters							
	Salinization							
Release of urban and domestic wastes	Eutrophication			●			●	●
	Development of water-borne diseases							
	Organic pollution							
	Persistent organic pollutants							

* A: human health; B: water availability; C: water quality; D: carbon balance; E: fluvial morphology; F: aquatic biodiversity; G: coastal zone impacts.

SUSTAINABLE MANAGEMENT OF WATER USE

Water is an important geofactor. On a global scale, this is particularly true of drinking water. In some marginal-yield sites, in particular in the transitional zone from semi-arid to arid regions, the availability of water for human beings and the environment is a critical factor that limits anthropogenic regional development and agricultural success.

The sustainability of food production depends on sound and efficient water use and such conservation practices as irrigation development and management with respect to rainfed areas, livestock water supply, inland fisheries and agroforestry. Achieving food security is a high priority in many countries, and agriculture should not only provide sufficient food for rising populations, but also save water for other uses. The challenge is to develop and apply water-saving technology and management methods and enable communities to adopt new approaches for both rainfed and irrigated agriculture.

Water has now become an increasingly scarce and extremely valuable resource, without which sustainable development is impossible. Everything humanly possible should be done to encourage sustainable use of fresh water.

The amount of water present on earth is estimated at about 1.41 billion km^3, of which only about 2.5% is fresh water. The overwhelming part, the other 97.5%, is sea or brackish water which is unsuitable for human use.

Table 2.3 shows the continent-wise renewable water supply distribution. South America, with an average of 583 mm, has the largest amounts of fresh water available, while Africa only has an average of 139 mm.

Table 2.3 Renewable water supply of the earth (after Wolff, 1999)

Region	km^3/a
Europe	3110
Asia	13190
Africa	4225
North America	5960
South America	10380
Australia	1965
Total	**38830**

Virtually all human activities degrade the natural quality of the water. Beyond a certain degree of pollution, water cannot be used. The following are the major causes of pollution (see Wolff, 1999):

1. Discharge of untreated or insufficiently treated industrial and domestic waste water (bacterial, organic and heavy metal pollution);

2. Improper inadequate storage of domestic and industrial waste (pollution by seepage water);
3. Unsuitable irrigation which can cause salinization of groundwater and surface water.
4. Excessive use of fertilizers and inefficient use, unsuitable storage and disposal of pesticides (discharge of nitrate, pesticides, herbicides, etc.);
5. Deforestation (pollution of outlets with nutrients and suspended matter); and
6. Overuse of the groundwater (salt water contamination of groundwater bodies in coastal areas, subsidence and consequent changes in the behaviour of surface runoff, rising salt content and, ultimately, the drying up of lakes).

The assured sustainable supply of good-quality water is a formidable challenge for many countries, especially in the developing world, because that is where, in many cases, the existing water shortage has become a serious multi-faceted problem. The areas of concern include, *inter alia*, the rising costs of tapping new water reserves, the wasteful use of those already available, soil degradation in irrigated areas, the overuse or exhaustion of groundwater resources, the release of harmful substances into the water with adverse effects on the health of water users and high subsidies for providing water, which means an unsustainable use of water resources.

Humans cannot live without water. Water is also an indispensable raw material for many industries. The quality of groundwater tends to be more vulnerable to risk in arid regions than it is in humid regions. As a rule, the sustainable utilization of this resource is not so much at risk through the withdrawal of groundwater as such, as it is through its improper or non-optimal use and the failure to regularize its disposal (Vierhuff, 1999).

Water Wastage and Overuse

Wasteful use of water is widespread in many places and there is need to introduce economies in agricultural, municipal and industrial water use. The efficiency level of water use in irrigation in many developing countries ranges between 25 to 40%. Water losses in the cities of developing countries usually exceed 50%. These high losses could be turned into a great savings potential in water use and, thus, a clear improvement in sustainability. There exist many possibilities of water saving. Overuse of groundwater is also widespread in many parts of the world, especially in dry regions.

Groundwater exploitation is some Arabian countries greatly exceeds the rate of renewal. In these countries, medium or long-term sustainable development is not possible.

Future Strategies

Supply management and demand management may be used to help meet the challenges ahead. The former strategy seeks to identify, exploit and utilize hitherto unexplored water resources. Supply management, besides optimization of operations and of the maintenance of water supply plants, is taken to include the reallocation of water resources among the various consumer and user sectors. In contrast, demand management involves using special incentives and water-saving mechanisms to encourage economical and efficient use of water as a resource. Water needs to be regarded as an economic asset and both the strategies may be pursued judiciously in the coming years. However, with economic growth, increasing competition for water resources and the increase in the value of water, demand management is gradually becoming more important than supply management.

Striking a good balance between supply and demand by means of supply management does not necessarily depend on exploiting new water resources. It can also be done by better management of the water resources made available.

Reallocation of water made available from one user sector to another (e.g. from irrigation to other use sectors) is one of the most effective ways of combating critical shortages.

Because water losses in all consumption sectors are quite high, an increase in the efficiency of water use is a most effective measure in demand management. One good example is the use of drip irrigation with dense row crops. Compared with flood irrigation, drip irrigation results in water savings of 50% or more.

On a global scale, the supply of fresh water is quite sufficient to meet mankind's present and future needs. Regionally and locally, however, there are considerable deficits in providing good quality fresh water. This shortage is expected to spread and worsen in future with the population increase and resultent urbanization. Conflicts between countries provinces, etc., within countries and between the sectors of a country will intensify, if they do not succeed in setting up a sustainable and workable water management plan (Wolff, 1999).

Figure 2.7 and Table 2.4 shows the rationale, implementation and problems in integrated watershed management.

Freshwater Management

A worldwide movement of local governments committed to achieving tangible improvements in the sustainable use of freshwater resources is gaining momentum. The objective is to protect environmental flows of water, reduce water pollution, and improve the availability and efficiency of water and sanitation services.

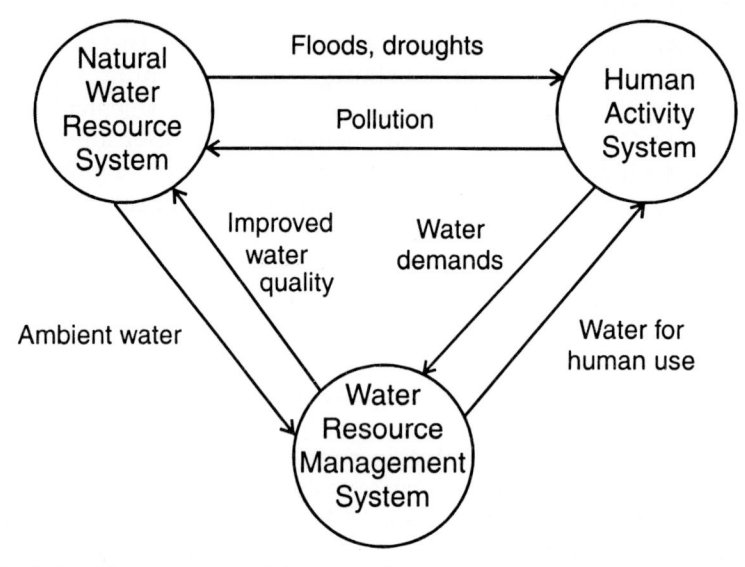

Fig. 2.7 The concept of integrated water resource management (after Jones, 1997).

Freshwater management is a very important environmental, social and economic issue facing local governments all over the world. Water supply, sanitation, urban drainage and water pollution issues are all intimately intertwined with basic rights and needs of citizens, industries as also the freshwater environment.

ONE Putting your house in order	Effective water resources management within the municipal corporation's sphere of operations.	• Development and implementation of an Integrated Water Resource Management System.
TWO Communities working together for sustainable water management	Community participation in water resource management activities.	• Implementation of a participatory multi-stakeholder planning process that is responsible for the development of targets for water quality improvements in the municipality. • Creation of a registry of local water management efforts. • Periodic evaluation of targets.
THREE Sustaining water resources through watershed planning	Establishment of mechanisms for water resources planning and management on a watershed basis, typically involving a number of municipal jurisdictions.	• Establishment of a regional multi-stakeholder forum to develop a framework in which to address water management issues within the local watershed area. • Creation of a regional watershed management strategy.

Fig. 2.8 A sustainable water code for local governments.

Table 2.4 Integrated watershed management: rationale, implementation and problems (After Jones, 1997).

Rationale (major aims)	Implementation (stages of development)	Problems (common causes of failure)
1. Treating watershed as a functional region with interrelationships between water and land management	1. Establishing watershed management objectives	1. Lack of local participation
2. Evaluating the biophysical linkages of upstream and downstream activities	2. Formulating and evaluating alternative resource management actions, involving institutional structures and tools for implementation	2. Weak technical support and guidance
3. Enabling planners/managers to consider all relevant facets of development	3. Choosing and implementing a preferred course of action	3. Inadequate management
4. Strong economic logic in integration	4. Evaluating performance, i.e. degree of achievement of specific objectives, by monitoring activities and outcomes	4. Delays in key inputs, e.g. financial
5. Allows ready assessment of environmental impacts		5. Fragmented governmental control structure
6. Can be integrated into other programmes, e.g. forestry and soil conservation		6. Ignoring downstream interests
		7. Inappropriate institutional arrangements
		8. Political boundaries unrelated to catchment boundary

Cities need to be proactive in managing their local water resources (see Fig. 2.8). There is a great need for: (a) interdisciplinary and multi-jurisdictional solutions; (b) communities to work together improve the local freshwater environment; and (c) a variety of strategies and approaches to freshwater management within the municipal corporation, the city, and the regional watershed area.

Rainwater Harvesting

Rainwater can be collected on the roofs of buildings and then stored underground for later use. Rainwater harvesting increases water availability and checks the falling water table. Rainwater harvesting is a simple, environmental-friendly method. It improves groundwater quality and helps meeting growing demand for water, particularly in urban areas. It prevents flooding of roads and streets. Recharging of groundwater not only arrests the declining water table, but even raises the table significantly. Groundwater recharge from a house with a 100 m^2 roof top can be around 50,000 litres in a year, sufficient for a period of four months for a family of four members. For New Delhi, the additional recharging of groundwater through rainwater harvesting has recently been estimated (by the Central Ground Water Board) to be around 76,000 million litres per year, which can help meet 4% of the Indian capital's annual requirement.

Many megacities (e.g. Delhi) are under severe strain and if drastic remedial action is not taken soon, their collapse may not be too far. More than half the urban population in India is managing without sanitation. Four-fifths of beds in city hospitals are occupied by patients of air- and water-borne diseases caused by wastes. Waste segregation needs to be mandated by law and enforced strictly.

The dust load in many Indian cities is the highest in the world Delhi, Mumbai and Calcutta are easily the world's noisiest cities today. Most Indian rivers are hardly distinguishable from vast urban gutters (see *The Times of India*, New Delhi, January 2000).

AQUATIC RECOVERY FROM ACIDIFICATION

Rates of acidic deposition from the atmosphere (acid rain) have decreased throughout the last two decades across large portions of North America and Europe. Many recent studies have attributed observed reversals in surface-water acidification at national and regional scales to the declining deposition. To test whether emissions regulations have led to widespread recovery in surfac-water chemistry, Stoddard et al., (1999) analysed regional trends between 1980 and 1995 in indicators of acidification (sulphate, nitrate and base-cation concntrations, and measured alkalinity) for over 200 lakes and streams in eight regions of North America and Europe. Dramatic differences in trend direction and strength for the two decades were noted. In concordance with general temporal trends in acidic deposition, lake and stream sulphate concentrations decreased in all regions with the exception of Great Britain; most of these regions exhibited stronger downward trends in the 1990s than in the 1980s. In contrast, regional declines in lake and stream nitrate concentrations were rare and/or very small. Recovery in alkalinity, expected wherever strong regional declines in sulphate concentrations have occurred, was observed in all regions of Europe, especially in the 1990s, but in only one region (of five) in North America (Stoddard et al., 1999).

PROTECTION AND MANAGEMENT OF TROPICAL LAKES

Although tropical limnology has only recently progressed beyond the stage of exploration (Melack 1996; Talling and Lemoalle 1998), the need for application of limnological knowledge is as pressing at tropical latitudes as it has been in the temperate zones. As many tropical lakes are accounted for by natural river lakes or reservoirs, degradation of water quality in rivers has direct negative effects on these lakes. Regulation of rivers, which occurs as a result of river impoundment, can potentially damage river lakes. Tropical lakes are more sensitive than temperate lakes to increases in nutrient supply and usually show higher proportionate changes in water quality and biotic communities in response to eutrophication. Tropical lakes are especially prone to loss of deep-water oxygen. Therefore, in order to maintain ecological stasis, they require more stringent regulation of organic and nutrient loading than do temperate lakes.

A nutrient contaminant must be more strongly oriented toward nitrogen, the most probable limiting nutrient in tropical lakes, than for temperate lakes. But phosphorus control is also important. Nitrogen management may be more feasible in the tropics because of the higher temperature, which is critical for efficient denitrification. Planktonic and benthic communities in the tropics bear a close resemblance, both in composition and diversity, to those of temperate latitudes; there is no parallel to the latitudinal gradient in biodiversity that characterizes terrestrial ecosystems. Foci of biodiversity, which require special attention, include the endemic species of ancient lakes and the diverse fish communities of very large rivers. The latter are an especially valuable untapped economic resource, but are facing much impairment due to hydrological regulation and pollution of rivers. Effective management programs for tropical lakes should focus on interception of nutrients, protection of aquatic habitats from invasive species, and minimization of hydrological changes in rivers to which lakes are connected. In the absence of protective management, tropical lakes are sure to decline in their utility for water supply, production of commercially useful species, and recreation (Lewis, 2000).

Two Kinds of Latitudinal Contrasts

The two very different kinds of latitudinal contrasts among lakes are physiographic and climatic. The former contrasts relate to the relative abundance of lakes per unit of land mass, and to the proportions of different kinds of lakes. Climatic contrasts relate to the fundamental climatic variables associated with latitude.

Globally, lakes cover approximatley 1% of land surface area. The distribution of lake area with latitude is uneven (Fig. 2.9). Lake surface area is highest at temperate latitudes and moderate at tropical latitudes. Approximately, 90% of lake surface area is accounted for by the 250 largest lakes in the world (Herdendorf, 1990). Fig. 2.9 primarily reflects the distribution of only 250 lakes. There are, approximately, 10 million small lakes on the earth (Wetzel, 1992). Protection and management of the world's largest lakes presents a difficult but important problem. Lakes of small to moderate size are also important. These lakes do not contribute significantly to the global surface area and yet, are of critical importance to human populations throughout the tropics (Lewis, 2000).

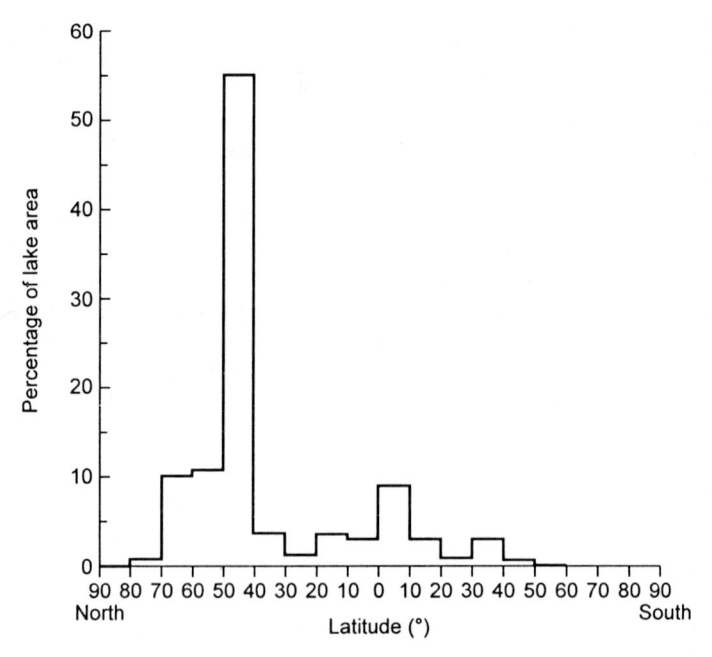

Fig. 2.9 Latitudinal distribution of lake surface area (after Herdendorf, 1990).

Climate varies with latitude, but some important latitudinal trends become masked by variations among individual lakes and variations in climate within regions at a given latitude.

The mean annual solar irradiance at tropical latitudes exceeds that of temperate latitudes, and is more uniform across the months of the year. Of direct importance for limnological contrasts, however, are the latitudinal trends in maximum and minimum irradiance over an annual cycle at a given location. The maximum and minimum annual irradiance are determined, in a large part, by the seasonal tilting of the Earth, but also to some degree by regional variations in the attenuation coefficient of the atmosphere (as determined by moisture or dust for example). The single-most important basis for understanding latitudinal trends in lakes is that the chief contrast between temperate and tropical lakes lies in the season of minimal irradiance (hemispheric winter). The season of maximum irradiance (hemispheric summer) differs much more modestly across latitude. An identifiable season of negative heat budgets is characteristic of tropical lakes, and its timing is predictable. Except in rare cases, it coincides with the hemispheric winter.

Seasonality in Tropical Lakes

One useful way of classifying lakes is on the basis of frequency of complete mixing. Lakes in temperate latitudes are most often dimictic, i.e. they have a seasonal period of ice cover, fall and spring mixing, and a season of stable stratification coinciding with maximum heat content. Although the dimictic status of temperate lakes is fundamental, it is not universal.

Tropical lakes are basically warm monomictic in the same sense that temperate lakes are fundamentally dimictic. The occurrence of a cool season coinciding with the hemispheric winter ensures that a mixing season at that time is very likely. The difference in temperature between the top and the bottom of the water column is quite small (approximately 2°C). But because water density responds markedly to changes in temperature when the temperature is high (Fig. 2.10), there is sufficient density difference between the top and bottom of tropical lakes to sustain seasonal stratification, except in very shallow lakes.

Deviations from the fundamental mixing type occur both in the tropics and at temperate latitudes. Lakes that are not very deep tend to show intermittent mixing throughout the year rather than seasonal mixing, i.e. they are warm polymictic for the same reasons that shallow temperate lakes are cold polymictic (Fig. 2.11). Polymixis will be continuous (i.e. daily) for very shallow lakes or discontinuous (multiple but not daily mixing events) for lakes with a mean depth of 5-10m, depending on area.

88 Environmental Technology and Biosphere Management

The deepest tropical lakes show the characteristic tropical seasonality, that is, deep mixing on a seasonal cycle. But sometimes wind strength is not high enough to move the entire water mass. This also applies to the deepest temperate lakes. Such lakes verge on meromixis (failure to mix completely), but a strong vertical exchange during the cool season prevents most of them from developing the chemocline so characteristic of truly meromictic lakes (Lewis, 1983). Truly meromictic lakes are found occasionally in the tropics, as they are at temperate latitudes.

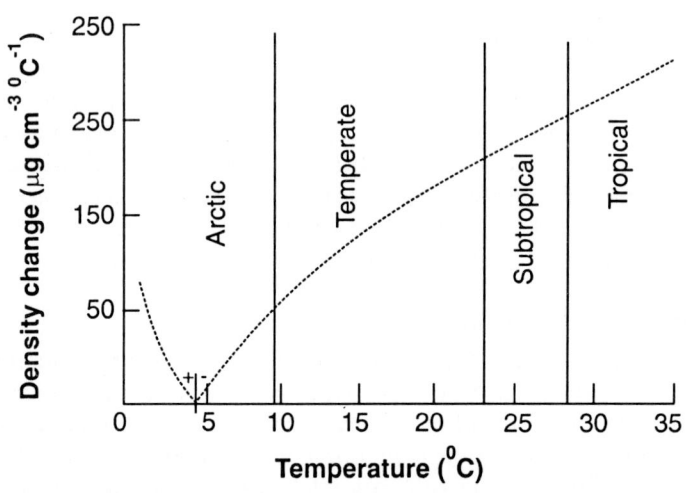

Fig. 2.10 Change in the density of water that accompanies a change in temperature of 1°C (after Lewis, 2000).

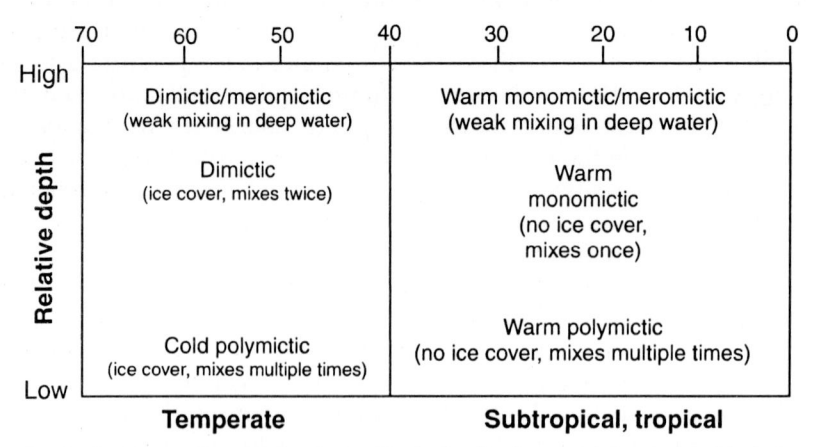

Fig. 2.11 Illustration of the latitudinal distribution of lake types based on mixing (after Lewis, 2000).

It appears that the potential for production of phytoplankton biomass on a given nutrient base is higher in tropical lakes than it is in temperate lakes (Lewis, 1974). So, tropical lakes may be more responsive to eutrophication than temperate ones.

Temperature

In general, the relationship between water temperature and lake metabolism is determined by a Q10 of approximately 2.0, i.e. metabolic rates are likely to double with a 10°C increase in temperature, provided that some other factor does not strongly suppress the rate.

In summer, the upper water column of temperate lakes at low elevation reaches temperatures that approach those typical of the tropics, because of the relatively low sensitivity of maximum irradiance to latitude. But the annual average temperature of the mixed layer is much lower for a temperate than for a tropical lake. Metabolic processes dependent on temperature tend to be steadier and stronger in the upper mixed layers of tropical lakes. Consequently, the nutrient cycling caused by regeneration of inorganic nutrients through microbes in the mixed layer occurs more rapidly in the upper zones of tropical lakes during stratification.

The temperature of the hypolimnion shows an even more significant latitudinal contrast than temperature in the upper water column. The hypolimnion temperature in a tropical lake is only as low as the seasonal minimum air temperature, that is, in the vicinity of 24°C for lakes located within 10° of the equator at low elevation. Such high temperatures are conducive to high rates of microbial metabolism. Nutrients tend to be regenerated more rapidly and completely, and oxygen is removed much more rapidly than would be the case at the lower temperatures characteristic of temperate lakes (Lewis, 2000).

Oxygen

Oxygen in the deep water of tropical lakes is critical for their protection and management. Three factors act against the retention of oxygen in the deep waters of tropical lakes: (i) the long duration of stratification in tropical lakes, which will typically last 10-11 months rather than 6-9 months, as in temperate lakes; (ii) the weaker ability of water to hold oxygen at high temperatures than at low temperatures (approximately 8 mg L^{-1} tropical vs. about 12 mgL^{-1} at 45° latitude during mixing; and (iii) higher rates of microbial metabolism at high temperatures prevailing in the deep waters of tropical lakes. These three factors magnify the influence of any organic enrichment of deep waters, so undesirable effects of anoxia caused by eutrophication or direct organic enrichment of waters are more serious and more quickly realized in tropical lakes rather than temperate lakes. Any tropical lake that sustains oxygen in deep waters as a natural condition should be protected with special care if deep-water oxygen is to be preserved, because small increases in oxygen demand often lead to the elimination of deep-water oxygen in such as lake. Indeed, oxygen conservation is an even more important management principle for tropical lakes than it has been for temperate ones.

Primary Production and Photosynthesis

If primary production were exclusively under the control of solar irradiance, one would expect it to show a slight suppression near the equator caused by high humidity (cloudiness) and a steady decline beyond 30° caused by short days (Fig. 2.12). But primary production is also influenced by temperature. When the effect of temperature is superimposed on that of solar irradiance, the expected latitudinal trend becomes even more striking (Lewis, 2000).

Primary production is also influenced by the availability of nutrients, which are more difficult to predict than for solar irradiance or temperature. Indeed, persistent nutrient depletion tends to nullify the effects of temperature and irradiance; a nutrient-limited tropical lake sustains no more net primary production than a temperate lake with a similar degree of nutrient limitation. Work done under the International Biological Programme showed, however, that primary production does decrease as a function of latitude (Fig. 2.12). Measurements of the production of tropical lakes have verified the general trend, i.e. that annual primary production is highest at low latitudes (Lewis, 1996). This implies that tropical lakes are favoured by high solar irradiance, high temperature and a high efficiency of nutrient use, which allows primary production to reflect the favourable irradiance and temperature.

Under nutrient-limiting conditions, the ability of a lake to sustain primary production is determined by recycling mechanisms. Recycling mechanisms within the mixed layer occur more rapidly at low latitude because of sustained high temperature. On the basis of nutrient-cycling efficiency alone, tropical lakes need to be considered more sensitive to eutrophication than temperate lakes, i.e. the potential for eutrophication to degrade water quality (phytoplankton biomass, deep-water oxygen, etc.) is highest at tropical latitudes.

Table 2.5 compares the general characteristics of tropical and temperate lakes.

Nutrients

Phytoplankton, attached algae and macrophytes require some 20 elements for the synthesis of protoplasm but the growth rate of a given species population at a particular time will be either unresponsive (unlimited by nutrients) or responsive (limited by nutrients) to the addition of nutrients.

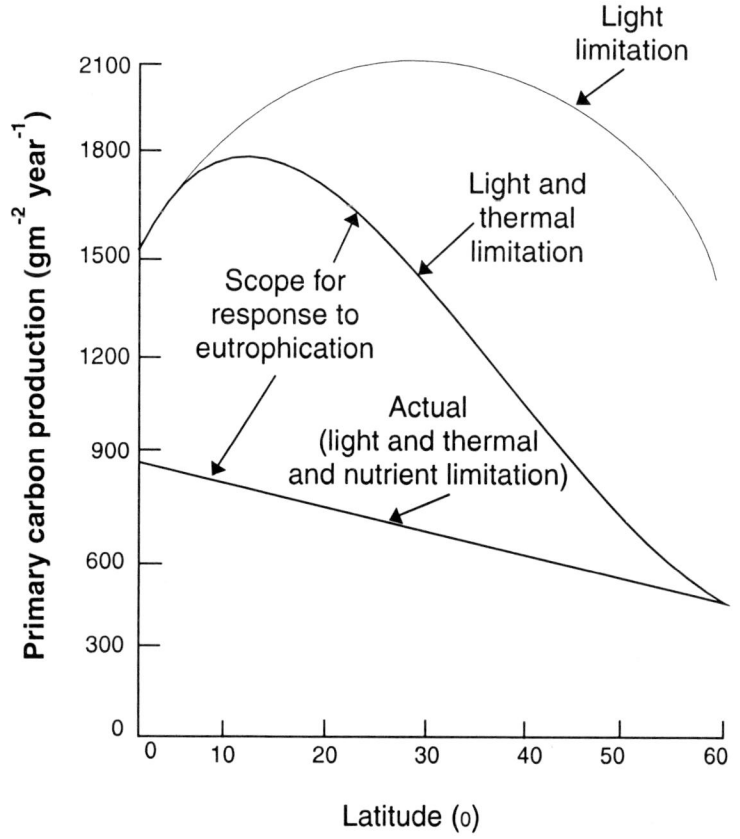

Fig. 2.12 Latitudinal trends in annual net primary production of lakes under three conditions: optimal solar irradiance with optimal temperature and nutrient saturation (top line); optimal solar irradiance with actual temperature (middle line); and observed rates (i.e. reflecting actual solar irradiance, temperature, and nutrient limitation) (after Lewis, 2000).

If the population does not respond to the addition of nutrients, it is growing under conditions of nutrient sufficiency. Often, however, the addition of nutrients causes a growth response. The response to the addition of a mixture of nutrients can be traced exclusively or mainly to a single nutrient in the mixture (the limiting nutrient). A complication occurs when different species are limited by different nutrients, but usually, the community responds as if limited by one nutrient (Lewis, 2000).

Tropical lakes, like temperate ones, often show nutrient limitation of autotrophs. There may be latitudinal trends, however, in the persistence or frequency of nutrient limitation. As at temperate latitudes, tropical

Table 2.5 Some characteristic features of tropical lakes, their consequences for the functioning of these lakes, and management implications (after Lewis, 2000).

Features	Consequences	Implications
Natural lakes not abundant Glacier lakes scarce among natural lakes	Reservoirs of high relative importance River lakes are the predominant natural type	Management of reservoirs is of high priority Status of rivers dictates welfare of most lakes
Predominantly warm monomictic	Predictable annual mixing season except in shallow lakes	Seasonal anoxia likely in deep water; seasonal cycle in water quality
High hypolimnetic temperature, long stratification season	High probability of hypolimnetic oxygen depletion	High vulnerability to eutrophication or organic loading
Recurrent changes in thickness of mixed layer	Recurrent nutrient enrichment of mixed layer	High efficiency of nutrient use; strong response to anthropogenic nutrient loading
Nitrogen limitation of autotrophs predominates over phosphorus limitation	Nitrogen pollution creates problems	Denitrification of waste critical; phosphorus removal also important
Invertebrate predators favoured by anoxia	Herbivores consumed mostly by invertebrate predators	Low fish production per unit primary production; eutrophication can decrease fish production
Planktonic and benthic biodiversity fairly similar to that in temperate lakes	Analogies with temperate lakes stronger than for terrestrial environments	Biodiversity protection challenging but less so than in terrestrial tropics

lakes that stratify seasonally are likely to show the most severe nutrient limitation. Such lakes, whether tropical or temperate, show a progressive seasonal separation of nutrient supply from nutrient demand as free nutrients accumulate in the lower water column, while growth is limited primarily to the upper water column. Recurring cycling of nutrients between the growth zone and deeper water allows tropical lakes to be more efficient in the use of nutrients than temperate lakes. High temperatures in deep water further increase this effect by rapidly regenerating the available forms of nutrients from their unavailable forms through microbial action.

Experience with the inland waters of temperate latitudes points to the significance of phosphorus and nitrogen as limiting nutrients for autotrophs (Welch, 1992). Other macronutrients (i.e. essential elements present in mgL^{-1} quantities or more) are generally present almost universally in amounts sufficient enough to satisfy the needs of autotrophs. Only inorganic carbon and silicon stand out as possible causes of limitation in addition to nitrogen and phosphorus.

Limitation by inorganic carbon differs fundamentally from that of nitrogen and phosphorus. It occurs only when photosynthesis rates are very high, i.e. only under full sun, and only in lakes rich in autotrophs. The atmospheric CO_2 pool relieves any incipient inorganic carbon limitation in most situations. Likewise, a silicon limitation is of much narrower significance than limitation by phosphorus and nitrogen. It affects mostly diatoms and may be less likely in the tropics than at temperate latitudes because of higher rates of rock weathering in the tropics.

Nutrient limitation in tropical lakes comes from deficiencies of phosphorus or nitrogen, as is the case in temperate latitudes. Phosphorus is widely believed to be the dominant limiting nutrient for lakes in temperate latitudes. Phosphorus and nitrogen limitation are rather closely balanced in temperate waters that are not subject to anthropogenic waste disposal. It appears that nitrogen limitation is more important at tropical latitudes than it is at temperate latitudes (Talling and Lemoalle, 1998). Phosphorus limitation does occur in some lakes and in other cases, the balance between phosphorus and nitrogen limitation seems to be rather close, as it is for many natural waters at temperate latitudes. In general, however, there are more instances of clear nitrogen limitation than of clear phosphorus limitation for tropical lakes.

Even more important to the balance between nitrogen and phosphorus in temperate lakes may be denitrification, which is stimulated by anoxia and high temperatures in sediments. Anoxia is more likely to occur and is more persistent in tropical lakes than in temperate ones, and sediment temperatures are higher.

The management implications of nutrient limitation dominated by nitrogen are quite complex. A high likelihood of nitrogen limitation for tropical lakes suggests that human disposal or mobilization of inorganic nitrogen is more dangerous in the tropics than at temperate latitudes: it is likely to provoke a trophic response even in the absence of additional phosphorus loading. At the same time, phosphorus control is also essential. Control of eutrophication requires management of phosphorus as well as nitrogen.

Removal of phosphorus is critical to the maintenance of beneficial uses for tropical lakes. Nitrogen removal may prove to be more feasible at tropical latitudes than it has at temperate latitudes. This is typically accomplished by processes that facilitate denitrification in wastewater treatment plants.

Containment of both nitrogen and phosphorus needs to be incorporated into wastewater treatment plants for the tropics. The importance of this principle is magnified by the high likelihood of connections between tropical rivers and the dominant types of tropical lakes, which include reservoirs and river lakes. Widespread nutrient pollution of running water in the tropics produces undesirable effects on the usefulness of tropical lakes (Lewis, 1996, 2000).

Biodiversity

The ancient lakes in the tropics are centres of biodiversity for both aquatic vertebrates and invertebrates. Some individual lakes of the African rift valley support hundreds of endemic fish. However, aside from the few ancient lakes, the composition and diversity of lacustrine biotas in tropical lakes appear to be similar to those of temperate lakes (Lewis, 1996). The morphotaxa of phytoplankton are largely cosmopolitan. Many of the dominant species that occupy tropical lakes are indistinguishable, even to the experts, from the taxa that are dominant in temperate lakes. There are some exceptions. For example, *Asterionella* is common in temperate phytoplankton assemblages, but not so in the tropics.

The diversity of individual phytoplankton communities appears to be no greater than it is at temperate latitudes. Proportional representation of taxa is also very similar, although tropical latitudes often show a higher proportion of blue-green algae and a lower proportion of golden brown algae.

The zooplankton communities of tropical lakes are also very similar across latitudes. Cladocerans, copepods and rotifers all are major components and constitute approximately the same proportion of biomass, productivity, and species diversity at tropical and temperate latitudes, although individual lakes vary a great deal at any given latitude. There is little endemism among the zooplankton. The calanoids show a greatest amount of endemism but there is no latitudinal trend. In general, large cladocerans are much less abundant at tropical than at temperate latitudes. Overall, however, the tropical and temperate zooplankton are quite similar (Lewis, 2000). Benthic communities also differ little across latitude with respect to diversity, although species composition varies more latitudinally than it does for phytoplankton or zooplankton. Benthic communities of lakes may, in fact, be depauperate in the tropics because of deep-water anoxia at tropical latitudes.

Fish faunas, which are much more endemic than the communities already mentioned, are more difficult to characterize latitudinally, but clearly, do not present such striking latitudinal contrasts as do terrestrial plants, the impressive and very special biodiversity of fishes in ancient lakes notwithstanding. The fish diversity in river system is related to the size of their drainage basins.

The fish faunas of tropical lakes are more likely than those of temperate lakes to have a strong affiliation with riverine fish faunas. The dominant lake types of the tropics lie astride or beside large rivers. Therefore, the welfare of tropical lacustrine fish faunas is very sensitive to the welfare of rivers, which makes the protection of these faunas particularly difficult (Lewis, 2000).

It appears that the lacustrine fisheries of the tropics are much less productive than might be expected, possibly because lacustrine fisheries in the tropics (especially in reservoirs) are handicapped by the weak adaptation of tropical fish faunas to lacustrine conditions. Except for ancient lakes, where specialized fishes have adapted extensively to lacustrine conditions, tropical lakes often contain fishes that have strong riverine affinities.

According to Lewis, there may exist an important and widespread relationship between *Chaoborus*, fish production and anoxia in tropical lakes that are deep enough to stratify seasonally. When deep waters become anoxic, *Chaoborus* is much less vulnerable to fish predation because fishes are unable to reach *Chaoborus* populations that are resting within the anoxic zone during the daylight hours. As *Chaoborus* becomes increasingly abundant, it consumes the production of herbivorous zooplankton — the largest energy source for fishes. Thus, many tropical lakes may show low efficiency of consumer production because of deep-water anoxia. Eutrophication enhances this tendency.

Managers of tropical lakes should know that the greatest potential for increasing indigenous or sustained commercial yields of fisheries lies in the adjustment of fishing practices. The increase of primary production, as might occur through eutrophication, is less likely to magnify the production of harvestable consumers because the connection between primary production and harvest is indirect and difficult to manipulate.

Also, eutrophication may have the undesirable consequence of increasing the population density of the widespread invertebrate *Chaoborus*, which intercepts primary production before it can reach fish populations, thus potentially reducing fish production drastically.

CONCLUSIONS

It appears that challenges to the ecological integrity and utility of lakes are very similar both at tropical and temperate latitudes — eutrophication, organic loading, invasive species and hydrographic changes in rivers that affect river lakes. Intensification of these challenges for individual lakes can be traced to waste disposal, changes in land use, and diversion or impoundment of water. The effects of some of these environmental challenges often differ greatly between tropical and temperate lakes. Tropical lakes are likely to show a higher degree of adverse response to eutrophication or organic loading than would occur in temperate lakes. Nitrogen plays a larger role in eutrophication than it does at temperate latitudes, and hydrographic changes in rivers are more important because of the greater proportional importance of river lakes in the tropics (Lewis, 2000).

REFERENCES

Agarwal, A.K. Crippling cost of India's big dam. *New Scientist* (January 30, 1975).
Anonymous. *A Primer on Water: Questions and Answers.* 2nd ed. Environment Canada, Ottawa (1991).
Eckholm, E.P. *Losing Ground: Environmental Stress and World Food Prospects.* W.W. Norton, New York (1976).
Fernando C.H., Gurgel, J.J.S., Moyo, N.A.G. A global view of reservoir fisheries. *Int. Rev. Hydrobiol.* (Special Issue) 83: 31-42 (1998).
Förch, G. Water resources development and environmental planning. *Natural Resources Develop.* 35: 34-50 (1992).
Fujita, M. International symposium highlights innovative ways of findings water for cities. *UNEP IETC INSIGHT Newsletter* pp. 5-7 (Sept. 1999).
Grover, J.P. *Resource Competition.* Chapman and Hall, London (1997).
Herbers, H. Water as a critical factor of nutritional security — on the role of water as a life-sustaining substance. *Applied Geography and Development* 54: 28-45 (1999).
Herdendorf, C.E. Distribution of the world's large lakes. In : Tilzer, M.M., Serruya, C. (eds.). *Large Lakes: Ecological Structure and Function.* pp. 3-38, Springer-Verlag, New York (1990).
Huisman, J., Weissing, F.J. Light-limited growth and competition for light in well-mixed aquatic environments: an elementary model. *Ecology* 75: 507-520 (1994).
Huisman, J., Weissing, F.J. Biodiversity of plankton by species oscillations and chaos. *Nature* 402: 407-410 (1999).

Jones, J.A.A. *Global Hydrology*. Longman, Harlow (1997).
Layne, N. *The Natural Environment: A Dimension of Development*. National Audubon Society, New York (1976).
Lewis, W.M. Primary production in the plankton community of a tropical lake. *Ecol. Monogr.* 44: 377-409 (1974).
Lewis, W.M. A revised classification of lakes based on mixing. *Can. Jóur. Fish. Aquat. Sci.* 40: 1779-1787 (1983).
Lewis, W.M. Tropical lakes: How latitude makes a difference. In: Schiemer F., Boland, K.T. (eds). *Perspectives in Tropical Limnology*. pp. 43-61, SPB Academic Publishers, Amsterdam (1996).
Lewis, W.M. Basis for the protection and management of tropical lakes. *Lakes and Reservoirs: Research and Management* 5: 35-48 (2000).
Margat, J. Les eaux souterraines dans le monde. Similitudes et differences. *Proc. 21 Journées de Hydraulique*. Sophia Antipolis, France, IV, pp 1-13 (1991).
Melack, J.M. Recent developments in tropical limnology. *Verh. Internat. Verein. Limnol.* 26: 211-217 (1996).
Meybeck M. The IGBP water group: a response to a growing global concern. *IGBP Newsletter* 36: 8–12 (1998).
Quigg, P.W. *Water, the Essential Resource*. National Audubon Soc., New York (1977).
Ramalingaswami, V., Jonsson, U., Rohde, J. The Asian enigma. In Adamson, P. (Ed.). *The Progress of Nations*. UNICEF-Jahresbericht, Wallingford (1996).
Rowe, D.K., Dean, T.L. Effects of turbidity on the feeding ability of the juvenile migrant stage of six New Zealand freshwater fish species. *New Zealand J. Mar. Freshwater Res.* 32(1): 21-29 (1998).
Shiklomanov, I.A. The world water resources: how much do we really know about them? 25 year commemorative symposium UNESCO, IHD/IHP Paris, pp. 92-126 (1991).
Stoddard, Jeffries, D.S., Lükewille, A., Clair, T.A., Dillon, P.J., Driscoll, C.T., Forsius, M., Johannessen, M., Kahl, J.S., Kellogg, J.H., Kemp, A., Mannio, J., Monteith, D.T., Murdoch, P.S., Patrick, S., Rebsdorf, A., Skjelkvale, B.L., Stainton,, M.P., Traaen, T., H. van Dam, Webster, K.E., Wieting, J., Wilander, A. Regional trends in aquatic recovery from acidification in North America and Europe. *Nature* 401: 575-577 (1999).
Talling, J.F., Lemoalle, J. *Ecological Dynamics of Tropical Inland Waters*. University Press, Cambridge (1998).
UNDP. Deutsche Gesellschaft für die Vereinten Nationen. *Bericht über die Menschliche Entwicklung* 1995. United Nations Development Programme, Bonn (1995).
UNEP. *Saving Our Planet. Challenges and Hopes*. UNEP, Nairobi (1992).
UNESCO-WMO. The Dublin Statement. UN International Conference on Water and the Environment (ICWE) UNESCO, Paris (1992).
United Nations (Water Conference Secretariat). *Resources and Needs: Assessment of the World Water Situation* (E/CONF. 70/CBP/1) (July 2, 1976).
Vierhuff, H. Groundwater withdrawal in arid areas — Sustainable utilization or overexploitation? *Natural Resources and Development* 49/50: 31-41 (1999).
Villarroya, F., Aldwell, C.R. Sustainable development and groundwater resources exploitation. *Environ. Geology* 34 : 111-115 (1998).

Welch, E.B. *Ecological Effects of Wastewater.* Chapman and Hall, London (1992).

Wetzel, R.G. Gradient-dominated ecosystems: Sources and regulatory functions of dissolved organic matter in freshwater ecosystems. *Hydrobiologia* 229: 181-198 (1992).

WHO : *Weltgesundheitsbericht 1996 : Krankheit Bekämpfen, Entwicklung fördern.* Genf (1996).

Wilson, E.O. *The Diversity of Life.* Belknap Press, Cambridge, Massachussetts (1992).

Wolff, P. On the sustainability of water use. *Natural Resources and Development* 49/50: 9-28 (1999).

World Bank. *World Development Report 1993. Investing in Health.* New York (1993).

Chapter 3
Atmosphere and Climate Change

INTRODUCTION

Traditionally, quantitative approaches and issues such as landscape-scale economic modelling and integrated-assessment models have received more attention, but now a growing number of qualitative socio-political and ethical issues, including sustainable development, environmental justice, environmental education, social learning of global environmental risks, ecological vulnerability, property rights, public perceptions and environmental policies at various levels are attracting focus. Also, such diverse problems such as deforestation, population growth, human migrations, technological changes, land use and ozone depletion are being addressed. Many workers have become interested in the burning issue of climatic change. By and large, it is being felt that global warming may not be a shaky hypothesis as believed hitherto but a real phenomenon of serious consequences for humankind as well as for other organisms. Usable, accessible and intelligent knowledge is the foundation stone for a sustainable society.

As the climate system is global in scope, international cooperation is essential to understanding and predicting its future states under the impact of human activities, especially those causing increase in atmospheric concentrations of greenhouse gases and aerosols. Better estimates of the rate, magnitude and regional distribution of global warming are essential for sound policy and decision making.

The social aspects of the environment need to be studied just as strongly as its natural science dimensions. Interdisciplinary, comparative-historical, and policy-related research is warranted.

Medium-term (a few months to a year) climate forecasts are valuable for many economic sectors. Recent successes in understanding the *El Niño* Southern Oscillation (ENSO) phenomenon have generated the ability to predict the worldwide climate anomalies associated with it. It is quite easy to predict major climatic events such as droughts and floods in tropical and subtropical regions but the projections are not so easy for temperate latitudes.

The global climate system is much more than just the atmosphere. It involves interactions between the atmosphere, oceans, the land, snow and the biosphere. Several decisions are made every day in various countries based on some assumption about future climate. These decisions relate to designing of buildings, storm sewers, bridges, highways, planting of agro-crops and trees, the sizing of dams and reservoirs, fishery and wildlife management, tourism, energy management, etc. All this warrants a continuing need for innovation in using the climatic information derived from various meteorological stations.

The Earth has gone through at least three glacial-interglacial phases during the Quaternary. These phases affected the flora and fauna, resulting in their migrations. Animals which seek niches with special microclimates are probably less subject to the vagaries of weather and climate than plants. North-south oriented mountain chains like the Western Ghats in India and the Andes provided more convenient corridors for migration than the east-west oriented ranges like the Himalayas and the Pyrenees, which proved to be high barriers to cross from north to south and vice-versa (Meher-Homji, 2000). Hill ranges with higher elevation provided a convenient passage for ascension than smaller mountains (see Gupta and Pachauri, 1989).

Legris (1963) plotted the existing forest types of Peninsular India within a framework, keeping the mean temperature of the coldest month on one axis and a precipitation index (hydric balance) on the other. On this graph, he showed how the vegetation types would shift with a drop of 6°C or with a rise in precipitation by 500 mm—scenarios quite likely during glacial and integlacial phases, respectively.

Some models predict sensitivity of temperature to greenhouse gases within a range of 1.5 to 4.5°C and the rise in sea-level by about 20 cm within the next 5 to 10 decades. These figures of projected higher temperature probably refer to mean annual temperature. Higher temperatures—contrary to popular belief—promote the growth of tropical species, though the direct impact of a climate change of the projected magnitude may be of greater consequence for the boreal forests of higher latitudes than for the tropical forests (Meher-Homji, 2000). However, as high temperatures are generally associated with hot deserts, the misconception persists. According to Meher-Homji, it is the mean temperature of the coldest month that determines the distribution of many forest types in India. Many miscellaneous tree species of deciduous forests are common in the zones with the mean temperature of the coldest month above or below 15°C, the sal tolerating somewhat lower temperatures than the teak. It would be interesting to compare the floristic composition of a deciduous forest near Warangal (Andhra Pradesh) with the one in the vicinity of Varanasi — there is a difference in mean temperatures of around 3° to 4°C between the two stations.

However, it is not just temperature and rainfall measures that determine the vegetation types. The length of the dry period and the seasons of rainfall play an equally important role. For example, the distribution of *Shorea robusta* is influenced by timely arrival of rains when ripe seeds are available. The tropical dry evergreen forest is confined to the Coromandel coast where the rainfall regime is dissymetric (Meher-Homji, 2000) with the bulk of rains during the north-east monsoon (October-December).

Anthropogenic interference in the geo-biosphere impacts on the environment on a two-tier scale. At the global level, indication is of warming of the earth to the extent of 2 to 4°C over half a century or so. On the other hand, at a local scale, an operation like major deforestation or land-use change can bring about a shift of greater magnitude in terms of temperature and humidity over a much shorter time frame (Meher-Homji, 2000).

Though the climate may have warmed in the Quaternary, the current rise in the global mean temperature is of great concern because unlike in the past, the burgeoning human population today exerts heavy pressure on the ecosystems. The combined effects of the two forces may not permit the intrinsic ability of the species to adjust to the disequilibrium; the threat of mass extinction hangs like the sword of Democles.

Global warming is often blamed for the increase in the frequency, intensity and, more importantly, the havoc wrought by cyclones, droughts and fires. Though these phenomena also occurred in the past, the havoc caused by them was not so devastating then because of human disturbance on a much lower scale. For example, the current rate of destruction of mangroves has exposed the coastal zones to the full fury of hurricanes. Man-induced earthquake activity is also on the rise but is due to the impounding of water over the geological faults and is in no way related to the global warming.

Other anthropogenic activities such as deforestation result in immediate drastic changes in the microclimates providing conditions not conducive to restoration of forest over the degraded land. There is evidence that major deforestation affects the hydrological cycle, reduction in rainfall (mainly of conventional origin), and hastens soil erosion (Meher-Homji, 2000).

The International Geosphere–Biosphere Programme (IGBP) is focused on acquiring basic scientific knowledge about global environmental change and, particularly, the interactive processes of biology and chemistry of the Earth system. The major aims of global change research are to describe and understand the interactive physical, chemical and biological processes that regulate the total Earth system, the unique environment that it provides for life, the changes that are occurring in this system, and the manner in which they are influenced by human actions. The functioning of the planet and particularly the biogeochemical system of the planet needs to be understood in the context of natural variability as well as changes induced by human activities.

Measurements of air trapped in ice-cores and direct measurements of the atmosphere have shown that in the past 200 years, the abundance of CO_2 in the atmosphere has increased by over 30%, from a concentration of 280 parts-per-million by volume (ppmv) in 1700 to nearly 370 ppmv as we enter 2000. The concentration was relatively constant (roughly within +/-10 ppmv of 275) for more than 1000 years prior to the human-induced rapid increase in atmospheric carbon dioxide.

The Earth's biogeochemical system is intimately linked with the physical climate system. Carbon dioxide and methane are but two of the biogenic greenhouse gases. The marine carbon cycle plays an important role in the partitioning of carbon dioxide between the atmosphere and the ocean. The primary controls are the circulation of the ocean (a function of the climate system), and two important biogeochemical processes; viz., (1) the solubility pump and (2) the biological pump. Both of these create a global mean increase of dissolved inorganic carbon with depth, and so maintain atmospheric CO_2 at a level considerably lower—about a factor of three—than it would be otherwise (Moore, 1999).

The interplay between the circulation of the oceans and the biogeochemical 'pumps' (which themselves depend on ocean circulation), determine the sea surface pC and hence are the primary determinants (with atmospheric pCO_2 and sea surface winds) of the air-sea exchange rates of carbon dioxide.

An interesting coupling also occurs on land. The metabolic processes responsible for plant growth and maintenance and the microbial turnover are associated with dead organic matter decomposition, cycle carbon, nutrients, and water through plants and soil. All these cycles also affect the energy balance and control biogenic trace gas production (i.e. both are climate couplings). When we consider the carbon fixation-organic material decomposition as a linked process, it is seen that some of the carbon fixed by photosynthesis and incorporated into plant tissue is perhaps significantly delayed from returning to the atmosphere until it is oxidized by decomposition or fire. This slower carbon loop through the terrestrial component of the carbon cycle affects the rate of growth of atmospheric CO_2 concentration and can also impose a seasonal cycle on that trend (Moore, 1999).

The terrestrial ecosystems respond on long time scales in an integrated manner to changes in climate and to the intermediate time scale carbon-nutrient machinery. The loop is closed back to the climate system, since it is the structure of ecosystems, including species composition, that largely fixes the terrestrial boundary condition on the climate system.

Water also connects up the terrestrial biogeochemical system to the climate system. Water availability is, obviously, an important regulator of plant productivity and sustainability of natural ecosystems. In turn, terrestrial ecosystems recycle water vapour at the land-surface/atmosphere boundary and exchange numerous important trace gases with the troposphere.

Soil moisture is crucial in the formation of runoff and, hence, riverine flows. It is an important determinant of ecosystem structure and also a primary means by which climate regulates (and is regulated by) ecosystem distribution. Finally, adequate soil moisture is an essential resource for human activity. Consequently, accurate prediction of soil moisture is important for simulation of the hydrological cycle, of soil and vegetation biochemistry (including the cycling of carbon and nutrients), of ecosystem structure and distribution and of climate (Moore, 1999).

Much work has been done in sectoral (e.g. forestry, transportation) climate impact studies based on climate model outputs with a doubled CO_2 atmosphere. These studies helped identify potentially adverse as well as beneficial consequences under atmospheric climate change. Such studies can benefit several countries.

It seems fairly certain that climatic changes due to anthropogenic actions will become a reality in spite of concerted efforts at serious international actions under the United Nations Framework Convention on Climate Change. Society could improve its performance and adapt more effectively to normal climate fluctuations. There exists a need for developing climate adaptation and limitation strategies.

Many discussions about the effects of climate change have related to the possibility that biomass production might decrease in areas where rainfall is reduced. In other areas, where temperature and rainfall increase, biomass production is likely to increase. Thus, it could be expected that increased supplies of fuelwood, a renewable source of energy, might bring benefits from climate change to some areas, and increase costs where production rates are reduced. This is an important aspect of the costs and benefits of the mitigation vs adaptation dilemma of climate change. One crucial question: is what damage can be avoided by abatement policies and how do different policy approaches compare? There are uncertainties in the response of climate to increased levels of greenhouse gases. Regional differences cannot be predicted. The effects of responses to changes that might occur are not quantifiable and, so, the costs and benefits of different actions are difficult to estimate.

General estimates of the costs of damage on a global scale that might be caused by a doubling of atmospheric CO_2 concentration have been made by some scientists (see Chadwick, 1994). These have been set as high as 2% of GNP for industrial countries and may well be twice as high for developing countries. But there exist several regional and sectoral variations.

In the context of a possible increase in certain areas in biomass production resulting from climate change induced by greenhouse gas emissions, a costing study showed an overall benefit from the effects of climate change on forestry (about US $10 billion). To determine if adaptation to human induced climate change is a more cost-effective strategy than mitigation, it is necessary to come to a decision, both in the south and north, before facing up to the now famous impeccable economic logic of differential per capita wages across the world (Chadwick, 1994).

Use of biomass and other renewable energy sources is a crucial factor in responses to climate change mitigation and adaptation.

Developing suitable strategies to mitigate global climate change as well as promote sustainable development is a daunting task for most developing countries. These countries must consider mitigating the effects of climate change as they strive to improve their economy and improve the quality of life for the vast majority of people. There is a strong interaction between energy and development; so developing countries will require substantial increases in high-quality energy consumption in the coming decades.

Increasing high-quality energy supplies from fossil fuels is not only unsustainable, but also limited because of the associated economic and environmental costs. High debt burden, low capacity for internal capitalization and relatively poor economic performance reduce most developing nations' access to external funds. Widesprad use of carbon-intensive fuels cannot be allowed in the light of the environmental implications. Hence, new routes of energy growth should not only satisfy the economic and environmental concerns, but also promote more sustainable and more equitable development. Past development paths in many developing countries have led to serious environmental degradation and wide income disparities. New routes should aim at controlling management of all natural, human, financial and physical assets.

As a result of good progress in technology and cost reductions in the energy sector, several mitigation options are now available which promote development and increase energy supplies. These options can be divided into three areas: improved energy efficiency, economic instruments, and wide use of renewable energy technologies. There is need to evolve strategies that combine all three options.

In Asia, rapid development is taking place presently. Africa and Central and South America are also developing rapidly, increasing the mobilization of carbon, phosphorus, nitrogen and sulphur. In 1990, Asia consumed about 45% of the nitrogenous fertilizer supplied on a global basis which, in 2020, is projected to rise to about 65%.

Natural and anthropogenic changes have impacted global biogeochemical cycles and attempts are being made to describe and understand the interactive physico-chemical and biological phenomena that regulate the total Earth system, the unique environment that it provides for life, the changes that are occurring in this system, and the manner in which they are influenced by human actions. Biogeochemical cycles are now being looked at from several angles, especially in Asia, with special reference to processes that mobilize, transform, transport and sequester carbon, oxygen, nitrogen, phosphorus and sulphur (CONPS) as well as environmental processes that are affected by the increased concentrations of active CONPS species. The Earth's biogeochemical cycles of CONPS show much natural variability. These cycles are being increasingly affected by such human activities as combustion, agriculture and industry. These activities have speeded up the mobilization of CONPS from inert (e.g. gaseous nitrogen) and sequestered (e.g. fossil carbon) forms into chemical forms that can impact critical processes of our biogeochemical environment, such as ecosystem productivity and atmospheric energy absorption and photochemistry.

Till recently, most anthropogenic changes in the natural cycles of CONPS took place in the industrially-advanced countries, but in recent years, combustion, agriculture and industry in Asia have developed to the extent that mobilization rates in some Asian countries now rank among the highest in the world and are already exerting considerable impact globally. Current trends point to a fairly high growth in Asian energy use, agriculture and industrial activity in the coming few decades, leading to further changes in natural cycles.

Striking land-use changes are occurring in Asia. The total land area under forests and wetlands declined by over 45% from 1880 to 1980. Human activities in southern Asia have converted high-biomass land cover to low-biomass land categories in the past century. The changes in land use resulted in an estimated contribution of about 29×10^6 T of carbon to the atmosphere during the last century. Land-use changes determine the fluxes of greenhouse gas emissions between the biosphere and the atmosphere. Land use changes occur mostly on local and regional scales and have global significance through their cumulative effects. Biomass burning strongly impacts atmospheric chemistry and aerosols, and could play a significant role in processes leading to global warming.

Human activities have strongly affected the structure and functioning of ecosystems in Asia for millennia. Cropping systems, domestication of plants and animals and building of large irrigation systems have changed the ecological landscapes considerably.

The changes in land use and mobilization rates have naturally changed biogeochemical cycles. In the terrestrial carbon budget, increases in the atmospheric concentration of CO_2 are attributed to two primary forcing agents, viz. (1) combustion of fossil fuels, and (2) land cover conversion. The former is a product of industrial activities while the latter is primarily due to the clearing of tropical forests. The former is well known but the latter is not. When estimates of sources and sinks are combined in global carbon models and are then compared with observed increases in the atmosphere, a global budget cannot be reconciled as there is some missing fraction.

Methane, an important greenhouse gas, is particularly significant in Asia where it is released from rice paddies to the atmosphere through ebullition and rice plant-mediated diffusion. The growth of irrigation systems and increased production of wetland rice has increased methane fluxes to the atmosphere from paddy fields. There is some hope that our present knowledge of processes controlling methane fluxes may help develop mitigation technologies in tune with sustainable increases in rice production.

Unfortunately, one of the major concerns from the viewpoint of global change is the serious situation in respect of Asian rivers. About 13% of global fluvial discharge is presently dammed in the world. In the present century, about 66% of the world's total stream flow to the ocean shores is estimated to be controlled by dams. The Nile river has already been completely dammed and several other rivers in the world will face the same fate as the Nile (see Deekshatulu and Nash, 1995).

On the huge agricultural continent of Asia, modelling work has shown that the growing seasons of some crops could lengthen due to warming, but will be accompanied by a harmful decrease of moisture. Changes of interannual variability, as well as the longer-term trend of water resource in Asia, are significant under global forcing. Historical records as well as modelling work bring out a general aridity trend in Northern China and Central Asia.

There has been considerable progress in modelling of global terrestrial primary production. Geographically referenced net primary productivity (NPP) and gross primary productivity (GPP) and their corresponding seasonal variation are key components in the terrestrial carbon cycle as they are needed to understand living ecosystems and their effects on the environment. Productivity is also the crucial variable for the sustainability of human use of the biosphere by providing food and fibre (see Deekshatulu and Nash, 1995).

WATER VAPOUR

Many aspects of the properties, distribution and roles of water vapour in the atmosphere and the climate system are not well understood. Water vapour is very hard to measure accurately, even in the lower atmosphere. The best current capacitive radiosondes are considered to be quite good sensors if not contaminated, but they have a dry bias at low levels, and are too moist at low humidities, where a temperature-dependent correction factor has to be applied. No single technique or instrument is appropriate for use at all latitudes. Individual instruments reach an accuracy on the order of 5%. Intercomparisons between different instrumental techniques show even larger discrepancies. No technique available today has proven to be an absolute standard, and future calibration and characterization will be required for high-quality measurements. Inexpensive balloon borne frost-point hygrometers may be one way to overcome some of these difficulties.

Atmospheric Radiation and Chemistry

Water vapour has several roles in atmospheric chemistry. It is the major source of the hydroxyl radical (OH) — the major oxidizing agent and cleanser in the atmosphere. Additionally, ice provides absorbent sites for polar molecules, and a matrix for chemical reactions. Liquid water provides sites for aqueous chemistry and is a solvent, resulting in wet removal processes and acid rain. In the stratosphere, water is created by the oxidation of methane and is responsible for ozone destruction in the upper stratosphere, while water vapour chemistry is responsible for processes that cool the mesopause by 9K/day. These strong links between the hydrological cycle and atmospheric chemistry mean that future climate changes are tightly linked with atmospheric chemistry (Gille, 2000).

Upper Troposphere–Lower Stratosphere (UT-LS)

Airborne measurements of water vapour in the UT-LS have been used to infer that the Lagrangian mean transport in the stratosphere, later termed the Brewer-Dobson circulation, consists of rising motion through the tropical tropopause, poleward drift, and sinking at high latitudes. Recent measurements have highlighted the importance of water vapour as a tracer of atmospheric motions.

Water vapour poses a particularly challenging problem because it decreases by 4 orders of magnitude between the surface and tropical tropopause. In response to CO_2 increases, all current climate models predict a larger temperature increase in the upper tropical troposphere than at the surface, and an increase in UT specific humidity of ~40%. The stratospheric temperature decreases in response to increased radiative cooling (see Gille, 2000).

The Tropical UT

Water is injected into the stratosphere through the tropical tropopause; possibly convective overshooting inserts water vapour and cirrus into a tropopause layer, suggested to lie between roughly 150-50hPa. Entry into the stratosphere occurs away from the overshooting region, where radiative heating leads to upward motions greater than those of the Brewer-Dobson circulation. Clouds appear to be a source of moisture in the UT in the tropics.

The Extratropical UT

New measurements are proving useful in understanding distributions and processes in the extratropical middleworld, the UT and stratosphere. There are strong seasonal variations, with higher specific humidity in the summer than the winter, but with relative humidities about 10% lower in the summer than the winter, and about 30% lower in the subtropics than in mid-latitudes. These changes are associated with the seasonal march of the subtropical and polar jet streams, where strong latitudinal boundaries occur (see Gille, 2000).

WATER VAPOUR AND CLIMATE

Water vapour plays a fundamental role in the chemistry, dynamics and the radiation budget of the atmosphere. In conjunction with UV-B photolysis of ozone, water vapour is the precursor for the hydroxyl radical which controls the oxidative capacity of the atmosphere. Transport and release of latent heat provides a major contribution to the energy budget of the global and regional atmosphere (Kley and Russell, 1999).

The importance of water vapour extends to all atmospheric regions — troposphere, stratosphere and mesosphere, but its distribution, climatology and the variability of its concentration in the Upper Troposphere

(UT) and stratosphere are not adequately known, particularly in the altitude region just below and above the tropopause.

Water vapour is increasing in the lower mesosphere at a higher rate than expected from methane oxidation, while in the upper mesosphere, its long-term variations may be attributed to a 11-year solar cycle variability in the photo-dissociation rate by solar lyman-alpha irradiance. The causes of the long-term cooling, the effects of episodic volcanic warming and the water vapour increase in the lower mesosphere are unknown.

Water vapour is the greatest contributor to the atmospheric greenhouse effect — even more so than carbon dioxide. Unlike CO_2, which contributes to the anthropogenic forcing of the greenhouse effect, the action of water vapour is a feedback rather than a forcing, because in a static system, thermodynamic arguments dictate that a temperature increase of a water body causes an increase of the water vapour partial pressure above that body. However, the Earth-atmosphere system is not static and it is not clear if evaporation to the air above a water body is governed by thermodynamics or by dynamics.

According to Lindzen (1990), increased convection in a warmer climate leads to a drying rather than a moistening of the UT. Lindzen's cumulus drying hypothesis is equivalent to a strong negative feedback by water vapour in the enhancement of the greenhouse effect caused by increasing CO_2.

According to Harries (1997), because the effects of water vapour on the radiative balances of the earth are very large, even small errors in the spectroscopic parameters can produce large uncertainties in the prediction of climatic change.

Serious difficulties arise from uncertainties about quantification of the water vapour feedback (Kley and Russell, 1999). There is a woeful paucity of data on water vapour concentration and relative humidity in the UT and LS. Indeed, no coherent picture of the global UT-LS water vapour distribution exists today to allow quantification of the water vapour feedback relative to the effects on the radiative forcing from a doubling of CO_2. Indeed, it is the water vapour issue that represents the single largest uncertainty in the prediction of global and regional warming caused by the increase of CO_2 concentration.

HOLISTIC APPROACH AND BIOSPHERIC PROCESSES

Although the phrase 'global environmental change', like 'risk' or 'sustainable development' is difficult to define precisely and adequately, four conditions are associated with the notion. A typology of global change, for which different risk management strategies would be appropriate, can be generated:

1. Rapid global environmental change, e.g. stratospheric ozone depletion is caused primarily by human activity superimposed on the underlying biogeochemical processes.
2. The effects of these changes manifest globally in various ways, meaning that they have direct implications, or that the cost of trying to alter these changes is so high as to affect the present or future global economy.
3. The rate of change is so rapid that it can be identified within a human lifetime. Yet, sadly, most institutions do not adapt well to this time frame when planning for the future.
4. The scale of change may be potentially irreversible — or at least so expensive to remedy as to be economically crippling (O'Riordan and Rayner, 1991).

These conditions should be considered in relation to the idea of the wholeness of earth's physical and social processes—referred to as 'biospheric' processes. From this viewpoint, the complex interactions among biogeophysical processes constitute a self-correcting global organism on different scales and locations of time and space.

The biosphere may be taken to be virtually timeless. No doubt, some perturbations do take place occasionally and phase states change, but the principles of homeostasis still hold. This suggests that states maintain fluctuating dynamic equilibrium over ages, altering catastrophically on occasion, but retaining the basic objective of evolving life on earth. The biospheric linkages between the organic and the inorganic do not break. No matter how much of the organic is lost, the scope for life restoration via the metamorphosis of the inorganic to new organic forms still remains. Catastrophes may be viewed as a cleansing process, rather like a liver diet where only fruit is consumed for a week to eliminate unwanted residues in an overworked organ (O'Riordan and Rayner, 1991).

Many examples are known in which extreme, dramatic environmental perturbations cleanse and heal diseased ecological states. For example, fire is essential to maintain heathland, moorland, grassland and the conifer Douglas fir. Complete destruction of vegetation by irregular extreme condition purifies this seed-bed, preparing the ground for successful recolonization. Likewise, many rivers do their real erosive and sediment removal work in the 0.1% of the time they are in high flood, leaving 99.9% of the time to reorder the mess created the flood. Conceivably, this may also be the case with global environmental change: a foreseeable peril of such great magnitude on such a relatively short time frame, where holistic issues of social justice and resource entitlement must be faced if the crisis is to be avoided.

Typology of Global Change

O'Riordan and Rayner proposed the following four types of global environmental change, where each level warrants a different scale of alteration and character of societal response.

1. Biospheric catastrophe

This is the cumulative onslaught of perturbations sufficient to cause a human-induced catastrophic gaian 'flip' or change of phase, e.g. a so-called 'runaway' greenhouse effect. Some have suggested that slight warming of the northern hemisphere may result in the rapid release of methane stored in tundra peat bogs. Consequent changes in ocean temperature differentials might induce a catastrophic reversal in the direction of global oceanic circulation, with further climate repercussions.

2. Climate perturbation

Many analysts equate climate change with global environmental change, but climate perturbation is only a subset of global environmental change. Nevertheless, climate change is specific enough to be regarded as a risk arena in its own right firstly because much atmospheric chemistry and global kinetic energy are locked up in climate patterns and, secondly, because climate oscillations spell significant economic repercussions.

3. Undermining basic needs provision

Basic needs (e.g. access to sufficient fertile soil, water, and energy to provide food and shelter for survival together with enough surplus to allow for education culture and enjoyment) go beyond mere survival. They apply to sustainable survival. Hence, a manageable surplus for reciprocal use and for personal and social development must be incorporated. Global change occurs if such basic needs are not met for such a high proportion of the population that the cumulative outcome is degradation of ecosystems whose health is essential for the health of the planet.

4. Micropollutants

The least understood feature of global environmental change may be the cumulative, biogeochemical impact of persistent and toxic micropollutants arising from dispersed (and uncontrollable) sources via undetectable concentrations. These can accumulate unpredictably (chaotically) in air and ocean currents, in rainfall, and through food chains in such a way as to become extremely hazardous. Examples can be given of the many volatile organic compounds, traces of heavy metals and the polychlorinated compounds. Their long-term consequences cannot be predicted.

Three general types of response strategies for global environmental change are prevention, adaptation and sustainable development. Preventive response is based on the notion that nature is fragile and it is morally wrong to abuse it. Adaptive strategy, in contrast, rests on the idea that nature is quite robust and it is wrong to curtail development. The concept of sustainable development implies that nature is resilient but only within limits, and it is morally imperative to preserve choice.

Those who believe in sustainable development feel that it is possible to avoid global catastrophe by careful stewardship of the limited opportunities that nature provides for controlled growth (Adams, 1990). The ethical imperative is to preserve choice for future generations.

Prevention is the desirable strategy where conceptual uncertainty is endemic, e.g. in biospheric catastrophe. For a holistic risk assessment of climate change, practical responses must be based on a combination of both prevention and adaptation.

The basic-needs-plus level constitutes the core of sustainable development. There is a strong link between non-sustainability (basic-needs-minus) on the one hand, and population growth on the other. There is no doubt that a rechanneling of resources from military spending to basic-needs-plus (e.g. education, ethnic and gender rights) will result in net gains for sustainable development and global stability.

When we come to the issue of multiple micropollutants and the possible implications for biospheric stability, prevention may be the best option.

Ethics and Environmental Policies

The issue of ethics and environmental policies has become a central concern of both environmentalists and policy makers. The global moves towards the establishment of international agreements and covenants, as well as the need to focus on longer-term arrangements, have catalyzed a reconsideration of underlying value systems that aid or constrain action. At the local, regional and national levels, the political process is increasingly being called upon to respond to appeals for environmental equity. The assessment and adjudication of moral and ethical claims and counter-claims are becoming more and more appropriate tasks.

Environmental ethics may be envisaged as operating in the contexts of ordinary human activity, economics, technology and politics. Any ethic that is to accomplish changes in social policy should promote 'the integrity of nature'. The notion of sustainable development has strong ethical implications.

The process of environmental decision-making should be based on friendly negotiations among parties involved in disputes. There is also the need to balance power and the diverse sources of access to information. However, the issue of whether or not negotiation should be the central guiding or structural principle in environmental decision-making processes, rather than only one step in such process, is debatable.

The discipline of economics has contributed significantly to the foundations of environmental policy. Examples include issues of economic efficiency, benefit-cost analysis, multiple-objective planning and evaluation, and the application of economic analysis to the elucidation of problems that require the design of innovative institutions. To overcome certain common problems, it has been argued that one could rely on commonly-shared beliefs or moral norms. These norms, which once used to be based on tradition and custom, are being expressed through public opinion. To be guided by this opinion, however, practitioners need a suitable framework within which to operate, and through which they can evaluate the claims being made upon them. The problem is, consequently, how to educate decision makers and industrial managers towards a higher level of environmental concern, and how to restructure laws and decision making systems with a view to facilitating this shift. During the last few years, there has been substantial discussion on major trends in economic theory, including considerations of economic rationality, economic ethics and the mechanics of green taxes. The issue of the ethics surrounding the application of various discount rates to future environments has been hotly debated and the concept of the 'fiduciary trust' for the protection of non-economically valued objects and places has been introduced as an alternative model for consideration.

The environmental question is becoming a crucial ethical issue of our times, and involves many critical aspects of our lives, ranging from the well-being of the individual to the destruction of many global ecosystems. This issue requires that we reconsider our current ways of thinking and acting; and it points towards possible major changes in our social, economic and political structure. We must reconsider our approaches and methods that in the long term have a profound effect on social and political systems, and there is need for the analysis and application of ethics to environmental policies.

THE ROLE OF THE STRATOSPHERE IN THE CLIMATE SYSTEM

For long, it was believed that the stratosphere, which represents about 10-20% of the atmosphere in terms of mass, plays only a minor role in climate change. Recent evidence shows that the stratosphere is a sensitive component of the climate system and can affect the troposphere through various coupling mechanisms.

The temperature structure of the stratosphere represents a balance between radiative and dynamical heating. Largely, radiative transfer in the stratosphere is a clear-sky process and is well understood, but a significant exception occurs in the aftermath of a large volcanic eruption.

The radiative equilibrium state tends to be dynamically stable nearly everywhere. There are no rapid small-scale adjustments to be considered (as there are in the troposphere). The one place where stability is not guaranteed is in the tropics, where the seasonal migration of the solar heating maximum about the equator leads to a process of quasi-horizontal inertial adjustment (Dunkerton, 1989).

Dynamical heating is associated with vertical (diabatic) motion, which is partly caused by radiative heating, but is mainly caused by the breaking and dissipation of certain waves emanating from the troposphere. In the extratropical stratosphere, the wave-induced forces are almost exclusively westward and drive air poleward (Holton et al., 1995). The resulting dynamical heating is negative in low latitudes and positive in high latitudes.

The Role of Transport and Mixing

The distribution of radiatively active trace gases in the stratosphere constitutes an important part of climate forcing, including the shape of the tropopause itself. This distribution is controlled by dynamics in two ways: (1) through the temperature distribution which affects chemical reaction rates; and (2) through transport and mixing by winds.

The tropopause is higher in the tropics than in the extratropics. A critical factor affecting tropospheric climate is the height of the tropopause, but this issue, so basic to climate, is poorly understood. Considerations of radiative-convective adjustment suggest that the tropopause should be identified with the top of the region of moist convection. There are, however, important feedbacks from the stratosphere itself (Shepherd, 2000).

Coupling Mechanisms between the Stratosphere and Troposphere

There are three principal mechanisms by which the stratosphere can affect tropospheric climate. The first is through radiative transfer. This may be either by changes in the amount of solar radiation that reaches the surface (e.g. after a volcanic eruption), or by changes in the amount of downwelling longwave radiation emitted by the stratosphere (e.g. because of stratosphere ozone depletion). The impact depends on the vertical, latitudinal and seasonal structure of the changes in the radiatively active substances, particularly in the vicinity of the tropopuase (Hansen et al., 1997). The second and third mechanisms are based on the basic dynamical fact that tropospherically forced waves propagate up, while zonal-mean anomalies propagate down. Thus, the second mechanism is that the stratosphere can affect the 'upper boundary condition' of the troposphere by affecting the propagation characteristics of tropospheric waves. The possibility of wave reflection at the tropopause has obvious implications for regional climate perturbations.

The third mechanism involves the downward propagation of zonal-mean anomalies (Baldwin et al., 1994).

The changes in stratospheric ozone over the past two decades have been important in forcing climate change. They have led to significant changes in the temperature in the lower stratosphere and the troposphere (Karoly, 2000).

Besides the recent decreases in stratospheric ozone, there have been significant increases in tropospheric ozone over the last century. While the observational network is sufficient to identify these increases, it is not adequate to describe the horizontal distribution of these changes in tropospheric ozone.

Major volcanic eruptions lead to increases in the stratospheric aerosol load, which influence the radiative forcing of climate through scattering and absorption of solar radiation. Direct measurements of stratospheric aerosol optical depth have been made only for the past two decades with reasonable global coverage through satellite observations. The radiative forcing associated with historic eruptions before the satellite era is more uncertain (see Andronova et al., 1999). The eruption of Mt. Pinatubo in 1991 caused the most recent major increase in volcanic aerosols.

MAN-MADE CLIMATE CHANGE

The climate of a planet mainly depends on its size and its distance from the sun. For the Earth, these parameters made it possible for water to exist in all three phases and, hence, made life possible on earth. This life has, over the last three billion years, created an atmosphere whose constituent parts relevant to climate (greenhouse gases) only account for 0.3% of its mass but nevertheless, lead to a greenhouse effect of 30°C, with CO_2 being the most important greenhouse gas.

During the last two centuries, industrialized countries have produced a 30% increase in atmospherice CO_2, doubled the concentration of methane and created CFCs which are new and potent greenhouse gases. Global warming alongside with a change of circulation in atmosphere and oceans have been the result. Table 3.1 summarizes potential climate change impacts for different systems.

The natural greenhouse gases CO_2, CH_4 and N_2O as well as certain CFCs, with their long lifetimes, are increasing. Although the concentrations of CO_2 and CH_4 either did not, or only marginally increased in 1992 and 1993, this does not indicate a change of the trend but rather that the dynamics of their sources and sinks have not yet been fully understood. CO_2 absorption and release by oceans and the vegetation, for example, vary from year to year, while a sink for CH_4, UV–B radiation,

Table 3.1 Potential climate change impacts for various systems

Systems	Potential impacts
Forests/terrestrial vegetation	Migration of vegetation. Reduction in inhabited range. Altered ecosystem composition.
Species diversity	Loss of diversity. Migration of species. Invasion of new species.
Coastal wetlands	Inundation and migration of wetlands.
Aquatic ecosystems	Loss of habitat. Migration to new habitats. Invasion of new species.
Coastal resources	Inundation of coastal development. Increased risk of flooding.
Water resources	Changes in supplies. Changes in drought and floods. Changes in water quality and hydropower production.
Agriculture	Changes in crop yields. Shifts in relative productivity and production.
Human health	Shifts in range of infectious diseases. Changes in heat-stress and cold-weather afflictions.
Energy	Increase in cooling demand. Decrease in heating demand. Changes in hydropower output.
Transportation	Fewer disruptions of winter transportation. Increased risk for summer inland nagivation. Risks to coastal roads.

Source : Smith, J.B., Mueller-Vollmer, J. Setting Priorities for Adapting to Climate Change, contractor paper prepared for the Office of Technology Assessment, March 1992.

increases due to stratospheric ozone depletion. In 1994, the concentration of CO_2 and CH_4 increased again with the average rate of the 80s. The Pinatubo eruption can also be accounted for these fluctuations, since the increase in sulphuric acid aerosols in the stratosphere causes greater stratospheric ozone depletion resulting in increase in UV–B radiation, and it induced a cooling of the Earth's surface in 1992 by about 0.25°C.

Another important greenhouse gas—ozone—has shown a marked decrease in the lower stratosphere as well as an increase in the troposphere of the northern hemisphere, leading to a complicated pattern of radiative forcing.

Aerosols are formed as a result of the burning of fossil fuels, or via atmospheric reactions of the emitted gases SO_2, NO_x and NH_3. This can reduce visibility. Man-made aerosols decrease irradiation, increase acidic deposition, and cause unintended nitrogen fertilization of landscapes.

The aerosol effect of the climate cannot be deduced from the overall greenhouse effect, because aerosols are distributed unevenly in the atmosphere and across the continents and neither accumulate nor are effective at night. The effect of aerosols is smaller than that of the greenhouse gases. Estimates of the aerosol effects are uncertain, especially for those resulting from biomass burning.

Four main and important reservoirs of CO_2 are atmosphere, soils, vegetation and oceans. How fast can the latter three absorb additional amounts of CO_2? A few answers are:

1. Seven billion tons of carbon (GtC) are released into the atmosphere annually from biomass burning and burning of fossil fuels, of which in the 80s, 3.2 ± 0.2 GtC have remained in the atmosphere annually, thereby increasing CO_2 concentration by almost 0.5% annually;
2. Of the 3.5 ± 0.2 GtC which leave the atmosphere, the oceans only absorb 2.0 ± 0.8 GtC per year;
3. 1.5 GtC are assumed to be absorbed by terrestrial CO_2 reservoirs, implying that the destruction of the biosphere, especially in the tropics, which accounts for 1.6 ± 1.0 GtC carbon release annually, is partly compensated. This compensation occurs through afforestation and sustainable forest use in industrialized countries (0.5 GtC) through humus formation and N–fertilization of forests (1.4 ± 1.5 GtC);
4. The modelling of the carbon cycle shows that even for a CO_2 concentration of 750 ppm in the twenty second century (which means more than doubling the pre-industrial 280 ppmv), we have to reduce CO_2 emissions below today's rates. To achieve CO_2 stabilization at 450 ppmv, the emission rate in the next 100 years has to be reduced by at least 50%, and still we will have to live with an increase of the average global temperature of 2°C.

Who will be affected by climate change and to what extent? Some partial answers are the following:

Expected regionalization of climate changes
Stronger monsoon circulations, general increase of precipitation in higher latitudes as well as in mid latitudes in the winter, less warming in areas where oceans mix (Iceland and around Antarctica).

Expected altered variability
Increased precipitation variability from year to year, especially in the tropics, increasing periods between precipitation events.

Expected new extremes
Every change in mean values will cause new extremes; increasing variabilities increases the frequency of extremes. Precipitation intensities increase with warming.

Anthropogenic climate change is highly complex, in view of the influence of anthropogenic aerosols on the radiation balance, the negative greenhouse effect due to increased stratospheric ozone depletion, as well as the greenhouse effect of the increase in tropospheric ozone.

According to some recent calculations of the German Max Planck Institute for Meteorology in Hamburg, the 25% increase in CO_2 concentrations concomitant with the increase of other greenhouse gases have up to now resulted in a temperature increase of 0.5 – 1.0°C. An increase of the average global surface temperature of 0.7°C since 1880 has been measured. So far, it was not possible to directly link this temperature change to the increase in greenhouse gases, since the changes still fell in the range of the natural variability of the climate system. New model calculations based on improved methods for detecting human impact have shown that the probability for a temperature change caused by natural factors is less than 5%. Conversely, the probability for this change to be caused by an anthropogenic increase in greenhouse gases is greater than 95%.

GLOBAL CLIMATE CHANGE

Many governments and industries at the highest level are greatly concerned about the broad implications of climate variability and change for human well-being, ecosystems and economic development. Foundations have been laid for the changing views on anthropogenic climate change, and possibilities for increasingly confident seasonal and annual climate predictions have materialized. It is very important to know as precisely as possible, the extent and rate of climate change brought about by human activity.

The emerging challenges of global climatic change, environmental hazards and the sustainability of natural resources have led to a cultural shift in emphasis in research priorities. As a result, the priorities attracting greater notice are: to develop more effective ways of discovering and assessing such natural resources as give us minerals and energy; to improve our understanding of the factors and processes involved in global environmental change as a basis of securing and improving the living conditions on the planet; and to enhance our understanding of natural, physico-chemical, biological and geological processes and concepts by comparative analysis around the globe.

The issue relating to sustainability of resources have come in sharp focus. The problem is not simply how to find and sustain economic minerals, but extends to a broad area including toxic rock-derived substance in soil, water and foodstuffs. Geochemistry has thrown much light on such problems and is increasingly involved in investigations of the epidemiology of human and animal diseases in the developing world. Studies of rock and soil geochemistry have revealed both natural and human-induced deficiencies of vital elements (e.g. iodine) or the presence of toxic substances that have demonstrable effects upon the earth's life-support systems, including food quality and the geography of disease.

CO_2 is a major player in global climatic change. A close relationship exists between fluctuations in the atmospheric carbon dioxide and methane and global climatic changes as expressed in temperatures, sea levels and ice volumes, for example. The reservoir of carbon held on the surface of the earth's landmasses has been an important factor in the carbon budget over geological time. Rock weathering, particularly the dissolution and precipitation of carbonate-rich rocks such as limestones, involves transfers of carbon between rock, water and atmosphere. Precipitation of carbonates in deposits such as stalagmites in caves, a process that releases carbon dioxide into the atmosphere, can provide a climatic 'calendar' of decadal resolution.

Our understanding of the reasons for natural shifts in desert margins has also improved, as a part of a process of characterizing both natural and human-induced changes. Because a significant proportion of the world's population lives in the arid and semi-arid drylands, this improved understanding of shifts in desert margins is of obvious importance for future dryland zone socio-economic development. It has been established that during the last major warm stage (about 80,000 to 130,000 years ago), the southern margin of the Sahara Desert was almost 500km farther to the north, grasslands, lakes and marshes being widespread where today there are only sand dunes. It was inhabited by antelopes, elephants and giraffes, with crocodiles, hippopotamuses and fish. In contrast, during the last major cold phase (80,000 to 10,000 years ago), the desert margin was about the same distance south of the present margin. These spectacular shifts took place over only one or two thousand years, during which periods there were migrations of the human population. Such documented sequences may provide a model for both cooling and warming phases based on changes in the greenhouse gas budget. The changes were also accompanied by a rising sea level during warming, and a declining sea level as the climate turned glacial.

Considerable new knowledge has been generated about the world's deposits of tin/tungsten, coal, phosphorites and bauxite, and the ores of copper, lead and zinc. New maps have been made as a basis for mineral exploration in Africa, East Asia and South America and also new geochemical maps produced with a bearing on animal and human health, soil fertility, water supplies, irrigation and waste disposal. New light has been thrown on the processes of desert advance and desertification; improved understanding of the complex relationship between terrestrial and extra-terrestrial events and the evolution of life forms. Integrated data sets on stability and instability in the coastal zones and the present trends in sea level change have been produced.

Table 3.2 lists some communities/species vulnerable to be affected by global climate change.

Table 3.2 Some species and communities most likely to be affected by global climate change (after Peters and Darling, 1985)

Species type	Predicted effects
Peripheral populations	Populations near edges of the species' range are more likely to be affected by range shifts due to global warming.
Geographically-localized species	Local endemics are unlikely to have populations in areas of suitable habitat after a climatically-induced range shift.
Genetically-impoverished species	May lack the variability necessary to adapt to changing climatic or habitat conditions.
Specialized species	Generally less tolerant of environmental change and often require a narrow range of conditions.
Poor dispersers	Would have difficulty in spreading their distributions into newly-created suitable habitat during climate change.
Annual species	Reproductive failure in one year can lead to local extinction. Such failure is more likely during climate change in annual than in perennial species.
Montane and alpine communities	Species distributions would shift to higher elevations during global warming, occupying smaller areas with smaller populations, making them more vulnerable to extinction.
Arctic communities	Temperature increases in the arctic are expected to be greater than in equatorial areas, so the arctic communities may undergo greater stress.
Coastal communities	Upwelling patterns in coastal areas are likely to be altereed, and sea levels will rise, flooding coastal habitat. These events will stress coastal communities, which may not be able to follow rising waters inland due to human development.

THE ACTIVE CARBON CYCLE

Much of the earth's carbon is stored in rocks and sediments. A relatively small fraction is 'active' and keeps circulating among the carbon reservoirs of the oceans, atmosphere and terrestrial and marine biospheres. This small active fraction is, however, vitally important for the maintenance of life on earth. For it is within this pool of carbon that a series of biological, physical and chemical cycles constantly re-distributes the element, effectively transferring energy from sunlight into fuel for the biosphere. These processes may be called the 'active carbon cycle'.

Since the advent of the industrial and agricultural revolutions, beginning over 220 years ago, the flow and distribution of carbon between the various active carbon reservoirs has changed. It took from about 1750-1970 for atmospheric CO_2 to increase by about 45 ppm. However, it has taken only a few decades for it to increase by the same amount. This large increase in growth rate has been due to the dramatic escalation in fossil fuel usage in the second half of the twentieth century. There has been a relatively rapid addition of fossil carbon to the atmosphere in the form of carbon dioxide and other carbon-containing gases. These gases are emitted by a wide variety of processes, including the burning of coal and oil, large-scale clearing of forests, widespread ruminant animal farming and certain changes in wetlands (Lowe, 2000). This additional carbon affects the basic physical and chemical properties of the atmosphere and is thought to be the cause of current and projected climate change.

Seasonal Cycle

In many places (such as Baring Head, New Zealand), a seasonal cycle is superimposed on the long-term trend in atmospheric CO_2. Its range is quite small, only about 1 ppm, but this is highly variable. The seasonal cycle is caused by a combination of seasonal changes in the rates at which CO_2 is transferred between the atmosphere, oceans and terrestrial and marine biosphere reservoirs.

The seasonal cycle is affected by photosynthesis as well as seasonal changes in the transfer of atmospheric CO_2 into the oceans caused by changes in sea-surface temperature and growth of marine organisms. An additional effect in the Southern Hemisphere occurs due to the transport of CO_2 from the Northern Hemisphere. As most of the earth's terrestrial biosphere is in the Northern Hemisphere, seasonal changes in photosynthesis there lead to a much larger annual cycle in atmospheric CO_2. This cycle is transported into the Southern Hemisphere and, although becoming dispersed, still modifies the seasonal cycle in atmospheric CO_2 observed at Baring Head.

Fossil fuel usage adds about 6 billion tonnes of carbon to the atmosphere each year. Of this, about one half is taken up by the oceans and the biosphere. The remainder stays in the atmosphere, leading to an increase in the concentration of atmospheric CO_2. (Note: 1 GtC = 1 billion tonnes of carbon).

One serious scientific challenge is to determine the impact of the additional CO_2 released by the combustion of fossil fuels. The processes involved are very complex. Only about half of the CO_2 released by fossil fuel combustion remains in the atmosphere. The remainder must be absorbed by the other reservoirs of the active carbon cycle, but which ones, and how much goes into each? It is crucial to understand the changes in the partitioning of CO_2 between the reservoirs of the active carbon cycle. Storage of excess CO_2 in the deep oceans, for example, could prove safer for humanity because that CO_2 may be less likely to re-enter the atmosphere over a short time-scale. Conversely, carbon stored in terrestrial ecosystems may be only temporary—a possibility coming from projections for the degradation of forests as a result of global warming. Besides, forests are vulnerable to future human intervention through forest clearing along with other land management practices (Lowe, 2000).

One way of differentiating between the oceanic and biospheric sinks of CO_2 is through measurement of O_2 in the atmosphere. Combustion of fossil fuels reduces atmospheric O_2. These changes are too small to be of concern for air-breathing animals (including people). But the rate of decrease and small seasonal variations in O_2 levels also enable us to separately estimate the CO_2 uptake by the oceans and by vegetation. At Baring Head, such measurements have been made and it is the first clean-air site in the world at which continuous O_2 measurements are being made in collaboration with the University of California, San Diego (Lowe, 2000).

There is uncertainty as to where the CO_2 from current fossil fuel emissions is going, and the effects of future climate change increase the uncertainty for future prediction. Some oceanographic studies suggest that the Southern Ocean is already changing in ways that will reduce its uptake of CO_2. But uncertainties in the fate of fossil fuel emissions are far less serious than the uncertainties in projections of how people might behave in future. Recent studies of scenarios for social and economic development and their implications for CO_2 emissions reveal that the historical links between population, GDP and emissions may not hold in future. The rate of transfer of technology from developed to developing countries, subsidies in overseas energy markets and the worldwide awareness of environmental issues have emerged as key factors. It appears that understanding the nature and implications of human-induced disturbances to the active carbon cycle is going to be one of the greatest scientific and environmental challenges of the twenty-first century (Lowe, 2000).

GREENHOUSE GASES AND CLIMATE CHANGE

The following account is extracted from recent reports of the Intergovernmental Panel on Climate Change (IPCC).

There is now much scientific evidence that the greenhouse gas emissions emanating from human actions will affect the climate system. There is also scientific evidence that the current warming of the global atmosphere is being caused by these human actions.

Several developed countries have agreed to limit their emissions in 2000 to their 1990 level. It seems both desirable and feasible, for some countries at least, to take suitable actions on climate change, including:

1. Adoption of measures that will stabilize emissions of greenhouse gases at 1990 levels by 2000.
2. Developing a strategy to achieve further reductions beyond 2000. Further reductions will require actions beyond the energy efficiency measures that will dominate the recommended emission reduction strategies for the next few years. This next level of action will very likely include restructuring of the transportation system, fuel switching and development of renewable energy, especially in electricity generation, and use of economic instruments to encourage emission reductions.

There has been a tendency to overlook some measures that could reduce greenhouse gas emissions at relatively low cost, and be implemented by 2000. Such measures include co-firing coal-fired generating stations with natural gas, stronger measures in the transportation sector, and mandatory implementation of energy-efficiency measures in government buildings.

Studies in Europe have indicated that limiting greenhouse gas emissions yield substantial indirect economic and environmental benefits.

After 2000, both additional energy efficiency measures and several new initiatives may include:

1. Restructuring of the transportation system and land use planning, including development of a sustainable transportation strategy that would reduce greenhouse gas emissions from that sector in the long run;
2. Fuel switching and development of renewable energy, especially in electricity generation.
3. Use of economic instruments to manage greenhouse gas emissions. Emissions fees and tradeable permits enable environmental goals to be achieved at low cost.
4. Additional energy efficiency measures in the industrial, commercial and residential sectors; and
5. Measures to reduce non-energy emissions, such as limiting methane emissions from landfill sites and enhancing carbon dioxide sinks such as forests.

RECENT ASSESSMENT REPORTS OF INTER-GOVERNMENTAL PANEL ON CLIMATE CHANGE

The salient issues discussed by the Intergovernmental Panel on Climate Change (IPCC) are summarized below.

Are Human Activities Changing Climate?

1. Increases in greenhouse gas concentrations lead to a positive radiative forcing of climate, tending to warm the surface and to produce other changes in climate.
2. The atmospheric concentrations of carbon dioxide, methane and nitrous oxide have grown significantly since pre-industrial times; by about 30%, 145% and 15%, respectively. These trends can be attributed largely to human activities, mostly fossil fuel use and agriculture.
3. Tropospheric aerosols resulting from the combustion of fossil fuels, smelting and biomass burning give rise to a negative radiative forcing over particular regions.
4. Global mean surface temperature has increased by about 0.3 to 0.6°C since the late nineteenth century.
5. Recent years have been among the warmest since 1860, i.e. in the period of instrumental record.
6. A trend towards higher frequency of extreme rainfall in recent decades was evident in Japan, the U.S.A., former Soviet Union and China.
7. There is evidence of an emerging pattern of climate response in the observed climate record to forcings by greenhouse gases and sulphate aerosols. The evidence comes from the geographical, seasonal and vertical patterns of temperature. Taken together, these results point towards a detectable human influence on global climate.

What Future Changes are Likely Without Limitations of Green-house Gases?

1. All the IPCC scenarios of future greenhouse gas and aerosol emissions (even one which assumes very little population growth between the present and the year 2100) imply increases in GHG concentrations relative to pre-industrial levels by 2100 (e.g. CO_2 increases range from 75 to 200%; methane (CH_4) increases to between about three and six times the pre-industrial level).
2. The increasing realism of simulations by coupled ocean-atmosphere climate models has enhanced our confidence in their use for projection of future climate change.

3. From the mid-range IPCC emission scenario, assuming the 'best estimate' value of climate sensitivity and including the cooling effects of projected future increases in aerosols, models project an increase in global mean surface temperature relative to the present of 2.0°C by 2100 with a range of estimates from 1°C to 3.5°C. In all cases, the average rate of warming would probably be greater than any seen in the last 10,000 years. Models project an average increase in sea level of about 50 cm by 2100 with a range of estimate from 15 to 95 cm (taking into account the aerosol effect).
4. Models suggest an increase in the probability of intense precipitation with increased greenhouse gases. In some areas, there is also likely to be an increase in the probability of dry days and the length of dry spells (consecutive days without precipitation).
5. Numerous factors today limit our ability to model future climate change. In particular, there are inadequacies in:
   (a) Estimates of future emissions of greenhouse gases, aerosols and aerosol precursors; and
   (b) Representation of climate processes in models (especially those associated with clouds);
   (c) Observations of climate (improved systematic measurements).

Will There Be Significant Impacts?

Human-induced climate change constitutes an important stress. Climate change will have a significant impact directly upon the environment and society, and also some indirect impacts all over the world. While there will be some beneficial effects of climate change, there will be many adverse effects, some potentially irreversible.

Quantitative estimates of the impacts of climate change on ecological and socio-economic systems are difficult because of uncertainties regarding climatic change at the regional scale; several of the following studies have identified potential effects:

1. In North America and East Asia, the number of heat-related deaths would increase several-fold in response to two climate change scenarios for 2050. In large cities, this would represent several thousand extra deaths annually. Climate-related increases in malaria incidence are estimated by one model to be of the order of 50-80 million additional cases annually, relative to an assumed global background total of 500 million by 2100. These would occur primarily in tropical and subtropical populations currently at the margins of endemically-infected areas. Higher temperatures in urban environments would enhance both the formation of secondary pollutants (e.g. ozone) and the health impact of certain air pollutants.

2. Climate change is expected to occur rapidly relative to the speed at which forest species grow, reproduce (in boreal forests), and re-establish themselves. An average global warming of 1-4°C over the current century would be equivalent to shifting isotherms poleward approximately 160-640 km or an altitude shift of 150-650 meters. Entire forest types may disappear, and new ecosystems may take their places.
3. Increased fire frequency and pest outbreaks are expected to decrease the average age of biomass and carbon store of some forests.
4. Overall, climate change will produce small to moderate effects on global agricultural production. Subtropical and tropical areas show negative consequences more often than temperate areas. People who depend on isolated agricultural systems in semi-arid and arid regions face the greatest risk of increased hunger due to climate change. Many of these at-risk populations live in Africa; South, East, and Southeast Asia; and tropical areas of Latin America, as well as some Pacific island nations.
5. Many coastal zones, estuaries and small islands are particularly vulnerable to direct effects of climate change and sea-level rise. Current estimates of global sea-level rise represent a rate two to five times of that experienced during the last century.
6. Fairly small changes in temperature and precipitation can have large and nonlinear effects on runoff and lake levels in arid and semi-arid lands. This can affect water supply, demand and hydro-power production. Even in those areas where models project a precipitation increase, higher evaporation rates can potentially lead to reduced runoff. More intense rainfall would increase flooding. A 50 cm sea level rise would increase the number of people world-wide subject to coastal inundation from about 45 to 90 million.

What Can Be Done and at What Cost?

The following insights can be gained from analyses of economics in the literature on climate change:
1. Estimates of aggregate net damages from a 2-3°C global warming tend to be a few percent of world GDP, with higher damages to developing countries, especially small island states.
2. The risk of aggregate net damage due to climate change, consideration of risk aversion and the precautionary principle provide rationales for GHG mitigation actions beyond 'no regrets'.
3. Energy efficiency gains of up to about 30% of current energy use can be realized, over the next three decades, with net economic benefits to zero net cost (an example of 'no regrets' measures).

4. The range of estimates from macro-economic models of costs in OECD countries of stabilizing emissions at 1990 levels range from minus 0.5% of GDP (an economic benefit) to plus 2% of GDP. Other models which emphasize technological options indicate that costs of reducing emissions by 20% in developed countries over the next few decades are either negligible or even beneficial.
5. There are potentially significant offsets to mitigation costs, known as 'environmental double dividends' and 'economic double dividends'. Environmental double dividends accrue in view of the fact that actions to reduce GHG emissions also reduce other pollutants and precursors to smog. This, along with other secondary benefits of mitigation action, results in economic benefits which can offset over 30% of the mitigation costs. If the reduction of emissions is achieved through a carbon tax or energy tax, 'economic double dividends' can also accrue through using the tax revenue to offset other more distortionary taxes and to increase investment.
6. Upto 30% of global energy related emissions could be offset by additional carbon sequestration in forests for a period of 5 to 10 decades.
7. Individual countries that adopt mitigation policies can choose from among a large set of potential policies and instruments, including carbon taxes, tradeable permits, deposit refund systems as well as technology standards, performance standards, product bans, direct government investment, and voluntary agreements. Public education on the sustainable use of resources has an important role in modifying consumption patterns and other human behaviour.

The Longer Term Possibility

1. Without mitigation actions, cumulative GHG emissions are expected to be, in the twenty-first century, three to ten times of those of the past 130 years.
2. Stabilization of emissions by some countries at 1990 levels would, if continued beyond the year 2000, result in an 8 to 12% reduction below intermediate emission projections.
3. The ultimate objective of the U.N. is a stabilization of greenhouse gas concentrations in the atmosphere at a level that would prevent dangerous anthropogenic interference with the climate system.
4. To achieve stabilization at levels between 25% greater (450 ppmv) than at present and to more than twofold of pre-industrial values (750 ppmv) will warrant going much beyond stabilization of emissions. In fact, stabilization at 750 ppmv CO_2 equivalent would require reductions, over the coming decades, to two-thirds of present

CO_2 emissions, while achievement of 450 ppmv CO_2 equivalent would require global reductions to one-third of present emissions.
5. In view of the long lifetimes of most greenhouse gases in the atmosphere, the earth is expected to warm further and sea level to continue to rise, even if measures were implemented which could immediately stabilize emissions or even concentrations. So, besides mitigation measures, countries should also strive to adapt to some degree of climate change.
6. To some extent, emissions can be reduced by switching from coal, to oil, to natural gas (natural gas has twice as much energy as coal per unit of carbon in the fuel) and large resources of natural gas exist in many areas.
7. Renewable energy sources are sufficiently abundant to technically provide all of the energy needs of the world even at levels expected over the current century.
8. The removal and storage of CO_2 from power-station stack gases is a feasible option to reduce GHG emissions. However, it can increase the production cost of electricity.
9. Producers and consumers need to adapt cost-effectively to constraints on GHG emissions.

CLIMATE CHANGE AND CARBON DIOXIDE

Diverse human enterprises have now made it possible for humans to alter the Earth on a global scale. One important index of human activity is the rate of utilization of energy, which has increased greatly in the last century because of rapidly increasing human population, coupled with increasing per capita energy consumption. Accelerating human activity may already have caused globally significant environmental changes, especially a possible human alteration of the Earth's climate, which is essentially the summation of weather and its variability. Weather varies on long time scales and, hence the climate is variable (Keeling, 1997). Long-term weather records have even provided evidence of significant variability over decades, possibly associated with climatic change (NRC, 1996). This short-term variability makes it quite difficult to separate subtle climate changes that might be caused by accelerating human activities.

Climatic change is chiefly caused by exchanges of energy, momentum, and chemicals between the atmosphere, the oceans and land surfaces. Oceanic and atmospheric circulation, turbulent mixing, photochemistry and radiative transfer all have roles here. Although these processes are mainly natural, a few of them especially those involving the greenhouse gases, are susceptible to human influence.

These greenhouse gases, mainly carbon dioxide, but also including others such as methane, nitrous oxide and halocarbons, enter the air mainly as byproducts of the combustion of coal, natural gas and petroleum. To a lesser degree, they come through other industrial and agricultural activities. The rates of emission of greenhouse gases into the air are roughly proportional to the global rate of energy consumption arising from human activity. With increasing human population and per capita energy consumption, concentrations of these gases have risen proportionately to the product of both increases. As they build up, these gases trap radiation upwelling from the Earth's surface leading to rising temperature at the Earth's surface unless some compensating process cancels out this tendency.

Carbon dioxide is particularly important as a greenhouse gas because it is indisputably rising in concentration. These processes, pertinent to CO_2 changing levels, include its (CO_2) interactions with the chemically-buffered carbonate system in seawater and with vegetation because of its vital role in photosynthesis. The sum total of all processes affecting the carbon on the Earth and, hence, controlling the concentration of atmospheric carbon dioxide, is called the 'carbon cycle', whose understanding is necessary in order to know how human activities may affect carbon dioxide. The pathways of carbon through the global carbon cycle are understood generally but the actual rates of change of the fluxes between the atmosphere, land, and ocean are not very clear. The annual anthropogenic carbon input to the atmosphere between 1980 and 1989 was estimated to include 5.5 ± 0.5 GtC (thousand million metric tons of carbon) from fossil fuel combustion and 1.6 ± 0.6 GtC from land–use change, making a total of 7.1 ± 1.1 GtC. Of this annual input, 3.3 ± 0.2 GtC remained in the atmosphere whereas 3.8 GtC were removed. Oceanic uptake, related to carborate buffering probably accounts annually for about one-half of the removal. Regrowth of northern hemisphere forests accounts for perhaps 0.5 ± 0.5 GtC. The removal mechanisms of the remaining carbon, 1.3 ± 1.5 GtC per year, remain uncertain (Keeling, 1997). This residual term is the so-called 'missing carbon sink'. Also, there is a feedback mechanism, whereby climate change may itself change carbon cycle. Significant warming from increasing greenhouse gases can change the rates of uptake of carbon dioxide by the oceans globally and also alter gas exchange with vegetation. But very little is known about such feedback mechanisms.

Some current issues of interest relating to the carbon cycle and the CO_2 missing sink are the following: (i) the extent to which climate is changing owing to both natural causes and human activities; (ii) whether these changes, in part, are long-term manifestations of increasing carbon dioxide; and (iii) how the oceans, terrestrial plants and soils, and atmosphere function in general as a necessary basis for exploring the first two topics.

The world's sea level is expected to rise as a result of global warming. Unless pre-emptive actions are undertaken, this can destroy millions of people. Over 60% of the world's people live in coastal communities, and most productive agricultural zones are found in or near coastal areas.

Mangrove forests along the intertidal belts of tropical and subtropical coastlines can help mitigate, if not completely check, such potentially catastrophic consequences. Indeed, had there been no deforestation in the coastal areas of Orissa state, the fury and havoc caused by the severe cyclone in October/November 1999 could have been avoided or minimized.

Sea Level Rise Scenarios

The earth's surface temperature is expected to increase by 0.5°C to 2.0°C by 2050. A 1°C rise in global temperature is projected to be very likely by 2030. A 3°C increase is predicted by the end of the twenty-first century.

These upward changes in global temperature may melt the polar ice caps and cause thermal expansion in the oceans. This, in turn, will raise the sea level.

The world's sea level has risen by 20-30 cm in the past century. There is an annual sea level increase of 2-7 mm. A 20-cm sea level rise is projected to be probable in the early part of the present century. A 30- to 100-cm rise is expected by the end of the millenium.

A 30- to 50-cm rise in sea level will seriously diminish the habitability of areas in low-lying regions such as the Netherlands. In Bangladesh, the overloading of the Ganga-Brahmaputra rivers from the Himalayan headwaters will cover most of its accreted land including its capital city, Dhaka.

Huge water bodies, such as the Nile of Egypt, the Mekong of Indo-China, the Indus of Pakistan and the Yangtse and Huang Ho of China, will devastate their banks and coastlines (Palis et al., 1999).

A 100-cm sea level rise will swamp a large portion of Jakarta, Indonesia, inundate much of the river deltas of Vietnam and may well displace 140 million people in Bangladesh and China.

Higher water level will narrow the boundaries between salt and freshwater back land in river estuaries, resulting in salt water intrusion of about one kilometre towards the interior of a flat estuary, greatly mpairing porductivity in agriculture and fishing. Water supply for domestic and industrial uses will also be adversely affected.

Sea level rise and increased salinity will cause a landward migration of coastal and wetland vegetation. Plant species with adaptive features will overcome the stressful environment while plants with limited adaptability may become extinct (Ramachandran, 1993).

Mangrove Forest and Sea Level Rise

A mangrove's adaptive features point to a capability to cope with gradual sea level rise. The factors that determine the response of mangrove ecosystems are: (i) changes in sedimentation; (ii) changes in erosion rates; and (iii) inland migration.

Mangrove trees such as *Rhizophora* protect shorelines and riverbanks from erosion through their extensive root system that traps eroded soil and sediment.

Peat soil is formed wherever there is a densely-populated mangrove stand. The mangrove plants have all the time to adapt to rising sea level.

Litter production leads to peat formation and contributes to the gradual increase in the height of the compiled soil. Estimates of total litterfall in eleven locations worldwide range from 0.2 to 1.6 kg of dry matter/sq.m/year (FAO, 1994).

Table 3.3 shows the three types of common mangrove habitats, the rate of soil, peat and beach barrier buildup and likely effects on mangrove community.

Table 3.3 Types of mangrove habitats and prediction of effect by sea level rise (after Palis et al., 1999)

Type of Mangrove Habitat	Effect on Soil/Peat	Development	Effect on Plant Community
Estuary or delta	RP > SL		Advance
	RP = SL		Stagnation
	RP < SL		Retreat
Back marsh or lagoon	BB >/= SL	P>/=SL	Expansion
		P<SL	Reduction-extinction
	BB<SL	P>/=SL	Retreat to inland
		P<SL	Reduction-extinction
Tidal flat	Front of scarp	P>/=SL	Maintenance
		P<SL	Extinction
	Front of lowland	P>/=SL	Expansion to inland
		P<SL	Reduction-extinction

where
RP : Sedimentary rate (sedimentary rate by river transportation+accumulative rate of mangrove peat)
SL : Rate of sea level rise
BB : Growth rate of barrier or beach ridge
P   : Accumulative rate of mangrove peat

Mangrove plants can withstand the impact of a moderate sea level rise such as maintaining state, expansion to inland, retreat to inland and stagnation of the stand. Mangroves play an important role in coastal protection through the reduction of coastal erosion from undesirable land-based influences.

Threats to the 17,000 sq km of mangrove areas worldwide endanger the buffering function of over 600,000 km of coastline. The ability of mangrove ecosystems to mitigate the damage posed by sea level rise will substantially diminish if their current state is not maintained or improved. There is an urgent need to conserve the remaining mangrove forests and halt the inward march of degraded coastal lands.

It is generally agreed upon that future changes in climate are likely to have a profound impact on global forests. Developing countries may be particularly at risk from the potentially adverse effects of climate change because of their heavy dependence on forests and other natural resources.

Increased concentrations of CO_2 may raise plant productivity and improve water use efficiency. Higher temperatures increase transpiration. Ecological phenomena of succession and migration may also be affected by changes in temperature and moisture regimes. Climate change could trigger outbreaks of pests, fire and disease. In short, radical upheavals in the composition, distribution and productivity of forests may occur.

Millions of people in developing countries directly depend on forests and farming systems for their livelihood. Any change in the structure and distribution of forests can have serious consequences for such people.

Option to Reduce Atmospheric CO_2 Levels

There are several alternatives available to reduce atmospheric CO_2 levels. The best environmentally-acceptable options relate to biomass. However, biomass options form only a part of an overall strategy that ought to include energy efficiency, alternative fuels and emission control technologies. On the available areas of land, it is more desirable to grow trees or other types of biomass for energy production rather than just planting trees to act as a carbon sink, providing the biomass is produced in a sustainable manner. Such a strategy can make considerable impact on atmospheric carbon levels and also provide many ancillary benefits into the bargain, for instance, restoration of degraded lands, energy security, foreign exchange savings, job creation and provision of electricity and other forms of energy to rural areas, thereby helping to promote development. Moreover biomass production does not conflict with food production; with agroforestry and integrated farming systems it could even enhance agricultural output.

Since bioenergy yields an income, it also constitutes a way of paying for CO_2 mitigation and land restoration. To derive maximum benefits, bioenergy production and its use must be modernized. Some developing countries have gone in strongly for modernization of biomass for energy as they are so heavily dependant on it.

The situation was relatively simple over 20 years ago, when the greenhouse gas syndrome first emerged in public focus as disappearing forests, chemically contaminated rivers and lakes, a shrinking biodiversity base and expanding deserts. Sustainability, rather than exploitation, and conservation instead of extermination emerged as the watchwords of the future. Unfortunately, these new feelings often came to be exaggerated in emphasis and such irresponsible statements as 'unprecedented change; or the greatest risk for mankind second only to thermonuclear war' projected doomsday scenarios readily accepted by the gullible in the 1970s.

At the other extreme are those who point at those early exaggerations and dismiss climate change as a figment of the scientist's imagination.

The true position lies somewhere between the above two extreme viewpoints. The Intergovernmental Panel on Climate Change (IPCC) assessments, made in 1990, 1992 and 1995, represent the consensus viewpoint of thousands of the world's experts. The Panel's opinion combines the best predictions of future climates tempered with the full range of uncertainities, unknowns and imponderables which qualify them.

The probable consequences of greenhouse gas accumulation in the atmosphere are global warming, climate change and sea level rise. Reasonable estimates of the average global change can be made and should suffice to make us cautious. Much of the world's low-lying coastline and several islands could face pressures that will be very expensive to counter if unregulated greenhouse gas emissions continue. Many of the protests voiced against stringent emission controls come from those regions and industries which are the largest fossil fuel producers and energy consumers. Environmentally, enhancement of forest CO_2 sinks, adoption of energy efficiency techniques, the introduction of simple pollution controls, and the development and use of renewable energy sources make good economic sense.

According to the IPCC, if we wish to achieve the objective of global atmospheric stabilization at today's levels, immediate reductions of over 60% in the net emissions from human activities of long-lived gases are needed. There are serious difficulties and expense in achieving this goal. The very fabric of societal behaviour would have to change and the pace of future development would have to be greatly slowed or even halted.

Some people urge minimal response based on the degree of uncertainty of how the climate system will actually respond to the atmospheric changes. Such uncertainties must be removed by intensifying scientific research in due course of time. For environmentalists, the responsibility is clear; uncertainity cannot be a ground for adopting risky policies until there is a clear signal of harm done. Polluters are guilty until proven innocent. It is for the emitters of greenhouse gases to justify claims of minimal damage.

EL NIÑO

Wind and Sea

Like two people talking to each other in a dialogue, the tropical Pacific Ocean and the overlying atmosphere influence and react to one another, as illustrated schematically in Fig. 3.1. The strength of the easterly surface winds along the equator controls the amount and the temperature of the water that 'upwells' to the surface along the equator. The upwelling of cold water, in turn, determines the distribution of sea-surface temperature, which affects the distribution of rainfall, which determines the strength of the easterlies, and so on.

The mood of the dialogue changes from year to year. *El Niño* events are marked by weak easterlies over the central Pacific, reduced upwelling and above-normal sea-surface temperatures and rainfall over the central and eastern Pacific, as well as above-normal barometer readings over Australia.

Coastal Upwelling

Surface winds affect the temperature and chemical properties of the surface water along the coast of Peru and southern Ecuador. The southeasterly winds that blow along the coast tend to drag the surface water along with them. The Earth's rotation deflects the water toward the left, away from the coast. Water 'upwells' from below to replace it.

The temperature and chemical properties of the upwelled water depend upon the strength of the easterlies far to the west, in the central and western equatorial Pacific. In the absence of wind, the 'thermocline' (the layer that divides the warm surface water from the colder water below) would be nearly flat. When the easterlies are strong, they drag the surface water westward, raising the thermocline nearly all the way up to the surface along the South American coast and depressing it in the western Pacific.

The cold water below the thermocline is rich in chemical nutrients. Whenever the easterlies in the central Pacific are strong, the thermocline along the South American coast is so shallow that upwelling and stirring by the wind brings nutrient-rich water to the surface. In the presence of sunlight, photosynthetic microalgae (phytoplankton) use up the nutrients in their growth. These algae would soon use up all the nutrients were they not continually being replenished by upwelling.

During *El Niño* the easterlies along the equator become weak and the thermocline along the South American coast plunges several hundred feet,

Atmosphere and Climate Change **135**

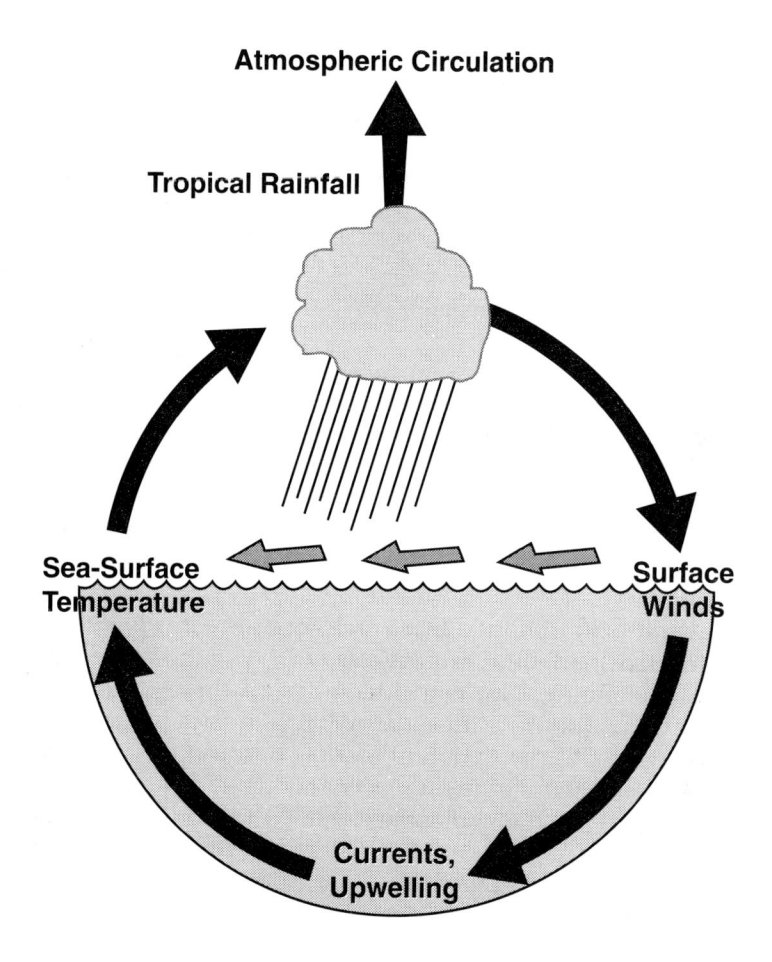

Fig. 3.1 Interactions between the tropical Pacific Ocean and the overlying atmosphere.

preventing nutrient rich water from reaching the surface. Phytoplankton production declines, reducing the food supply for zooplankton which graze on them. Ultimately, anchovies, sardines, sea birds and many other animals at higher levels of the marine food web are adversely affected.

Clouds and Winds

The easterly winds along the equator produce local upwelling, which takes cool water to the surface. When the easterlies are blowing strongly, the band of cold water along the equator chills the air above it, which becomes too dense to rise high enough for water vapour to condense to form clouds and raindrops. Consequently, this strip of ocean remains markedly free of clouds and the rain in the equatorial belt becomes largely confined to the extreme western Pacific, near Indonesia, as shown in the upper panel of Fig. 3.2. But when the easterlies weaken during the early stages of an *El Niño* event, the upwelling slows down and the ocean warms up. The moist air above the ocean also warms and becomes buoyant enough to form deep clouds which produce heavy rain along the equator (the lower panel, Fig. 3.2). The change in ocean temperatures thus causes the major rain zone over the western Pacific to shift eastward.

Related adjustments in the atmosphere cause a further weakening of the easterlies in the central Pacific. In this way, the dialogue between wind and sea in the Pacific can intensify as each partner sends back a stronger message. Small perturbations in the ocean and atmosphere can amplify one another until eventually a full-fledged *El Niño* develops. It is often difficult to identify the subtle change in the ocean-atmosphere system that initiates a transition into or out of *El Niño* conditions.

Australia, Brazil, Ethiopia, India and Peru are already successfully using computer modelling predictions of *El Niño* in connection with agricultural planning. Not coincidentally, all these countries lie at least partially within the tropics. Tropical countries have the most to gain from successful prediction of *El Niño* because they experience a dispro-portionate share of the impacts and again coincidentally, they occupy the part of the world in which the accuracy of climate prediction models is highest. But for many countries outside the tropics, such as Japan and the United States, more accurate prediction of *El Niño* will also benefit strategic planning in areas such as agriculture, and the management of water resources and reserves of grain and fuel oil.

Encouraged by the progress of the past decade, scientists and governments in many countries are working together to design and build a global system for: (1) observing the tropical oceans; (2) predicting *El Niño* and other irregular climate rhythms; and (3) making routine climate predictions readily available to those who require them for planning purposes, in the same way as weather forecasts are made available to the public today. The ability to anticipate how climate will change from one year to the next will lead to better management of agriculture, water supplies, fisheries, and other resources. By incorporating climate predictions into management decisions, humankind is becoming better adapted to the irregular rhythms of climate.

Atmosphere and Climate Change 137

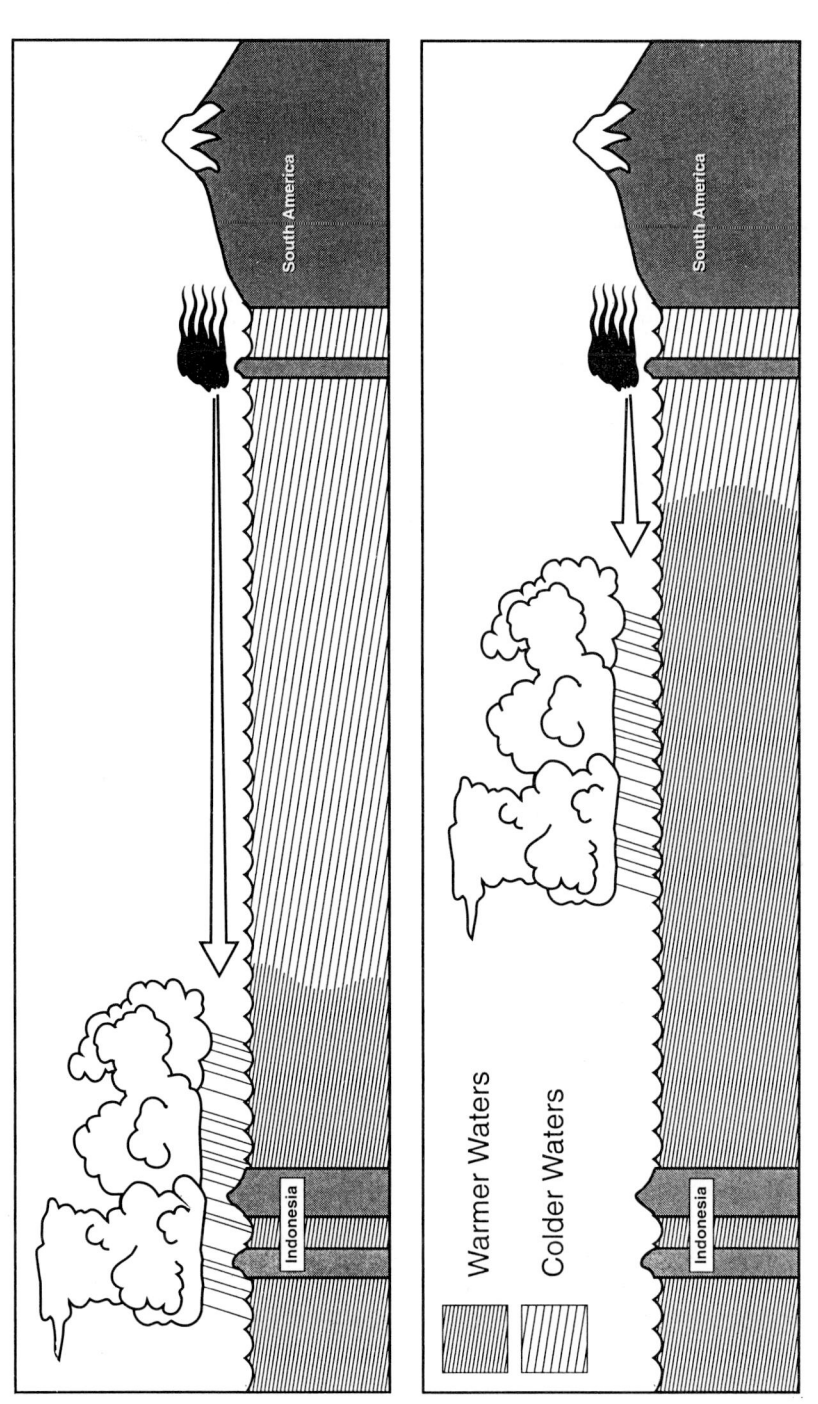

Fig. 3.2 Linkages between clouds, wind, upwelling and El Nino. (Source: UNEP poster)

SUSTAINABLE FORESTRY FOR CARBON MITIGATION

Combustion of fossil fuels significantly increases the CO_2 concentration in the earth's atmosphere. The current carbon emission from fossil fuel burning and cement manufacturing in India is estimated at about 0.272 Giga tonnes (Gt). This increased CO_2 can be sequestered through massive afforestation programmes. The annual rate of afforestation in India today is about 2 million hectares (mha.), sufficient to sustain the growing needs of increasing population, but has only marginal CO_2 sequestration potential. Analysis of recent Indian land-use/land cover statistics generated by remote-sensing techniques has revealed that the areas under non-forest degraded lands and forest degraded lands are about 93.7 mha. and 35.9 mha., respectively, totalling to 129.6 mha. If even a reasonable managed forest productivity of 5.5 tonnes per hectare per year could be attained on only 40 mha. of the available surplus degraded land in India, through programmes such as short rotation forestry plantation, carbon mitigation of about 3.3 Gt may be achieved in the coming half a century with an annual reduction of about 0.07 Gt of carbon (Singh and Lal, 2000).

Carbon emissions are strongly related to economic growth and standard of living. Developing countries suffer from poverty to eliminate which they need to grow. Therefore, India cannot reduce its energy consumption. A potential option for mitigating the increasing CO_2 in the atmosphere is the enhanced cycling of carbon by the terrestrial biosphere through massive reforestation or sustainable afforestation programmes.

The world's forests contain about 830 peta gram (Pg) carbon in their vegetation and soil, with about 1.5 times as much in the soil as in the vegetation (see Singh and Lal, 2000). Forests store and recycle much of the earth's carbon. Managed forests can potentially conserve and sequester carbon and, thus, mitigate emissions of CO_2 by an amount equivalent to 11-15% of the fossil fuel emission. About 60 to 90 Gt of carbon emission could be reduced or sequestered by slowing deforestation, establishing plantations, agroforestry and forest regeneration between the period 1995 to 2050 (Singh and Lal, 2000).

According to FSI (1998), the forest area in India is about 76.52 mha., but the actual forest cover is only 63.34 mha. This is classified into reserved, protected and unclassed forests each of which covers fractional areas of about 54.44%, 29.18%, and 16.38%, respectively. Almost 58% of the Indian forest is dense (with crown density more than 40%), 41% is open (crown density less than 40%) and the rest is covered by mangroves (about 0.78%). The growing stock of the country has been estimated to be about 4740 million cubic meters with an average volume of 74.42 m^3/ha. Almost 50% of forest area is reserved with a view to preserving the unique diversity of Indian flora and fauna.

In India, plantations are being established both in the forest area and in the village farmlands to fulfil the growing demands for fuelwood by the local people. The share of fuelwood consumption for total energy generation has declined from 30% in 1980 to 20% in 1994. Short rotation plantations are expected to be able to fulfil the demands of industrial wood and fuelwood requirements in the coming years.

Renewable energy technologies exploit energy from the sun, wind, water and plants. Their development and use can not only decrease CO_2 emission, but also reduce our dependence on oil imports, improve air quality and create rural employment. Eventually the supply of a substantial portion of India's energy needs could be met from renewables thus minimizing carbon emissions. Some potential renewable energy options are shown in Table 3.4.

Table 3.4 Renewable energy potential and achievements (after TERI, 1998).

Source	Approximate potential	Status (as on 31 March 1988)
Biogas plants	12 million	2.71 million
Improved wood stoves	120 million	28.49 million
Biomass power and gasifiers	17,000 MW	29.50 MW
Biogas-based cogeneration	3500 MW	84 MW
Solar photovoltic	20 MW/km^2	32 MW
Solar water heating systems	35 MW/km^2 (30 mm^2 collector area)	0.38 mm^2 collector area (13.3 MW)
Wind power	20,000 MW	970 MW
Small hydro power (up to 16 megawatt (MW))	10,000 MW	155.38 MW

CO$_2$ Mitigation Options through Activities in Forestry Sector

Forestry activities provide a variety of options to reduce the increasing CO_2 concentrations in the atmosphere. There exists a great potential for expanding energy plantations in India. Some of the basic requirements for the sustainable energy plantations include land, technology and plant species with high biomass yields. The available wasteland from non-forest degraded and forest degraded land is 93.7 mha. and 35.9 mha., respectively, totalling about 129.6 mha. Almost 50% of this land is, however, highly degraded and not suitable for afforestation.

The mean productivity at different age groups of plantations of high-yielding species seems to increase up to a certain age and then decreases (Table 3.4). These findings offer some choice in selection of species and rotation time for plantation projects.

Besides electricity, biomass from energy plantations can provide heat and liquid fuel. Biomass-based power generation in India has proven to be economically viable and is already being practised in some villages (Singh and Lal, 2000).

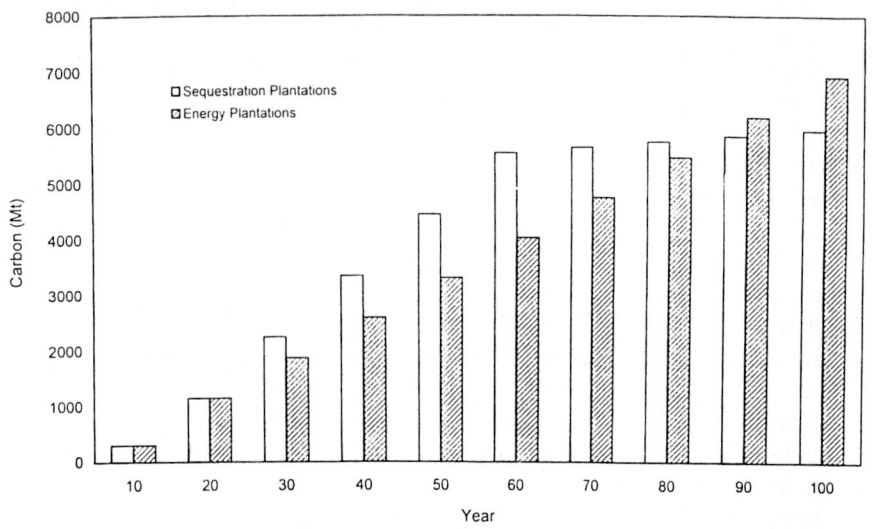

Fig. 3.3 Carbon sequestered/substituted through plantations (after Singh and Lal, 2000).

If used only for carbon conservation and sequestration, the amount of carbon stored by plantations in India would be the same during the first 20 years but they would absorb more carbon thereafter (this land would sequester approximately 0.11 Gt annually) than the energy plantations but only up to about 60 years of plant age (Fig. 3.3). The amount of carbon stored would slow down with the maturation of these plantations. Therefore, for long-term carbon mitigation, energy plantations are the best options as they provide for energy along with reducing carbon emissions from fossil fuels (Singh and Lal, 2000).

BIOLOGICAL IMPACTS OF GLOBAL CLIMATE CHANGE

Global climate change has strong potential consequences for the Earth's living systems (biota). All life on Earth is affected by the surrounding chemical and physical climate. Organisms tolerate certain ranges of temperature, have certain requirements for moisture and nutrients, and respond to the chemical composition of the surrounding air or water. The rate and the magnitude of global environmental changes can severely stress the capability of organisms to adapt. Humans have a vested interest in this topic because they are important members of the biological community and because they depend on other living systems.

There are many uncertainties concerning the biological consequences of climate change — about exactly how the biota might respond to temperature perturbations, sea-level changes, changes in rainfall patterns, increased ultraviolet radiation, or an altered atmospheric chemical composition. The range of possible responses is quite large and provides even greater motivation to study the topic carefully.

The forces of global change can potentially alter physical and chemical climate conditions, which determine where organisms can live. In order to understand how climate change might affect living systems, it is necessary to know how biota is linked to climate both spatially and temporally. The 'spatial scale' extends from millimeters to kilometers to continental to global. The times needed to cause significant change at any level can vary from relatively short, in the case of individuals, to centuries and more at the level of biomes and the biosphere. When we discuss the effects of climate change on the biota, we may be concerned with any of the above levels of organization. Studies of effects at the ecosystem or biome level differ greatly from research at the level of individual organisms because of the differences in temporal (time) and spatial scales. The different types and scales are interdependent and complementary in the search to understand the biotic response to change.

The climate plays a determining role in the distribution patterns of biota in so far as it determines the physical and chemical attributes of the abiotic environment.

Terrestrial vegetation illustrates the connection between climate and the biota. There exists a remarkable association with the climate zones (tropical, temperate, boreal, etc; see Fig. 3.4). Both climates and biomes show strong latitudinal zones proceeding from pole to equator. The broad terms for vegetation zones (such as boreal forest, tropical forest, temperate forest), in fact, carry climatic undertones, showing that climate variables influence plant distribution.

Temperature is undoubtedly an important climatic factor that governs the biology of animals and plants; it directly affects the rate of most biological processes (e.g. photosynthesis, respiration, digestion, excretion).

Moisture is also an important factor limiting the distribution of organisms. Water, which accounts for up to 90% of the weight of most living organisms, is essential for proper plant and animal physiology.

This dependence on water links the biota to several climate-related variables, such as relative humidity, rainfall amount, and the distribution of rainfall through the year.

All organisms are sensitive to the chemical composition of the air or water they live in.

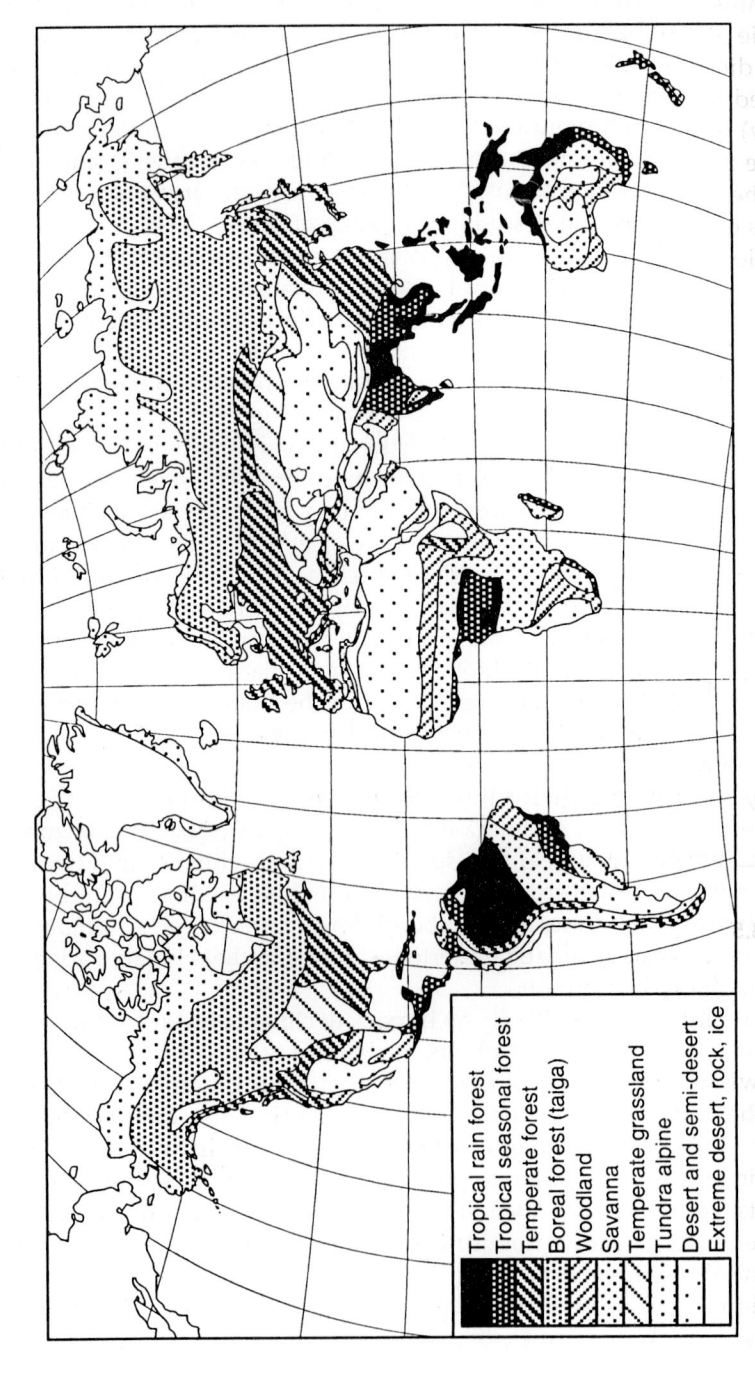

Fig. 3.4 A map of the world showing where the major biomes occur. From Bolin, B., B.R. Döös, J. Jäger, and R.A. Warrick, eds., Scope 29: *The Greenhouse Effect, Climate Change, and Ecosystems*. Wiley, N. York (1986).

Atmosphere and Climate Change 143

Organisms require certain chemicals, oxygen being the most obvious example for all except the anaerobic microorganisms. Plants require carbon dioxide. Beyond these essential needs, other chemicals can only be tolerated at certain levels. For water-dwelling (aquatic) species, the pH (acidity) of the water is a critical variable. Acid rain, currently a problem in some regions, could alter the pH of lakes or coastal areas enough to affect the health and survival of aquatic plants and animals. Pollution of streams or coastal waters by such toxic metals as mercury and lead also has biological consequences (see Fig. 3.5).

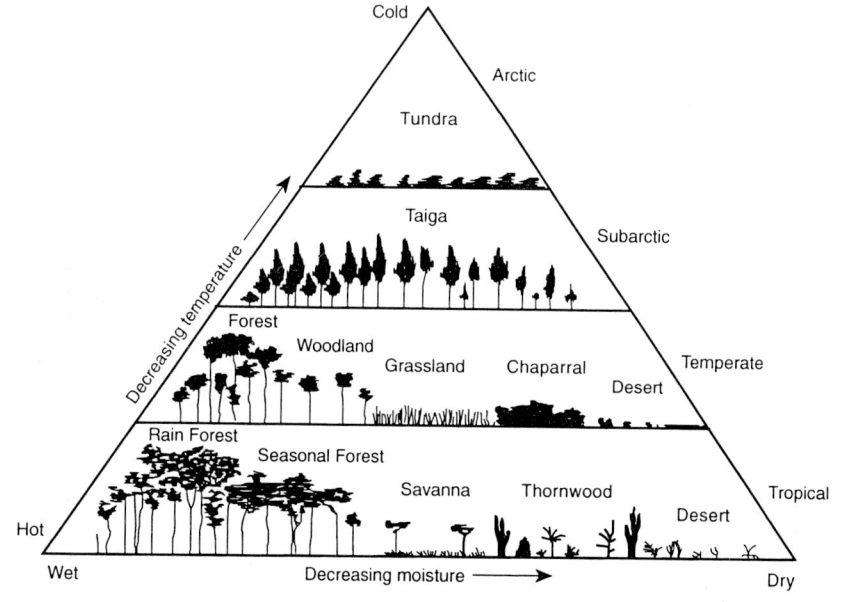

Fig. 3.5 The relationship between climate and biomes. Temperature and precipitation are the major factors determining what biome occurs where. Agter Arms, K. *Environmental Science*. Saunders College Publishing, Philadelphia (1990).

Likewise, organisms that depend on air have limited tolerance for some atmospheric gases (Table 3.5). Today's air pollutants such as sulfur dioxide (SO_2) and ozone (O_3) affect plants, animals and people by interfering with basic physiological processes. These gases are the by-products of industrial processes that have increased with the advancement in human standard of living. Some toxic gases such as phosgene are being employed as chemical weapons because of their rapid and serious disruptions to human physiology. The occurrence of gaseous oxidants is now becoming a global problem.

Table 3.5 Major air pollutants and their effects

Pollutant	Principal adverse effects
Particulate matter	Increased respiratory illness and mortality; visibility impairment; materials damage
Sulfur dioxide*	Increased respiratory illness and mortality; plant and forest damage; materials damage
Carbon monoxide	Asphyxiation; mortality in cigarette smokers; impaired functioning in heart patients
Ozone	Increased respiratory symptoms and illness; plant and forest damage; materials damage
Nitrogen oxides*	Increased respiratory illness; leads to ozone formation; materials damage
Lead	Neurological malfunctioning; learning disabilities; increased blood pressure

* Results in acid rain which can damage human health, aquatic systems, plants and animals. (Modified from Kumar, 1995).

People influence the Earth in many different ways, but only a few global scale examples are known. Three of these have important implications for living organisms are: (1) increases in greenhouse gases such as carbon dioxide; (2) stratospheric ozone depletion; and (3) the increase in oxidants (such as ozone) in the lower atmosphere (the troposphere). These seemingly diverse changes share two basic properties. First, each of these changes is being caused by chemicals that people have been adding to the atmosphere (carbon dioxide, chlorofluorocarbons, nitrogen oxides, volatile organic compounds). Second, each of these anthropogenic changes can potentially affect the fundamental physiological functions of life forms, either directly or indirectly.

The atmosphere and oceans are the Earth's greatest storage tanks of carbon, with a sizeable amount residing in carbonate rock and fossil fuels. The cycling of carbon begins with atmospheric CO_2 entering plants during photosynthesis and becoming incorporated into carbohydrates. Plants, during respiration, release some of the carbon back into the atmosphere and soil as CO_2. Some of the plant carbohydrates are eaten by animals, where they can be stored or where respiration can return them to the atmosphere as CO_2. Microorganisms break down dead plant and animal matter in their own respiration to return the carbon to the atmosphere as CO_2 (Fig. 3.6). Methane (CH_4) and carbon monoxide (CO) are also important carbon-containing gases. Their natural and human sources are shown in Fig. 3.6. Two significant sources of CO_2—the burning of fossil fuels and of tropical forests by humans—have recently been added to the carbon cycle.

Atmosphere and Climate Change 145

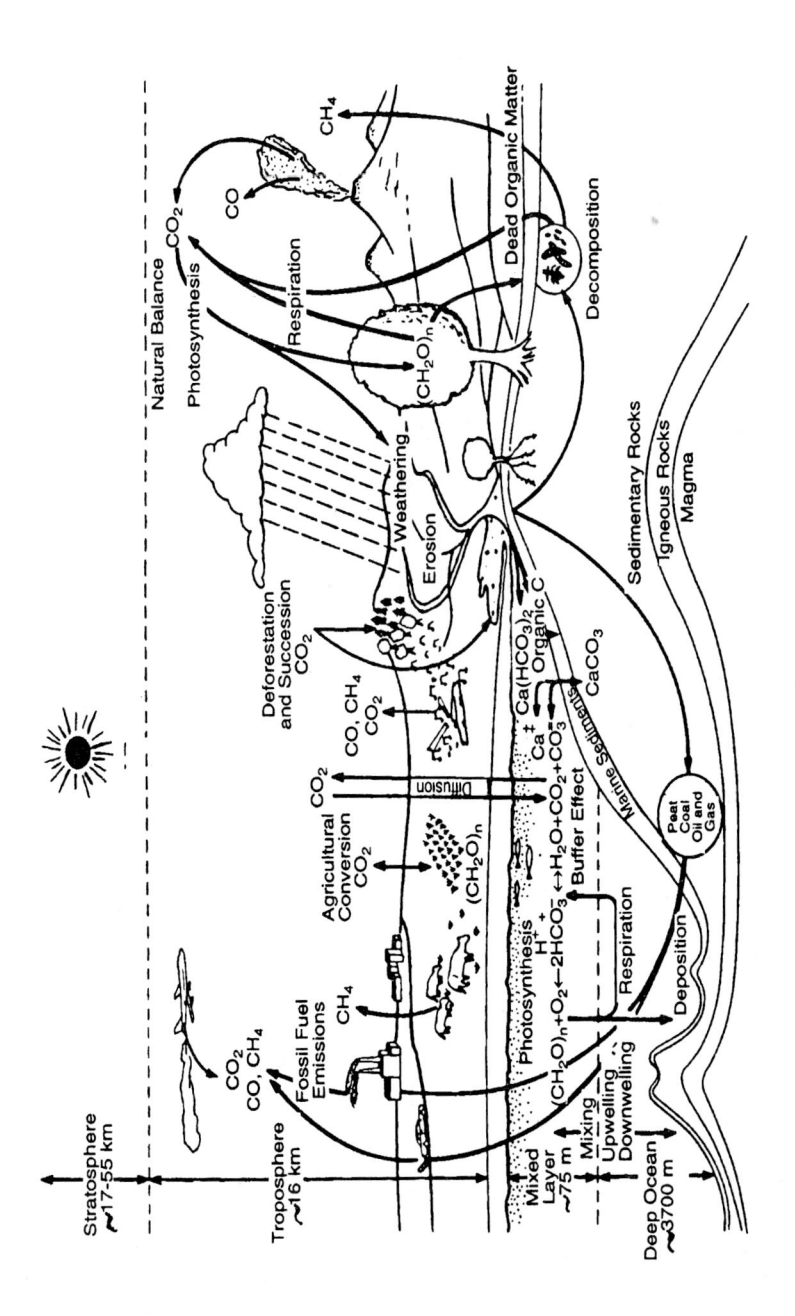

Fig. 3.6 The carbon cycle. From Trabalka, J.R., *Atmospheric Carbon Dioxide and the Global Carbon Cycle.* U.S. Department of Energy, Washington, D.C. (1985).

146 Environmental Technology and Biosphere Management

In spite of several uncertainties, the topic of CO_2 fertilization is important because it is a possible 'feedback' between climate change and the biosphere. It has been hypothesized that increased atmospheric CO_2 would stimulate photosynthesis; increased photosynthesis; would consume greater amounts of atmospheric CO_2 than would otherwise occur, and atmospheric CO_2 levels would then decrease. Thus, the plants would be acting like a sponge, soaking up some of the excess CO_2 that people are adding to the atmosphere (see Fig. 3.7). This would amount to a sort of 'negative feedback'. Scientists are now debating as to whether CO_2 fertilization occurs in the real world and whether it could be significant enough to buffer some of the additional CO_2 that will be added to the atmosphere in the coming decades (see Ennis and Marcus, 1996).

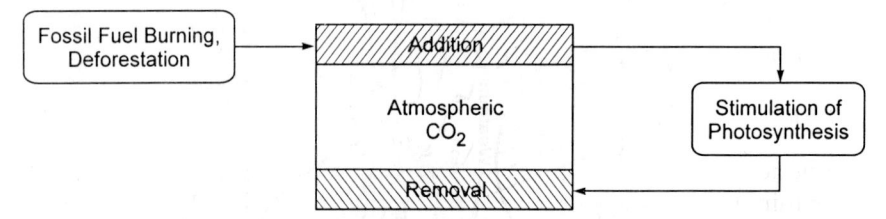

Fig. 3.7 Example of the CO_2 fertilization feedback. Additional CO_2 in the atmosphere stimulates plants to consume more CO_2 in photosynthesis, thus removing CO_2 from the atmosphere. This feedback is stabilizing (negative) (after Ennis and Marcus, 1996).

Some other GHGs such as methane, nitrous oxide (N_2O), and chlorofluorocarbons (CFCs) are not expected to have direct effects on biota. But ozone does have strong direct impacts on plants and animals. O_3 differs from the other GHGs because its effects on biological organisms, and its importance in the chemistry of the troposphere, arise from the fact that it reacts so readily with other molecules.

Indirect Effects of Greenhouse Gases

The increase in CO_2 and other GHGs is likely to change the physical climate, which, in turn, will impact living organisms in diverse ways. These are 'indirect' effects because the greenhouse gas molecule itself is not what the biota is responding to.

Global climate models indicate that temperature and rainfall are expected to change as a result of the buildup of GHGs. Both increases and decreases are possible, depending on which region of the world is being considered. Precipitation may increase at high latitudes, while the middle latitudes of the Northern Hemisphere may experience less summer rainfall and soil moisture. Not only averages but annual patterns of temperature and precipitation may change (Ennis and Marcus, 1996).

For people, the effects of climate change on world food production are of serious concern. Water and heat stress can be controlled to some extent by irrigation and the choice of heat-resistant crops, so adjustments may possibly be made to minimize the impacts of a changing climate.

Several aspects of human health may be influenced by global warming and associated changes in humidity. The temperature of the human body must remain constant within a very narrow range (37°C to 37.5°C). Various thermoregulatory mechanisms ensure maintenance of this basic temperature. People having circulatory problems are particularly susceptible to heat stress. Studies have correlated heart and respiratory diseases, birth defects, infant mortality and the survival and transmittance of many airborne pathogens with temperature, humidity, and other meteorological variables (Ennis and Marcus, 1996).

GLOBAL WARMING

The notion of commensurability is the idea that social realities, amongst other things, can be indexed by a unitary measure of value. Economic cost-benefit analysis takes commensurability to its logical extreme, propounding that, for instance, the question whether to construct another airport in a megacity can be decided by applying a strict commercial analysis, attributing costs and prices to the whole range of environmental and human factors. A particularly difficult aspect of cost-benefit analysis is that, since any larger investment project will almost certainly create benefits for some people and disbenefits for others, the analysis must include invidious interpersonal comparison of utility — and while it might be equitable for those who benefit from such a project to compensate those who are disadvantaged by it, this in practice rarely happens (Pollock, 1999). The application of cost-benefit analysis to the global warming caused by GHGs is an even more complicated and a less satisfactory approach. It appears that despite the anxieties of the climatologists and earth scientists, the costs of our present rate of consumption of fossil fuels are bound to become enormous.

Neoclassical economics applies commercial logic to the analysis of human welfare and envisions an ideal world in which commercial interests and individual interests are synonymous. In this (ideal) world, the quantities of the goods and services which are produced and the prices at which they are traded are determined equally by the preferences of consumers and by the technology of production.

In describing the choices faced by an individual consumer, it is usually assumed that the consumer is bound to accommodate himself to the rates of exchange which are offered by the market because as an individual, he is unable to alter these aspects of the economic environment. Although a neoclassical economist might eschew the notions of society and of social choice, he can show how the aggregate of individual choices reaches an optimal accommodation with the processes of economic production.

Macroenvironmental Issues and Cost-benefit Analysis

Economists have attempted to assess the likely impact upon the global economy of the global warming due to the anthropogenic emissions of carbon dioxide and other GHGs. The methods used have been those of cost-benefit analysis and the nature of the analysis is global.

In the case of any action taken to limit the emissions of GHGs, the effect has to be considered in global terms. One important issue in applying cost-benefit techniques to the problem of global warming concerns the length of time which is necessarily spanned by the analysis: at the very least, the effects of current economic activities upon the lives of several human generations must be taken into account. Another issue is the highly uncertain nature of the effect of the emissions. We are unlikely to be able to ascertain the precise effects much in advance of the time when they actually materialize; and moreover, the processes of climatic change are liable to acquire some momentum which will make it impossible to reverse them quickly. In short, while we face great risks, we remain ignorant of their precise nature (Pollock, 1999).

Most of the cost-benefit analyses of global warming so far have focused on advanced industrialized countries, and a full assessment must include other regions. Frankhauser (1993) estimated the losses from the doubling of atmospheric CO_2 to be 1.4% of the gross national product of OECD countries and 1.5% for the rest of the world.

The serious anomaly about cost benefit analysis is the disregard of the circumstances in which the majority of the world's population is constrained to live. The most dramatic effects of climatic change will be experienced in areas where people are already miserably impoverished. It is amazing that anybody should think of applying a cost-benefit analysis on a global scale without taking account of the distribution, or rather, the maldistribution, of incomes.

VULNERABILITIES AND POTENTIAL CONSEQUENCES

The climate changes expected from increased atmospheric concentrations of GHGs are expected to have wide-ranging effects — many of them adverse — on ecological systems, human health and socio-economic sectors. People in developing countries are usually more vulnerable to climate change because of limited infrastructure, poverty and greater dependence on natural resources.

Climate change will impact human health in a variety of ways. Warmer temperatures increase the risk of mortality from heat stress. CO_2 concentrations of 550 ppm (double the pre-industrial level) could make such events 6 times more frequent. Children and the elderly tend to suffer disporportionately with both warmer temperatures and poorer air quality.

Some diseases, e.g. malaria, dengue and yellow fever, encephalitis, and cholera thrive in warmer climates. These are likely to spread due to the expansion of the ranges of mosquitoes and other disease-carrying organisms and increased rates of transmission. This could result in 50 million to 80 million additional malaria cases per year worldwide by 2100.

Rising sea level erodes beaches and coastal wetlands, inundates low-lying areas, and increases the vulnerability of coastal areas to flooding from storms or intense rainfall. By 2100, sea level is expected to rise by upto 90 cm.

Disruption of the Water Cycle. Among the strongest effects of climate change are intensification and disruption of the water cycle. This will produce more severe droughts in some places and floods in others. Areas of greatest vulnerability are those where quality and quantity of water are already problematic, such as the arid and semi-arid regions of the world. Climatic shifts will force some species to migrate northwards or to higher elevations in order to stay in the appropriate climatic zone.

Water scarcity in the Middle East and Africa is likely to be aggravated by climate change, which could increase international tension among countries that depend on water supplies originating outside their borders.

Changing Forests and Natural Areas. Climate change could markedly change the geographic distributions of vegetation types. The composition of one-third of the Earth's forests would undergo major changes as a result of climate changes associated with a CO_2 level of 700 ppm.

Effects on Agriculture and the Food Supply. Climate markedly affects crop yields. A CO_2 concentration of 550 ppm may increase crop yields in some areas by 30 to 40% but it will decrease yields in other places by similar amounts, even for the same crop. A warmer climate would reduce flexibility in crop distribution and increase irrigation demands. Expansions of the ranges of pests could also increase vulnerability and result in greater use of pesticides. Despite these effects, total global food production is not expected to be altered substantially by climate change, but negative regional impacts are quite likely to occur in some areas. Agricultural systems in the developed countries are highly adaptable and can probably cope with the expected range of climate changes without dramatic reductions in yields. It is the poorest countries, already subject to hunger, that are the most likely to suffer significant decreases in agricultural productivity (OSTP, 1997).

CLOUDS AND ATMOSPHERIC RADIATION

The scientific issues surrounding climate and hydrological systems are of great concern and there is need to improve cloud and radiation parameterizations, e.g. the role of clouds and of cloud radiative feedback.

Observed radiative fluxes and radiances in the atmosphere, spectrally resolved and as a function of position and time should be related to the temperature and composition of the atmosphere, specifically including water vapour and clouds, and to surface properties. A wide variety of situations ought to be sampled so as to cover a good range of climatologically relevant possibilities.

Parameterizations that can be used to accurately predict the radiative properties and to model the radiative interactions involving water vapour and clouds within the atmosphere need to be developed and tested with the objective of incorporating them into general circulation models (GCMs) (see DDE/USDE, 1996). The following key scientific issues need to be resolved in order to achieve these objectives.

1. What are the direct effects of temperature and atmospheric constitutents, particularly clouds, water vapour and aerosols on the radiative flow of energy through the atmosphere and across the Earth's surface?
2. What is the nature of the variability of radiation and the radiative properties of the atmosphere on climatically relevant space and time scales?
3. What are the primary interactions among the various dynamic, thermodynamic and radiative processes that determine the radiative properties of an atmospheric column, including clouds and the underlying surface?
4. How do radiative processes interact with dynamical and hydrological processes to produce cloud feedbacks that regulate climate change?

Each type of cloud system results from different formation processes. The structure and optical properties of a cloud are determined by both large-scale and cloud-scale circulations (via vertical motion on these scales). Cloud-scale circulations are influenced by large-scale circulations, radiation and surface properties and, in turn, influence the large-scale circulation and surface properties (via convective and turbulent fluxes). Cloud structure and cloud optical properties greatly impact radiative fluxes. Radiative and turbulent fluxes affect surface properties, and the large scale circulation is affected by convective and radiative heating.

The Earth's climate is very sensitive to radiation. While the Solar Constant is 1360 W m^{-2}, a mere 4 W m^{-2} change in radiation flux at the tropopause results in strong changes in the Earth's equilibrium surface temperature and rainfall patterns. An extra 10 W m^{-2} at the ocean surface, if not countered by fedback effects, would raise the sea-surface temperature 1°K in just one year (assuming a 75-m deep ocean mixed layer). The precision (5-10% or tens of W m^{-2}) with which meteorologists usually measure and calculate radiation fluxes in the atmosphere, is not good anough for climate studies. Within the entire global change research programme, there is need to bring atmospheric radiation measurements, and hence models, to the level of high quality needed in climate studies.

Atmospheric radiation is a transport problem whose temporal and spatial scales have usually not been taken into account while planning past field programmes. First, the transport is near-instantaneous. The entire radiation field adjusts almost instantaneously to changes in physical properties, hence measuring those properties at times as little as several minutes from the times of radiation observation can be of no use. Second, radiation spatial scales have been factored into the design of past field programmes only with great difficulty. A broadband flux measurement typically sums photons originating all the way from the visible horizon to the immediate neighborhood. Radiation time and space scales are smaller, usually by many orders of magnitude, than those accounted for the present climate models. These scaling considerations greatly challenge the designing of a radiation observation programme in conformity with the underlying assumptions built into radiation models.

All the physics underlying the radiative transfer equation — conservation of energy, Maxwell's equations, Lambert's Law, Planck's Law, Mie scattering — are well known, and the equation itself merely keeps track of photons as they are scattered, absorbed and emitted. But some other aspects, such as far-wing and continuum absorption and ice crystal scattering, are not well established and merit much further study. The chief difficulties in solving the radiative transfer equation occur in handling: (a) multiple scattering; (b) polarization; and (c) more than one spatial dimension. One major practical difficulty is knowing the input variables to the radiative transfer equation: temperature; surface boundary conditions; the absorption line parameters and spatial distribution of all radiatively-active gases; and the scattering properties and spatial distribution of cloud and aerosol particles. Integrals over space and time further complicate the matter (see DDE/USDE, 1996).

Plane parallel geometry, the assumption of horizontal homogeneity in a vertically non-homogeneous atmosphere, is the main plank of current radiative modeling. With this geometrical idealization, theoretically rigorous treatment of multiple scattering can be achieved for typical water clouds and aerosols for all wavelengths of the spectrum and for Rayleigh scattering. The results are used as reference point for comparing the performance of certain more approximate radiative transfer treatments.

Plane-parallel geometry with horizontally homogeneous layers is quite convenient for mathematical analysis and modeling but real clouds show significant spatial non homogeneity; their radiative effects can be modeled using certain three-dimensional radiative transfer techniques. These calculations have shown that spatial inhomogeneities decrease cloud reflection and absorption while increasing direct and diffuse transmissions.

Clouds play a dual role in the atmospheric general circulation and the global climate. First, they owe their origin to large-scale dynamical forcing, radiative cooling in the atmosphere and turbulent transfer at the surface. Secondly, they provide an important mechanism for the vertical redistribution of momentum and sensible and latent heat for the large scale, and affect the coupling between the atmosphere and the surface as well as the radiative and dynamical-hydrological balance.

In the currently available diagnostic cloudiness parameterization schemes, relative humidity is the most often used variable for estimating total cloud amount or stratiform cloud amount. But the prediction of relative humidity in GCMs is usually poor, and the predicted relative humidity can deviate greatly from that actually observed.

One requirement of a cloud parameterization is that it should represent the full lifecycle of clouds; i.e. their formation, persistence, and decay. For clouds forming under strongly advective conditions, the domain traversed by a cloud during its lifecycle can be thousands of kilometers, much larger than that of a single grid cell of a GCM. Evaluation of such a cloud parameterization can only be achieved by a model spanning a domain much larger than that of Single Column Model (SCM); otherwise, the lateral boundary conditions for the cloud variables will control the simulation of the cloud.

Climate model dynamics are driven by external and internal forcing. The primary forces affecting the thermal field are long wave radiative (LWR) heating, short wave radiative (SWR) heating, and convection (cumulus). These forcing effects cycle through the thermal field to the motion field by nonlinear transfer. The model dependent variables, especially temperature, moisture and clouds, evolve in time in a Global Climate Model and thereby determine the subsequent forcing. If the dependent variables are not accurately calculated in space and time, such inaccuracies can introduce errors in climate prediction.

Mace et al., (1996) developed an operational data processing and analysis methodology that enables continuous examination of the influence of clouds on the radiation field and to test new and existing cloud and radiation parameterization (Fig. 3.8).

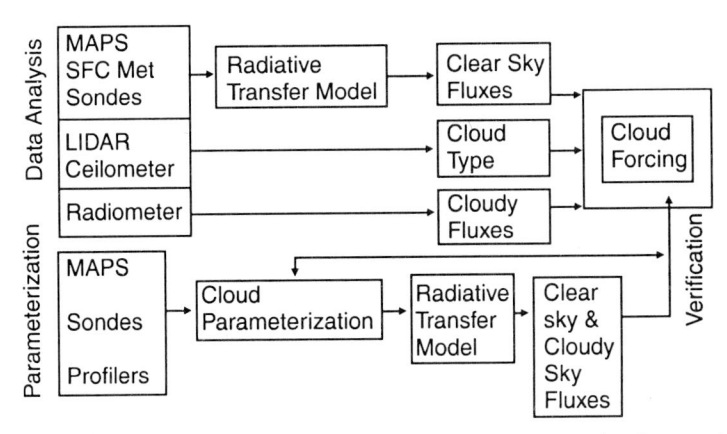

Fig. 3.8 An operational methodology for cloud parameterization testing. The upper portion depicts the data analysis procedure, while the lower portion shows parameterization forcing and evaluation (after Mace et al., 1996).

Satellite measurements have shown that global effects of clouds on solar and infrared radiation are quite large. Effective treatment of cloud formation and cloud properties is crucial for reliable climate prediction. Clouds are some of the most important moderators of the earth's radiation budget but one of the least understood. The effect clouds have on the reflection and absorption of solar and terrestrial radiation is greatly affected by their shape, size and composition. Physically accurate parameterization of clouds is essential for any general circulation model to give meaningful results.

A good understanding of radiation transport through clouds is basic to studies of the earth's radiation budget and climate dynamics (see Buch et al., 1996). The transmission through horizontally homogenous clouds has been studied thoroughly using discreet ordinate radiative transfer models, but the applicability of these results to general problems of global radiation budget is limited by the plane parallel assumption and the fact that real cloud fields show both vertical and horizontal variability, on all size scales. To understand how radiation interacts with realistic clouds, Gautier et al., (1996) used a Monte Carlo radiative transfer model to compute the details of the photon-cloud interaction on synthetic cloud fields. Synthetic cloud fields, generated by a cascade model, were shown to reproduce the scaling behaviour, as well as the cloud variability observed and estimated from cloud satellite data.

Comparison of remotely-sensed meteorological parameters with in situ direct measurements is a challenging task because the sampling volumes for these comparisons are very difficult to match precisely. Estimations of vertical profiles of cloud by sampling a cloud at various altitudes at different times (Matrosov et al., 1996).

Cloud-radiation feedback is a crucial factor limiting the usefulness of GCMs for progress in climate change research. It is also a chief uncertainty in judging the impact of GHGs on climate simulations. Many GCMs are very sensitive to the treatment of clouds and cloud radiative properties. A better understanding of cloud-radiation feedback on the large-scale environment is absolutely essential for improving the use of cloud processes in CGMs. There is more absorption of solar radiation than estimated by current atmospheric GCMs and that the discrepancy is associated with cloudy scenes. It appears that this may be an artifact of stochastic radiative transport (Byrne et al., 1996).

In the light of the currently available findings/data, the following general statements can be made in the context of radiation effects in relation to climate (see DDE/USDE, 1996).

Continuum absorption is just the sum of thousands of far-wings of lines, and, thus, one cannot change one without changing the other. Aerosols may be important at the 10 W m^{-2} level in the longwave. Aerosol optical depth is almost unknown in the longwave since all observations are made in the shortwave. Anthropogenic aerosols may be regionally counteracting greenhouse warming. The indirect (Twomey) effect of aerosols on cloud albedo may be substantial for oceanic clouds. The contribution of solar influx beyond 4 microns is about 5-10 W m^{-2} and is important in understanding systematic offsets between model calculations and accurate UAV (unmanned aerospace vehicles) longwave flux measurements.

Atmospheric sphericity is important for shortwave radiation for solar zenith angles between 80 and 90 degrees. Longwave radiation may also experience spherical-geometry effects in the upper troposphere and stratosphere, where long ray paths are possible.

Horizontal flux divergences can be significant (compared to vertical ones) in the shortwave when the sun is away from the zenith. No attempts to measure horizontal fluxes have been made. Horizontal variability appears to be unimportant except for water vapour. Water vapour may have strong horizontal inhomogeneity on scales which are important for radiation.

Deep inside clouds, where a radiation diffusion regime is established, horizontal net fluxes are small compared to vertical ones. But horizontal flux divergences near the edges of clouds, or spanning cloudy/clear regions, can be significant in the shortwave. They may also be significant in longwave situations, for example in a partially cloudy upper troposphere.

Liquid water clouds have variable emissivities, depending on total liquid water path and its horizontal distribution. They must therefore scatter, and if the scattered radiation is significantly warmer or colder than the locally emitted radiation, there will be a significant effect on radiative fluxes.

Cloud top temperature can be quite different from that of the surrounding air. Most clouds, not just boundary layer clouds, help create and maintain a temperature inversion at their tops which significantly alters the longwave radiative transfer.

Ramanathan et al., (1995) pointed to observations which show an anomalous absorption of cloudy skies in comparison with the values predicted by usual models (homogeneous atmosphere) and which thus introduce serious uncertainties for climate change assessments. These observations question the manner in which GCMs have been hitherto tuned by relying only on studies of the radiation/dynamics relationship on an arbitrary scale (often considered as 'characteristics', when really the interactions between clouds and radiations occur over a wide range of scales (Naud et al., 1996). These observations also suggest that homogeneous models may not be relevant in relating the highly variable properties of clouds and radiation fields. Byrne et al., (1996) proposed a simple model of broken clouds and measured an increase in the value of the photon mean free path in comparison with the value calculated for a homogeneous atmosphere. According to them, photons diffused by a first cloud can circulate horizontally in a 'clear sky' region and be reflected in the opposite direction by another cloud and cloud can thus be 'trapped' between two clouds (Fig. 3.9). This phenomenon increases significantly the mean free path of photons and thus increases the total absorption of the layer.

CLOUDS AND AEROSOLS

Tropospheric aerosols (particularly anthropogenic sulfate aerosols) contribute to the radiative forcing by exerting a cooling influence on climate (-1 to -2 W/m^2), which is comparable in magnitude to greenhouse forcing, but opposite in sign.

Aerosol particles affect the Earth's radiative budget either directly by scattering and absorption of solar radiation by themselves or indirectly by

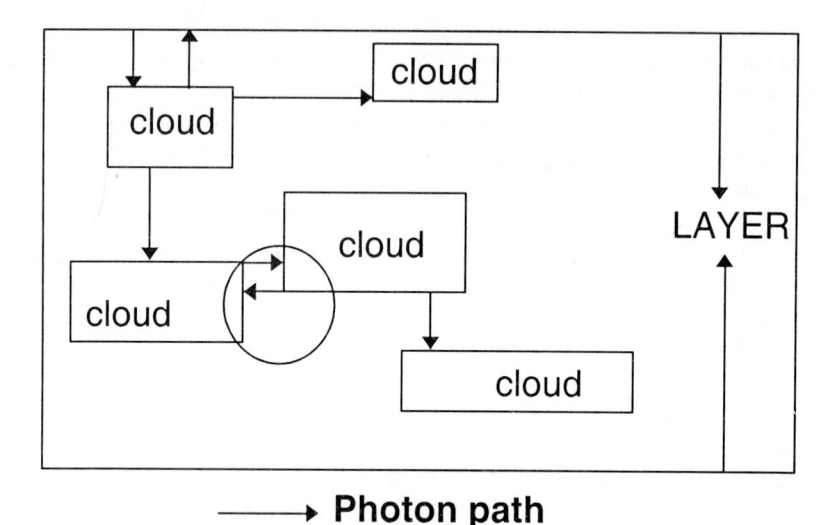

Fig. 3.9 Trapping of photons between two or several clouds increases the global absorption (after Byrne et al., 1996).

altering the cloud radiative properties through changes in cloud microstructure. Some marine cloud layers and their possible cooling influence on the atmosphere as a result of pollution are specially interesting in view of their high reflectivity, durability, and large global cover.

Aerosols influence the absorption of shortwave radiant energy in the troposphere both directly through enhanced scattering of sunlight in the absence of clouds and indirectly via their basic role in controlling cloud microphysical properties. Estimates of the magnitudes of both effects are, however, uncertain.

The cloud drop size distribution near cloud bases is initially determined by aerosols that act as cloud condensation nuclei and the updraft velocity. Chuang and Penner (1996) developed parameterizations which relate cloud drop number/concentration to aerosol number and sulfate mass concentrations, and used them in a coupled global aerosol/general circulation model to estimate the indirect aerosol forcing.

In order to gain an understanding of the global effects of aerosols on clouds, one must first understand the global concentrations of the different aerosol components or types. This involves the development of a source emissions inventory for aerosols and for gas-phase species that form aerosols. Natural and anthropogenic emissions of gaseous sulfur species (which form sulfate in the aerosol), soot or black carbon emissions, emissions of particulate organic carbon and of gas phase non-methane hydrocarbon species which produce aerosols during photo-oxidation, nitrogen oxides (which form nitrate in the aerosol), and ammonium are important in this context.

Chuang and Penner (1996) worked on paarmeterizing the effects of anthropogenic sulfate-containing aerosols on initial cloud drop number concentration. They refined the treatment of the aerosol size distributions by specifically calculating the processes by which aerosol sulfate is produced in the atmosphere, condensation of H_2SO_4 on pre-existing particles, and the formation of sulfate by in-cloud oxidation of SO_2. The pre-existing particles with a prescribed size distribution are assumed to mainly consist of organic matter, black carbon, and natural sulfate. The resulting anthropogenic sulfate-containing aerosol size distribution is then used in a detailed microphysical model to parameterize the cloud drop number concentration for use in the climate model.

Their simulations have indicated that current concentrations of anthropogenic sulfate have direct and indirect effects, are comparable in magnitude, and at least locally tend to mask the warming effect of increased greenhouse gases.

It has been well recognized that the Earth's radiation balance is highly sensitive to changes in aerosol characteristics. The direct effects of aerosols on the earth's radiation come from their single scattering albedo and hence their composition. A potentially more important effect of aerosols, however, is their impact on cloud microphysical characteristics. Because aerosol concentrations vary, both naturally and because of human activities, it is important to include both effects in GCMs.

Extremely small (submicron) particles with the longest lifespan are included in almost all atmospheric processes (Rosenberg, 1983). These have special importance among the great variety of sizes of particles present in the atmosphere. It is these submicron particles which mainly determine the optical state of the atmosphere in the visible spectral range and cause the absorption of infrared radiation (Panchenko, 1988). Also, since they are the products and participants in all aerosol-to-gas transformations, they absorb large amounts of various chemical compounds and move them to large distances.

Investigation of the processes of the spatial-temporal variability of aerosol particles for different climatic zones of the earth makes up the experimental base for studying their effect on climatically- and ecologically- significant factors and for estimating their unfavourable tendencies. The increasing anthropogenic loading of the earth's atmosphere underscores the urgency for aerosol research. Regardless of how perfect the analytical and numerical methods of solving radiation problems may be, success in forecasting climatic change would be determined mainly by the reliability of the experimental data on optical parameters of the atmosphere and of the description of their variability under the effect of external factors (Panchenko et al., 1996).

The June 1991 volcanic eruption of Mt. Pinatubo (Philippines) is perhaps the largest, best documented global climate forcing experiment in recorded history. The time of development and geographical dispersion of the Pinatubo aerosol has been monitored and sampled by satellite, airborne, and ground based instruments using visible, near-infrared (IR), and thermal wavelength measurements. The principal information regarding the optical depth, particle size, and vertical and geographic distribution of the aerosol is derived from the solar occultation measurement at 385, 453, 525, and 1020 nm taken by the Stratospheric Aerosol and Gas Experiment (SAGE) II aboard the ERBS satellite (McCormick et al., 1995). Balloon-borne measurements (Deshler et al., 1992) identified the aerosol composition as concentrated sulfuric acid and also characterized the particle size distribution. The variance (width) of the particle size distribution has also been recorded (see Lacis et al., 1996). The Pinatubo signature on the global energy balance could also be recorded in the Earth Radiation Budget Experiment (ERBE) measurements of top of the atmosphere (TOA) solar and thermal fluxes.

With the availability of the above data as well as the spherical shape of the sulfuric acid droplets and the refractive indices of sulfuric acid, Mie scattering has been used to calculate the aerosol radiative properties at all wave-lengths of the spectrum for a broad range of particle size distributions, thus making possible the accurate determination of the radiative forcing generated by the Pinatubo aerosols (Lacis et al., 1996).

Based on preliminary estimates of the Pinatubo aerosol loading, Hansen et al., (1992) accurately predicted the impact on global climate due to the Pinatubo aerosol: predicted the decrease in the global surface air temperature by 0.5°C by the end of 1992 with recovery to normal by 1995 has turned out to be remarkably accurate.

Clouds and Precipitation

Atmospheric moisture distribution is directly related to the formation of clouds and precipitation and affects the atmospheric radiation and climate. Several remote sensing systems can accurately measure precipitable water. Some clouds produce precipitation, whereas other produce small amounts of drizzle which has some impact on both cloud macrophysical properties (spatial coverage, depth and liquid water content) and microphysical properties (droplet size distributions, effective radii).

Drizzle production is related to the number and size of cloud condensation nuclei (CCN) as well as to cloud dynamics and the ability of clouds to support droplets within their bounds and allow for repeated collision-coalescence cycles.

WORLD'S WEBS

Morais (1999) discussed the opportunities and challenges of addressing global change research under a highly diversified and ever changing social and scientific kaleidoscope with a view to finding a panacea for the world's many problems. Reviewing the major steps of history of science may help us to better understand where we are as well as the obstacles or progresses that shaped particular world views.

Practical solutions are what societies and good governance eagerly strive for. For scientists, this provides the social context for aiming to explain the nature of today's problems. Such needs lie at the core of the Global Environmental Change Programmes (see Fig. 3.10).

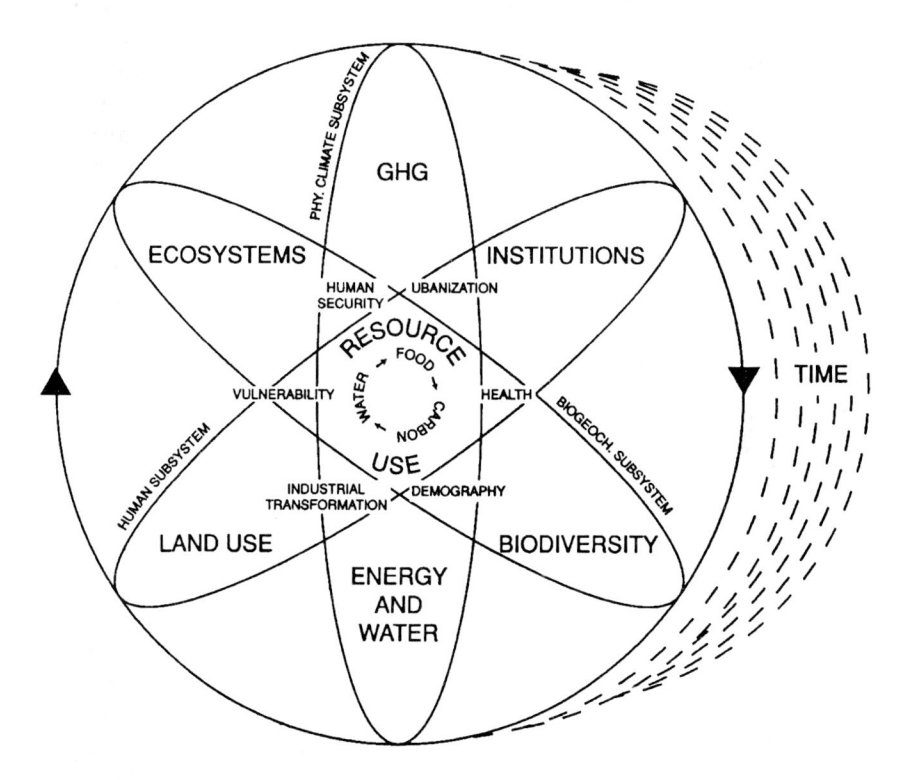

Fig. 3.10 Global Change Subsystem Web (in time) (after Steffen, 1999).

For social sciences, the basic problem is to explain events in terms of human actions and social situations. Social science models aim at reconstructing 'typical social situations', which in the natural sciences realm would be labelled as 'initial conditions'. However, typical social institutions are well embedded in physical bodies as illustrated, among other examples, in the type of human habitat or socially constructed systems impacting biophysical changes, namely at the level of production, distribution and consumption of goods and services at larger spatial scales. Such 'indicators', expressing a rationality principle, are reflected in accumulated societal knowledge and cultural behaviour, criteria which are being integrated in new global modelling efforts and of crucial importance for their calibration and validation where historical output is the best approximation to a 'real system' (Rotmans, 1998; Morais, 1999).

PREVENTION OF CLIMATE CHANGES

In order to enable people as well as nature to adapt to a warmer global climate, the temperature rise should be limited to no more than 0.1°C per decade and no more than 2°C above the preindustrial level. The current trend points towards a global temperature increase of about 0.3°C per decade, with the limit for the total temperature increase being reached in about 2030. The low-risk limit that entails a total temperature increase of no more than 1°C above the preindustrial level has already been crossed.

With a view to meeting the environmental goals and preventing the risks of more serious effects of a temperature increase, it is estimated that annual global carbon dioxide emissions need to be cut by 50 to 60% within the coming half a century. This means that the emissions of the industrial countries have to be reduced more if there is to be some allowance for an industrialization in the developing countries.

The GHGs that should be the focus of response strategies include carbon dioxide, methane, nitrous oxide and the stable fluorine compounds, fluorocarbons (FCs) and hydrofluorocarbons (HFCs). The possible contribution to the greenhouse effect (radiative forcing) caused by CFCs (chlorofluorocarbons) as well as of carbon monoxide (CO), nitrogen oxides (NOx) and volatile organic compounds (VOCs) is somewhat uncertain.

The growth (increment) of forests currently accounts for a net uptake of the order of some 40 million tonnes of carbon dioxide per year. It is estimated that this accumulation of carbon can continue for another two to six decades, depending on the harvest rate.

There is considerable potential for reducing carbon dioxide emissions through efficiency improvements and conversion to other energy forms, such as biofuels, within both the energy and transport sectors. Many efficiency-improving measures within the energy sector are profitable if one compares with the cost of new electricity and heat production. Unfortunately, however, this entire potential is not being exploited. In the transport sector also efficiency-improving measures often prove to be cost-effective.

Fairly good prospects exist for being able to stabilize carbon dioxide emissions at the 1990 level from the year 2000 until the nuclear power plants are phased out. It appears that the effects on the gross national product and gross national income will be small in this time perspective if the carbon dioxide tax is set lower in energy-intensive industries than in the rest of society (or if the rest of the world adopts similar strategies).

In order to keep emissions low when the nuclear power phaseout begins, it will be necessary to put very strong policy instruments into effect. The changes in the energy system that emerge as being the most cost-effective for combining a phaseout of nuclear power with a stabilization of carbon dioxide emissions include vigorous energy saving, biofuel-based CHP (combined heat and power), and wind power.

Within the transport sector also far-reaching measures need to be adopted, including stringent fuel-efficiency standards for transport modes; introduction of alternative biofuel-based vehicle fuels, and extensive investments/expansion of public transport (SEPA, 1992). SEPA (1992) proposed implementation of the following response strategies as soon as possible to stabilize carbon dioxide emissions by the year 2000 and to initiate a long-term restructuring:

Energy sector: CO_2 tax on electricity generation; raised CO_2 tax and unchanged energy tax for non-energy-intensive industries; energy tax on CHP; electricity conservation measures, including standards for domestic appliances; R&D on biofuel technology; and examination of energy use in energy-intensive industries.

Transport sector: New rules for company car benefits and travel deductions; programme for expansion of public transport system; environmental impact description of traffic plans and infrastructure investments; introduction of electric cars; emissions limits for light vehicles; and environmentally adapted speed system.

REFERENCES

Adams, W.M. *Green Development.* Routledge, London (1990).

Andronova, N.G., Rozanov, E.V., Yang, F., Schlesinger, M.E., Stenchikov, G.L. Radiative forcing by volcanic aerosols from 1850 to 1994. *J. Geophys. Res.* 104: 16,807-16,826 (1999).

Baldwin, M.P., Cheng, X., Dunkerton, T.J. Observed correlations between winter-mean tropospheric and stratospheric circulation anomalies. *Geophys. Res. Lett.* 21: 1141-1144 (1994).

Bolin et al., Scope 29: *The Greenhouse Effect, Climate Change, and Ecosystems.* Wiley, N. York (1986).

Buch, K.A., Sun, C.-H., Thorne, L.R. Cloud classification using whole-sky imager data. In Proc. 5th Atmospheric Radiation Measurement Science Team

Meeting, March 19-23, 1995, San Diego, California, pp. 33-37. Published by U.S. Dept. of Energy, Washington, D.C. (1996).

Byrne, R.N., Somerville, R.C.J., Subasilar, B. Broken-cloud enhancement of solar radiation absorption. *J. Atmos. Sci.* (submitted) (Cited in Naud et al., 1996).

Chadwick, M.J. Climate change: The mitigation/adaptation dilemma. *Renew. Energy for Develop.* (SEI Newsletter) 7 (3): 1-2 (1994).

Chuang, C.C., Penner, J.E. The role of aerosols in cloud drops parameterizations and its application in global climate models. pp. 49-51 In Proc. 5th (1996).

DDE/USDE. Science Plan for the Atmospheric Radiation Measurement Program (ARM). Report DOE/ER-0670T. U.S. Dept. of Energy, Office of Energy Research, Washington D.C. (1996).

Deekshatulu, B.L., Nash, S. Natural and anthropogenic changes: impacts on global biogeochemical cycles. *Global Change Newsletter* No. 24, pp. 5-7 (Dec., 1995).

Deshler, T., Hofman, D.J., Johnson, B.J., Rozier, W.R. Balloonborne measurements of the Pinatubo aerosol size distribution and volatility at Laramie, Wyoming, during the summer of 1991, *Geophys. Res. Lett.*, 19: 199-202 (1992).

Dunkerton, T.J. Non linear Hadley circulation driven by asymmetric differential heating. *J. Atmos. Sci* 46: 956-974 (1989).

Ennis, C.A., Marcus, N.H. *Biological Consequences of Global Climate Change.* Univ. Science Books, Sausalito, CA. (1996).

FAO. *Mangrove Forest Management Guidelines.* FAO Forestry Paper 117, Rome (1994).

Fosberg, M., Gash, J., Odada, E., Oyebande, L., Schulze, R. Freshwater resources research in Africa. *IGBP Newsletter* 40: 9-15 (1999).

Frankhauser, S. The economic costs of global warming: some monetary estimates. In *Costs, Impacts and Benefits of CO_2 Mitigation.* Kaya, Y., Nakicenovic, N., Nordhaus, W.D., Toth, F.L. (eds) IASSA, Laxenberg (1993).

FSI. (Forest Survey of India). Ministry of Environment and Forests, India, 1998 p. 56.

Gautier, C., Lavallec, D., O'Hirok, W., Ricchiazzi, P., Yang, S.R. A study of Monte Carlo radiative transfer through fractal clouds. In Proc. 5th Atmospheric Radiation Measurement Science Team Meeting, March 19-23, 1995, San Diego, California. pp. 105-109. Published by U.S. Dept. of Energy, Washington, D.C. (1996).

Gille, J. The Chapman water vapour conference. *SPARC (Newsletter), WCRP* 14: 21-23 (2000).

Gupta, S., Pachauri, R.K. *Perspective from Developing Countries.* Tata Energy Research Institute, New Delhi (1989).

Hansen, J., Lacis, A., Ruedy, R., Sato, M. Potential climate impact of Mount Pinatubo eruption. *Geophys. Res. Lett.* 19: 215-218 (1992).

Harries, J.E. Atmospheric radiation and atmospheric humidity. *Q.J.R. Meteorol. Soc.* 123: 2173-2186 (1997).

Holton, J.R., Haynes, P.H., McIntyre, M.E., Douglass, A.R., Rood, R.B., Pfister, L. Stratosphere-troposphere exchange. *Revs. Geophys.* 33: 403-439 (1995).

Karoly, D. Stratospheric aspects of climate forcing. *SPARC (Newsletter)* 14: 15-16 (2000).

Keeling, C.D. Climate change and carbon dioxide: An introduction. *Proc. Natl. Acad. Sci. USA.* 94: 8273–8274, (1997).

Kley, D., Russell, J.M. Plan for the SPARC water vapour assessment (WAVAS), *SPARC Newsletter* No. 13. pp. 5-8 (July, 1999).

Lacis, A.A., Carlson, B.E., Mischenko, M.I. A general circulation model (GCM) parameterization of Pinatubo aerosols. In Proc. 5th Atmospheric Radiation Measurement Science Team Meeting, March 19-23, 1995 San Diego, California. pp. 151-154. Published by U.S. Dept. of Energy, Washington, D.C. (1996).

Legris, P. *Inst. Fr. Pondicherry Tr. Sect. Sci. Tech.* 6: 1-589 (1963).

Lindzen, R.S. Some coolness concerning global warming. *Bull. Am. Meteorol. Soc.* 71: 288-299 (1990).

Lowe, D. Atmospheric CO_2 and the active carbon cycle. *Water and Atmosphere (NiWA)* 8(1): 14-16 (2000).

Mace, G.G., Ackerman, T.P., George, A.T. A comparison of radiometric fluxes influenced by parameterized cirrus clouds with observed fluxes at the Southern Great Plains (SGP) Cloud and Radiation Testbed (CART) site. In Proc. 5th Atmospheric Radiation Measurement Science Team Meeting, March 19-23, 1995, San Diego, California. pp. 185-186, Published by U.S. Dept. of Energy, Washington, D.C. (1996).

Matrosov, S.Y., Heymsfield, A.J., Kropfli, R.A., Snider, J.B. Comparisons of cloudice mass content retrieved from the radar-infrared radiometer method with aircraft data during the second international satellite cloud climatology project regional experiment (FIRE-II). In Proc. 5th Atmospheric Radiation Measurement Science Team Meeting, March 19-23, 1995, San Diego, California, pp. 189-192, Published by US Dep. of Energy, Washington, D.C. (1996).

McCormick, M.P., Thomason, L.W., Trepte, C.R. Atmospheric effects of the Mt. Pinatubo eruption. *Nature* 373: 399-404 (1995).

Meher-Homji, V.M. Climate changes: projections and prospects. *Current Science* 78: 777-778 (2000).

Moore, B. International geosphere-biosphere programme: A study of global change, some reflections. *IGBP Newsletter* 40: 1-3 (1999).

Morais, J. World's manifold webs. *IGBP Newsletter* 40: 7-8 (1999).

N.R.C. *Natural Climate Variability on Decade-to-Century Time Scales.* National Academy Press, Washington, D.C. (1996).

Naud, C., Schertzer, D., Lovejoy, S. Fractional integration and radiative transfer in a multifractal atmosphere. In Proc. 5th Atmospheric Radiation Measurement Science Team Meeting, March 19-23, 1995, San Diego, California, pp. 227-229, Published by U.S. Dept. of Energy, Washington, D.C. (1996).

O'Riordan, T., Rayner, S. Risk management for global environmental change. *Global Environ. Change* 1: 91-108 (1991).

OSTP. Climate change—State of knowledge. Office of Science and Technology Policy, Washington, D.C. (1997).

Palis, H.G., Alan, J., Castillo, A., Lat, C.A. Can mangrove forest cope with sea level rise? *Canopy International* 25 (2): 6-7 (1999).

Panchenko, M.V. Relative humidity and IR absorption by submicron aerosol. *Atmos. Opt.* 1: 25-29 (1988).

Panchenko, M.V., Zuev, V.E., Belan, B.D., Terpugova, S.A., Pol'kin, V.V., Tumavok, A.G. Airborne studies of submicron aerosol in the troposphere over West Siberia. In Proc. 5th Atmospheric Radiation Measurement Science Team Meeting, March 19-23, 1995, San Diego, California, pp. 241-243, Published by U.S. Dept. of Energy, Washington, D.C. (1996).

Peters, R.L., Darling, J.D.S. The greenhouse effect and nature reserves. *BioScience* 35: 707-717 (1985).

Pollock, S. Discounting the future—the cost of global warming. *Interdisciplinary Science Reviews* 24: 195-201 (1999).

Ramachandaran, S. Sea level rise and its impacts on coastal ecosystems. In Swaminathan, M.S., Ramesh, R. (eds.). *Sustainable Management of Coastal Ecosystems*. Swaminathan Research Foundation, Madras (1993).

Ramanathan, V., Subasilar, B., Zhang, G.J., Conant, W., Cess, R.D., Kiehl, J.T., Grassl, H., Shi, L. Warm pool heat budget and shortwave cloud forcing : A missing physics? *Science* 267: 499-503 (1995).

Rosenberg, G.V. Appearance and development of atmospheric aerosols — kinetically caused parameters. *Atmos. Oceanic Phys.* 19: 21-35 (1983).

Rotmans, J. Global change and sustainable development: Towards an integrated conceptual model. In Schellnhuber, H.-J., Wenzel V. (eds.) *Earth System Analysis*. Springer-Verlag, Berlin, pp. 421-453 (1998).

SEPA (Swedish Environmental Protection Agency). Strategies to prevent climate changes. Summary. SEPA, Solna, Sweden (1992).

Shepherd, T. On the role of the stratosphere in the climate system. *SPARC* 14: 7-10 (2000).

Singh, R., Lal, M. Sustainable forestry in India for carbon mitigation. *Current Science* 78: 563-567 (2000).

Smith, J.B., J. Mueller-Vollmer, Setting Priorities for Adapting to Climate Change, Contractor paper prepared for the Office of Technology Assessment, Washington, D.C. March 1992.

Steffen, W. Global change science in the next century, a personal perspective. *IGBP Newsletter* 40: 4-5 (1999).

TERI (Tata Energy Research Institute). *Energy Data Directory and Yearbook*. Delhi (1998).

Trabalka, J.R., Atmospheric Carbon Dioxide and the Global Carbon Cycle. U.S. Department of Energy, Washington, D.C. (1985).

Chapter 4
Energy and the Environment

INTRODUCTION

Energy is undoubtedly a principal component of economic development, but its production and consumption exerts major short and long-term impact on the environment. Efforts need to be intensified to design sustainable energy policies and to develop technologies for production and use of energy in an economically-efficient and environmentally-sound manner.

In the past several decades, the human race has greatly expanded its perception of mankind as a part of the biosphere. Early in the twentieth century, the chief emphasis of many scientists was on surveying the air, water, land and biota of the continents and oceans. The earth was the object of mapping, description and research intended to illuminate various processes. However, before long, the emphasis shifted to the modes and conditions of development. To cope with rising population, advancing technology and expanding demands, the attention shifted toward development plans and to ways of economically using resources in the short run. In the early 1950s, the focus was mainly on conservation and utilization of natural resources, river basins and agricultural and forest use. By the time of the Stockholm Conference in 1972, serious questions started being posed about threats to the integrity of the biosphere. Global environmental monitoring systems and environmental impact assessment methods received much attention.

By the 1992 Rio Conference, people all over the world were perceiving the Earth as something more than a globe to be surveyed or developed for the welfare of some, or to be protected from threats to its sustain-ability, both natural and human. There is now a general awareness, sharing and pooling of efforts to understand and maintain the earth as the one material and spiritual home for one global human superfamily.

Current global environmental problems are being viewed as relating to various sectors of the environment, to particular regions, to sectors of human society and to the adjustment of human society to environment.

Energy input is crucial in virtually all areas of social and economic activity. For developing countries, improving the access to high quality energy services is vital for meeting basic human needs and for raising standards of living. But the production, conversion and use of energy often can cause negative environmental effects. These impacts range from the adverse health effects caused by the indoor use of biomass fuels, through urban air pollution in big cities, to changes in the concentrations of greenhouse gases in the atmosphere, notably CO_2, with the consequent risk of climate change.

Environmental impact is not limited to air pollution from fuel combustion (see Fig. 4.1). The damming of rivers for hydropower utilization, coal mining, oil production and many other energy activities also affect the aquatic and terrestrial environments, leading to ecosystem damage (including biodiversity loss), and degradation of productive land. Economic development and policy decisions in the industrialized world, as well as in the developing world and transitional economies, must consider the environmental issues connected with energy if sustainable local, regional and global development is to be achieved.

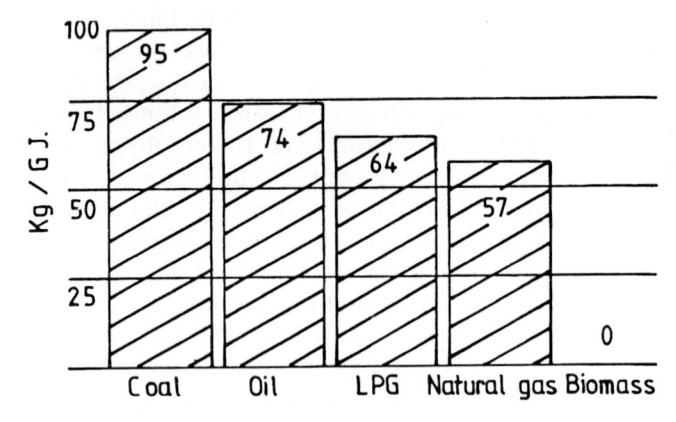

Fig. 4.1 Approximate CO_2 emission/GJ for different fuels. (*Source:* Danish Energy Agency)

Over two decades ago, the energy crisis was perceived as a major threat to Western society. Unusually low oil prices in the USA are firmly rooted in both politics and economics, and western nations have resorted to force to ensure that their supply lines to oil source from the Middle East are not interrupted. However, although cheap energy is the engine of economic progress and affluence, it involves high environmental costs. Western nations have now woken up and started thinking on how to check the environmental costs of oil use (e.g. oil spills and rising concentrations of greenhouse gases). To rein in the pollution, energy efficiency and revival of alternative energy sources in being advocated. Some new technologies are being developed, e.g. fuel cells and ceramics. For improving thermoelectric devices for refrigeration, materials with high electronic conductivity but low thermal conductivity—properties that normally tend to increase or decrease together—are being sought (Stone and Szuromi, 1999).

Alternative fuels are also being developed. However, use of hydrogen in fuel cell vehicles requires enormous expenditure to create the infrastructure to deliver the gas. Meanwhile, the present infrastructure could perhaps become much more energy efficient. Many resources that could be recycled, such as waste water or flare gas, often are not recycled today. If most people had been driving smaller, less gas-consuming cars the energy crisis would have been largely averted. Some other forms of transportation consume less gasoline, and their use needs to be encouraged. Advances in energy technology must go hand in hand with changes in our own habits of energy use (Stone and Szuromi, 1999).

Energy and power are the engines that drive the network of transportation, agriculture, health care, manufacturing and commerce. Yet some of the energy sources release sulfur oxides, heavy metals, greenhouse gases and other noxious pollutants that damage the quality of life. The use of fossil fuels makes us dependent on a non-renewable energy source on which we cannot depend forever (Holt, 1999).

In the USA, petroleum products are available in abundance at so low prices that no one needs to stand in a queue for buying gasoline for one's car. This contrasts sharply with the situation in some developing countries (e.g. India), where the prices of petrol are high and there are often long queues for filling the vehicles. Whereas in the underdeveloped countries, most people cannot afford cars and petrol, in the advanced countries, current system of energy use is unsustainable. Although fossil fuel supplies are limited, total energy use will rise rapidly in coming years with economic development gaining momentum in developing countries. Even current greenhouse gas emissions—let alone any greater emissions in the future—can produce significant environmental changes (see Fig. 4.2).

168 Environmental Technology and Biosphere Management

Per capita Carbon dioxide Emission

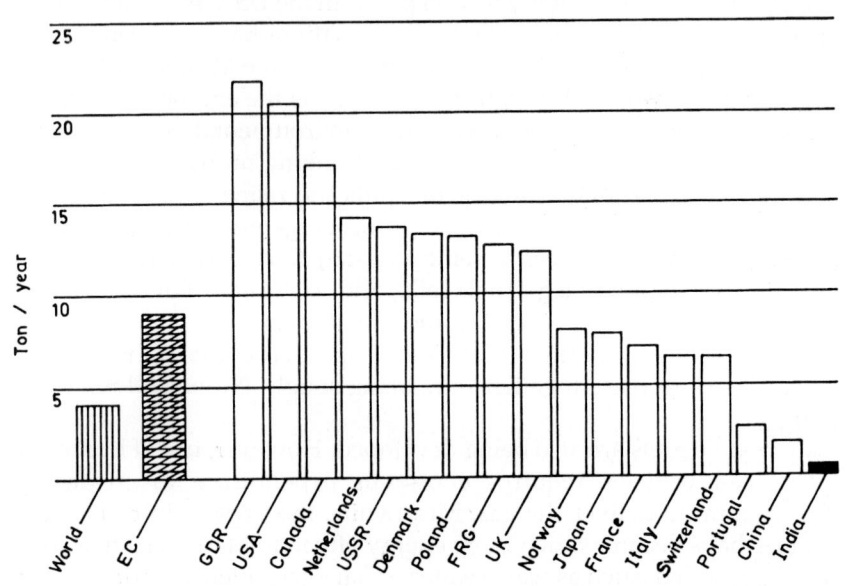

Fig. 4.2 CO_2 emissions per capita (tons/year) as a result of energy consumption in 1986 (Source: Danish Energy Agency).

Projected greenhouse gas emissions for the coming decades can potentially produce climate changes such as an increase of about 3°C in average global temperature, a one-half meter rise of sea level and an increased intensity of hurricanes and tropical storms. Some other adverse effects of fossil fuel pollution (e.g. smog, acid rain, water contamination from leaky fuel tanks, oil refinery emissions, and oil spills) have already materialized in many parts of the world in both industrialized and developing nations (Holt, 1999).

ENERGY RESOURCES, RESERVES AND AVAILABILITY

There is some confusion about the precise definition of the terms 'reserves' and 'resources'. The main difference is that for some energy sources, reserves mean those resources that can be economically recovered and for other energy sources, reserves are considered to be a separate category from resources. Hydrocarbon reserves are those hydrocarbons that are contained in a deposit and which have been demonstrated by drilling to be economically-recoverable by using existing available technology. The reserves present before production begins are the original reserves. Hydrocarbons that are demonstrated to be present but are presently technically and/or economically non-recoverable, are termed resources.

Resources also include those quantities that may exist in deposits that have not yet been demonstrated, which means that new discoveries may be expected.

Resources of nuclear fuels encompass those quantities for which there is a high uncertainty or which cannot be recovered for technical and/or economic reasons.

For hydrocarbons, resources do not include reserves. For nuclear fuels, reserves are termed Reasonably Assured Known Resources. Estimated resources (known and undiscovered) are termed Additional Resources I and II, respectively. For coal, resources in addition to reserves are termed 'additional resources'.

In 1993, the globally known recoverable amounts of non-renewable energy sources amounted to about 1105×10^9 tce (tce = ton coal equivalent) corresponding to about 32400 EJ (EJ = exajoule = 10^8). Coal is the dominant energy source. It represents 51.1% of the reserves, of which 44.6% is hard coal and 6.5% soft brown coal (Fig. 4.3). It is followed by conventional oil with 18.5% and conventional natural gas with 15.7%. Unconventional oil ranks fourth with about 11%. Uranium accounts for the remaining 3.7% of the reserves. Hydrocarbons as a whole (about 45%) have about the same proportion of reserves as hard coal.

The world demand for energy has been increasing more rapidly than population. Regions with a growing hunger for energy are mainly South, Southeast and East Asia.

At present, the fossil energy sources of oil, coal and natural gas cover about 90% of the world's consumption of commercial primary energy — a trend which is not likely to change much within the next one or two decades: Oil is – and will remain in the near future — the dominant energy source.

The global distribution of consumption of commercial primary energy is extremely uneven: Industrialized countries account for 23% of the world's population and consume 71% of the commercial primary energy. The population of Asia, however, uses less than a quarter of the primary energy commercially available. In contrast, Western Europe, which has about 6% of the world's population and consumes 17% of its primary energy, accounts for 30% of the output of the global economy (Anonymous, 1998).

Coal reserves are by far the largest (about 51%) among the non-renewable energy sources. They have a static lifetime of 185 years and are, thus, adequate to secure the present share of coal in the world's consumption of primary energy (26%) until far beyond the year 2000. The proved recoverable reserves are distributed in 79 countries, of which 65 produce about 3×10^9 tce in 1993. Due to its transport disadvantage, most coal is used in the producing country, and only 11% is offered on the

170 Environmental Technology and Biosphere Management

Resources

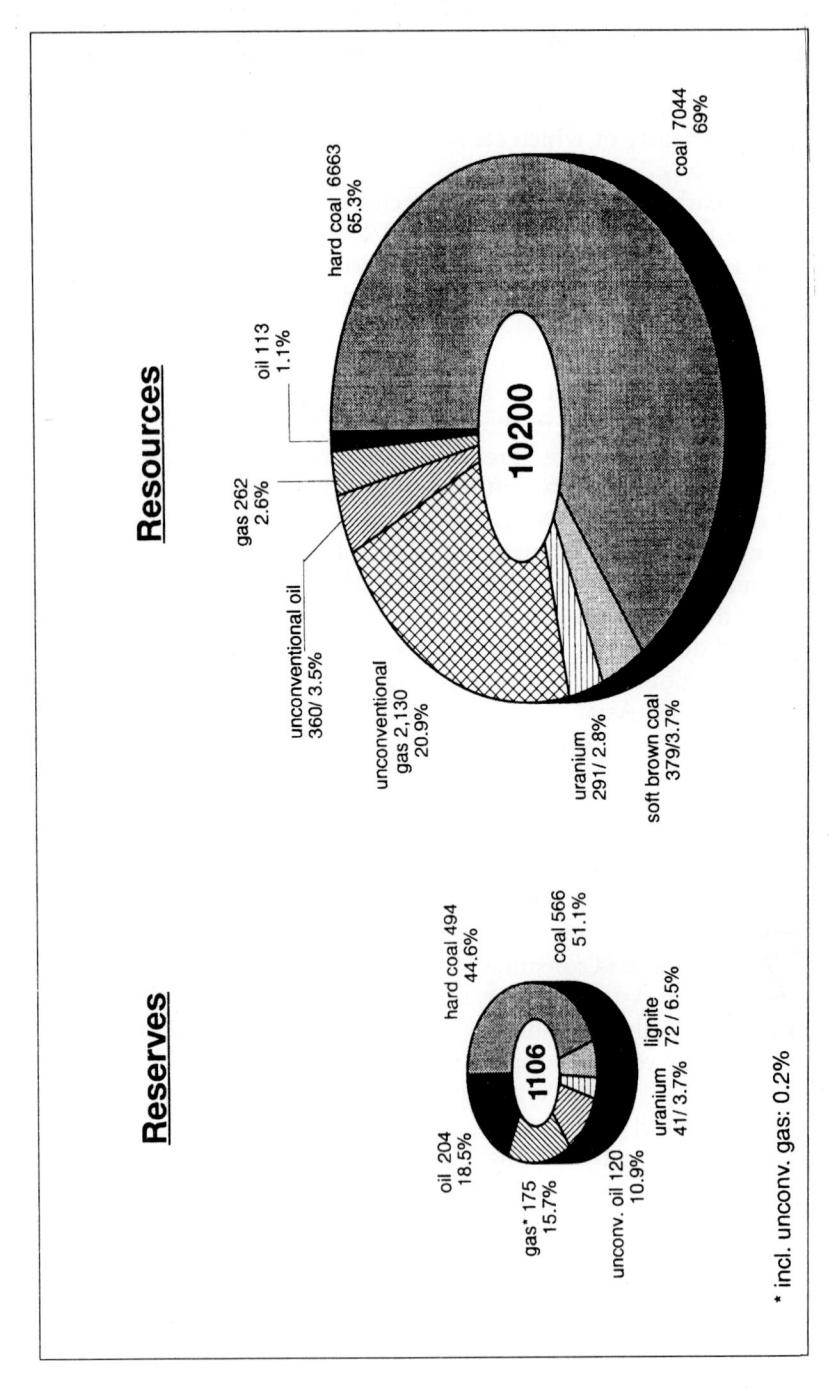

- hard coal 6663 / 65.3%
- coal 7044 / 69%
- oil 113 / 1.1%
- gas 262 / 2.6%
- unconventional oil 360 / 3.5%
- unconventional gas 2,130 / 20.9%
- uranium 291 / 2.8%
- soft brown coal 379 / 3.7%

Total: 10200

Reserves

- hard coal 494 / 44.6%
- coal 566 / 51.1%
- oil 204 / 18.5%
- gas* 175 / 15.7%
- unconv. oil 120 / 10.9%
- uranium 41 / 3.7%
- lignite 72 / 6.5%

Total: 1106

* incl. unconv. gas: 0.2%

Fig. 4.3 The world's nonrenewable energy reserves and resources in bill. tce (after Anonymous, 1998).

world market. The bulk of coal production is consumed by the electricity industry (coal is used for 45% of the electricity generated in the world). Besides power generation, coal is used by the iron and steel industry as well as for heating.

Nuclear fuels accounted for about 7% of the global consumption of primary energy in 1994 and contributed to about 18% of the generation of electricity. About 82% of the nuclear power generation capacity falls in the OECD countries and 32% in the EU.

Figure 4.4 illustrates the ultimate global recovery of conventional and unconventional natural gas sources. Conventional natural gas accounts for only a small percentage of the ultimate recovery of all natural gas.

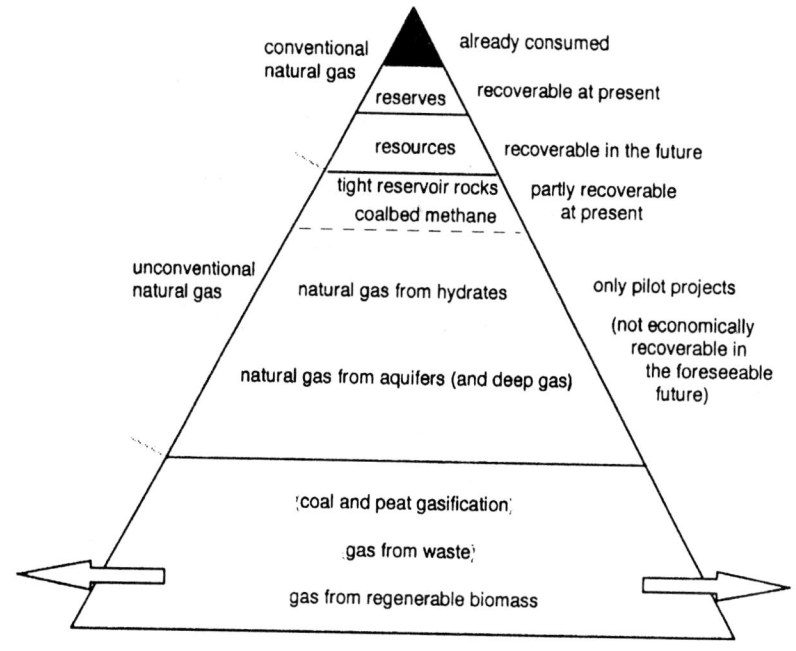

Fig. 4.4 Schematic representation of the ultimate global recovery of conventional and unconventional natural gas (after Anonymous, 1998).

Coal

The demonstrated recoverable coal reserves amount globally to 566×10^9 tce. They have a static lifetime of 185 years and will, therefore, last far ouer the twenty-first century. This makes coal by far the most important source of fossil energy. The proved recoverable reserves account only for 7% of the total coal resources, which amount to about 7600×10^9 tce. This enormous potential attests to the supreme economic significance of coal as a source of energy.

Coal reserves are available in 79 countries of the world, with the USA accounting for more than 35% of the proved recoverable reserves (about 195 x 10^9 tce), followed by the People's Republic of China, accounting for 13% (76 x 10^9 tce), and Australia, accounting for 10% (58 x 10^6 tce). Other countries with major reserves are the Republic of South Africa, India, Germany, and Poland, which together account for about 27% of the global reserves (145 x 10^9 tce).

A distinction based on their energy content can be made between low-calorie soft brown coal, which cannot be transported long distances owing to its low value, and high-caloric hard coal, which is traded on the world market. The proved recoverable reserves (72 x 10^9 tce) of soft brown coal represent only 13% of the global coal reserves, with the 494 x 10^9 t of hard coal reserves, accounting for the remaining 87%. Because hard coal makes up the major portion of the total reserves, its distribution among various countries is identical with the distribution of total coal. The largest reserves of soft brown coal are in the USA (16 x 10^9 tce = 22%), followed by Australia (14 x 10^9 tce = 19%), Germany (13 x 10^9 tce = 13%) and the People's Republic of China (7 x 10^6 tce = 10%). These four countries account for 69% of the global reserves of soft brown coal.

The distribution of proved recoverable coal reserves according to region and economic group are shown in Table 4.1.

Coal has a share of 26% of the global consumption of primary energy in 1993, and thus ranked second after mineral oil. The global share of coal in the generation of electricity was 45%, thus ranking first (Anonymous, 1998).

Total global resources of uranium in 1993 were estimated at 27 x 10^6 t, which figure includes speculative resources of about 11 x 10^6 t U and other resources of about 7 x 10^6 t U, which are of no economic significance at present (see Table 4.2). The known resources amount to about 6.2 x 10^6 t of uranium.

The resources are unevenly distributed with respect to geographic region and economic grouping of countries, due to differences in geological conditions and exploration.

The distribution of uranium resources in terms of 10^9 tce is given in Table 4.3. The conversion is based on OECD/NEA 1993, p. 306 (1 t of uranium = 14000-23000 tce, with the lower value being used of the conversion).

In 1994, global consumption amounted to about 55000 t U (Anonymous, 1998).

Energy and the Environment   173

Table 4.1 Coal reserves according to region and economic groups (in 10^9 tce) (after Anonymous, 1998).

	Total coal		Hard coal		Soft brown coal	
Regions and economic groups	Proved recoverable reserves	Total resources	Proved recoverable reserves	Total resources	Proved recoverable reserves	Total resources
World	566	7608	494	7157	72	451
Western Europe	45	592	29	558	16	34
Eastern Europe	54	3874	39	3852	15	22
Europe	98	4467	68	4410	30	57
Asia	137	988	127	904	10	84
Africa	60	229	60	229	<1	<1
North America	202	1265	185	1043	17	222
Latin America	9	59	9	57	<1	2
Australia/Oceania	58	601	44	515	14	86
EU countries	42	588	29	557	13	31
Asia (FMEC)	49	205	47	196	2	9
Asia (SCE)	89	783	80	708	9	75

Deviartions between totals and sum of individual entires caused by rounding effects
(*) includes soft brown coal from Russia
FMEC Free Market Economy Countries
CPE Centrally Planned Economies

Abbreviations:

Gcal	gigacalorie (10^9 cal)
GW	gigawatt (10^9 watt)
J	joule (0.239 cal)
LNG	liquefied natural gas
LPG	liquefied petroleum gas
MW	megawatt (10^6 watt)
MWe	megawatt (electric)*
MWth	megawatt (thermal)*
Nm^3	cubic meter of gas at STP (standard temperature and pressure: 1013 millibar and 0 ^{0}C)
t	metric ton
tce	ton coal equivalent
toe	ton oil equivalent
TW	terawatt (10^{12} watt)
* MWe =	electric capacity of power plant
MWth =	thermal capacity of power plant, which is transferred to the steam generator

Definitions of condenstate, light oil, heavy oil, oil sand, etc.

Table 4.2 Uranium resources and production figures (after Anonymous, 1998).

		Europe	North America	Latin America	Africa	Asia	Australia	World	Western industrial countries	Eastern industrial countries	Developing countries	OECD	EU
Uranium production 1945–1993	1000 t U	604	606	3	309	257	56	1835	889	698	248	742	80.4
World	%	32.9	33.0	0.2	16.8	14.0	3.1	100	48.4	38.0	13.5	40.4	4.4
Reserves													
Reasonably assured reserves (RAR)													
up to US$ 80/kg U	1000 t U	118	390	168	432	351	631	2090	1225	59	807	1081	48
	%	5.6	18.7	8.0	20.7	16.8	30.2	100	58.6	2.8	38.6	51.7	2.3
up to US$ 130/kg U	1000 t U	154	765	173	570	562	707	2932	1712	23	1197	1608	115
	%	5.3	26.1	5.9	19.4	19.2	24.1	100	58.4	0.8	40.8	54.8	3.9
Known reserves (RAR + EAR I)	1000 t U	782	1216	275	1479	1385	1101	6238	3306	528	2404	2562	210
	%	12.5	19.5	4.4	23.7	22.2	17.6	100	53.0	8.5	38.5	41.1	3.4
Resources													
Estimated additional reserves II (EAR II)	1000 t U	177	1458	15	89	688	0	2427	1500	180	747	1468	10
	%	7.3	60.1	0.6	3.7	28.3	0.0	100	61.8	7.4	30.8	60.5	0.4
Other reserves (speculative and unconventional)	1000 t U	357	2060	697	7714	3604	3900	18332	7184	246	10901	6071	101
	%	1.9	11.2	3.8	42.1	19.7	21.3	100	39.2	1.3	59.5	33.1	0.6

Source: BGR-database

Table 4.3 Reserves and resources of uranium (10^9 tce) (after Anonymous, 1998).

Region	Reserves, recoverable at a price of up to US$ 130/kg U	Known reserves	Additional resources
Western Europe	1.7 – 3.0	3.3 – 5.5	1.5 – 2.5
Eastern Europe	0.3 – 0.6	1.6 – 2.6	0.6 – 0.9
CIS	6.4 – 10.5	21.8 – 35.9	15.5 – 25.5
Middle East	0 – 0	0 – 0	0.2 – 0.4
Asia (not incl. CIS)	1.5 – 2.4	3.5 – 5.8	49.8 – 81.9
Africa	8.0 – 13.1	20.7 – 34.0	109.2 – 179.4
North America	10.7 – 17.6	17.0 – 28.0	49.3 – 81.0
Latin America	2.4 – 4.0	3.9 – 6.3	9.9 – 16.3
Australia/Oceania	9.9 – 16.3	151 – 25.3	54.6 – 89.7
World	40.9 – 67.5	86.9 – 143.4	290.6 – 477.6
EU	1.6 – 2.6	2.9 – 4.8	1.5 – 2.5
OECD	22.5 – 37.0	35.8 – 58.8	63.6 – 104.4
Western industrial countries	25.9 – 42.6	46.2 – 75.9	1215 – 199.6
Eastern industrial countries	0.3 – 0.6	7.7 – 12.6	5.6 – 9.7
Former centrally planned economies	0.3 – 0.6	25.7 – 42.2	63.4 – 104.2
Developing countries	14.8 – 24.3	33.4 – 54.9	163.4 – 268.4

1t U = 14000–23000 tce; the lower value was used for the calculations

RENEWABLE ENERGY

Demand for fossil fuels in the long term, especially for oil and gas, may deplete global reserves to the point where prices could rise greatly. However, it is estimated that their global resources are probably sufficient for some 50 years at present and projected consumption levels, and within a tolerable price range of less than $40 a barrel for oil. Nevertheless, renewable energy sources are of interest in the short term for four main reasons:

1. Many oil-importing developing countries have heavy burdens of international debt which are aggravated by the cost of imported oil, and these countries are looking for ways to reduce this burden.
2. Whereas the industrialized countries have extensive electricity supply grids, networks of oil and gas pipelines, and other means of

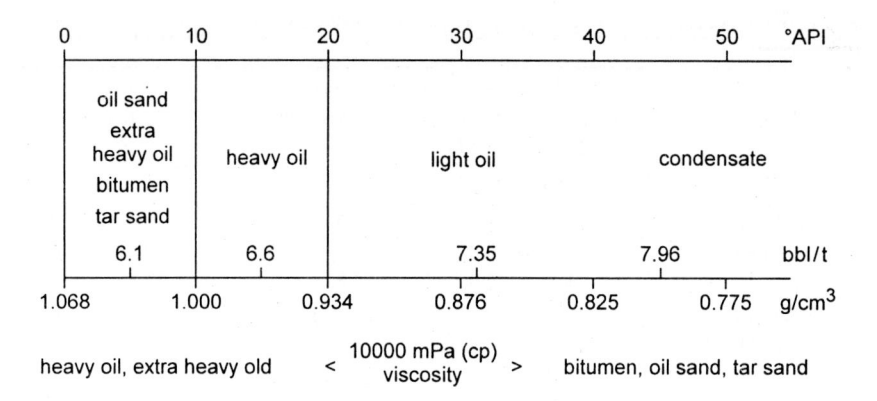

```
0         10        20        30        40        50    °API
|---------|---------|---------|---------|---------|------
|  oil sand  |           |            |                 |
|  extra     |           |            |                 |
|  heavy oil | heavy oil | light oil  |   condensate    |
|  bitumen   |           |            |                 |
|  tar sand  |           |            |                 |
       6.1       6.6          7.35           7.96        bbl/t
|---------|---------|---------|---------|---------|------
    1.068      1.000      0.934    0.876      0.825    0.775  g/cm³

                         10000 mPa (cp)
heavy oil, extra heavy old  <  viscosity  >  bitumen, oil sand, tar sand
```

Conversion factors*

1 t of oil	= 1 toe = 7.35 bbl = 1.50 tce
	= 1270 Nm³ of natural gas = 44.0×10^9 J
1 t LNG	= 1400 Nm³ of natural gas = 1.10 toe
	= 1.65 tce = 48.3×10^9 J
1000 Nm³ of natural gas	= 35.31 cuft = 0.79 toe = 1.18 tce
	= 0.71 t LNG = 34.6×10^9 J
1 tce	= 0.67 toe = 850 Nm³ of natural gas
	= 29.3×10^9 J
1 EJ (10^{18} J)	= 34.1×10^6 tce = 22.8×10^6 t of oil
	= 28.9×10^9 Nm³ of natural gas = 278×10^9 KWh
1 t of uranium (nat.)	= 14000 – 23000 tce, without or with recycling of reprocessed uranum and plutonium, respectively = 0.4–0.7×10^{12} J
1 kg of uranium (nat.)	= 26 lb U_3O_8

* Because they are natural products, fossil energy sources are subject to variation; the specific energy contents given here represent average values, which might, in some cases deviate considerably.

distribution, most developing countries have large rural populations and many urban areas with no electric supply. In many cases, renewables such as photovoltaics, offer a possible way ahead.
3. Populations in developing countries which rely on fuelwood for upto 70% of their domestic energy are facing acute shortages of biomass to burn and are damaging their soil and agricultural base by cutting and burning woodlands which exposes precious land resources to erosion and desertification.

4. When fossil fuels are burnt, environmental problems arise at several levels, e.g. global warming, acid rain (from sulfur and nitrogen oxides); photochemical smog in urban areas (from motor vehicle exhausts); soot, dust and particulate matter; oil spills and environmental damage related to the transportation of fossil fuels; and the threat to the ozone layer, which is partly related to the burning of fossil fuels.

It is the environmental threat which has aroused fresh interest in renewable sources of energy in the developed countries.

Except for large-scale hydropower, the existing market for renewable energies is small and fragmented, projections from past trends have proved unreliable, and only a few technologies are mature in the sense that reliable equipment is commercially available.

Until the early 1970s, renewable energy research and development was greatly neglected. The 1974 oil crisis prompted several efforts aimed at the rapid development of renewable energy options. Some of these technologies have become economically viable, while the marketability of others remains uncertain.

The renewable energy situation in the early 1990s may be summarized as follows:

Solar

Flat-plate solar collector systems for space heating and hot water became technically viable in several parts of the world. Hot water systems were in widespread use in areas with low space heating requirements, but proved economical only where they were integrated into the roof structure of new buildings.

Photovoltaic collectors were developed and found an increasing number of niches in the marketplace.

Wind

Horizontal axis wind turbines were perfected and gained a fair market position in Denmark, California and some other areas. In windy regions, they were cost competitive with fuel-based alternatives for bulk electricity production up to 20% of demand.

Wind energy can potentially provide 10% of the world's electricity requirements by 2020, create 1.7 million jobs, and reduce global emissions of carbon dioxide by more than 10 billion tons, according to a study commissioned by Greenpeace, the European Wind Energy Association and the Forum for Energy and Development (see *Acid News*, Dec. 1999, p. 13).

The study demonstrates that altogether, about 1.2 million megawatts (MW) of wind power could be installed worldwide by 2020, producing more than Europe's present total consumption of electricity. As much as one-fifth of that capacity could be placed in Europe, creating a quarter of a million jobs there.

The 1.2 million MW is equal to about one fifth of world electricity consumption in 1998 (the conservative assumption of the study is that the global consumption of electricity will have doubled by 2020).

In order to achieve the 10% aim, governments should set firm targets for production, remove inherent barriers within the electricity sector and halt subsidies to fossil fuel and nuclear power, while introducing a range of legally enforceable mechanisms to promote wind energy.

In 1998, wind power was the world's fastest-growing source of energy, with an average growth of 40% between 1994 and 1998. The greatest increases were in Denmark, Germany and Spain.

Hydro, Biomass

Hydro and biomass technologies have so far had the largest share of renewable energy use.

If the use of biomass is modernized and much higher conversion efficiencies are achieved, it can provide an abundant source of renewable energy. Indeed, in the long term, all of humankind's energy needs can be met in environmentally benign ways.

The burning of biomass was quite widespread in developing countries and in forest-rich industrialized countries, despite adverse environmental effects. Biomass use includes food production (which is basically an energy use).

Besides photovoltaics, biomass and hydro are the most promising future renewable energy sources. Their potential contribution to world energy supply is substantial, so also their potential to stimulate sustainable development and to displace unsustainable energy systems which are causing serious environmental degradation. They are unlikely to be used at scales large enough to realize those potentials, however, without government action on both national and international levels.

Biogas

Biogas production operates at two levels. The first is in relatively primitive, labour-intensive, domestic and other facilities, e.g. in Asia. The second is in fully automated plants in industrialized countries, e.g. Denmark. Several technically and economically viable installations emerged in both areas (Sørenson, 1991).

In the early 1990s, renewable energies commanded 25% of the total energy picture, including non-energy uses such as those in the timber, paper and pulp industries (Fig. 4.5).

Energy and the Environment 179

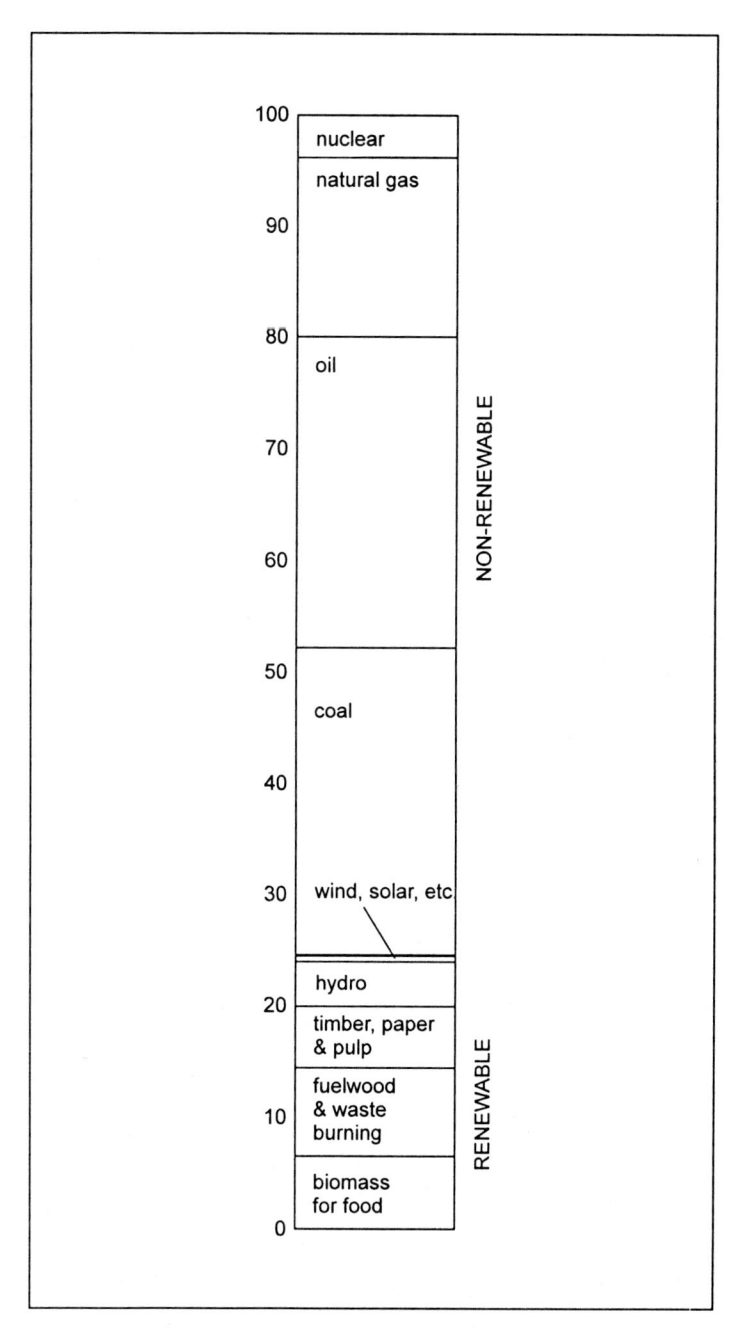

Fig. 4.5 Global energy use in early 1990s (after Sørenson, 1991).

ENERGY AND DEVELOPMENT

No economic growth, social mobility and cultural expression can be achieved without energy. Energy fires industries, fuels agriculture and keeps houses warm or cool. It enables us to cook food, pump water, communicate and travel. The energy for these various activities is so abundant and easily accessible in the industrialized world that consumers have taken it for granted, pausing only in recent years to consider their overdependence on fossil fuels and the implications this has for the global environment. Concern at global warming related to the buildup of 'greenhouse' gases such as CO_2, NO_x and SO_2 in the earth's atmosphere, has driven a search for cleaner fuels and energy-efficient technologies.

Driven by population growth and rising standards of living, energy consumption in developing countries grew from 14% of the world total in 1973 to 22% in 1987, and it might exceed 33% in the year 2020. Over the next two decades, the population of these nations is likely to increase greatly and their energy demand could at least triple. The mix and match of energies which they consume will have a profound bearing on climate and development at home and abroad.

It is noteworthy that about 33% of the developing world's commercial energy consumption is from coal, with the two most populous nations—China and India—burning most of it.

Although the energy field is rich in alternatives, clean and renewable sources such as solar, wind, tidal and geothermal are not universally available and the technologies to exploit them most efficiently are accessible mostly in the private sector of the industrialized world. Many, if not most, developing countries cannot easily change their domestic fuel consumption habits, or afford to import solutions involving the use of advanced technologies developed abroad.

Brazil is a country with a well balanced energy supply mix. Its three main energy sources—electricity, biomass and petroleum—represent 92% of the total primary energy consumed nationally, 60% of this energy being renewable.

Hydro Electricity has become increasingly important in Brazil's energy picture, representing 16% of the total in 1973 and more than 39% in 1990. Coal and natural gas play a small role in Brazil's energy supply mix. Coal use is restricted to the southern regions, where the mines and thermoelectrical stations are located. In the early 1940s, biomass represented 70% of the nation's total primary energy production in the form of fuelwood or charcoal. In 1973, fuelwood generated 50% of the energy used nationally but by 1988, its share had fallen to 22%.

The production of alcohol (ethanol) from sugar cane has a major impact on the energy picture. Ethanol's share of energy supply increased from 0.5% in 1973 to 4.6% in 1988 and bagasse generated twice this amount.

In addition to ethanol's environmental advantage alcohol combustion releases only about 5% of the CO_2 produced by fossil fuels), alcohol is competitive with gasoline when oil prices rise above $30 a barrel. Technological progress suggests that its production costs could be further reduced. Alcohol may be made even more competitive with gasoline by using bagasse for electrical energy co-generation. Indeed, the Brazilian experience with biomass clearly brings out the possibility of establishing sustainable patterns of growth through an appropriate mix of energy sources, the application of modern technologies and energy conservation.

Exploitation of solar, wind and other forms of renewable energy can be encouraged if fossil fuel prices in the western countries (e.g. USA) were not so low. Thousands of 50-meter tall turbines, each with three blades, have already been installed in some wind farms in California. Energy from these wind farms is supplied to over 100,000 families. Some companies, such as Pentagonia Inc., have chosen to pay extra to go green (Brown, 1999).

California customers are increasingly turning to renewable energy from the wind, sun, rivers, geothermal vents and even corn stalks. In Pennsylvania in 1999, 300,000 residents dumped their old electricity suppliers, often switching to renewable alternatives. In Texas, electric companies have been asked to add 2000 megawatts (MW)—the equivalent of two large coal-fired plants—of new renewable resources over the next decade, the largest provision of its kind in any state. At least 20 more American states are planning to deregulate their markets soon, giving renewable companies a chance to compete. Even though natural gas is very cheap, a growing number of people are willing to pay extra to go green, if it means avoiding pollution that may degrade health or cause climate change (Brown, 1999).

For renewable energy to become affordable and cheaper, better technologies will be needed.

Renewables are being favoured in the light of concerns about climate change and also because power plants have to cut GHG emissions substantially under the Kyoto climate change treaty.

Renewable energy is catching on in some countries where there is customer desire for green power.

Japan, Germany and the Netherlands have, for decades, striven to free themselves of oil dependence. In the USA, there was an announcement of an ambitious goal to produce 20% of the country's energy from renewable resources by the year 2000, beginning with some wind turbines. But low natural gas and coal prices prevented wider use of alternative energy sources. In 1997, renewable energy accounted for just 8% of total U.S. energy consumption, compared to 24% for natural gas. The renewable energy that makes the largest contribution of U.S. energy production is hydroelectricity whose future is uncertain.

Wind and solar power are leaping ahead worldwide: Global wind and solar power capacities, in megawatts, have been growing by roughly 22% and 16% a year, respectively, since 1990, according to the Worldwatch Institute, in Washington, D.C. In 1998, the world's consumption of coal — chief source of energy for electricity — declined 2% partly because China cut subsidies to its coal producers.

Fueling the global gains of wind and solar impressive technical achievements, such as more efficient turbine design, have greatly improved the global benefits of wind and solar power; wind power costs about 5 cents per kilowatt-hour (kWh), less than a tenth of the 1980 price, and PV power averages less than 20 cents per kWh.

Solar cells are being sold for items like highway signs, roofs and radios in many countries. Japan—where electricity is relatively expensive —is expected to lead solar PV sales worldwide in the near future. Since 1998, the world's wind energy capacity has grown more than 35%, topping 10,000 MW in 1999 which is double the amount of 3 years ago.

Solar Homes

In Amersfoort, (Netherlands), a project called Niewland to build 500 houses with roofs covered with photovoltaic (PV) panels has been launched in 1999. When finished these homes are expected to draw 1.3 megawatts of energy from the sun, enough to supply about 60% of the community's energy needs with the rest coming from the power grid. This project is the world's largest attempt at 'building-integrated photovoltaics' (BIPV). The attempt is all the more noteworthy because Amersfoort gets much less sunshine than the world average of 1700 watts per square metre: Nieuwland's homes should be bathed in about 1050 watts worth of energy per square metre, from which they may glean as much as 128 watts per square metre, thanks to nifty PV cells that respond best to light reflected by clouds. However, each kilowatt-hour from the solar panels will cost about four times more than electricity supplied by the grid, but the costs may come down later with further expansion.

THE FUTURE OF RENEWABLE ENERGY

There is no doubt that renewable resources can potentially provide all of society's energy needs. But actual exploitation of renewable energy sources mainly depends on their costs as well as on the perceived risks of using fossil fuels. The energy crisis of the mid-1970s generated the perceived risk of running out of conventional fossil fuels and stimulated developing renewable sources and energy conservation measures, including higher vehicle fuel economy and energy-efficient buildings and homes (Turner, 1999). But these programmes suffered a setback when supply once again met demand. In the 1980s when the risks associated with pollution increased, the focus shifted to avoid or remedy environmental damage from fossil fuel extraction, processing and transport and catalytic converters to burn fossil fuels more cleanly were employed. More recently, the risks associated with CO_2 emissions and global warming have again enhanced interest in renewable energy. However, it is not possible to continue to burn fossil fuels and sequester the produced CO_2 efficiently enough to prevent or control global warming because the processes of concentrating and burying or transforming the CO_2 are themselves energy intensive.

Thus, if global warming issues prompt us to minimize fossil fuel use, are there, in fact, practical renewable alternatives? Can a sustainable energy system be developed that would supply a growing population with energy without harming the environment providing energy for the present without compromising the ability of future generations to meet their needs? The crucial issue today is whether governments should promote building, manufacturing, and implementing renewable energy technologies, or should they instead develop and build CO_2 sequestration technologies. By developing and implementing renewable energy technologies and manufacturing capabilities, we can build a sustainable energy infrastructure.

As the USA is the largest user of energy in the world (about 50% of total consumption), any change in global energy use would require a change in US production and consumption of energy. The major renewable energy systems include photovoltaics (PVs) (or solar cells), solar thermal (electric and thermal), wind, biomass (plants, trees), hydroelectric, ocean, and geothermal (Turner, 1999). Solar cells directly convert the sun's energy to electricity with no moving parts. Solar thermal systems generate heat. Wind energy represents the nearest term cost-competitive renewable energy source. Wind presents a dual-use technology: The land can still be used for farming and forestry. Biomass power ranges from burning wood chips in power plants to burning biogas from waste treatment plants to the generation of methanol and ethanol, which may be used as fuels. The ocean is Earth's largest collector of solar energy — ocean thermal platforms have a large potential for electricity generation (Turner, 1999).

Figure 4.6 shows expected world consumption of primary energy.

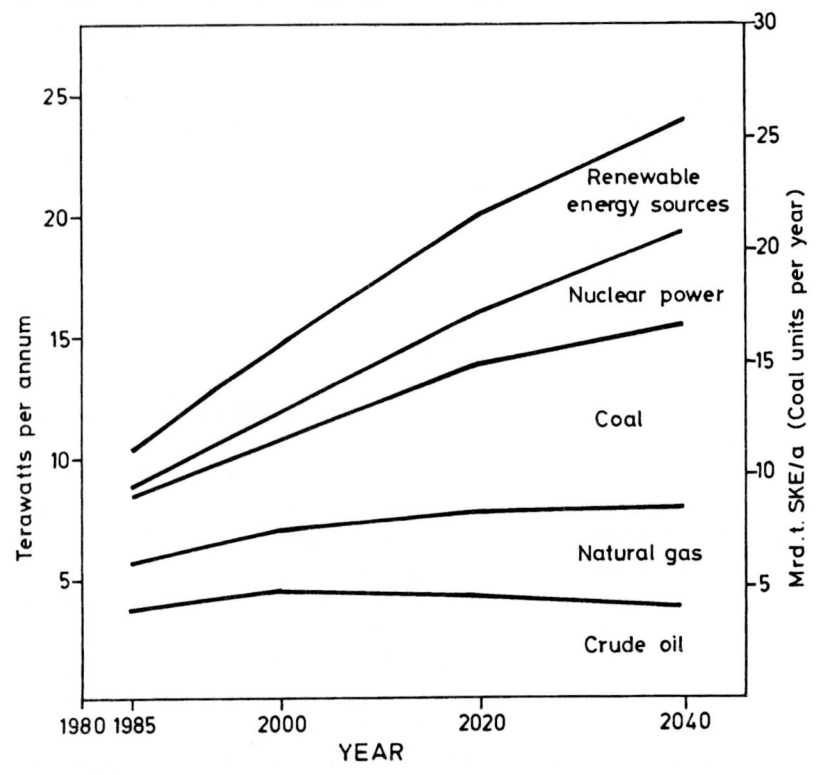

Fig. 4.6 Expected world consumption of primary energy (after Hiller, 1991).

Two important pathways can lead to a renewable-based energy infrastructure while reducing CO_2 emissions and oil imports. One of these focuses on the electricity-producing sector, whereas the other focuses on the transportation sector.

Intermittency and H_2 Use

The missing links required for a sustainable energy system are suitable storage scheme for the renewable energy and an energy carrier to replace gasoline and other fossil-derived energy carriers. Energy storage technologies include H_2, batteries, flywheels, superconductivity, ultracapacitors, pumped hydro and compressed gas. According to Turner (1999), the most versatile energy storage system and the best energy carrier is H_2, which can replace fossil fuels for transportation and electrical generation when renewable energy is not available. H_2 is transportable by gas pipelines or can be generated on site, so any system that requires an energy carrier can use H_2. The conversion of the chemical energy of H_2 to electrical energy by a fuel cell produces only water as waste.

Today, H_2 is mostly manufactured from steam reforming of natural gas. But it can be generated by solar energy (Fig. 4.7). Wind energy and PV systems coupled to electrolyzers may possibly be the most versatile of the approaches and may be the major H_2 producers of the future. Unfortunately, although these systems are commercially available, they are very expensive (Turner, 1999).

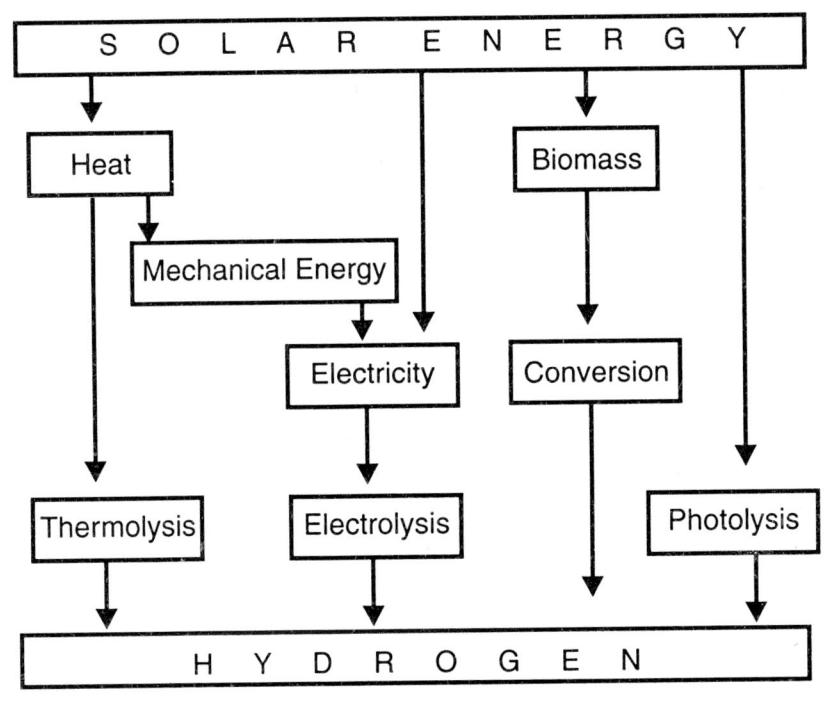

Fig. 4.7 Sustainable paths to hydrogen (after Turner, 1999).

The photolysis systems are direct conversion systems including photoelectrolysis and photobiological systems, and are based on the fact that visible light has a sufficient amount of energy to split water.

Transportation

Fuel cell vehicles have the potentiality of being highly efficient, reaching an equivalent fuel economy of over 42 kilometers/l, with a performance and range equivalent to current vehicles (see Ogden et al., 1998). When powered with H_2 generated from renewable resources, these vehicles have no emissions. The major problem with these vehicles is the non-existence of any H_2-fueling infrastructure. Attempts are underway to develop a fuel cell vehicle that uses an on-board chemical process plant to convert gasoline to H_2. Miniaturizing converters is problematic. In this context, H_2 as the energy carrier is a far better option. On-board H_2 can allow a vehicle to maximize the energy-efficient aspects of fuel cells and the environmental attributes of H_2. The near-term approach would be to build stationary converters at current compressed natural gas filling stations. Because it would run continuously, putting the same small-scale converter on the ground would spread the cost over a larger number of consumers and would minimize development costs. Initially, the H_2 may be used in modified internal combustion engines and the exhaust from a H_2 internal combustion engine would be much cleaner than that from gasoline engines. This engine technology is already quite well developed. This approach allows a H_2 infrastructure that deals with safety issues to be built simultaneously with the development of fuel cells. Fuel cell vehicles would be available when the costs of fuel cells can be brought down somewhat for automotive applications. Renewable energy can supply all of the H_2 necessary for cars (Turner, 1999).

Research and Development Issues

The crucial element in the above infrastructure scenario is energy *storage* and its matching with the energy generation system. Hydrogen storage systems are particularly important. Hydrogen stored as compressed gas or as a hydride works well for stationary application; however, H_2 storage for transportation is problematic because of its low volumetric density. Research is needed for the development of compressed gas tanks made of lightweight advanced composites and of new H_2 absorbents such as carbon nanotubes. Some additional research is also needed for improvement of fuel cells and electrolyzers.

THE ENERGY EVENTS OF THE 1970s

In the 1970s, it became apparent that industrial society was being adversely affected through pollution and excessive use of natural resources. It was even predicted that oil resources would be exhausted by the year 2000. These worries were highlighted at the 1972 UN Conference on the Human Environment in Stockholm.

The energy issue became a crisis during the 1973 Middle East war, which enabled OPEC (Organization of Petroleum Exporting Countries) to impose a steep hike in oil prices, which ended the era of cheap and abundant oil. Many people believed that oil resources were running out. Finding of suitable substitutes for oil became a top priority for many countries. Technologies for the manufacture of synthetic crude oil, the exploitation of tar sands and oil-shales, and the harnessing of such unusual energy sources as the ocean thermal gradients started attracting heavy grants in several countries. France launched a strong nuclear programme. The high price of oil also stimulated much interest in new and renewable energy sources. The Iranian revolution of 1979, which saw the disruption of the country's oil production, further dealt another severe blow to the already fragile economies of these oil-importing developing nations which could not afford the greatly increased prices of oil.

Fuelwood was another major energy concern of the 1970s. There was a parallel between the industrial world faced with vanishing oil resources and the disappearance of forests in the developing world—it was on these forests that a large number of people depended for their energy needs. Indeed, fuelwood cutting was a major threat not just to household energy supplies but also to the whole agricultural system on which people were dependent for their food. A massive programme to promote tree-growing was suggested. Some experts also linked fuelwood consumption with general environmental degradation and food production.

The major component of development assistance in the 1970s was given for providing substitutes for oil and increasing wood-fuel supplies. The decade of the 1980s, however, witnessed the failure of many of these energy assistance programmes (Foley, 1991).

Several social forestry programmes to promote tree-growing by rural people were launched. They differed from traditional forest department activities focused on commercial production of timber or pulpwood. In India, programmes in various states included community woodlots, farm forestry, 'vanamohotsav', planting on waste lands, and plans to provide landless peasants with land on which they could grow trees. Peri-urban plantation projects, designed to provide woodfuel supplies for urban dwellers, were launched in many African countries. In several cases, however, there was little interest in growing trees for fuel because the costs were so high that the fuelwood produced was uncompetitive with wood from natural forests. By the mid-1980s, the peri-urban approach had been discredited.

Attempts made by individual farmers to grow trees for fuelwood also did not make any significant impact. By and large, in many African countries, local people resisted attempts to grow trees for fuel. But in India, 'farm forestry' was adopted substantially and fairly widely. Billions of eucalyptus trees were planted. However, the aim of the farmers was not to produce fuelwood but to supply the market for poles and pulpwood. While these programmes were financially profitable, they nevertheless attracted much criticism on social and environmental grounds. Rather than using marginal or degraded lands, as was the government's intention, much of the area planted was high-quality farmland which had previously been used for cash or food crops.

Some programmes to improve the efficiency or design of rural cooking stoves were also launched, but most of them failed to make any strong impact as the technical performance of some heavy home-built stoves turned out to be poor and the material from which they were made cracked easily from the heat. Consequently their energy performance declined and after sometime the stoves saved little, if any, energy (see Fig. 4.8).

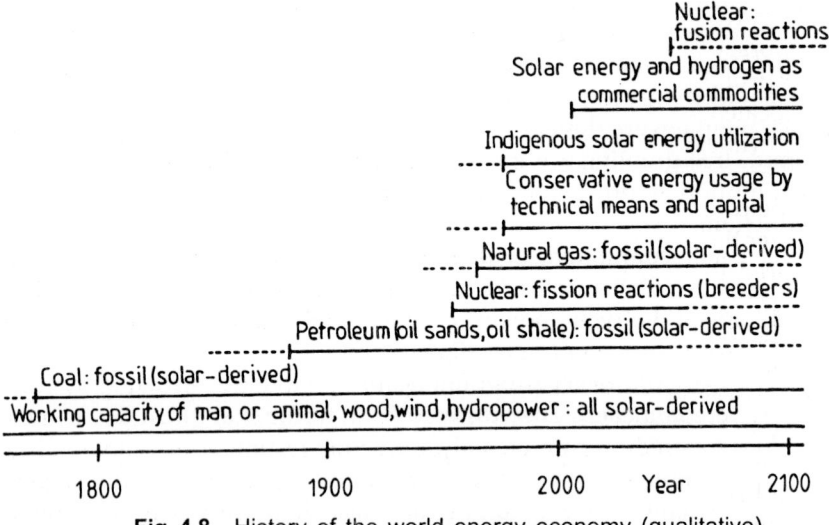

Fig. 4.8 History of the world energy economy (qualitative). (Source: based on information from the UN)

In the 1980s, lighter, portable stoves made of metal and/or ceramic material became popular and also emphasis shifted from the rural to the urban areas. Such stoves have been widely introduced and become very popular in Kenya, Sudan, Tanzania and some other African countries, as also in Sri Lanka.

Large stoves for use by institutions such as schools and hospitals have also been promoted. Well designed institutional stoves are conducive to substantial woodfuel savings compared with the use of an open fire.

Change in Energy Use

For about two decades now, the major ingredient in energy developments has been change. International tensions, conservation, technological advances and changing consumer preferences have disturbed old, predictable patterns of energy supply and demand.

Solar energy has tended to develop more slowly than many had hoped. Coal faces environmental problems. Nuclear energy development also has stagnated.

In the USA, in 1984, total imports rose to nearly 5.2 million barrels a day—an increase of 7.7%. In India, most of the export earnings go to finance the import of fossil fuels.

Globally, there has been some oil 'glut', but the word 'glut' is misleading. It gives the impression as if the world is choking on oil. The glut simply means that some foreign countries can produce more oil than they can sell at a given price. That surplus could turn out to be small and temporary. Free world production is about 43 million barrels a day. There is a surplus of about 10 million barrels. But about 8 million barrels of it are located within the Persian Gulf. If there's major trouble there, the glut could vanish quickly.

A combination of factors usually brings about any surplus. The glut can be caused partly because of higher prices. During the 1980s, people all over the world cut back on energy use. The world-wide recession also lowered the demand for oil. Then the high price of dollars—all oil is traded in dollars—cut back on consumption in some countries. And, of course, the higher energy prices encouraged producers all over the world to step up their efforts to find new oil supplies. Now the glut is helping to keep prices down. So, on the whole, it may be good for consumers and many industries. But it is wiser to work on long range solutions to our basic energy problems. We need to make sure we'll have enough energy 10 and 20 years from now.

Domestic energy sources need to be trapped and developed. There are long lead times in all energy industries. We need to start today to find and produce energy for a decade from now. The pace of new exploration needs to be increased.

It is extremely risky to depend on imports to supply what we'll need. A Mideast crisis or war can disrupt oil supplies. Terrorist activity could result in supply problems.

Oil is the most widely used energy resource in the economy. It can be used for transportation, electricity generation or industrial production. Oil goes into everything—from automotive fuel to farm fertilizers, plastics and paints. And because of that versatility, oil imports are needed to make up the difference between domestic energy demand and energy production. Although in some places, power companies have switched from oil to coal,

coal creates its own problems such as air pollution and acid rain. Therefore, oil will remain the major energy resource for many years to come.

Besides coal and nuclear power, synthetic fuels and solar energy will contribute significant amounts in the decades ahead but petroleum will probably continue to be the major energy source. Of course, we can count on new energy developments to solve our problems. These new developments will relate to solar, geothermal, winds and tides. We will need all we can produce because our population is increasing, of course, and that means more cars, more houses and more work places, all needing energy.

Policies are needed that will lead to more domestic energy development. Increasing conservation is very important. But energy-saving, by itself, won't solve our problem. We'll need more energy supplies. And there are ways to encourage more development. Onshore and offshore oil exploration deserves high priority.

Renewable Energy Technologies

Figures 4.8 and 4.9 illustrates historical aspects of the world energy economy.

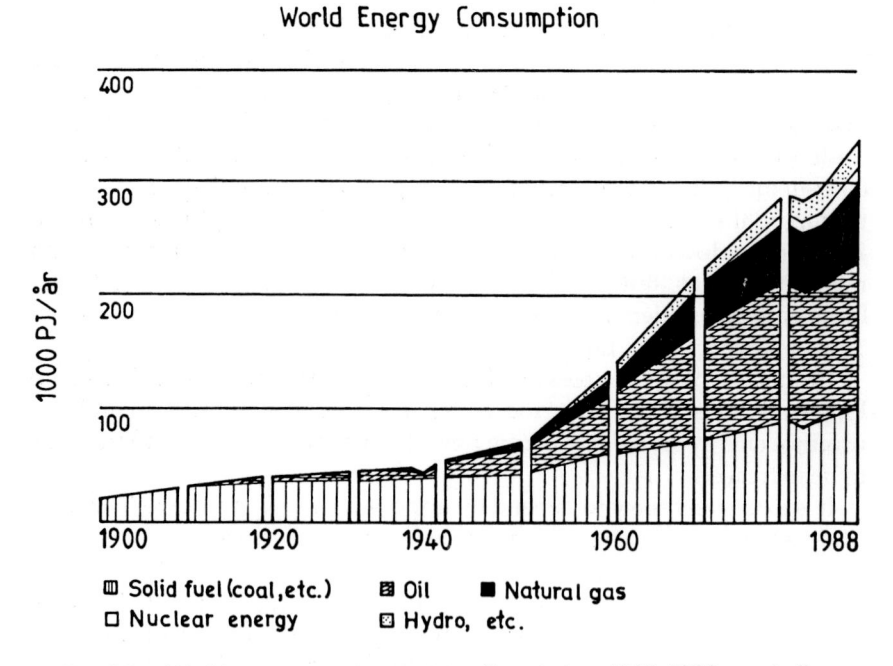

World Energy Consumption

- ▥ Solid fuel (coal, etc.)
- ▧ Oil
- ■ Natural gas
- ☐ Nuclear energy
- ▨ Hydro, etc.

Fig. 4.9 World gross energy consumption during 1900-1988, excluding non-commercial fuels, e.g. wood in developing countries. (*Source:* based on information from the UN).

During the 1970s and 80s, the renewable technologies promoted included solar energy, small hydro, biomass gasification, biogas, wind power, residue briquetting and others (Kristoferson and Bokalders, 1988). Unfortunately, the success ratio in renewable energy projects was quite low. Some projects did not work at all; few could survive after the departure of the foreign project staff who installed them. Their impact on petroleum imports was very small, if any.

In solar energy, the main emphasis was on photovoltaics for water pumping and electricity generation (McNelis et al., 1988). The solar cells themselves were quite durable but the ancillary equipment frequently failed under the harsh operating conditions of the rural Third World. Spare parts and competent technicians were not available and, in many cases, the equipment had to be simply abandoned. Over 2000 solar pumps have by now been installed, whose average peak output is around 1 KW, making them primarily suitable for drinking water supplies. Some 1000 solar refrigerators have been installed, generally powered by photovoltaic units with a peak output of around 100 watts (Foley, 1991). Many small lighting kits have also been installed, typically powered by panels with a peak output of upto 100 watts, and suffice for a few fluorescent tubes and bulbs. Most of these programmes have been heavily subsidized and yet their adoption by the general public has fallen short of expectation.

Some complex, flat plate solar collectors have been used in urban areas. They are unsuitable in the rural areas, especially where there are no skilled technicians or piped water supply. Solar driers and solar cookers are also quite cumbersome, awkward, expensive and quite impracticable.

Small hydro projects have achieved some success in Nepal, where many locally made turbines have been substituted for traditional wooden water mills. There have also been several successful small hydro installations in mission stations, commercial farms and small enterprises in some countries.

In the inland delta of the river Mali, a Chinese-designed rice husk gasifier has worked successfully in spite of the fact that rice husk gasification is technically difficult. When the oil prices fell in the mid-1980s, most commercial firms engaged in making gasifiers abandoned the technology.

In the Philippines, a major dendrothermal programme was launched in 1980 with the objective to use fast growing *Leucaena leucocephela* trees, grown in special plantations, as fuel for small power stations. The programme has, however, failed (Foley, 1991).

China appears to have been the first country to start a large-scale biogas programme during the 1960s. Over 7 million digesters have been built. The second largest programme has been in India where at least 300,000 have been installed.

There are two main types of biogas digesters. The Chinese type has a fixed brick-built dome, in which the pressure increases with the accumulating quantity of the gas produced; this digester has a tendency to leak, and it appears that many of the digesters are now defunct.

The second type of digester was developed in India and has a masonry with a floating steel cover which maintains a constant pressure. It is very costly. Only rich farmers can instal it. The steel cover tends to corrode.

Some other renewable energy technologies proposed include wind power, residue briquitting, small steam engines, hybrid wind-diesel systems, and ocean thermal gradient power. Of these, wind power has had some success, mainly for pumping, in some countries.

The Power Sector

During the 1960s, several developing countries started highly ambitious programmes for developing the power sector with high emphasis on hydro-electric power stations which utilized indigenous energy resources. A few examples of such major projects included Aswan in Egypt, Akosombo in Ghana, Bhakra-Nangal in India and Tarbela in Pakistan. But as the financial, socio-economic and environmental costs of these large hydro projects started being recognized, public opinion turned against them.

However, smaller hydro projects and oil- and coal-fired power stations continued to be supported through the 1970s and 1980s. India happened to be the largest recipient of World Bank lending for rural electrification, with Brazil coming next. Malaysia and Thailand also received substantial funds. In 1985, the power sector was estimated to account for about one fifth of Third World debt (Foley, 1991).

Unfortunately, the technical performance of many of the projects has not been upto the mark, and the overall financial performance of utilities has undergone deterioration. Also, there were weaknesses in management and financial control.

Indeed, despite the inputs of enormous sums of financial aid into the power sector during the 1980s, the national power utility in many countries was worse off at the end of the decade than it was at the beginning (Foley, 1991). Some factors which have led to the extremely bad state of electric supply situation in India have been heavy pilferage and illegal electric connections, corruption of electric supply staff, heavy losses, defective and poor-quality fixtures and equipments, inadequate repairs and maintenance and political reasons (e.g. ignoring widespread thefts of electricity by slum dwellers with a view to getting their votes, etc.). Some industrial establishments have, in collusion with corrupt linemen, freely drawn electricity from the poles thus inflicting heavy financial losses on the electricity boards or companies concerned (see World Bank, 1990). Politicians should be made to realize that a financially-viable electricity utility, which provides a reliable electricity supply, and charging what it really costs to provide, is much more likely to promote development than the inefficient, subsidized, populist and near-bankrupt power utilities which so commonly exist today. The long term sustainability of projects requires that donor agencies should withhold funding from those utilities which cannot bring tariffs up to a realistic level.

According to Foley (1991), for the electric power sector, projects need to be funded only if they are likely to lead to the balanced and sustainable development of the whole power sector. However, there are many opportunities for providing funds for new generating plant, expansion of the transmission systems, rehabilitation of the existing system, and for the provision of technical and managerial assistance.

The rural electrification sector in many Third World countries is also growing. Electricity is, of course, essential if the rural areas have to develop properly. But the provision of an electricity supply, in itself, may not be enough to promote rural development. Electricity is only useful when there is a demand for the service it makes possible. Rural electrification programmes should, therefore, be started only in those areas which have already attained some degree of economic development which enables them to benefit from an electricity supply.

It is the belief of several economists that prices are the best indicator of energy scarcity or lack, and that governments should intervene as little as possible. Many scientists believe that some useful information about economic systems can be gained by applying ecological systems analysis to the interrelations between natural and socio-economic systems. There appears to be a relationship of energy to fundamental economic processes and the ways that changing energy regimes impact economic processes. An important tool in this approach is to compare the fiscal return of an economic process such as agricultural production, with the amount invested.

If, in fact, new agricultural technology continues to be energy intensive and if the supply of fossil fuel continues to be finite (see Figs. 4.10 and 4.11), then the best strategy for developing countries is to invest in incentives for population control. All other strategies may only delay the inevitable population/land pressure by a few decades.

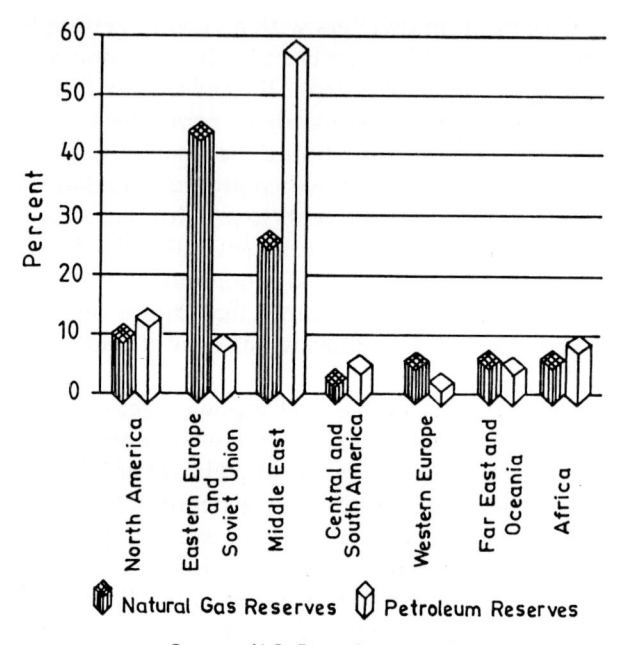

194 Environmental Technology and Biosphere Management

Fig. 4.10 Comparison of world's petroleum and natural gas reserves as in 1987 (Source : USDA).

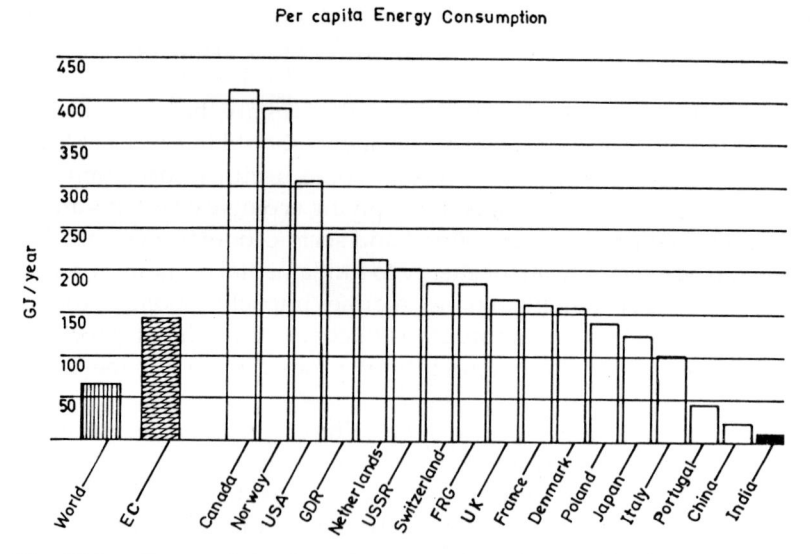

Fig. 4.11 Gross per capita energy consumption in 1986 (in GJ/year) in several countries (Source : Danish Energy Agency).

BIOENERGY

Well-managed biomass energy systems can form part of a matrix of energy supply (see Fig. 4.12), which is environmentally sound and conducive to sustainable development. The overall impacts of bioenergy systems are usually less damaging to the environment than those of conventional fossil fuels. They produce many, albeit local and small impacts on their surrounding environment. This contrasts with fewer but much larger impacts, distributed over greater areas, for fossil fuels. These qualities make the environmental impacts of bioenergy systems more controllable and more reversible and, hence, more benign. Another merit of bioenergy is that its use often results in certain beneficial side effects, both locally and globally.

Among renewable energy resources, biodiesel fuel made from rapeseed is of special importance in Europe. Economical, technological, ecological and toxicological arguments have been advanced implying that, at present, biodiesel is at best just a 'niche' product that can only compete with traditional fossil diesel fuel because of significant tax incentives. Given the present state of knowledge in these very different areas, the crucial issue is whether the competitiveness, and thus marketability, of biodiesel can be enhanced by biotechnological manipulations of the rape plant (Martini and Schell, 1998).

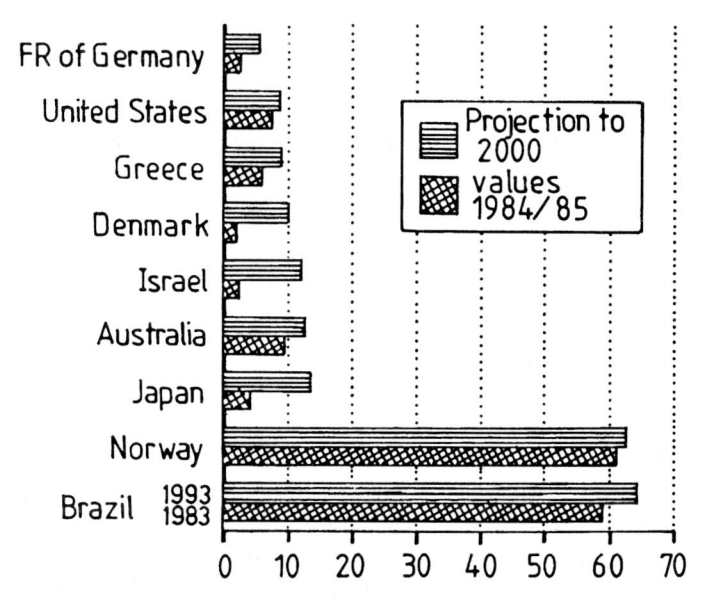

Fig. 4.12 The renewables share of total primary energy supply for selected countries (after OECD, 1987).

Of course, all bioenergy systems, like any other energy system, do have some environmental impacts. While some of these are environmentally more preferable than others, it cannot be stated that there exist energy systems without any impacts at all. For bioenergy, the impact depends more on the manner in which the whole system is managed than just on the fuel or the conversion technology. The impacts vary both in quality and in quantity, and our goal should be to strive for a situation where the total environmental impact is as low as possible, in tune with the prevailing socio-economic realities.

The term 'environmental impact' implies either an adverse or a beneficial impact resulting from the production, conversion and use of energy. Environmental impact includes harm to human health, to natural or man-made ecosystems, to man-made structures or to the socioeconomic system. Some impacts reversibly damage the object they act on. For example, if someone falls off a roof, he may break a leg, but after some weeks, the damage would be reversed following appropriate health care. In contrast, deforestation is often irreversible, since the exposed soil may not be able to support new vegetation, and hence the topsoil erodes fast.

Various kinds of environmental impact are not always independent of each other. A small emission may result in a reversible effect. However, if the level of emission rises above a certain threshold value, the effect can become irreversible. A small amount of toxic emission may only produce on occupational hazard, but after a certain level, the emission can also become a public health problem.

The aim of environmentally-conscious energy development is to make considered choices between different energy systems, with a view to minimizing their environmental impacts, while being conscious of social and economic realities. However, most environmental impacts have different aspects, which are measured by different, often incomparable measures. Rarely can one express all concerns in just one value, such as monetary costs. For instance, how to decide between one irreversible impact and many reversible ones? How to decide between a potential health impact (and hence a cost) today and one for people of future generations, especially when conventional economic analysis is heavily weighted against present costs and future returns? Should we opt for an energy system with an occupational or public health impact? By going in for a well-managed bioenergy system for home heating, one is implicitly promoting some health risk to individuals in the home asf well as generating some health risks outdoors for the general public with air pollutants produced from biomass combustion. But still, biomass energy compares favourably with the widespread environmental impacts of fossil fuel-based air pollution and climate change (Pasztor and Kristoferson, 1990).

Bioenergy systems also can have some potential benefits, including those accruing from avoiding an adverse impact of a strongly polluting energy system, by substitution. For instance, replacing a well-managed coal-fired boiler with a well-managed wood-fired, one can spell improvements in air quality. In some cases, besides the avoidance of adverse effects, some positive effects also accrue. Thus, a well-managed energy forest, besides providing wood for the above boiler, would also provide green space, absorb carbon dioxide, improve soil, increase wildlife, and provide valuable forest products.

Traditional Biomass Resources

In the developing countries, especially rural households and urban slums, traditional biomass resources for energy needs are fuelwood, charcoal and agricultural and animal wastes. In villages, most of these resources are either collected by the users or come from non-commercial markets, in marked contrast to urban and industrial traditional biomass resources, which are bought and sold in commercial markets.

Traditional biomass systems have typically high levels of site specificity. In the past, wood fuel has been blamed as the primary cause of deforestation and consequent desertification in developing countries, but recent information from critical, site specific studies does not support this idea. No doubt, many of the felled trees do end up as fuel, but the important driving force for the cutting of the trees and, consequently, for the disappearance of forests, is the need to open up land for agriculture and grazing, followed by other commercial uses of trees, including commercial wood fuel requirements by urban and industrial users. Indeed, rural wood fuel requirements rarely cause deforestation. Therefore, the 'second energy crisis' due to wood fuel shortage needs to be looked at, not merely as an energy crisis, but rather, as a subset of crises in land-use and development patterns (Pasztor and Kristoferson, 1990).

Further, rural collection of traditional wood fuels rarely causes much environmental damage. Several traditional practices of tree pruning and combining tree growing with food producing units (agroforestry) are, in fact, environmentally sound, and may even result in increased wood and crop yields. It is the exploitation of forests and woodlands by commercial loggers, as well as the expanding agricultural and grazing interests, that produce the adverse impacts. Forest soils cannot sustain agriculture when the trees are removed, and become completely exhausted and barren within a few years. This, however, is not an energy issue but an agricultural one.

Also, it is the charcoal and fuelwood requirements of urban and industrial users that usually result in strong and adverse environmental impacts.

Traditional Charcoal Conversion

Being cleaner, charcoal is more popular than fuelwood in many developing countries. It also has a higher energy content per unit weight and is more easily available than fuelwood. Most charcoal conversion is done in rather inefficient, traditional earthen kilns. Whole trees, branches, and some agricultural residues can be used as feedstocks in these kilns. Nevertheless, one typically needs more than twice as much wood for producing charcoal to deliver the same final energy as with fuelwood. Also, charcoal use is almost entirely commercial, mainly in urban and industrial areas. While a lot of the wood which is being converted to charcoal comes from forests cleared for agriculture, most often the producers cut trees specifically to produce charcoal. This directly contributes to deforestation.

The health impacts of charcoal use are much less than of fuelwood, although invisible carbon monoxide production during incomplete combustion can be a problem.

Agricultural Residues

Agricultural residues include woody crop residues, cereal remains and green crop residues as well as crop processing residues. Dry residues are best for use as fuel, whereas green residues are better as animal fodder or manure. The latter can also be used as feedstock in biogas plants. Residues are widely used as domestic fuel in developing countries. In certain areas of China and Bangladesh, residues provide as much as two-thirds of household energy.

The wide variety of agricultural residues implies a proportionate variety of potential environmental effects. The removal of residues may have an impact by intensifying soil erosion and rapid water run-off. Also, the combustion of the residues themselves may result in specific hazards.

Increased removal of residues from the soil is usually strongly correlated with declining soil fertility and intensified erosion. The enhanced erosion, in turn, results in the loss of soil nutrients, water-holding capacity, soil organic matter and soil biota. In addition, residue removal increases the rate of water run off sometime upto even hundredfold.

Crop Residues

Crop residues are plant materials left in the field or in agro-based industries after harvesting the main crop produce. The materials occur in the form of straw, stalks, husk, leaves, fibres, roots and other parts of plants. Reliable data are not available about the quantity of crop residues used either as fuel or as fodder but a rough estimate for non-fodder crop residues for the year 1991 was about 130 Mt, nearly half of which was used up in the sugar industry.

Dungcakes

The total production of dung from an estimated bovine population of about 275 million is around 1000 million tonnes per annum on a wet weight basis (dry dung around 200 Mt). About 100 million tonnes of dung is assumed to be used for fuel purposes.

Biogas

Most wet biomass can be converted into a combustible gas by anaerobic fermentation in a biogas digester. Typically, animal or human wastes or wet plant material are collected, mixed with water and fed into an airtight container. The gas which is produced can be collected and the resulting sludge can be used as an excellent, high-quality fertilizer. Biogas digesters come in various sizes from small, single family units up to large, industrial scale facilities. The fermentation destroys some harmful bacteria and pathogens present in animal and human wastes. The technology of biogas, therefore, is generally an environmentally sound way of managing animal and human wastes, which otherwise, would either present important health hazards or whose disposal would by quite costly. Emissions from the combustion of biogas are essentially harmless, though the small amounts of hydrogen sulphide present may cause eye irritation. Rarely does an explosion of the biogas occur in critical mixtures with air.

Producer Gas

Producer gas is generated by pyrolysis or the partial combustion of biomass, such as wood or various crop residues. The main types of gasifier include up-draught, down-draught, cross-draught and fluidized-bed gasifiers. Their environmental impacts are fairly similar.

The gas can be utilized either for direct combustion in a boiler or for operation in an internal combustion engine. Both of these, in large and small sizes, can operate using producer gas. The biomass feedstocks can be obtained directly from an energy plantation, or from commercial or non-commercial biomass markets.

The supply of biomass feedstocks to the gasifier process will have different environmental impacts, depending on the actual source of the biomass. Following gasification, the gas mixture will have carbon monoxide, hydrogen, carbon dioxide, water vapour, methane, nitrogen and traces of condensable organic vapour. The gas, if mixed with air, can cause explosion. The high content of carbon monoxide makes the gas quite toxic, and chronic exposures may lead to chronic health problems or even to death but, suitable design and safety devices can reduce the risks considerably.

Combustion of producer gas mainly yields carbon dioxide and water vapour, with some nitrous oxide, which causes acidification and health effects analogous to those of fossil fuels.

Alcohol Fuels (Ethanol)

Ethanol can be produced from different sugary and starchy plants. The main processes are based on sugarcane (as in Brazil) or on corn (as in the USA). The basic processes involve fermentation and distillation, which require large energy inputs. The alcohol produced may be used either neat or in various blends with gasoline (gasohol).

Raising of the feedstock needed for alcohol production has much environmental impacts analogous to those of commercial agriculture or production of other energy crops.

Production of alcohol fuels entails two chief environmental impacts. In the distillation considerable energy—either from fossil fuels or from sugarcane—is needed. Several air pollution related problems arise and need to be controlled. The stillage waste (a by-product) has very high biological and chemical oxygen demands. Dumping this into local water bodies has caused serious environmental problems. It is better to recycle this stillage as a fertilizer in the sugarcane fields.

Modern Wood Fuel Resources

The large-scale harvesting of trees specifically for energy purposes has been increasing in many parts of the world. In Brazil, steel and cement industries obtain their charcoal from energy forests particularly planted for that purpose.

The effects of short rotation forestry mostly depend on previous land use. If plantations of carefully-chosen species are made on poor soils or abandoned arable land, the forest gradually causes soil *improvement*, and also proves good for wildlife. In contrast, when an existing natural forest is cleared to make space for fast-growing new species, the effect can be negative since biodiversity is reduced and rare species are threatened. When burnt, biomass fuels cause various forms of air pollution broadly similar to but less severe than that caused by fossil fuels. Indoor air pollution caused by unvented fires is a most critical local impact. It seems that the burning of wood is less hazardous for health than dirty, low-quality coal, but more hazardous than clean, high-quality coal. Another important local effect is the damage to man-made structures through various constituents of photochemical smog. One of the causes of this is sulphur: coal has 1-5%. Dung and other residues can contain even more sulphur. Since sulphur causes acidification, substitution of fossil fuels by biofuels reduces this problem considerably.

Climate change, induced by various greenhouse gases, e.g. CO_2, is the most important global effect of air pollution. For a given production of energy, coal produces somewhat more of CO_2 while oil produces somewhat less CO_2 than wood. If biomass combustion is accompanied by an equal amount of regrowth or replanting, the overall effect of the wood cycle on the CO_2 balance of the atmosphere would be negligible. Also, large biomass plantations serve as additional sinks for the CO_2 produced by the combustion of fossil fuels.

RATIONAL USE OF ENERGY IN DEVELOPING COUNTRIES

There is now a great demand for energy-saving concepts in the face of the impending climatic changes. The need of the hour is to drastically reduce carbon dioxide emissions. This can be achieved by more efficient use of fossil fuels. Rational-use-of-energy (RUE) is also economically desirable. Simple efficiency measures in the industrial sector in some developing countries have given encouraging results. RUE constitutes a comprehensive tool in environmental protection and conservation of natural resources, in improving energy and raw materials productivity in industry and hence in raising productivity.

Energy-saving does not just mean cutting down fossil fuels. It also means reducing emissions from industry and other uses like building and households. Consequently, the often severe environmental pollution from large complexes can be reduced leading to a decrease in the greenhouse effect. The rational use of fossil fuels is a core area in efforts to reduce environmental pollution.

The energy saving can be as much as 15 to 25%, depending on the type of industry. Modern, efficient technologies may even achieve 75% energy saving in some cases. An economically-feasible reduction potential of 20% of the industrial sector in several developing countries can result in savings of at least 150 million toe or about 470 million tons of CO_2. Considering the energy mix—the combined use of conventional and renewable sources of energy—it seems feasible to reduce energy costs in developing countries by a few billion dollars.

Sustainable development means producing goods differently and more efficiently, not producing less. The following factors can aid minimize environmental pollution from industrial production:

1. The knowledge and capability to produce on an environmentally sound, resource-saving and cost-effective basis;
2. Improving the efficiency of the input factor;

3. Substituting pollugenic raw materials and production methods by renewable, environmentally sound, clean technologies;
4. Modernizing of inefficient, pollutant production systems;
5. Structural change on an inter- and intrasectoral basis.

Transition to alternative energy technologies or better production plants can give the greatest savings (20-80%): combined heat- and power generation plants, process, re-engineering and production changes (see Schutt, 1996).

In some ways, the mid-1990s were a dark period for the world energy system. Oil consumption reached the record levels of the late 1970s, with demand in some countries growing at rates as high as 10 percent per year. Even the use of coal has increased in many countries, pushing emissions of carbon dioxide to more than 6 billion tons annually. Emissions are increasing particularly rapidly in China and India. The United States and Canada have not been able to hold carbon dioxide emissions steady as per the recommendates of the Rio climate convention. Today, three major forces of change bear down on the world energy economy—new technologies, industry restructuring, and stricter environmental policies—all of these are likely to be intensified by incipient climate change.

The conditions for successful RUE policies have recently improved in many countries: Liberalizing imports has raised competitiveness. Companies can no longer just transfer their higher energy costs to the customer but are forced to adopt suitable measures to keep prices low and competitive.

More stringent environmental policies also force change. Competitiveness, higher energy prices and environmental protection considerations are all conductive to RUE programmes in several countries.

Industries are likely to invest in energy saving in the face of growing domestic markets, greater competition, higher environmental demands, cost-oriented energy prices and functioning capital markets. Competitive industry is of vital importance because on their own the other factors could become detrimental to industrial production.

RUE is one of the most effective instruments available today to rapidly and significantly lower local, regional and global emissions from industry (see Schutt, 1996).

In the face of growing shortage of natural resources and continuing population growth, we must change our energy consumption habits.

A new culture of handling energy needs to be ushered in. But till then, renewable energies and rational use of energy will remain the most promising instruments in exploiting the potential to reduce energy consumption (Schutt, 1996).

CO-GENERATION POWER PLANTS AND EFFICIENT ENERGY UTILIZATION

Combined heat and power generation using co-generation power plants is now an established technology in some European countries. Over 100 such plants have been installed in Germany with an electrical output between 100 KW and 10 MW.

In terms of energy, it is not sensible to generate electricity from heat, which entails high losses, and then to transform it back into heat again. Generating electricity from power stations which are usually operated by municipal companies and local industry means lower turnover for large electricity generating companies. From a macro-economic viewpoint, combined power and heat generation is quite cost-effective and environmentally-sound.

Co-generation power plants match heat and power needs. Renewable energy sources (wind, solar radiation, biomass, water power) can be integrated into the concept of combined heat and power generation.

The operating management company aims to provide a reliable supply of heat, power and cooling using low energy inputs, causing low environmental pollution on an economically viable basis, while simultaneously bearing all the operating and investment risks.

This technology is being used in the commercial and industrial sectors and has potential applications in the following groups on the basis of their different heat and power characteristics:

1. Industrial companies
2. Commercial companies
3. Public buildings and utilities/hospitals
4. Heat supply to housing/local heating

Industrial Companies

Several kinds of industries require power and heat simultaneously—an ideal set-up for the use of co-generation power plants. Providing process steam for technological purposes is a typical task for the co-generation power plant. Process heat needed in summer usually requires some parallel cooling capacity. These requirements are met with by absorption coolers driven by heat from co-generation power plants. A gas-turbine plant with waste-heat boiler may be installed to provide an independent power supply for a paper mill. The hot exhaust gases from the turbines generate the process steam in the waste-heat boiler that is needed for paper manufacture.

The gas turbines can continue power generation when the local mains power supply fails.

Public Utilities

As public buildings and utilities must be equipped with emergency power supply systems, co-generation power plants provide a low-cost option when the daily demand for power and heat occurs simultaneously. The base load to be covered by the co-generation power plant has to be high in relation to the total demand.

The co-generation power plant takes over the function of the emergency power supply, at the same time generating energy-saving and environmentally-friendly power and heat (steam and hot water) for a hospital's requirements, in a co-generation process.

ENERGY WASTAGE IN THE STEAM BOILER

The steam boiler is undoubtedly a most widely used energy technology. Combustion of oil, coal, natural gas and biomass generates steam which is used as heat in several manufacturing processes. Small, medium and large-scale industries use thousands of steam boilers of various sizes. A medium-sized boiler (15 tons steam per hour) uses one ton of heating fuel per hour and emits pollutants such as: carbon monoxide 2.7 tons; sulphur dioxide, 50 kg; nitrogen oxides, 10 kg; dust and soot, 1.3 kg; and hydrocarbons, 0.4 kg.

In-company organizational measures alone have the potential to save from 15 to 50% energy in individual cases. At 4000 operating hours, one medium-sized steam boiler could save 600 tons of oil per year with corresponding reductions in emissions (1620 tons CO_2). The company would reduce its energy costs by thousands of dollars. Improved maintenance and regular control of old equipment could also save about 10% energy. The use of modern machinery converting to natural gas can reduce SO_2 and dust emissions by up to 90% (Schutt, 1996).

PHOTOVOLTAIC SYSTEMS

Decentralized rural photovoltaic power supply can become an environmentally friendly option for developing countries.

Many villagers in developing countries do not have access to commercial energy supplies and their energy consumption is mostly fuelwood and charcoal use for cooking purposes.

Cheap biomass-based energy prohibits substituting this source by fossil fuels or electricity and would also not be a rational-use-of-energy strategy.

In areas of low user/consumption density, grid extensions and diesel-powered village grids for household electrification are economically unfeasible or unaffordable.

PV-technology has emerged as a genuine alternative to conventional grid-based electricity for rural areas. Compared to other renewable enrgy sources, solar radiation has a low regional variability. The most important comparative advantage of PV-systems over conventional systems is their adaptability for small consumers, facilitating decentralized basic electrification concepts for dispersed settlements with low consumption densities. Individual PV-systems are modular and can be stacked incrementally to match the user's demand, making the energy costs proportional to the capacity installed.

Introducing selected PV-applications on the commercial market is a feasible option in developing countries, but the lack of product know how and market transparency tend to prevent the development of existing market potentials.

In the context of any major rural household electrification programmes, three different technical concepts requiring different approaches need to be considered, viz.,

1. The central PV-powered village grid.
2. The individual solar home system.
3. PV charging stations for batteries (see Posorski, 1996).

Solar Home Systems (SHS) can satisfy the typical electricity demands of rural households and open up a new decentralized approach to rural household electrification. SHS are technically mature, but the electronic devices (charge regulator and ballasts) must be of high quality because they determine the lifetime of the battery and fluorescent lamps (Posorski, 1996).

Rural Drinking Water Supply

Photovoltaic driven pumps (PVP) are now becoming popular in Argentina, Brazil, Indonesia and Philippines.

Photovoltaic water pumps can be put to good use for rural drinking water supplies. They do not require much maintenance and provide a good alternative to diesel driven pumps.

In many villages, water is usually drawn by hand pumps and maintenance-intensive diesel-driven pumps. Due to breakdowns and lack of fuel, the pumps often remain out of order. By contrast, photovoltaic pumping systems require no fuel and are often a very reliable solution to the problem of drinking water supply.

PVP Standard System

A photovoltaic generator provides the electricity needed for operating a submersible motor-pump unit. An inverter converts the direct current provided by the PV generator into the requisite three-phase alternative current and serves as the system controller (Fig. 4.13). As in any PV system some energy storage is required for night-time operations and to bridge periods of low sunlight.

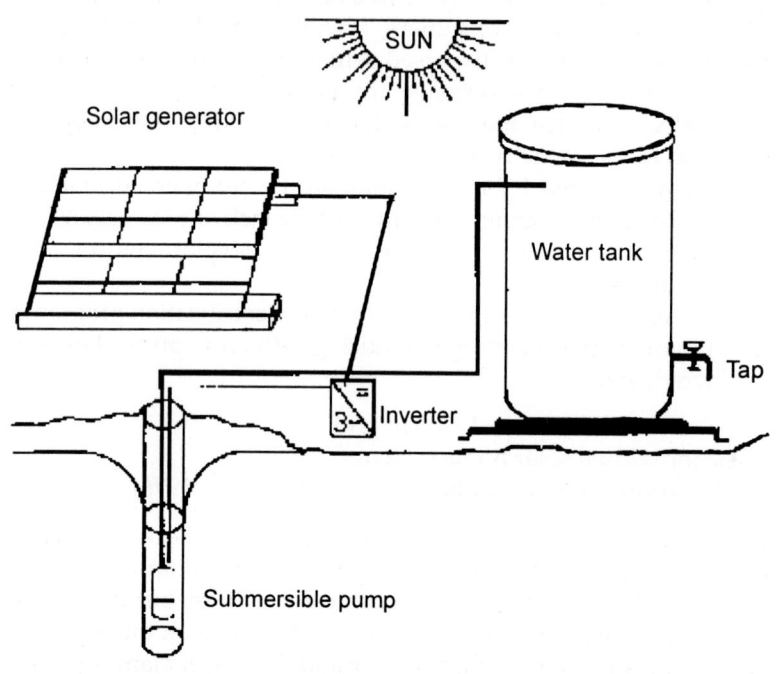

Fig. 4.13 The working of a photovoltaic pumping system.

In PV pumping systems, the delivered water is stored in a high level reservoir which feeds the water by gravity to public water taps or for watering points for cattle.

For pumping heads of 30 m and daily insolation levels above 5 kWh/m^2, a typical, medium-size standard system with an installed generating capacity of 1.6 kWh can deliver some 30 m^3 per day. That amount of water is sufficient for communities with populations up to 1200 (Hahn, 1996).

In the PV standard pumping system, the inverter has been the weakest link but now some improvements are being made in its design and operation. All inverters should have an effective lightning barrier. Within the scope of the PVP Programme, many inverter failures were caused by lightning. In particular, the level switch terminal needs to be protected because the levels switch cable acts like an aerial, especially at long distances, between inverter and water storage tank.

On a clear day, a well designed PVP plant can achieve an overall efficiency of about 4%. The use of so-called tandem systems enhances the system efficiency. The special feature of this system is that it has a second pump in the well and an electronic control system that either divides the power from the PV generator between both pumps, or supplies it all to one of them, depending on how much power is available. Consequently, this arrangement supplies more water on overcast days than a comparable standard system could.

The popularity and commercial diffusion of PV products in developing countries depends upon good maintenance and availability of spare parts.

PV systems have already been commercially disseminated in some developing countries, e.g. power supplies for relay stations and remote radio stations. But these conditions do not apply to most potential PV-users in rural areas.

LOCAL AND REGIONAL ENERGY AND ENVIRONMENTAL ISSUES

Local and regional environmental effects associated with energy, such as air pollution and land degradation, are already presenting serious problems both in developing and industrialized countries. With the projected large growth in energy consumption in the developing world, these local and regional effects will cause increasing damage to ecosystems, agricultural land and crops, and human health, in the coming years.

Local, regional and global environmental effects cannot be seen in isolation. Often the same root causes contribute to environmental degradation at the different levels. Thus, addressing the local environmental threat can help alleviate the global problem.

Local and regional energy related environmental issues involve some of the major environmental problem areas such as urban air pollution, indoor air pollution, disturbance and occupation of land, and electromagnetic fields. Some areas such as coal, oil and transport have multiple environmental effects. There is no doubt that local and regional environmental issues should be considered along with the global ones. Indeed, the immediate policy priorities focus on local and regional issues, especially for developing countries and countries in transition. In recent years there has been a general willingness to see problems in a more integrated manner and to expand interaction and collaboration aimed at the transfer of technological and financial resources.

One of the chief sources of GHG emissions in developing countries is the burning of biomass. Combustion of biomass releases the greenhouse gas CO_2 to the atmosphere, but this does not necessarily mean a net release of CO_2. The carbon that is lost to the atmosphere in the CO_2 may be returned by subsequent regrowth of vegetation. There is, however, a net emission of other greenhouse gases such as methane and nitrous oxide. Other trace gases are also emitted, resulting in acidification and indirect greenhouse effect. Besides, the total amount of smoke particles (aerosols) emitted is of the same magnitude as that of sulphate particles arising from the global SO_2 emissions of fossil fuel burning. These smoke particles influence the radiative properties of clouds and the Earth's radiation balance, and may also disturb the hydrological cycle in the tropics.

Savanna burning is one of the largest biomass sources of atmospheric pollutants. Globally, the savannahs cover around 1800 million hectares and it is estimated that about 3700 million tonnes of dry matter are burned annually. This is more than 40% of the total amount of biomass burned annually and about three times larger than the amount of biomass burned annually in tropical forests. The emissions of CH_4, N_2O and other gases are perhaps proportional to the emission of carbon from the biomass burning. Today, the best estimate of annual global emissions from savanna burning is 8.2 Mt CH_4 and 0.1 Mt N_2O. Almost 95% of these emissions originate from developing countries.

Grassland, bush and woodland fires play an important role in tropical agroecosystems. They are used for diverse purposes and have intended as well as unintended effects on ecosystems properties and evolution. Changes in the extent, timing and physical properties of fires, partly correlated with population growth and other elements of social change, have an important bearing on human livelihoods, production systems and biodiversity at both local and regional levels.

More information and data are needed on the effects of fire on plant nutrient uptake, ecology and diversity of higher plants and soil invertebrates and also on soil microbiological processes, plant nutrient release and emission of methane and nitrogen oxides. There is also need for macro-scale analysis of the distribution in time and space of fire, of emissions of greenhouse gases and of natural and human controls of fire distribution.

COAL INDUSTRY

Opencast mines, by and large, are fully mechanized and current capacities can be as high as 10 Mt per annum. Substantial increases in coal output in the future are expected to come from opencast mining. This can create problems of land availability and environmental degradation.

The introduction of large-scale heavy earth-moving machinery (HEMM) and adaptation of the technology has been quite satisfactory in opencast mines but the low-capacity utilization of equipment has to be tackled if productivity is to be increased. There has been no increase in production from the underground mines in the last 2 decades.

Productivity of the underground mines is very low due to surplus manpower and lack of mechanization.

The average quality of coal produced has declined over the years. Today, more of inferior coal is being produced rather than superior coal. The quality has deteriorated due to mining of thick inter-banded coal seams by opencast technology where shale and stone become mixed up with the coal due to deployment of heavy equipments. Selective mining of coal seams is not done because of the high cost of operations.

Since the quantities of coal that are projected to be consumed in the future would be substantial, it is imperative that clean coal technologies are adopted not only to improve the quality of coal being supplied but also to improve its efficiency of utilization. With regard to utilization, the strategy should be twofold: (1) Upgrading existing technologies and retrofitting of existing equipment and devices. On the question of retrofits, the problem is largely one of creating institutional arrangements that would bring about investments in technology upgradation. There are over 10,000 boilers which need to be replaced or retrofitted with efficient combustion systems and which can potentially result in coal savings to the extent of at least 5 million tonnes annually. (2) There is need for a more pro-active approach, which would result in evolution of coal technologies that are also environmentally benign. The various components of the clean coal technology programme would comprise coal beneficiation, coal gasification, pressurized fluidised bed combustion and integrated gasification combined cycle.

The environmental impacts of coal mining may be regarded as a chain reaction where the initial effect of coal extraction produces other impacts which pass through the reaction chain like a series of impulses. For example, the initial effect of opencast mining is the removal and deposition of waste overburden material causing land degradation at the surface. Damage to land adversely affects water courses, vegetation, animal communities and man-made structures. The intensity and impact of the environmental degradation depends on the method of mining and beneficiation, scale and concentration of mining activity, geological and geomorphological setting of the area, nature of deposits, land-use pattern before the commencement of the mining operations, and the natural resources existing in the area. Opencast mining operations pollute the environment much more severely than does underground mining. Some of the potential impacts of mining projects on the environment include the following: disturbance of land surface due to excavation, stacking of waste dumps, tailing ponds, haul roads; disturbance of the ecosystem and its flora, fauna and aquatic life, destruction of water sources; disruption in water regime and drainage system; pollution of land and water due to erosion from waste dumps, tailing ponds and wash-off from workshops; silting of water stream and water reservoirs; dust pollution due to blasting, excavation, stockpiles and tailing ponds; noise pollution and ground vibration due to blasting, movement of heavy earth-moving machinery, crushers and beneficiation plants, and destruction of aesthetic and recreational values of landscape.

According to the Worldwatch Institute, a global phase-out of coal is both necessary and feasible.

Coal's share of world energy peaked at 62% in 1910 and has already declined to 23%. While coal's market price is at an historic low, its enviromental and health costs have never been higher. Drastic curtailment of coal use at the earliest possible will be essential to slow down climate change during the next century. Coal is the most carbon-intensive of fossil fuels and it accounts for 43% of the world's annual carbon emissions.

Two main polluting ingredients of coal smoke are particulates and sulphur dioxide, which cause 500,000 premature deaths and millions of new respiratory illnesses each year in urban areas the world over. In rural areas, coal smoke from cooking accounts for as many as 1.8 million deaths annually.

One effective measure towards 'de-coalizing' is to reduce the subsidies that are encouraging its use in several countries. Happily, China has more than halved its coal subsidy rates since 1984—a move that contributed to a 5.2% drop in Chinese coal consumption in 1998. There is great potential for reducing coal use in other countries by eliminating subsidies.

Land Degradation

The most serious impact of mining operations is undoubtedly on land-use pattern in the area. Earlier, most of the coal mines in India were underground mines generally restricted to Jharia and Raniganj coalfields. However, after nationalization, coal production has been increasing rapidly and in the last two decades, the incremental production has come from opencast mines. The emphasis on opencast mines has led to degradation of land and displacement of people.

Figure 4.14 shows the world stocks of coal (and other fossil energy sources).

ENERGY DEMANDS

As emissions from energy are the single largest contributor to global warming, a study of the future trends in energy consumption, and the potential for emissions reduction, is crucial to any attempt to minimize global warming.

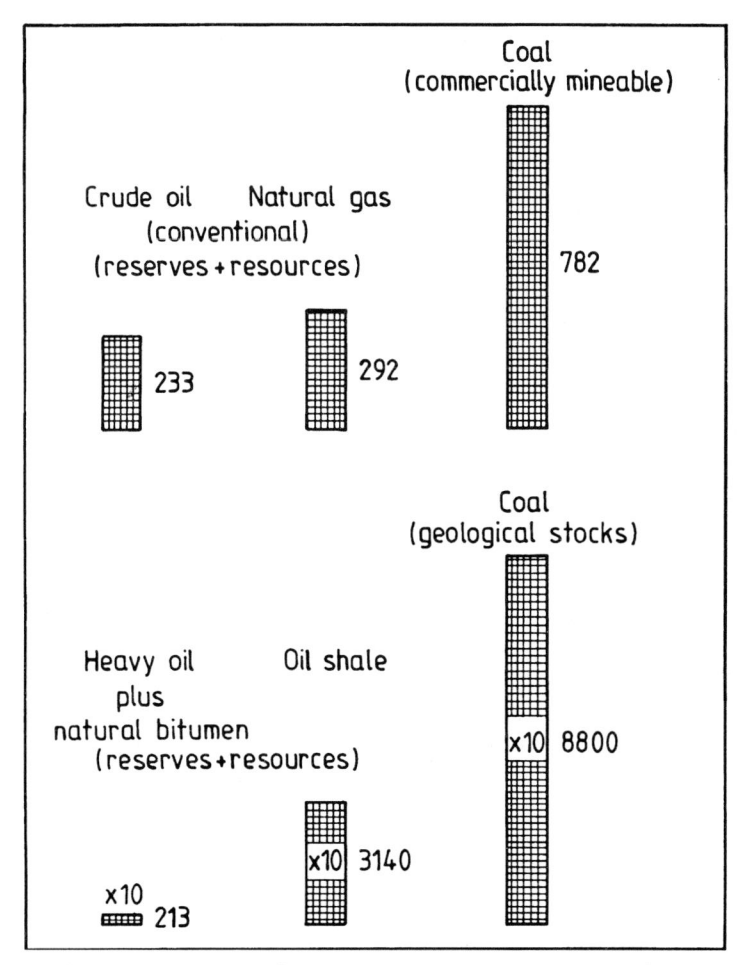

Fig. 4.14 World stocks of various fossil energy sources (incl. coal) (1 kg SKE = 7000 kcal = 8.14 kWh = 29308 kJ) (after Hiller, 1991).

Energy-related activities contribute both directly and indirectly to the generation of carbon dioxide (CO_2) and other potent greenhouse gases (GHGs). CO_2 emissions from fossil fuel combustion account for about one half of the radiative forcing caused by GHGs. Methane emissions from natural gas leaks and coal mines are also significant.

The greatest possibility for reducing emissions is through improving energy efficiency. This will require earnest efforts on the part of consumers, manufacturers, energy-supply companies and governments. Also, developing country analysts should understand and pursue end-use approaches which clearly delineate the technical and attainable potential that energy efficiency offers to reduce emissions (Sathaye, 1995).

The differences in incomes and urbanization levels among industrialized countries are usually relatively small as compared to those in the much more diverse group of developing countries, whose high diversity means that the mix of industrial output and transport structure and urban and rural dwellers influences energy consumption differently in different countries. Further, government policies that affect energy consumption also differ from country to country. Therefore, country or region-specific analysis of human activities and their linkages to energy consumption is crucial for proper interpretation of historical trends and for developing sensible scenarios of energy demand.

Nevertheless, despite the diversity of energy use in various developing countries, their commonality of activities does allow classification of their energy demand; each class of activities uses energy technologies which are also common across countries.

The End-use Approach

Useful energy means the amount of energy necessary for accomplishing an end-use. The energy provided to the end-users is termed the final or delivered energy. The delivered energy required for cooking, for example, will vary with the fuel and cook stove to satisfy the same useful energy demand. Since fuel wood stoves are less efficient, more wood is needed as compared to kerosene to cook the same amount of food.

The provision of the delivered energy involves transformation of primary energy sources such as coal, oil and natural gas. During the transformation and transportation, certain amount of energy is lost.

Five important sectors, viz., industry, transportation, residential, commercial and agriculture, can be studied for analyzing energy demand. Measures of energy intensity (energy consumption per unit of activity) for end-use within each sector help in understanding the changes in structure of that sector and other factors that influence its energy intensity.

Sathaye (1995) defined the end-use approach by the following equation:

$$\text{Energy Consumption} = \Sigma_n \text{ Activity Level} * \text{Energy Intensity}$$

$$\text{Energy Intensity} = \frac{\text{Energy Consumption}}{\text{Activity Level}}$$

This equation states that energy use in a sector or economy is the sum of activity times the energy intensity of each activity.

Activity defines the purpose for which energy is used. Thus, e.g. in transport, it can be the passenger kilometers travelled by car; for cement industry, it may be the physical output of the industry.

Structure denotes the mix of activities in each sector. In industry, it can be the percentage of value added contributed to by its components — manufacturing, construction, mining and utilities. Structural change is one of the two factors that can lead to changes in energy use, the other factor being the intensity of energy use.

Energy Intensity refers to the energy used to perform a particular activity.

Table 4.4 gives examples of the types of indicators used for analysis of various end-use sectors.

Table 4.4 Some examples of indicators used for analyzing various end-use sectors (after Sathaye, 1995)

End-Use Sector	Indicators
General	Population (urban and rural); Fuel price, income distribution
Residential	Urban and rural population and number of households; Level of electrification
Heating	Fuel use per household per day
Cooking	Fuel use per capita per meal
Refrigerators	Unit electricity use per month
Transportation	Ton km by mode, distance travelled, number of vehicles per capita
Industry	Physical output and value added
Energy-Intensive (Steel, Cement)	Fuel and electricity use per ton of output
Non-Energy-Intensive	Fuel and electricity use per value added
Mining, Commercial	Fuel and electricity use per value added
Agriculture	Fuel and electricity use per value added

The end-use approach allows the estimation of long-term energy demand and supply and the consequent GHG emissions. It is fairly simple and transparent but is data-intensive. The approach involves separating energy demand according to its major end uses. Energy use is linked to the service provided by an end-use. The approach begins with the most detailed level of end-use for which data are available and combines the detailed energy use into a more aggregated energy projection at the sectoral level. The sectoral energy use is added to yield the total energy demand by fuel type. The supply of primary energy is estimated on the basis of the demand for individual fuels and electricity for each major end use in each economic sector (Sathaye, 1995).

For each end-use, the future energy demand depends on a series of driving forces. For example, factors such as distances travelled per car, fuel use per km (fuel intensity) and levels of car ownership determine the energy demand for automobiles. The availability of fuels and their relative prices as influenced by government policies determine the choice of fuels.

A FLAME-FREE FUTURE

In the past millennium, the progression of fuel usage has been from wood to coal to oil. Today, methane gas appears to be the preferred clean fuel of the major electricity generators. But we are moving towards hydrogen as the ultimate clean power source of the future, with fuel cells as the electrochemical conversion devices, producing electricity and heat at the point of need from hydrogen delivered through pipes (Kendall, 2000). As most hydrogen is still derived from hydrocarbons, making it expensive, it is not storable and prone to explosion. These drawbacks are slowing the advent of the hydrogen economy. In the meantime, is it possible to use the hydrogen stored naturally in hydrocarbons, such as propane/butane (C_3H_8/C_4H_{10} or camping gas), to produce clean power? Nature achieves this easily enough through biochemical routes. Park et al., (2000) described a fuel cell that mimics this trick of nature (Fig. 4.15).

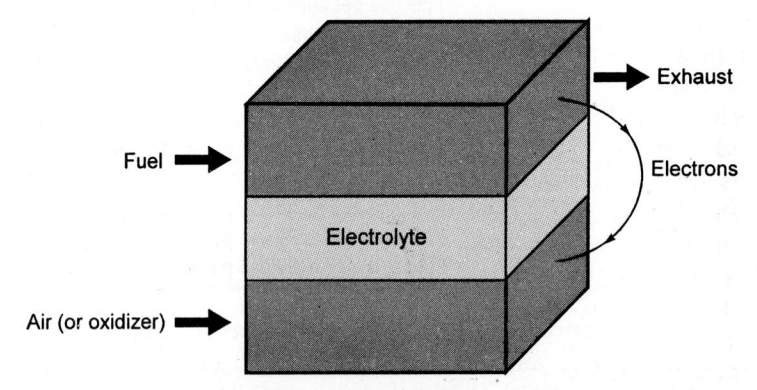

Fig. 4.15 A fuel cell that directly oxidizes the hydrogen stored in hydrocarbons found in several fuels (after Kendall, 2000).

The electrochemical processes involved in hydrocarbon oxidation reactions are not understood. Although humans oxidize about eight gigatonnes of hydrocarbon fuels annually, most of this is burnt in crude, dirty and wasteful flame processes in engines and burners. These processes may well be banned in the near future. Even today, the trend is apparent. Smoking is frowned upon; fires in forests are not permitted; dirty vehicles are penalized; and in some German cities (e.g. Hamburg), a new regulation prescribes limits of ten ppm of nitrogen oxides as the upper level of effluent from fossil fuel burners. This limit is difficult to achieve because ordinary flame burners give off roughly ten times this amount of nitrogen oxides.

Therefore, we should emulate nature and oxidize the hydrocarbon fuel catalytically, while extracting the electrons and useful energy directly through a membrane. Unfortunately, our knowledge and expertise of such processes is extremely meagre even today.

There is now evidence that methane can be converted directly by a solid-oxide fuel cell, without any flame, using cerium oxide as the electrode catalyst. A typical solid-oxide fuel cell uses a hard ceramic material such as ziroconia (which conducts O^{2-} ions) as the electrolyte, and has nickel-based anodes. Park et al., (2000) proposed that copper with cerium or samarium oxide may prove a better anode catalyst for direct oxidation of more complex hydrocarbons that exist in kerosene or diesel. The challenge is to find pathways through which ethane and butane, or even aromatics such as toluene, may be reacted without fouling up the process through damaging side reactions that tend to produce tar or carbon (Kendall, 2000).

A fuel cell works much like a battery. But unlike a battery it does not run down or need recharging, as long as it has a supply of fuel. Hydrogen is the ideal fuel because it reacts with oxygen from the air to produce an electric current and water, but pure hydrogen is expensive and prone to explosion. Park et al., (2000) developed a fuel cell that directly oxidizes the hydrogen stored in hydrocarbons found in regular fuels. They used a porous anode catalyst which mediates the chemical conversion of hydrocarbons without the undesirable carbon formation that usually fouls up the reaction.

Park et al., used a copper catalyst intimately in contact with cerium oxide to produce a porous anode material. When this operated on pure butane for 48 hours, there was no sign of carbon deposition, but considerable evidence of complete chemical conversion and electron transfer through a zirconia electrolyte membrane. When the copper-ceria-samaria catalyst was used, even toluene showed reasonable reactivity and electron conversion. This is encouraging because aromatics are usually prone to rapid graphite deposition, which destroys the catalyst (Kendall, 2000).

Conceivably, even gasoline or diesel fuel may one day be catalytically converted to energy and heat in a clean and powerful manner. It is, however, easier to use partly-oxidized fuels such as methanol or formic acid because these work in aqueous solutions with catalytic electrodes. Such fuels react with oxidizing species at lower temperatures without carbon formation and may be ideal power supplies for computers or mobile phones (Büchi, 1999). However, the costing and economics of the newer devices need to be worked out.

ENERGY AND ENVIRONMENTAL RECOVERY

In Europe, barring a few exceptions, two decades of effort have not been very successful in significant environmental recovery. In the coming years, the environment will face even greater pressure through the growth of sectors such as transportation, manufacturing and other industries, recreation and tourism.

Despite some improvements in terms of more efficient and cleaner technologies in industry, ever-increasing production and consumption will means ever-greater pressure on resources. The problems of emissions and waste will, in turn, become more serious. Economic development has in fact already wiped out some of the successes of environmental policy, such as that from the EU directives on air quality. There is immediate need of integrating environmental policy with public policy as a whole.

No doubt energy is now being used more efficiently but overall use may well increase in Europe by 15% between 1995 and 2010. Travel by car is likely to increase by 30%, and road freight carriage by 50%. All this will make it difficult to achieve the EU's aim of reducing the emissions of climate-affecting gases; instead of decreasing by 8% between 1990 and 2008-12, as anticipated in the Kyoto Protocol, they may actually increase by 6%. It also appears unlikely that the proportion of renewables in the energy mix will be doubled from the present 6% to 12% as proposed by the Commission. These warnings from the European Environmental Agency (EEA), are based on the assumption of business-as-usual, with no fresh legislation for environmental improvement. A few of the positive emerging trends are reduced emissions of substances causing acidification, eutrophication, and depletion of stratospheric ozone. The EEA forecasts, on the other hand, increased emissions in cases that are already difficult to handle, such as climate-affecting gases and waste.

The situation in 2010 is expected to be worse as regards climate and waste. The proportion of ecosystems where depositions are exceeding the critical loads for acidification will, however, continue to drop—from the present 25% to 7% in 2010. Urban air quality is also expected to be better.

Although the general trend has been negative, a number of small, positive developments have occurred in several countries, e.g. the increasing installation of windpower, and the growing use of the bicycle for local transportation in some cities.

ENERGY USE AND POLLUTION IN ASIA

Unless strong measures are taken to control emissions of air pollutants, in the coming few decades, northeast Asia will face a very difficult situation. Even today, the emissions of sulphur dioxide in China, Japan and North and South Korea are fairly high. The critical loads are being exceeded in several parts of these countries. The outlook would be much gloomier were it not for the large quantities of neutralizing dust particles that are blown in from the desert areas in the west.

The situation will become really serious if development activity continues rapidly and nothing is done to curb emissions. The use of energy is rising, particularly in the power generation and transport sectors, as a result of increasing electrification and a steadily increasing number of private cars

According to Streets et al., (1999), under to No Further Controls scenario (NFC), the emissions of sulphur dioxide may increase in northeast Asia from 14.7 million tons in 1990 to over 40 million in 2020 (Table 4.5). If this happens, the acid fallout and the expected concentrations of SO_2 in the air would seriously damage crops, natural ecosystems, and human health. Depositions in the hardest hit areas—parts of South Korea and the Chinese provinces of Sichuan and Jiangsu—are expected to exceed the limits for critical load by 100 to 200 kg per hectare annually.

Streets et al., suggested the following three ways to attack the problem.

1. Best Available Technology (BAT). This would be the most far-reaching solution. All important point sources of emissions would have to install state-of-the-art systems for flue-gas desulphurization, and all other users of fossil fuels must switch to low-sulphur types. This would help cut down emissions from the 14.7 million tons in 1990 to 4.7 million by 2020 (Elvingson, 1999).
2. Advanced Control Technology (ACT). Flue-gas desulphurization (FGD) for all new power plants (but not for existing ones), together with moderate fuel switching in other sectors. This would result in an increase of 40% in SO_2 emissions between 1990 and 2020.
3. Basic Control Technology (BCT). More modest emission-control methods, such as limestone injection in the flue gases to be used in all new power plants in China. The limestone technology achieves only a 50% reduction of SO_2 emissions, as compared with about 95% for FGD. This scenario would mean a 73% increase between 1990 and 2020.

Nitrogen oxides also contribute to acidification, but have not been considered in the above scenarios. Nitrogen oxides take part, too, in the formation of ground-level ozone, whose concentrations would be expected to increase markedly if nothing is done to check emissions.

There is also considerable transboundary movement of air pollutants from west to east.

The costs for improvement in northeast Asia according to the various scenarios are quite high but still they only cover technical measures. There exists a great potential for a more efficient use of energy, for instance, which can reduce emissions at little or no cost (Elvingson, 1999).

Table 4.5 Emissions of sulphur dioxide in Southeast Asia in 1990 and 2020 according to the various scenarios. Million tons of SO_2 (after Elvingson 1999; Streets et al., 1999)

Scenario	Year	NE China	Japan	N. Korea	S. Korea	Total
	1990	11.9	0.8	1.7	0.3	14.7
NFC	2020	32.5	1.1	5.5	1.4	40.5
BCT	2020	22.3	1.0	1.5	0.7	25.5
ACT	2020	17.4	1.0	1.5	0.7	20.7
BAT	2020	3.7	0.4	0.6	0.1	4.7

AIR POLLUTION CONTROL

In some countries, the levels of air pollution started falling some years ago but contamination of the atmosphere still remains a major problem. In 1994, air emissions of chemical pollutants in the United States were as high as 706,000 metric tonnes — a heavy burden upon the atmosphere, but in general, there has been a steady decrease in air emissions in recent years. There is some scope of improvement through application of new control technologies. In the world as a whole the problem may well be at least twice that of the US emissions. In this context, some uranium-oxide-based catalysts seem promising as they offer a potential treatment technology that can destroy volatile organic compounds in the gas phase at much lower temperatures and costs, and with greater destruction efficiencies, than can other catalysts or high-temperature thermal oxidation (Cooper and Holloway, 1996).

Emissions of air pollutants are of two kinds: those emanating from non-point sources such as losses during a manufacturing process other than through a smoke-stack and controlled emissions from the stack as a point source of pollution. Of the 520 million kg of industrial stack emissions in the United States, about 70% were made of organic compounds amenable to destruction using catalytic processes. Uranium oxide based catalysts would be applicable in the clean-up of stack emissions, and the comparatively low temperature (350°C) necessary for destruction would lead to lower energy costs.

These catalysts may also have some applications in areas other than stack emissions. For instance, cleaning up of contaminated land involves the use of soil vapour extraction (rather like a large vacuum cleaner) to remove the highly volatile compounds from soils or other contaminated media. To treat completely, a second process is needed to destroy the compounds that result from vapour extraction. Thermal oxidation has commonly been resorted to, but the uranium-based catalysts are probably much cheaper.

One possible pitfall in commercialization is the creation of toxic by-products during the catalytic process, but in the case of uranium-based catalysts, no reaction by-products seem to be formed for most of the pollutants examined.

It is not possible to use conventional three-way catalyzers to reduce the emissions of nitrogen oxides from diesel engines. Some alternatives are being tested for this purpose, however. One of these is the EGR (Exhaust Gas Recirculation) system devised by the Swedish company, STT Engineering AB. This system is claimed to be able to halve the emissions of nitrogen oxides from heavy vehicles—it can limit the emissions to 3.2 g NO_x/kWh. In this system, the flow of oxygen to the engine is reduced by returning some of the exhaust gas, so obtaining an ideal mixture for combustion. To prevent damage to the engine, the return exhaust gases must be cleaned by passing them through a particle filter. The system comes under the low-pessure category because the return gases are mixed with the intake air before the turbo unit.

The above system can be fitted to existing vehicles together with a particle filter and an oxidizing catalyzer. The particle filter not only enables the system to function, but also reduces the emissions of particles by over 90 per cent (Kageson, 1999).

TRANSPORTATION

A variety of forces affect the transport sector. These include such issues as environmental degradation, sustainable development, lifestyles, the automobile's inherent power of attraction, urban and regional development and vehicle technology. Till hitherto, no serious efforts have been made to create an environmentally sustainable transport system.

The environmental problems of the transport sector cannot be solved through technology alone. Emission-free automobiles are desirable and so also the development of suitable portable energy cells, using renewable fuel sources.

Figure 4.16 shows the social costs of transportation.

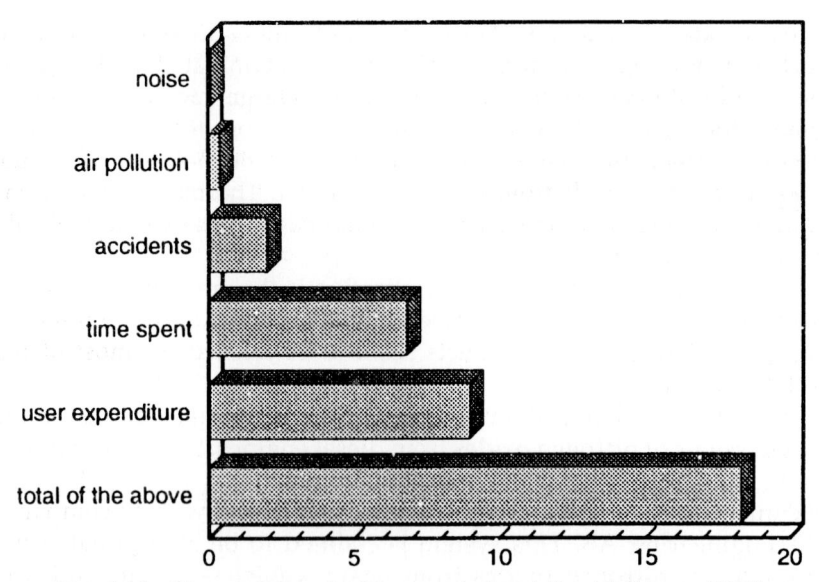

Fig. 4.16 The social cost of transportation (after Kullinger, 1994).

There is urgent need to make transportation more compatible with sustainable human development and proper management of natural resources.

In general, whereas wealth allows us to evade contact with the environmental hazards we create, poverty means that the problems stay close to home. The extravagant lifestyle of wealthy people is the principal source of the world's environmental problems. Large middle-income cities show the most extreme forms of urban environmental distress, and poverty and environmental degradation usually go hand-in-hand.

Diarrhoea, respiratory illness, acid rain, global warming and sustainable development are indeed interconnected. The diseases of urban poverty such as diarrhoea and respiratory illness reflect local environmental inadequacies, while the challenge of urban wealth is global sustainability.

THE ENERGY TRANSITION

A major shift from the use of traditional biomass fuels to 'modern' fossil fuels and electricity appears to be a basic feature of economic growth with its associated urbanization and industrialization. In the poorest developing countries biomass fuels account for 60-95% of total energy use, in middle income countries for 25-60%, and in high-income industrialized countries (with a few exceptions) for less than 5% (Leach and Gowen, 1987).

Leach (1992) discussed the substitution of traditional biomass fuels by modern energy sources in the household sector of developing countries. He demonstrated that this process is strongly dependent on urban size and, within cities, on household income, since the main constraints on the transition are poor access to modern fuels and the high cost of appliances for using them. Relative fuel prices are of lesser importance.

Table 4.6 shows the common urban fuel preferences and constraints.

Table 4.6 Urban household fuel preferences and constraints (after Leach, 1992).

'Ideal' goals:	Clean to use, Delivered to user. No storage. Versatile: e.g. good control of heat output (High efficiency reduces costs).		
Preference 'ladder'	Barriers to climbing the ladder		
	Equipment costs	Fuel payments	Access to fuels
Electricity	Very high	Lumpy	Restricted
Bottled gas/LPG (natural gas)	High	Lumpy	Often restricted, bulky to transport
Kerosene	Medium	Small	Often restricted in low income areas
Charcoal (may be higher in some cultures)	Low	Small	Good: dispersed markets and reliable supplies
Firewood	Low or zero	Small; zero if gathered	Good: dispersed markets and reliable supplies
Crop residues, animal wastes	Low or zero	Small; zero if gathered	Variable: depending on urban size, local crops and livestock holding

According to Leach (1992), minor, low cost interventions, or changes in market conditions can often greatly accelerate the 'natural' pace of the energy transition which occurs during the normal development process as a result of improving infrastructure for distributing modern fuels (the urban size and urbanization effects) and rising incomes to overcome the high costs of buying modern fuel appliances. The transition is occurring rapidly in many urban areas of the Third World but the situation and prospects for rural areas are much more uncertain.

ENERGY MODELLING, ECONOMICS AND ENVIRONMENT

Certain combined energy-economic models can evaluate energy technologies by their ability to reconcile economic growth targets with environmental constraints. This new generation of models makes it possible to evaluate the energy sector quite comprehensively and satisfactorily. For example, the effect of emission restrictions on the 'rebound effect' from energy conservation can be examined (see IEA/ETSAP, 1997).

National results with the combined models indicate that severe emission reductions lead to reduced energy demands, thus relieving the need for drastic technological change. But long-term emission reductions depend more upon new technology than economic policy. The reduction in economic growth due to emission restrictions in itself brings down small emissions only to a small extent, however.

When material flows are added to the energy flows in some of the models, the need for changes in energy technology to meet CO_2 emission restrictions is reduced because many changes in manufacturing materials and recycling can be made at lower cost. When emission restrictions are applied to the complete set of GHGs measured as 'CO_2 equivalents' through their Global Warming Potential, only methane (other than the CFCs already bound by the Montreal Protocol) is found to be important in industrialized economies.

When probabilities are assigned to various future emission reduction scenarios in stochastic programming, the best mix of near-term technologies to hedge against the uncertainty may prove to be different from that most suitable for any one of the assumed scenarios.

More significantly, when a common methodology is used, it becomes possible to make both side-by-side comparisons of national capabilities to reduce future emissions, and integrated multinational analysis to calculate the benefits of national activities implemented jointly. The widespread use of computer models such as MARKAL forms a good basis for assessment of options in an international context, in particular for activities implemented jointly. The value of a consistent methodology among nations needs to be stressed (IEA/ETSAP, 1997).

The United Nations Framework Convention for Climate Change calls for stabilization of GHG concentrations in the atmosphere at a level that would prevent dangerous anthropogenic interference with the climate system. For this stabilization to occur at today's levels, for example, carbon dioxide emissions from human activities will have to be reduced to less than half the current rate.

In the long run, an agreement must be reached on both the implementation of abatement measures and the manner in which costs are to be shared. Today, not all countries are able to evaluate the abatement measures most suitable to their situation. However, neither the ultimate climate stabilization target nor the strategy for achieving that objective needs to be fixed immediately. An important part of the initial response to the threat of global climate change is to build capability where it is needed.

Linking with Economic Approaches

In it simplest form, technology assessment can take the form of a side-by-side comparison that assumes the 'rest of the world' as given and constant. Up to a certain level, these simple comparisons are fairly satisfactory when concerned with isolated problems in the short term. But longer term assessments of a more structural than incidental nature can only be analyzed in more comprehensive frameworks. Mutual inter-dependencies between energy supplying and consuming sectors, or within subsectors thereof can then be accounted for.

By choosing suitable system boundaries, MARKAL has been successfully extended beyond national boundaries to take into account 'upstream' energy-consuming and polluting activities in so-called full fuel cycle analyses, and extended temporally for 'cradle-to-grave' life cycle analysis. MARKAL, pure and simple, has been used to evaluate the cost savings from 'activities implemented jointly' by three countries cooperating in reducing carbon dioxide emissions, and to evaluate the relative importance of reducing emissions of the many different greenhouse gases. However, no single model can provide all the answers. Therefore, the use of both bottom-up energy systems and top-down macroeconomic models is valuable in projecting energy futures.

For GHG gas abatement costing studies, the need for a hybrid approach that combines the essential elements of bottom-up technological models and top-down econometric models has been felt.

The inclusion of general economic aspects has expanded the analytical capabilities available to the participating nations and other users of MARKAL and other similar models.

An improved version, called MARKAL-MACRO, allows estimation of the economic value of energy technologies in a more comprehensive and satisfactory way than was possible with MARKAL alone.

Partial Equilibrium MARKAL

In MARKAL, emission restrictions are met by modifications of the energy system, such as changes in the fuel mix, new energy technology, and energy conservation. In MARKAL-MACRO, the necessary changes in the energy system to meet a specific emission cap are lessened by adjustments made with the rest of the economic system, both by reduced GDP and reduced energy needs. The effect of GDP on emissions is rather small, typically contributing just a few percent of the emission reduction. Therefore, a partial equilibrium model not representing the rest of the economic system but allowing demands to be reduced in response to higher energy prices usually suffices.

The partial equilibrium approach uses the traditional MARKAL linear programming model with the OMNI modeling language. The partial equilibrium approach makes use of the Equivalence Theorem (see Fig. 4.17) drawn from economics:

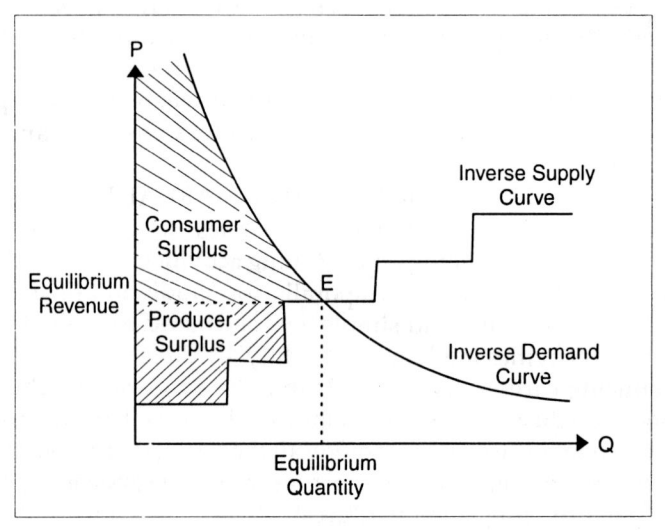

Fig. 4.17 Illustration of a supply-demand equilibrium (after IEA/ETSAP, 1997).

A supply-demand equilibrium is reached when the sum of the producers' and consumers' surpluses is maximized.

This illustration refers to a case where only one commodity is exchanged. Point E, which is the equilibrium point, is the intersection of the inverse supply and the inverse demand curves. ('inverse' because price is shown as a function of quantity, rather than the other way around). Consumer surplus, shown by one hatched area, is the difference between how much a consumer pays and the higher price he would have paid for smaller quantities. Producer surplus, shown by the other hatched area, is the difference between the price the producer received and the lower price he would have accepted at smaller quantities.

Note that at Point E, the area between the two curves is also maximized. The two hatched areas represent the sum of the producers' and the consumers' surpluses, sometimes called the net social surplus, which is a proxy for welfare.

The Equivalence Theorem is valid subject to the supply and demand curves meeting certain economic and mathematical conditions. One condition is that the area under the inverse supply curve is well defined, but that area is simply the value of the MARKAL objective function.

The objective of the MARKAL model is to minimize total energy system cost which is a linear function. The objective of the partial equilibrium model is to maximize net social surplus, which is a nonlinear function. Since the latter objective function is separable, however, it is easily linearized by piecewise linear functions. By so doing, the resulting optimization problem becomes linear again, and it may now be formulated entirely within MARKAL, by defining additional 'dummy' technologies, each representing a portion of the energy demand that is reduced due to its elasticity (see IEA/ETSAP, 1997).

Environmental Implications of Energy and Transport Subsidies

Interest in energy subsidies has increased by the possibility of imposing carbon taxes to reduce emissions of carbon dioxide. But the question arises, whether it is not more logical first to remove any subsidies that encourage the use of fuels that emit carbon dioxide? A subsidy is any intervention, or failure to intervene, that results in the prices of goods or services to producers or consumers, or the quantities produced or consumed, differing from those that would occur in a fully competitive market with all social and environmental costs internalized. Financial subsidies are rather easy to analyze, whereas other economic subsidies cannot be easily analyzed because of difficulty in quantifying their amounts and impacts. A 'net subsidy' may be defined as a financial transfer from the economy as a whole to, for instance, electricity producers and consumers. A 'cross subsidy' is a financial transfer from one subsector of the electricity market to another. Both kinds of subsidy perturb the market and make actual prices different from long-term marginal prices.

In the early 1990s, about 40% of the electricity market was found to be affected by financial subsidies. Net public support of the electric power sector by the economy was in the range of 15-20% of the value of production. Cross subsidies among different consumers and producers amounted to 20-25% of the market (see IEA/ETSAP, 1997).

Removal of subsidies from the power sector can generate the following likely effects:

1. CO_2 emissions decrease by a few per cent;
2. Control of NO_x emissions from stationary sources becomes slightly less expensive;
3. Average efficiency of electric production increases; and
4. Market share of independent producers increases.

The interaction of carbon dioxide emission control policies with the removal of electricity subsidies tends to be synergistic. When applied independently each of them improves the welfare of the system; when applied together, the welfare improvement is greater than the seem of the two independent applications.

Internalizing of External Costs

Full cost pricing means internalizing of external costs, or externalities. Externalities are unintended by-products of such commercial activities as power generation. Examples are impacts on health, the loss of environmental quality or recreational facilities. Externalities influence well-being but their precise quantification is quite difficult. Huge uncertainties exist in the valuation of externalities due to particulates, SO_2, NO_x and CO_2 sometimes by several orders of magnitude, depending primarily upon location and population density.

Adoption of full cost pricing by all or most countries can greatly help to regularize the international trade of electric power to the extent that it would equally reflect the full cost of production in each country. However, whereas full cost pricing is technically feasible, many serious hurdles remain, including the consensus about the acceptability and validity of damage cost estimates. The damage cost values in one country cannot be transferred to another.

Expansion of environmental cost adders explicitly to include carbon dioxide emissions would be an effective way to assure reduction of CO_2 emissions as well as those of the local and regional pollutants (IEA/ETSAP, 1997).

REFERENCES

Anonymous, Sources of energy in the nineties : Reserves, resources and availability. *Natural Resources and Develop.* 47: 7-43 (1998).

Brown, K.S. Bright future—or brief flare—for renewable energy? *Science* 285: 678-680 (1999).

Büchi, F.N. (Ed.) *Portable Fuel Cells Conference.* Bossel, Switzerland (1999).

Cooper, W.J., Holloway, L.A. Recipes for cleaner air. *Nature* 384: 313-314 (1996).

Elvingson, P. Pollution is likely to increase. *Acid News* 3: 12-13 (Oct. 1999).

Foley, G. *Energy Assistance Rivisited — A Discussion Paper.* Stockholm Environment Institute, Stockholm (1991).

Hahn, A. Photovoltaic Pumping Systems: Technically Mature and Reliable. *Gate* No. 2. pp. 21-25 (1996).

Hiller, H. Future combined energy systems using fossil energy sources. *Natural Resources and Develop.* 33: 67-91 (1991).

Holt, R. A responsible energy future. *Science* 285: 662 (1999).

IEA/ETSAP. New directions in energy modeling. Summary of Annex V (1993-1995). Internat. Energy Agency/Energy Technology System Analysis Programme, Report No. ETSAP-97-1, Petten, Netherlands (April, 1997).

Kageson, P. Reducing nitrogen-oxide emissions. *Acid News* 3: 19 (Oct., 1999).
Kendall, K. Hopes for a flame-free future. *Nature* 404: 233-234 (2000).
Kristoferson, L.A., Bokalders, V. *Renewable Energy Technologies: Their Applications in Developing Countries.* Intermediate Technology Publications, London (1988).
Kullinger, B. Transportation: visions and decisions. *SEI Environ. Bull.* (Stockholm) No. 2 pp. 2 (April 1994).
Leach, G. The energy transition. *Energy Policy* (Butterworth Heinemann) pp. 116-123 (Feb. 1992).
Leach, G., Gowen, M. *Household Energy Handbook.* World Bank, Washington D.C. (1987).
Martini, N., Schell, J.S. (Eds.) *Plant Oils as Fuels: Present State of Science and Future Developments.* (Proceedings of the Symposium held in Potsdam, Germany, February 16-18, 1997). Springer Verlag, Berlin (1998).
McNelis, B., Derrick, A., Starr, M. *Solar Powered Electricity: a Survey of Photovoltaic Power in Developing Countries.* Intermediate Technology Publications, London (1988).
OECD (and International Energy Agency). *Energy Policy and Programme of IEA Countries: 1986 Review.* OECD, Paris (1987).
Ogden, J., Kreutz, T., Steinbugler, M. *Technical Paper No. 982500* (Society of Automotive Engineers, Warrendale, PA (1998).
Park, S., Vohs, J.M., Gorte, R.J. *Nature* 404: 265-267 (2000).
Pasztor, J., Kristoferson, L.A. Bioenergy and the envirnment—the challenge. In Pasztor, J., Kristoferson, L.A. (eds.). *Bioenergy and the Environment.* pp. 1-28. Westview Press, Boulder, Colorado (1990).
Posorski, R. From Assistance Projects to the Market. *Gate* No. 2. pp. 14-20 (1996).
Sathaye, J.A. End-use energy modelling for developing countries. Working paper No. 2. UNEP Centre on Energy and Environment, Riso, Denmark (1995).
Schutt, W. Rational use of energy in developing countries. *Gate* No. 2 pp. 4-10 (April-June 1996).
Sørenson, B. Renewables proivde 25 p.c. of global energy. *Update* (Newsletter) No. 46, pp. 4. UN Centre for Sci. and Technol. for Development, UN., New York (Summer, 1991).
Stone, R., Szuromi, P. Powering the next century. *Science* 285: 677 (1999).
Streets, D.G., Carmichael, G.R., Amann, M., Arndt, R.L. Energy consumption and acid deposition in Northeast Asia. *Ambio* 28 (2): (March 1999).
Turner, J.A. A realizable renewable energy future. *Science* 285: 687-689 (1999).
World Bank. *Review of Electricity Tariffs in Developing Countries During the 1980s.* Industry and Energy Department Working Paper, World Bank, Washington DC (1990).

Chapter 5
Industry, Energy and Technology

INTRODUCTION

Many less developed countries face several serious problems in lifting their people above the poverty line. Development, industrialization and technological improvement are necessary for these countries. Properly directed, technological change can lead to a healthier environment as well as a more vibrant economy. There is need to do away with inflexible, analytical and legal requirements that only obstruct the introduction of improved technologies to solve environmental problems. Strong environmental demands encourage innovation, particularly among the most technologically dynamic firms. This is best exemplified by the American General Motors' development of the *Impact*, a high-performance, electrically powered car that embodies innovative propulsion systems and advances in materials science. Environmental considerations played no small part in the development of this model (Banks and Heaton, 1995).

Both regulators and industry, including those firms that pollute and those that produce environmental technology, ought to devise innovations that will reduce environmental hazards as well as the costs of control. This may be done by making enhancement of technological change, the central mission of environmental policy. Technologies and their diffusion may be the most important engine of progress in today's society. Private firms are society's main means of technological change.

Environmental policy should cease discriminating against new technology, thereby prolonging the commercial life of yesterday's products and processes.

Environmental technology is usually considered to comprise products and services developed principally or uniquely for purposes of environmental improvement. But this concept is rather narrow. Why, for instance, should nanotechnology—rarely if ever mentioned in this context—not be considered inherently environmental? Use of nanotechnology can decrease demands on natural systems and increase our ability to control the environmental consequences of production. Likewise, biotechnology can reduce the need for hazardous pesticides and new materials can give more function and have less environmental impact than the old staple inputs of the industrial age (Banks and Heaton, 1995).

Conceptualizing environmental technology as goods and services for uniquely environmental purposes may even be counterproductive. This thinking tends to focus technology development on 'end-of-pipe' remedial approaches that may limit or clean up pollution but do not wish it away. Unless environmental factors become integral design criteria for new technology and unless there is faster movement along new technological trajectories, the environmental consequences of explosive global population and economic growth cannot be overcome (NSTC, 1994). A technological transformation that reduces environmental damage per unit of output is essential to rule out environmental decline. Only those technologies that can co-optimize environmental and economic objectives early enough in the design process can accomplish this objective while preserving economic growth.

The key to sustainable and sound industrial growth is technology that produces little or no waste, coupled with careful management to maximize efficiency and safety. Considerable information on clean technology relevant to specific industries is now available. A computer-based information exchange system—the International Cleaner Production Information Clearing-house (ICPIC)—is operational.

In 1988, in response to devastating industrial accidents in Mexico, Bhopal and Basel, UNEP launched the Awareness and Preparedness for Emergencies at Local Level (APELL) programme. Its aim is to alert communities to industrial hazards and help them to develop emergency response plans.

The choice of appropriate technology is always situation-specific. A useful continuum of technology may be constructed from an environmental perspective. This continuum proceeds from small-scale changes to systemic changes and from technologies requiring short time horizons to those with longer time frames. It consists of the following five categories (Banks and Heaton, 1995).

1. **Pollution control and treatment:** Improved assessment, monitoring, control and remediation of pollution using modern methods.

2. **Process change:** Generic changes in production technology that augment quality, reliability, controllability and cost factors while improving environmental performance. Examples: computer-integrated manufacturing, new means of catalysis, new separation techniques.
3. **Product redesign.** Choices made early in the design process that reduce material inputs and eliminate toxic residues or disposal problems for consumer and industrial products. Examples: development of biodegradable plastics; the design of efficient, bright, and long-lasting light fixtures.
4. **New systems:** Major changes in infrastructure systems (energy, transport, communication, housing) that move the systems towards environmental sustainability. Examples: advanced urban public transport systems that move thousands of passengers per hour while reducing congestion and pollution.
5. **New technological fields:** Broad areas that present fundamentally different routes with the potential to be much more environmentally friendly. Biotechnology, new materials, miniaturization and information technology hold this promise.

Strong environmental regulations have, in fact, stimulated technological innovation in many cases. Environmentally-superior substitutes replaced ozone-depleting chlorinated fluorocarbons (CFCs) as soon as the phaseout was decided. Innovative firms replaced a monopoly producer after polychlorinated biphenyls (PCBs) were banned; and the combination of emissions control, new materials, and electronics technologies have transformed the automobile. Good firms may be expected to respond innovatively to signals such as regulation that shake them out of technological inertia; and strong regulations contribute to a technologically dynamic economy (Banks and Heaton, 1995).

Although regulatory commands are important, they do suffer from certain drawbacks. Economic incentive systems, in contrast, offer both a more cost-effective approach to cleanup and a continuing stimulus to innovation. Market-based approaches such as tradeable permits and pollution charges have a major virtue in providing constant incentive to innovate, particularly to those technologically dynamic firms likely to produce the most effective pollution control strategies. Till now, economic incentives have been employed only in the context of air pollution. They need to be expanded to water pollution as well.

One way to move faster on the trajectory of environmentally-beneficial perspective change is to make an in-depth srutiny of the environmental problems and solutions in particular industries, based on review by specially constituted groups, each concerned with some particular industry alone. Regulatory agencies should match their areas of primary focus with what occurs in the economy. For instance, there may be subunits concerned with the principal infrastructure areas such as transportation, communication and construction; other subunits concerned with emerging technologies (biotechnology, new materials); and still others with the major industrial sectors. The connections between environmental problems and technological solutions should be viewed through the same analytical perspective that firms use.

The issue of skill mix is also important. The regulatory process so far has been dominated by scientists, lawyers and economists. In the move toward technology, engineers and industrial managers need to be inducted for implementing technological improvement. Also there is need to acquire some understanding of, and anticipating, patterns of technological change across industry. These patterns, which are essentially visions of a more sustainable economy, should be the principal drivers of regulatory objectives. A focus on future technological possibilities is basically quite different from cost-benefit analysis, which continues in the same problem-by-problem mode as regulation traditionally has (Banks and Heaton, 1995).

CLEAN PRODUCTION

The concept of clean production signifies a procedural approach to production that requires that all the phases of the life-cycle of a product should be critically considered for prevention or minimization of all risks to humans and to the environment. It is a new approach to the problem of production: production processes, product cycles and consumption patterns which allow for the provision of basic human needs without disrupting and degrading the ecosystems. This is a goal-oriented definition; it does not give definite operational answer to several issues of technological, economical and social choice.

The term 'clean' has usually been used to mean several kinds of technological innovations including process-integrated preventive measures, end-of-pipe abatement measures and even remediation or clean-up measures (see Jackson, 1993). According to Jackson, clean production is an operational approach to the development of the system of production and consumption which incorporates a preventive approach to environmental protection. Environmental protection may be viewed in terms of the three principles which form the basis of the new clean production approach, viz.: (1) precaution, (2) prevention, and (3) integration. The precautionary principle warrants the reduction of anthropogenic inputs into the environment by redesigning the industrial system of production and consumption which has till hitherto relied on extensive throughput of materials.

The preventive approach dictates that we should give due consideration to change upstream in the causal network of the system of production and consumption. It is to focus not so much on environmental endpoints (as has commonly been the tendency so far), but on the production processes themselves, reducing the generation of potentially-polluting emissions from those processes and thereby reducing the risk of environmental damage at the source. The preventive approach makes possible a better understanding of the connection between economic activity and environmental damage as far upstream as possible; it also brings out clearly that the demand for products and services is the prime factor in the impact of anthropogenic systems on the environment. The preventive nature of clean production calls for the new approach to reconsider product design, consumer demand, and patterns of material consumption.

Traditional, end-of-pipe regulation commonly applies to specific environmental media. Process-integrated measures can, to some extent, reduce the generation of pollutants. Reduction in the need for emission into the environment of such substances can make possible an integrated protection of most or all environmental media.

It is also advisable to realize that selective reductions in waste generation do not necessarily always lead to reduced burdens on the environment. However, this limitation/drawback may be partly alleviated by ensuring that preventive clean production is also integrative in the following specific senses. Preventive clean production addresses: (1) all material flows, not just selected ones; (2) the whole life-cycle of a product from raw material extraction, through conversion and production, distribution, utilization or consumption, reuse or recycling, and ultimate disposal; (3) material flows into all environmental media and (4) the macroeconomic impacts of structural changes in the economy (Jackson, 1993).

The triple strategy of precaution, prevention and integration leads to two main operational pathways of clean production: (1) minimization of the environmental impacts of processes, product cycle and economic activities by reducing the material flow through those processes, cycles, and activities. In other words, improving the material efficiency of those processes. Material efficiency also includes energy efficiency, because energy supply depends on material flows. Direct dependence on material flows can often be reduced by increasing the use of solar energy. Material efficiency can be improved in production processes by avoiding leaks and spills, by closing internal material loops for auxiliary materials such as recycling acid streams. In product cycles, improving material efficiency means reusing, restoring, reconditioning and recycling products, and recycling of raw materials (Jackson, 1993). (2) The substitution of highly hazardous materials, products and activities with less hazardous ones.

In terms of consumption patterns, material-efficiency can be improved by designing products for longer service lives, and by reversing the throw-away mentality of the consumers.

Certain materials (e.g. some heavy metals) are highly toxic to human health and can cause disease or death in very short time-scales. Some other materials produce chronic (long-term) effects on human and animal health, acting as carcinogens, mutagens, or simply reducing the immuno-competence or suppressing the reproductive power of living organisms. Good examples are PCBs, DDTs and many other synthetic organic compounds.

Toxicity is primarily determined in terms of acute toxicity (lethal doses and lethal concentrations), carcinogenicity, mutagenicity, teratogenicity and longer-term effects. Most of the substances listed in Table 5.1 come under one or other of those categories.

Table 5.1 EC List of toxic and dangerous substances and materials requiring priority consideration (after Jackson, 1993)

Metals and their compounds	Others
Arsenic	Isocyanates
Mercury	Organohalogen compounds
Cadmium	Chlorinated solvents
Thallium	Organic solvents
Beryllium	Biocides and phytopharmaceutical substances
Chrome (VI), Lead	Tarry materials from refining and tar residues from distilling
Antimony	Pharmaceutical compounds
Phenols	Peroxides, chlorates, perchlorates and azides
Cyanides	Ethers
	Chemical laboratory materials
	Asbestos
	Selenium and compounds
	Tellurium and compounds
	Aromatic polycyclic compounds (with carcinogenic effects)
	Metal carbonyls
	Soluble copper compounds
	Acids and/or basic substances used in surface treatment and finishing of metals

Highly toxic substances are directly hazardous to human health, even in very small quantities.

Toxics-use reduction can occur both through substitution and material recycling, but the former is preferable to closing the cycles for such materials because recycling of hazardous materials does not preclude chance leakages and spillages from the cycle. Substitution of toxics for non-toxic materials is also more preventive than closing cycles provided that it does not pose any risks elsewhere.

Although many heavy metals are highly toxic, they have their biogeochemical cycles. It is these cycles that are subject to anthropogenic distortion which can spell danger of increased exposure to nonresistent species (including humans).

There are several synthetic substances which lack known biogeochemical cycles. This means that no level of emissions of synthetics into the environment can be deemed to be 'safe'.

The Dilemma

The avowed goal of clean production is the Utopian concept of zero pollution, which is virtually unattainable because all wastes are potential pollutants and some amount of waste, no matter how small, cannot be avoided. Even on thermodynamic grounds, recycling cannot eliminate all wastes.

No doubt, many people have claimed that it is possible to eliminate all wastes, but their definition of waste has been restricted or the analyses faulty and has taken into account only production process wastes. Even within this limited category, however, all wastes cannot be totally eliminated for thermodynamic reasons. Further, in this category of process wastes, analyses usually focus on certain listed hazardous wastes. While it is possible to eliminate a listed or regulated waste from a particular production process, the limits to conversion efficiency imply that there will always be *some* waste streams from the system of production and consumption. The following properties are indicative of hazard potential: (1) Toxicity (acute, chronic, carcinogenic, mutagenic, teratogenic, and pathogenic); (2) Corrosivity; (3) Flammability; (4) Persistence; and (5) Liability to accumulate or bioaccumulate.

Thus, it may be safely stated that not only it is impossible to eliminate all waste streams but also, society is currently producing a very broad spectrum of wastes which pose significant hazards to environment. These waste streams range from extremely toxic chemicals, such as heavy metal compounds or organohalogens, to almost completely degraded materials such as carbon dioxide. The 'lists' of potentially hazardous wastes only cover the tip of the iceberg. This does not mean, however, that we should throw up our hands in despair. We can certainly attempt to bypass the seriousness of the dilemma posed by the problem of clean production by recognizing that not all material flows into the environment actually constitute pollution. We can escape the dilemma by recognizing that not all wastes are actual pollutants even if they are potential pollutants. Also, there are still some scientific uncertainties as to the actual impact of various emissions, etc.

Technological process substitution is, in some cases, not the most effective or cost-effective long-term solution to environmental problems. The idea of a dynamic, preventive strategy is to take into account not only the product but also the process.

Whereas reactive, end-of-pipe strategies essentially involve media-specific actions designed to control the manner or rate of entry of some hazardous substance such as mercury into the environment, the preventive approach attends to the causative elements in the socioeconomic matrix with a view to finding solutions as far upstream as possible.

There is need to frame suitable policies about materials usage taking into account diverse factors, e.g. the nature of the materials, their flows through the economy and the environment, and the services which they provide to society. These policies should create good matches between material properties and materials use, taking account of environmental and health effects. A materials policy which follows the guidelines of clean production must include (at least) the following elements:

1. Phasing out dissipative uses of toxic materials.
2. Phasing out emissions of persistent, synthetic materials.
3. Reducing raw material extraction and consumption.
4. Ensuring sustainable use of renewable resources.
5. Optimising materials use with respect to product life.
6. Ensuring that full life-cycle analysis is applied to material choices.
7. Optimizing material flows with respect to natural material cycles (Jackson, 1993).

The clean production route of efficiency improvement makes possible reductions in emissions into the environment for all materials and hence, is conducive to a 'no-regrets' strategy for environmental management. Substitution of known toxics can also be a 'non-regrets' strategy in those cases where safe substitutes provide the same service to society and where the costs associated with emissions of toxics into the environment are quite high.

The disturbing dilemma here is how to determine materials choices, emission limits, and activity levels in those cases where no clear 'no-regrets' exists. this challenge needs to be viewed in terms of the 'dilemma of substitution', i.e. how to make choices about production processes, material usage, products, and economic activities which will diminish the burden of the economic system on the environment, without significantly jeopardising human welfare. The dilemma of substitution must take into account product life-cycle assessment, materials and economic policies on human lifestyle. Concern about the impact of human society on its environment centres on long-term, usually irreversible and potentially catastrophic risks of changes to the local, regional, and global ecosystems; the preventive strategy warrants that we reduce emissions emanating from anthropogenic activity. If this can be done by efficiency improvements or material substitution, there is no risk to our welfare. But the substitution of certain products or activities can sometimes pose the risk of losing certain services which had previosuly contributed to society's well-being. Here, there can be a trade-off between the environmental risk associated with continuing the activity level, on the one hand, and the risk of a loss of welfare on the other. Risk minimization in respect of potential loss of welfare constitutes a high priority for clean production strategies.

Decarbonized Energy

Ever since prehistoric times, energy and technology have been intimately intertwined. While materials (and the innovative products made from them) have changed since the Stone, Bronze and Iron ages, the hydrocarbon age is still with us. The interaction between materials and the products they have spawned through the ages has been greatly affected by the energy source. The burning of coal made railways possible. It was steam that helped usher in the age of electricity. Besides electronics that transformed the world economy, the petroleum-fuelled vehicle industry has been the other fundamental base of progress.

Starting with the Industrial Revolution, a basic change has been occurring—energy was being decarbonized from wood to coal to oil and gas, and so on. The curve of energy decarbonization directly overlays that of the growth of modern society. Extrapolation from that curve brings out that hydrogen, the ultimate energy source, will not begin its dominance for another 50 years. There are still reserves of oil to be used (although they are depletable). Some of the more familiar societal problems caused by fossil fuels have been pollution, climate change and a strategic dependence on oil that results in global conflict, inflation and the economic burdens of developing countries (Cassedy, 2000).

There is a solution that can dramatically shorten the above 50-year scenario. Based on hydrogen, it requires a complete systems approach to make it realistic and achievable in the short term as we already entered the hydrogen age a few years ago when nickel metal hydride batteries became the enabling technology for electric and hybrid vehicles—electric vehicles can achieve over 200 miles on a single charge and hybrids get 80 miles to the gallon of petrol (see Koppel, 1999).

Hydrogen can be used as a transition fuel in internal-combustion engines that emit no climate-change gases and in fuel cells that convert chemical into electrical energy. Whereas hydride batteries shuttle ions back and forth and are charged by electricity, fuel cells use up the hydrogen in generating electricity (Ovshinsky, 2000).

Fuel cells have attracted much attention but although these are necessary, they are not sufficient in the energy equation. Not only must hydrogen be produced in various ways, it must also be stored and transported safely, economically and practically, without the need for drastic changes in the energy infrastructure. The hydrogen economy must be systems-based with a simple universal means of utilizing it. One realistic approach that can usher in a new era is the use of thin-film, multijunction photovoltaics to obtain hydrogen by breaking up water (Ovshinsky, 2000).

The proton-exchange membrane (PEM) fuel cell—invented by Sir William Grove in 1839 and developed by General Electric in the 1950s—has been redeveloped by a Canadian team of creative individuals under the leadership of Geoffrey Ballard (see Koppel, 1999). Their success required excellent engineering and packaging rather than scientific breakthroughs.

But there are some limitations as well. The PEM fuel cell is fuel-neutral. The problems of storage have to be addressed. The fuel providers for the automotive industry are not likely to provide the required methanol infrastructure since they can only afford to change their infrastructure once. They know that pure hydrogen is required and that this will be very difficult to produce on board the vehicle. The use of gaseous and liquid hydrogen presents problems that limit these two forms to a transient place in the hydrogen economy (Ovshinsky, 2000).

A more promising approach is the onboard solid storage of hydrogen as a light-weight hydride capable of providing a vehicular range of over 300 miles. This also offers a solution to the important problems of infrastructure, as hydrogen in hydride form is at its lowest free energy and can be readily transported by conventional means.

Technological progress depends strongly on new materials. The use of hydrogen in a solid is made practical by atomic engineering, as various chemical, electronic and topological functions can be integrated in one material, through such means as catalysis, hydrogen-diffusion paths and acceptor sites (Cassedy, 2000).

LEATHER INDUSTRY

Hundreds of leather tannery units are actively engaged in processing and manufacturing of leather goods. They have a processing capacity of millions of pickled skins and cattle hide.

Leather products are really a mixture of art and industry. China, India and Iran have always been top-ranking producers in handicrafts as well as leather goods. Iran has transferred tanneries to leather towns especially set up in the major leather-producing regions such as Tehran, Tabriz and Mashhad.

Unfortunately, very few tanneries have their waste water plants in operation and some have plants under construction. The rest of the tanneries discharge their waste water into water bodies without any treatment. The need of good housekeeping and clean technology cannot be overemphasized.

An Eco-Label may prove helpful in encouraging proper effluent treatment and environmental protection. The leather industry itself should determine parameters, rather than have them imposed, and to establish independence and impartiality in regulating the scheme.

High priority needs to be given to such important areas of the leather and leather products technology as hide and skin improvement, economic utilization of slaughter houses and their by-products, introduction of cleaner technology and pollution control, and professional training and education. An integrated programme should be planned for extending technical assistance to the leather related industry sub-sector, taking into consideration such aspects as tannery effluent treatment, product development and marketing and assistance to manufacturing units to improve quality of final products.

TECHNOLOGICAL DYNAMISM IN ASIA

In a fast changing world, Asia has emerged and sustained its industrial and technological dynamism and competitiveness. Substantial growth of manufacturing value-added goods in several developing economies of the region can be achieved through structural change and building up of appropriate technological capabilities. The new international arrangements and globalization of manufacturing production are expected to strengthen the region's capacity for sustaining this dynamic process.

The outcome of the recent meeting of the Uruguay Round is likely to stimulate efforts aimed at improving the industrial and technological comeptitiveness of the region. The various provisions such as the Trade-related Intellectual Property Rights Agreement (TRIPs) and the Trade-related Investment Measures Agreement (TRIMs) may be expected to improve the global flows of trade, investment and technology. It may, however, be mentioned that various national-level means, including reoriented policies and institutional arrangements, are needed to be implemented so as to benefit from the global agreements of the Uruguay Round. Regional cooperation among countries is absolutely desirable in this regard.

The richer countries in Asian region should intensify their efforts in promoting outflows of investment resources and further enhance technology transfer. These measures not only lead to overall improvement of competitiveness of the Asia-Pacific region, but would also result in relocation of industries and a balanced development of industrial activities with appropriate use of scarce resources.

The crucial role of technology for enterprises to meet the competitivity challenge in an increasingly liberalized and globalized market context cannot be overemphasized. Availability of cheap labour in Asian developing countries is no longer a sufficient prerequisite for international comeptitiveness, because the advantage of cheap labour has been greatly eroded by increased automation and replacement of labour-intensive production processes. In this context, it is important to build up core technological competencies in enterprises, enabling them to make new or improved products. This requires not only the development of specific technological skills and modern machinery, but also the build-up of managerial capability for technology acquisition, adaption and innovation.

Three suggestions for improving technology access and the strenghthening of core competencies are listed below:

1. Promotion of foreign direct investment, enabling private sector enterprises to acquire capital, skills and technologies;
2. Promotion of generic technologies, particularly information technologies, biotechnology and new advanced materials, enabling enterprises to significantly transform products as well as production and management functions;
3. Better human resource development in defined technology areas, which may vary from country to country; and greater cooperation amongst enterprises and research and development establishments of the region for joint development of technologies.

ENERGY

Ever since humans discovered how to make and control fire, energy has been essential to their well-being. From the day man first choked in the smoke of the fire he had set up, energy production has been linked to pollution. The crucial issue is how to achieve the maximum benefit from energy at the least environmental cost.

All forms of energy involve some degree of risk to human health or the environment. Chernobyl shook confidence in nuclear power among a public already sceptical about its safety. Burning of fossil fuel causes local pollution and acid rain. It adds to the build-up of greenhouse gases, which may cause climate changes. Hydropower installations may displace local populations and disrupt the aquatic environment. Hundreds of millions of people suffer from respiratory diseases caused by the smoke from wood and dung fires.

In many cases, the most cost-effective means of reducing the pollution caused by energy use and production is conservation. Conservation reduces the degree of pollution caused by energy use and production. Electricity can be generated by burning waste in an incinerator. Both developed and developing nations should use energy more efficiently. Special guidelines on energy conservation for developing countries have been produced by UNEP, Nairobi.

The major energy problem for most people in developing countries is the shortage of fuelwood. About 3 billion people are now affected.

Projects in Senegal, Sri Lanka, Indonesia and the Philippines have demonstrated that windmills, biogas plants, mini-hydro turbines, solar photovoltaic cells and gasification can be used to meet the energy needs of local people (UNEP, 1990).

The key elements of a sound energy technology programme for a developing country such as India or China are shown in Fig. 5.1. The chief objective of such a programme is to develop a globally applicable cradle-to-grave methodology to analyze the true impact of electric power generation on the environment. This objective can be achieved by focusing on a new way of cooperation among industry, academic institutions and government.

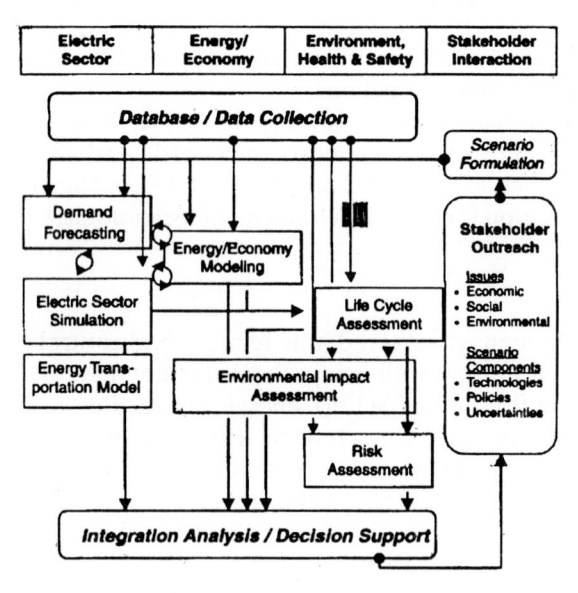

Fig. 5.1 Flow chart of a sound energy technology programme showing the relationships among the various tasks of the programme.

In China, energy intensity, i.e energy consumption per unit of economic activity, has been declining rapidly but still remains much higher than in the USA. Coal will continue to be the dominant fuel in China, causing sulfur dioxide air pollution and high emissions of carbon dioxide. If carbon dioxide emissions are controlled, sulfur dioxide is automatically reduced. However, this is an expensive process. Advanced coal technologies are about twice as expensive as the present inefficient coal technologies. The cost of reducing sulfur dioxide emission is modest and the Chinese are much more concerned with sulfur dioxide than with global warming.

FOSSIL FUELS AND THE ENVIRONMENT

Several environmental control technologies have been developed in concert with advanced fossil fuel conversion processes so as to eliminate the adverse health and ecological effects that are often by-products of energy conversion.

It is expected that the new and evolving conversion processes will turn out to be environmentally sound as well as efficient. Development of the novel technologies has been inspired by the following twin aims:

1. To safeguard health and the environment without unduly delaying the accelerated development and use of domestic energy resources.
2. To anticipate environmental impacts of energy conversion technologies and develop cost-effective environmental controls.

The overall objective in developing new control technologies is to integrate them properly into energy production processes as and when they are commercialized. Successful development of environmental controls in parallel with development of energy production processes can avoid the more costly and less efficient task of disposing of pollutant wastes after they have reached the environment.

Control technology development has been directed towards removing pollutants from both the discharge stream of an energy conversion process as well as the process stream itself. Even following development, however, application of these controls is often questionable. For instance, will the controls, in fact, reduce pollutants to levels that pose no threat to health or ecology? Are they sufficiently effective? How serious is the trade-off between lost energy and found environment?

One thing is certain, however—energy conversion processes do produce pollutants as by-products. Many of these are severe health hazards—some have impacts that may take years to manifest in clear-cut damage to health or the environment. Emissions standards must, therefore, be established and enforced to protect human health and key environmental relationships before these pollutants get out of hand.

Deriving these standards for today requires constant and careful evaluation of existing technologies. Deriving them for the future requires timely development of environmental control technologies for emerging fossil fuel processes, allowing for the identification, measurement and mitigation of these pollutants and their sources before widespread dispersion of both the pollutants and the concerned energy technologies can occur (EPA, 1977).

In the past two decades, happily, the focus has shifted to the development of new, cleaner processes that depart from the former practice of worrying about the mess after it has been created. Advanced fossil fuel environmental control technology is a key tool necessary to bring about these total process fuel conversions.

Processes/Control Technology

There are five major advanced fossil fuel processes and control technologies, viz.;

1. **Chemical Coal Cleaning**—four major processes. Methods of removing organic and pyritic sulfur and other contaminants.
2. **Synthetic Fuels**—gasification and liquification to provide synthetic fuels from coal. Reduction of gaseous, liquid and solid waste during and after conversion.
3. **Chemically Active Fluid Bed (CAFB)**—producing clean gas from sulfur-laden heavy residual oils.
4. **Oil Shale**—control of processing effects on the environment during recovery of oil from kerogenous shale.
5. **Liquid Fuels Cleaning**—methods of removing contaminants from liquid fuels before combustion.

Advanced fossil fuels will continue to be used to greater or lesser extent by different sectors of the economy. In the short term, for example, smaller users of conventional combustion who, for reasons of cost or location, cannot adopt gasification or liquefaction techniques may benefit from chemical coal cleaning technology. Large industrial users of gas and oil may soon turn to advanced fossil fuels when traditional fuel sources become critically scarce and prohibitively expensive.

The environmental effects of advanced fuel processes tend to be felt most significantly on the local or regional level—those areas where advanced processing plants and fuel sources are located.

The different technologies associated with these processing plants and fuel sources generate different types and amounts of pollution. Among the most hazardous pollutants are sulfur in its various forms, nitrogen oxides, phenols, aldehydes, ammonia, particulates, toxic trace metals, tars, oils and dust.

Some controversies has recently arisen on the approach to reducing emission of GHGs. According to Hansen (2000), the emphasis in the next 5 decades should be placed on reducing emissions of other GHGs such as methane, rather than carbon dioxide (CO_2). On the other hand, Manne and Richels (2000) believe that emissions of methane, a short-lived greenhouse gas, are not important in the near term if the goal is to minimize the long-term rise in global temperature. Methane is generally considered to be the second most important GHG, after CO_2.

Hansen et al., (1991) found that the rapid global warming in recent decades has been driven mainly by non-CO_2 GHGs such as CFCs, methane, and nitrous oxide. The climate forcing of CO_2 is less important because it is partially offset by aerosols such as sulfates and black carbon, that are also by-products of burning fossil fuel.

The growth rate of non-CO_2 GHGs seems to have declined in the past decade. If sources of methane and ozone precursors were reduced in the future, the change in climate forcing by non-CO_2 GHGs in the next 5 decades could be near zero. Combined with a reduction of black carbon emissions, mainly from diesel fuel and coal, and some tangible success in slowing CO_2 emissions, this reduction of non-CO_2 GHGs could lead to a decline in the rate of global warming.

The view of Hansen et al., is based on conclusions from a comparison of climate forcings of GHGs, other anthropogenic forcings such as changes in clouds and land cover, and natural forcings such as the sun and volcanic aerosols.

Manne and Richels (2000), on the other hand, view the problem from an economic perspective. Taking a maximum allowable increase in global temperature to be the goal of climate policy, they assess the importance of non-CO_2 gases by the incremental value of their emission rights—or price—relative to CO_2.

Because of the short life (about 12 years) of methane in the atmosphere, the value of methane reductions remains very low in the next 5 decades. Methane only becomes important later, when it remains in the atmosphere as the maximum acceptable temperature change approaches the assumed limits of 2°C or 3°C.

In contrast, nitrous oxide has a lifetime in the atmosphere of about 120 years, comparable to the 50 to 200 years of CO_2. The value of its emission rights is about double that calculated for CO_2, assuming the 100-year global warming potential, and it holds steady for the next century.

According to some climatologists, climate change damages may be sensitive to the *rate* of temperature change as well as the absolute temperature change. Too rapid a change, for instance, may not allow time for certain species of trees and even types of agriculture to 'migrate' with the climate.

Recent work of Manne and Richels on analysis of the importance of methane reductions supports that of Hansen if and only if near-term temperature changes are important.

Hansen's alternative scenario does not alter the desirability of limiting CO_2 emissions because the future balancing of forcings is likely to shift toward dominance of CO_2 over aerosols. The scenario calls for a moderate decrease in CO_2 emission rates in the coming half century, arising from the success in slowing CO_2 emissions. In the next 25 years, Hansen expects that this can be achieved by improved energy efficiency and a continued trend toward decarbonization of energy sources, such as increased use of gas instead of coal. In the longer term, attainment of a decreasing CO_2 growth will require the use of energy systems that produce little or no CO_2. Some renewable energy systems may be developed without concern for climate but it is important today to foster research and development in generic technologies at the interface between energy supply and use, such as gas turbines, fuel cells and photovoltaics.

ENERGY CHOICES AND TECHNOLOGICAL CHANGE

Energy choices have major implications for the environment. The seemingly unlimited availability of low-cost oil and gas up to the early 1970s not only fueled the automobile explosion, but also held down coal burning. Urban air benefited greatly, and acid rain did not spread as rapidly as it might otherwise have. Major commitments to coal and coal-based synthetic fuels can acidify the rain over wide areas and also be the cause of human-caused climate changes in the middle of the twenty-first century. This changes will certainly cause hardship for the underclass at least. Another major energy alternative—nuclear power—carries its own unresolved problems of safety, waste disposal and weapons proliferation.

In parts of the Third World, soaring oil prices have exacerbated the scarcity of fuelwood. Kerosene, traditionally the main wood substitute, has been pulled farther out of reach of the poor. Villagers in a few places have actually been forced backward in history as the high price or unavailability of kerosene requires them to switch back to wood for cooking their meals and warming their homes. It is not clear whether tree-planting programmes and alternative energy sources can be provided quickly enough to avoid massive landscape damage and social hardship. In this context, the increased use of renewable energy sources should reduce energy-related environmental damage. But biomass production and large-scale hydropower development can cause some damage to land and people if not properly managed. Increasing the efficiency of energy use remains the most environmentally-benign response to the energy challenge and one that has not yet been fully exploited anywhere.

In the absence of more stringent pollution-control regulations and of much stronger efforts to recycle materials and conserve resources, economic growth itself will inevitably mean increased degradation.

Technological change, which is the apparent source of many environmental problems, is sometimes also advocated as the key to their solution. Potential research breakthroughs that could make a difference are not difficult to identify. Cheaper solar power, cheaper and more effective ways to remove poisons from smokestacks and tailpipes, safer and more reliable contraceptives, hardier and faster growing trees are only some examples. New biotechnologies and applying genetic engineering techniques to agriculture and forestry, seem to hold great promise.

Even then, Still, few serious challenges can be resolved through technological progress alone. Appropriate social organization is very often the primary need. Higher-yielding crops can strengthen the battle against hunger but do not automatically mean more food for the hungry; for that, reforms in the distribution of assets, services, or employment will be needed. Better contraceptives are badly needed. Yet many among the underclass will not be interested in family planning unless their economic prospects can improve. The simple, low-cost method for treating the effects of severe diarrhoea will save many lives. But unless health-care systems are reorganized, many stricken rural children will not receive even this simple treatment. And only access to clean water, together with improved sanitary habits, will cut the awesome incidence of diarrhoea and other infectious diseases among the children of the underclass.

ENERGY MYTHS

In general, proven facts about energy are less well known and perhaps less interesting than the energy myths. So, energy myths have tended to endure. In particular, explanations of oil shortages and the workings of the energy industries have commonly been as much mythical as factual.

What are now recognized as energy myths originally served a purpose: as with the myths of old, they explained a puzzling phenomenon at a time when facts used to be scarce. Some of the myths about energy were—and are—harmless. But others have often led to false conclusions and harmful policies.

1. The Hoarding Myth

There is a widely held belief that gasoline shortages are caused by oil companies hoarding oil. Rigorous investigations in several countries, e.g. the USA, showed this myth to be unfounded.

2. The Monopoly Myth

There used to be a widely held belief that the oil industry is a monopoly—i.e. run by a few large companies that use their power to exploit consumers.

One of the reasons why people believe oil companies could influence prices by hoarding oil or shutting in wells is that they do not believe oil companies compete with each other. Nor do they accept the fact that if one oil company held back its oil, other companies would step in to sell their oil and gain a new market. Some collusion is suspected among oil companies.

Economists measure 'market concentration' to determine whether the firms in an industry could act in collusion. They usually consider market concentration as too low for effective collusion when the top four firms in an industry have less than a 50% share of the market. By this criterion, this myth is usually unfounded.

A powerful petroleum industry monopoly is mythical. The major oil companies usually cannot control energy markets.

3. The Exorbitant Profits Myth

Many of the early explorers sailed to the New World in the hope of becoming rich. They were encouraged by American Indians, who told them about cities where gold was so plentiful that the natives would give it away free.

For years, early explorers continued, despite repeated disappointments, to search for such cities of gold and their easy wealth. One of the most persistent energy myths is the belief that oil companies are like these cities of gold, grown vastly rich on excessive profits. Oil companies are not much richer than other industries; oil industries like any other industries, have profits that rise and fall.

Clearly, the oil companies are not fabulously wealthy modern cities of gold. Reasonable profits in the oil industry assure long-term investments in energy for the future.

4. The Myth That We May Run Out of Oil

Many people believe that the world will quickly run out of oil. There is no doubt that oil is a non-renewable fossil energy, but the actual position about oil may not be that bad. People have been predicting for years that we will run off the edge of our oil supplies—soon. But many such predictions have been proved wrong.

Many estimates of oil reserves are conservative and do not include vast amounts of known oil and gas resources that are now uneconomic to produce or difficult to extract. That oil and gas could become economic if prices, costs or recovery technology change in the future. The estimates also do not include the vast reserves of coal and oil shale that can, in time, be converted into synthetic oil and gas supplies.

Most edge-of-the-world predictions assume, firstly that oil consumption patterns do not change—that this world will go on using more and more oil each year. But in actual fact, energy consumption can change in a relatively short time as consumers have been found to reduce their consumption of oil significantly. Some of this reduction has been a result of increasing efficiency, as new technology reduces the amount of energy necessary to perform the same amount of work. Industrialized countries have become more efficient, producing more goods with less energy.

Doomsday predictions also assume that our present knowledge of oil reserves and our current state of oil production technology will not improve. But expertise in oil discovery and recovery has been increasing. New fields of oil and gas are being discovered as a result of technical methods that did not exist in the past and new recovery techniques squeeze more oil out of fields than was previously possible.

Exploration has shown that the end of our energy supplies is not an imminent danger in the near future.

The above energy myths resulted from explanations of events that seemed reasonable but later proved to be misunderstandings in the light of experience and more complete information. These misunderstandings led many people to believe that energy problems were not real and diverted attention from finding realistic, workable solutions to the serious energy problems confronting developing countries especially and the world in general.

As the world gains experience with energy markets and more information becomes available, workable solutions can be found easily. Some recent energy policy choices have had remarkable success. That success indicates that it is possible to turn away from myths and towards real solutions to real problems.

CONVERSION TECHNOLOGIES

Conversion technologies allow us to convert one energy form to another form, e.g. solar, hydro or nuclear energy to electricity. Technological advances have affected every aspect of the energy system, reducing losses and expanding options for energy systems. Lighting with photovoltaics has become better since the advent of compact fluorescent lamp technology which consumes less energy.

Although new technological and engineering breakthroughs transform scientific concepts into tangible products and processes, socio-economic criteria play a major role in selection of energy alternatives. Energy prices, howsoever arrived at, determine social choices. Much of the desirable changes such as improving efficiency of conversion or end-use devices

248 Environmental Technology and Biosphere Management

materialize only if the energy resources are correctly priced. Recently, financing energy expansion has become a major problem, especially in developing countries where capital is scarce.

Local environmental concerns have become important issues in making choices. Global environmental concerns call for reduction in use of fossil fuels to reduce emission of greenhouse gases to mitigate climate change.

Decarbonization of energy systems is likely to become a major criterion for choosing various elements of energy systems.

Pollution can be reduced mainly by reducing pollution intensity and energy intensity of GDP. When population level stabilizes, there is greater scope to reduce pollution and energy consumption. Power plant efficiencies—still about 20% to 30% in developing countries—have reached 40% to 60% in the developed countries for the new generation of power plants which recover and reuse waste heat (Parikh, 2000).

Energy and Equity

Distribution of energy across countries and income groups is far from equitable. Some three-fourths of energy resources are used by 25% of the world population in the developed countries. Within low income developing countries also, the rural poor use much less energy out of which the major fraction comes from locally-available biomass.

According to Parikh (2000), a good energy system for the future should include the following transitions from:

1. Low or medium efficiency to high efficiency;
2. Fossil-fuel based systems to decarbonized systems;
3. Environmentally damaging to environmentally friendly systems;
4. Inequitable to equitable consumption;
5. Public sector and government control to private sector undertakings; and
6. Highly centralized system to one with reinforcement distributed energy system.

ECOLOGICALLY-SUSTAINABLE INDUSTRIAL DEVELOPMENT FOR ASIA AND THE PACIFIC REGION

The rapid industrialization in the Asia-Pacific region as a whole has exerted strong pressures on environmental resources. The combination of industrial and economic growth, environmental degradation and population growth seriously threatens sustainable development. The major environmental challenges facing the region are related to atmospheric pollution and climate change, forest conservation, provision of adequate and clean water, clean and efficient energy development, preservation of biodiversity, marine and coastal pollution and the management and disposal of solid wastes.

Four of the important constraints that have limited the sustainable development of industry in the Asia-Pacific region include: (1) policy distortions especially relating to unrealistic low pricing of resources, notably water and energy; (2) weak institutional capacities for monitoring of environmental impacts and enforcement of regulations; (3) lack of cleaner and (energy) efficient technologies; and (4) lack of sufficient collaboration between government and private industry. To overcome these constraints, there is need for the application of clean production techniques and technologies; energy conservation; promotion of economic incentive policies for pollution prevention and the use of environmental management systems by industry; meeting increased ecolabelling requirements through the establishment of certification centres and accreditation bodies; and, environmentally-sound management of natural resources such as water and energy by industrial sectors.

There is a special need to improve the industrial situation in the least developed, land-locked and island developing countries, as well as economies in transition. A review of industrial progress in these countries as well as the current issues and challenges in industrialization suggests that these disadvantaged groups of economies need to strengthen their domestic efforts in effectively implementing relevant institutional facilities, so that the private sector might play significant roles.

It has been increasingly recognized that some Asian and Pacific regions are facing growing problems in the area of environment protection and optimum use of non-renewable resources, especially energy. In several countries, there is the need to direct industrial development towards the goal of poverty elimination. The small and medium-scale industrial sector faces the challenge of coping with global integration and international competition. Technology upgrading and improving competitiveness are key areas of concern.

Industrial growth, diversification and dispersal are the chief means for creating new jobs and incomes in less developed regions. The process of industrialization requires that the transition from low to high technology be properly programmed so as to minimize increasing umemployment and widening income gaps. As far as possible, the pace of industrialization in developing countries should increase in keeping with private sector development and increased export orientation. Market-friendly incentives can go a long way to correct market failures and shortcomings. Such incentives and supports should be provided at the local, state and international levels.

The future growth of the world economy, manufacturing and services is going to be mostly skill-based. To reduce their initial handicaps, developing countries will have to focus on the development of human resources as well as on new technological skills and innovations.

Technology is undoubtedly a core component in competitiveness and hence must be based on such integrated development efforts and capabilities as can be used in different market conditions. Some competitiveness in internal and external markets really constitute two sides of the same coin.

SOLAR ENERGY AND PHOTOVOLTAIC PUMPS

One potentially promising option for ensuring the food security for the growing world population in the new millennium is to intensify agriculture in the tropics and subtropics by encouraging responsible use of irrigated farming. The land under cultivation may be extended to include those areas which could not be exploited so far. In many developing countries, e.g. India, where there is serious shortage of electric power supply, it is extremely difficult or even impossible to run water pumps for irrigating crop fields. Therefore, combustion engines have to be used to generate electricity or to operate the pumps directly. That means that maintenance costs and the cost of procuring fuel will be high. Photovoltaically driven pumps and PV-systems for transporting water have proved to be less costly than are diesel pumps serving the same purpose (Posorski and Haars, 1994). Photovoltaic pumps constitute an economically attractive alternative for small farming enterprises in the developing countries; the unsteady performance of such pumps is a special factor when considering which method of irrigation should be used, however (Mayer and Müller, 1998).

Photovoltaic Pumping Systems (PVPs)

Recent years have seen an increasing use of PVPs for supplying water in the developing countries (Fig. 5.2). They have proven quite reliable and durable. The only demerit of solar energy, however, is that the energy supply is irregular and the energy source cannot be influenced directly. Thus, the energy consumer has to adapt to the irregularity with which the energy is supplied, or the energy has to be stored immediately so as to guarantee a reliable water supply. Storage of electric power in batteries is costly and not fail-safe. Thus, the transport of water depends directly on the amount of solar radiation existent, a water tank of corresponding size serving as an energy store.

PVPs consist of various independent components. Figure 5.3 shows a sketch of the commonly used components. The solar generator consists of solar modules with a defined combination of serial- and parallel-connected solar modules and converts the solar radiation into direct

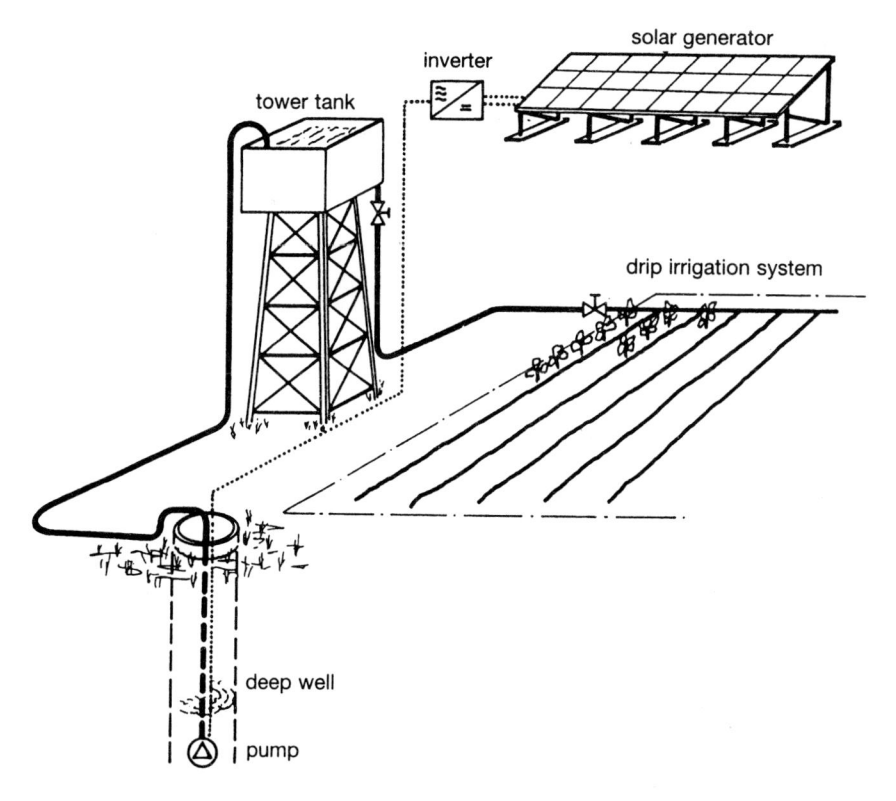

Fig. 5.2 Photovoltaically powered drip irrigation system (after Mayer and Müller, 1998).

current. The generator/load-matching device assures that the generator output power is optimally transferred to the load and, in DC/AC-systems it converts direct current into a three-phase alternating current. To operate the pumping system, submersible motors mounted directly on a rotary pump can be used. The transported water can be stored in a tank or, if the condition of the soil allow, fed into the irrigation system directly (Mayer and Müller, 1998).

There are certain ranges of capacity at which it proves profitable to use PVPs. The constant output of a human being amounts to approximately. 70 W and that of an animal amounts to approx. 300 W. Assuming that a pump's average effectiveness is 65%, the actual hydraulic pump capacity would be about 45 W for a human being and 195 W for an animal. In the countries of the Third World, manpower is generally less expensive than is the use of highly developed technology. The use of PVPs is generally economically justified for delivery heads over 5 m and power output of over 200 W. For power outputs of up to approximately. 10 kW, PV

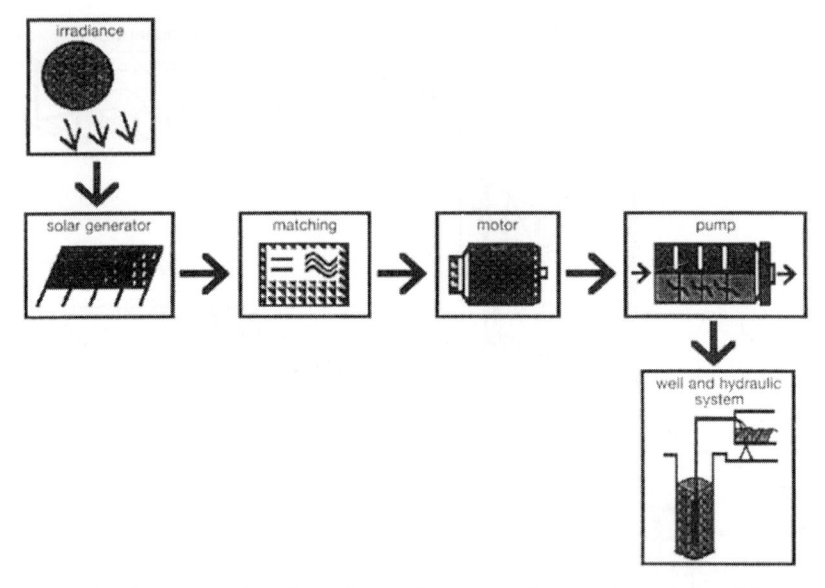

Fig. 5.3 Components of a photovoltaic pumping system (after Mayer and Müller, 1998).

generated electricity is more economical than that produced by diesel generators because PVPS break down less frequently than diesel generators. Fig. 5.4 shows the areas in which it is advantageous to use PVPs. The delivery heads typically range between 5 and 100 m (Mayer and Müller, 1998).

There are essentially three different categories of irrigation methods, viz., surface, sprinkler, and micro-irrigation. Each of these methods places different demands on the technology involved in transporting water as regards the delivery head and flow, which cannot always be met by PVPs to the same degree. Most surface irrigation systems in use in North Africa and Asia are integrated into the relief of the land and rely on the force of gravity to convey water. The water has often to be pumped from a supply canal to the fields. Typically, the delivery head is between 1 and 3 m. Various sprinkler irrigation procedures require a much higher delivery pressure. Mobile sprinkler systems such as those which are commonly used in the arid zones of Australia, North America and the Near East in the form of centre pivots or linear moves distribute the water through nozzles along a pipeline, which may be as long as 500 m. To obtain an optimal drip spectrum, operating pressures of 2 to 5 bar are essential. As regards operating pressure, micro-irrigation falls in between the two other methods of irrigation. Here, by means of drip elements or micro-sprinklers, the water is applied to the root-zone of the plants directly and in doing so, only a small amount of kinetic energy is needed to distribute the water.

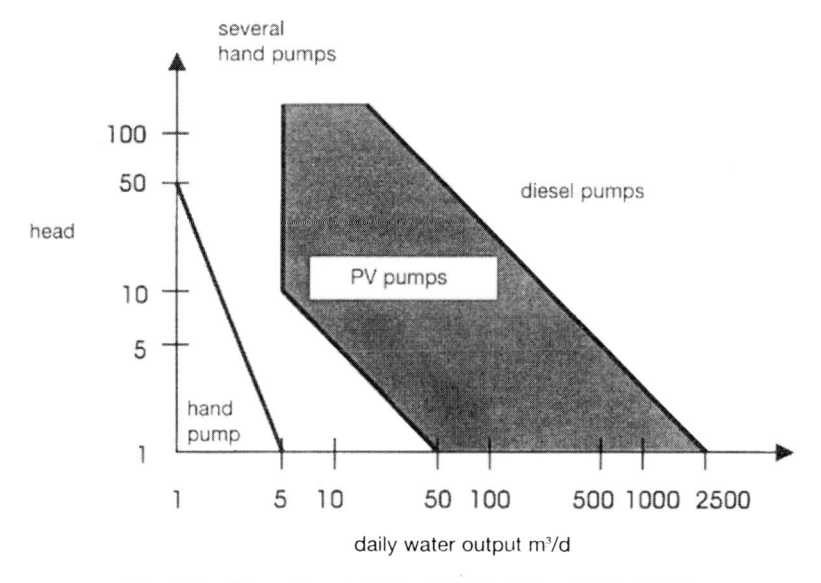

Fig. 5.4 Operational field of PVPs (after GTZ, 1992).

An operating pressure of 1 to 2 bar suffices to convey the water through the necessary filter systems and the hose net.

The effectiveness and suitability of the various methods of irrigation does not only depend on the delivery head, but also on the delivery flow.

Usually, the largest amount of water is required by surface irrigation. Sprinkling systems have an average degree of effectiveness of 0.8. The highest degree of effectiveness is obtained by drip irrigation.

In drip irrigation, the delivery flow is approximately. 90% less than for surface irrigation. For drip irrigation, the required hydraulic capacity is quite low. Since the costs of PVPS are determined largely by the capacity of the installed solar generators, it is preferable to use them in combination with drip irrigation methods (Mayer and Müller, 1998).

The daily water requirement correlates directly with radiation and thus harmonizes well with the performance of a PV-pump. Drip irrigation, as opposed to other forms of irrigation, allows for a continual supply of water so that the pump can be operated all day long and over the course of the entire vegetation period, thereby ensuring a high degree of utilization (Mayer and Müller, 1998).

PLANT OIL FOR USE IN ENGINES

In recent years, there has been some interest in using engine fuels from regenerative raw materials, residues and waste materials. In this context, particular attention has been given to biogas/dumpsite gas; producer gas; ethanol/methanol; and vegetable oil.

According to Bauer (1995), all engines will be hybrid fuel engines by the year 2005:

- By adapting to operating conditions or fuels by motor management, adjustment of the injection timing and injection process or the antiknock setting (in the otto engine); or
- By selecting out of various fuels the one that is best suited to the prevailing conditions.

Fuels from regenerative raw materials will have to be adapted to the extent possible to the substituted fuels based on mineral oils (e.g. ethanol for otto fuel and plant oil for diesel fuel) (e.g. by esterification or blending).

For industrialized countries with advanced distribution of modern diesel engines and pre-set structures for extracting and distributing fuels, normally it is the advantages of (rapeseed) methylester ((R)ME) which prevail, or, according to the technical properties of diesel fuels, the plant oil based fuels adapted to them.

Special engines for raw vegetable oil extracted using simple technology are gaining popularity.

Biofuels made from tropical and subtropical oil fruits are particularly advantageous.

In the case of vegetable oil, the oil-containing seeds are often produced, harvested and/or processed in the same region and are used on the one hand as fuel and, on the other as fodder, fertilizer, fuel for heating and drying. Sometimes, the whole plant is used.

In the tropical and sub-tropical developing countries, especially in remote inaccessible regions diverse biological fuels and lubricants are suitable for use in various engines for both mobile and stationary use (water raising and telecommunications). These uses must be balanced against each other and compared with traditional alternatives (such as draught animals/wind and water power) and with fuels based on mineral oil or other sources.

The local circulation economy, based on a small-scale technical solution for extracting and refining fuels, can certainly prove to be beneficial (Ackermann and Wolf, 1994).

Various kinds of advanced oil mills are now available and used for producing special oils while observing narrow tolerances regarding the crucial contents and properties. A number of fuels have been developed to meet various requirements.

Quality standards—at least for rapeseed oil—for use as fuel in plant oil motors have been laid down (see Table 5.2).

There are the following advantages in the use of crude vegetable oil : local, small-scale technical solution; circulation economy; hardly any

Table 5.2 Recommended quality standards for rapeseed oil to be used as fuel in internal combustion engines (Jansen and Steffen, 1991).

Specification	Unit	Requirement	Test method
Density at 15°C	g/ml	0.90–0.03	DIN 5157
Viscosity at 20°C	mm^2/s	max. 80	DIN 51562
Calorific value at 20°C	MJ/kg	min. 35	
Flashpoint	C	min. 55	DIN 51755
Cetane number	C	–39	DIN 51773
Neutralization no.	mg KOH/g	max. 1.5	DIN 51558
Saponification no.	mg KCH/g	max. 190	DGF CV 3 (77)
Iodine no.	G iod/100 g	max. 115	DIN
Free fatty acids	Weight–%	max. 1	DGF CIII 4 (53)
Total phosphorus	mg/kg.	30	DGF CVI 4
Sulphur content	Weight–%	max. 0.03	DIN EN 41
Residual coke	Weight–%	max. 0.5	DIN 51551
Ash content	Weight–%	max. 0.02	DIN EN 7
Water content	mg/kg	max. 1000	DIN 51777
Initial boiling point	C	–210	DIN 51751
Filter grade	μm	max. 5	

phosphatides in the oil; no chemical solvents; no distillation to retrieve the solvent; and high energy efficiency.

Engines can be run on plant oil. Prechamber and whirl chamber engines as standard engines are most tolerant to cold-pressed purified oils, but need a dual fuel system and, in cold regions, special heating devices for a cold start. Besides, the efficiency is about 15% lower than with direct injection engines. Special engines according to the duothermal method combine the advantages of the economical direct injection engines with the suitability for processed vegetable oils. The extra technical requirements regarding pistons, cooling and starting/heating devices, however, make them much more expensive than standard engines. Direct injection standard engines combine all the merits of technical advances in engines but impose high demands on the fuel, which in the case of plant oil can only be met by the esterification, which involves a higher technical, energy and financial expenditure.

TECHNOLOGY TRANSFER

Technological innovations and developments have been key elements in the competitiveness of nations, in the fields of economics, defence or politics. The selfish national interests have restricted the flow of information on new developments and limited their access to potential users. Commercial and legal considerations are invoked as restrictive factors in the transfer of technology, but most scientists have recognized the need for reciprocal exchange of information and also believe that all scientific and technological developments should be shared among nations. This has greatly contributed towards achieving the successful introduction of modern technologies in diverse fields.

'Technology' is the sum of knowledge, experience and skills necessary for performing a task, manufacturing a product and establishing an enterprise for this purpose. It is usually associated with patented processes and manufactured products often referred to as hardware, but an essential part of technology is the knowledge of how and why it works, which is referred to as the soft part of the technology, involving human resources and management processes (WMO, 1993). Transfer of technology should be considered to be the export of technologies from one country to another in various forms, including the provision of equipment, software, expertise and the training of local personnel.

It is basically necessary to ensure that there is a need for the technology before the transfer is envisaged. The suitability and adaptability of the technology and its capability of being absorbed also needs to be determined. Adequate facilities for the repair, maintenance and development of the technology should be accessible to the recipient. The transfer of technology should be accompanied by suitable training programmes for personnel, especially those of trainers, and those manufacturing the equipment or software. The use of locally available materials and tools should be encouraged.

Agricultural operations depend on the anticipated temporal and spatial rainfall patterns. Many meteorological and hydrological services maintain crop-weather calendars which are used to advise farmers and other agricultural interests of the best period for a specific operation, such as sowing. A problem of some interest to the developing countries is a choice between food and cash crops.

The importance of rainfall forecasts for agriculture cannot be overemphasized. As most of the agriculture in the tropics is rainfed, good rainfall forecasts assume great importance. In countries which depend on seasonal rains, such as the summer monsoons of Asia, there is a real need for forecasts on probable dates of the onset and stoppage of rains. Any error in the predicted date of onset leads to wastage of seeds and other resources. Any forecast for the total rainfall over a large country is not of much use, because rainfall and agricultural operations tend to be region-specific. Satellites and radars now provide information on rainfall. The need for such technologies in the developing countries is greater because their economies depend heavily on rain (WMO, 1993).

Management of Water Resources—Hydrological Forecasting

Hydrological forecasts are valuable for controlling river discharges and planning irrigation. For this area also, quantitative precipitation forecasts are valuable. The latter, in turn, need a good network of weather observations. Computers are used for data processing in hydrology. Improvements in computer software and storage through optical discs, assist hydrologists to speed up data processing. The benefits derived from the use of such technologies far exceed the investment costs because of the utility of hydrological forecasting in controlling river discharges and anticipating situations that lead to floods.

Bridging the Gap Through Technology Transfer

The chief aim of transferring technology from one part of the world to another is to reduce the existing disparities in levels of development. In meteorology and hydrology, the main objective is to ensure that comparable levels of services are provided to the developing countries with the overall goal of improving the scientific knowledge of atmospheric and environmental factors influencing human activities.

One of the potential contributions of technology transfer within the framework of international activities is the improvement of the level of meteorological and hydrological services of the developing countries with a view to providing better and effective inputs towards economic development (WMO, 1993).

Sustainable development will require the effective support of meteorological and hydrological services which provide vital information on the status of the environment and offer possible options for amelioration of changes that might be irreversible. Technology transfer will also depend on the energy options for the future. This, of course, depends on increasing the production of biomass and increasing the efficiency of biomass combustion. To reduce carbon dioxide emissions, a sustainable plan for biomass production will be needed, which implies better management of forests and non-renewable fossil fuels. Other energy options such as solar and wind energy, and more efficient use of renewable fuels will become necessary.

There are many different aspects of long-term planning for sustainable development which require an increase in the pace of technology transfer from the developed to the developing countries. The main requirements are for better exchange of knowledge on the environment, environmental data collection, research programmes in relevant scientific disciplines, and training and education.

In order to repay foreign debts, some African countries have encouraged agricultural and forest exports. This has indirectly contributed to the depletion of renewable resources such as forest, soil, water and biomass. Whereas unsustainable resource use at the household level means more time spent collecting firewood and potable water, that at the national level leads to declining yields and rural-urban migration.

In Africa, the Green Revolution technology involving breeding programmes for cereals and grains has in some cases led to ecological deterioration and economic decline. As a result, there is now a distinct shift to sustainable agriculture (which includes agriculture, forestry and fisheries), because sustainable development tends to conserve genetic, land and water resources, is environmentally viable and socially acceptable.

The most promising technologies for Africa and indeed, for other developing countries, are those which stabilize production while ensuring the conservation of natural resources, address farmer-identified problems and constraints, minimize the disruption of existing farming systems, are accessible and affordable for farmers, and are environmentally, socially and economically feasible to maintain in the long run (OTA, 1988).

NON-ENERGY BENEFITS OF INDUSTRIAL ENERGY EFFICIENCY

For too long, efforts to promote energy efficiency (E2) and pollution prevention (P2) have travelled on separate, parallel paths. Many E2 proponents considered only energy-savings aspects of their projects, and many P2 proponents did not include energy as a pollution source. Increasingly, however, the synergies between E2 and non-energy forms of P2 have become more apparent. E2 projects often have non-energy P2 benefits and P2 projects often save energy. In addition, both E2 and P2 projects often have non-environmental benefits, such as saving the direct costs of resources, reducing disposal costs, avoiding fines, minimizing adverse publicity, enhancing productivity, improving product quality, and improving workplace conditions. Companies attract praise by helping the environment, their employees, and their bottom line. When making a compelling E2/P2 case to businesses, all benefits—direct and indirect—must be taken into account to show how such projects impact the bottom line.

There is a strong need for industry to promote the non-energy benefits of energy efficiency investments. A broad range of productivity benefits can come from investments in energy-efficient technologies, including waste reduction and pollution prevention. These standard, widely-accepted analysis procedures are more credible to industry than the economic modeling done in the past because they are structured in the same way as corporate financial analysts perform discounted cashflow investment analyses on individual projects. Case studies of financial analyses, which quantify both energy and non-energy benefits from investments in energy-efficient technologies, are needed. Experience shows that non-energy benefits of energy efficiency projects often exceed the value of energy savings, so energy savings should be viewed more correctly as part of the total benefits, rather than the focus of the results (Pye and McKane, 2000). Quantifying the total benefits of energy efficiency projects can help companies to understand the financial opportunities of investments in energy-efficient technologies. Making a case for investing in energy-efficient technologies based on energy savings alone has not always proven successful. Evidence suggests, however, that industrial decision makers will understand energy efficiency investments as part of a broader set of parameters that affect company productivity and profitability.

Participation by big corporations can occur voluntarily if public good can be effectively linked to private profit potential. Most companies currently involved in sustainable development are corporate leaders. National governments need to work harmoniously with the private sector because corporations are absolutely essential to achieving any lasting improvements.

The roots of the problem—population explosion and rapid economic development in the emerging economies—are political and social issues that exceed the mandate and the capabilities of any corporation. And yet, multinational corporations may be the only organizations with the resources, the technology, the global reach and, ultimately, the motivation to achieve sustainability (Hart, 1997; Pye and McKane, 2000).

The foremost consideration in making a compelling case to business is the profit motive. Energy efficiency is generally not a primary driver in industrial decision making. Industry is much more interested in approaches whose impact on profit is more visible, such as productivity enhancements. Regardless of whether one's perspective is that energy efficiency is a byproduct of productivity gains, or that productivity gains are a byproduct of energy efficiency, it is very often the productivity gains that motivate industry to take action. Also, regardless of whether energy efficiency is the driver or the by product of a project, management must understand all the costs and benefits associated with an investment in efficiency in order to make wise decisions. Potential benefits beyond energy savings may include increased productivity; reduced costs of environmental compliance; reduced production costs; reduced waste disposal costs; improved product quality; improved capacity utilization; improved reliability and improved worker safety (resulting in reduced lost work and insurance costs).

When efficiency advocates understand the business decision-making perspective and can effectively communicate with management using financial and strategic arguments for energy efficiency, the case for energy efficiency becomes much stronger. Making business sense of energy efficiency reduces its perceived risk to management, which may, in turn, reduce the hurdle rate (or payback period) that a company requires of an energy efficiency investment. Since businesses make most decisions based on bottomline impact, it is logical to regard energy efficiency as part of overall 'efficiency' (e.g. process efficiency, enhanced productivity) to account for all the savings that a business will realize from energy efficiency projects (Pye and McKane, 2000).

REFERENCES

Ackermann, I., Wolf, J. Perspektiven einer dezentralen Pflanzenölgewinnung – *Landtechnik* 49. Jahrg. H. 5, 302/3 (1994).
Banks, R.D., Heaton, G.R. An innovation-driven environmental policy. *Issues in Science and Technology* pp. 43-51 (1995).
Bauer, P. Zehn Technologietrends für's kommende Jahrzehnt Nr. 8, 24. Febr. 1995 page 7 (1995).
Cassedy, E.S. *Prospects for Sustainable Energy: A Critical Assessment*. Cambridge University Press, Cambridge (2000).
EPA *Advanced Fossil Fuels and the Environment*. U.S. Environ. Protection Agency Report No. 600/9-77-013. Cincinnati, Ohio (1977).
GTZ : International program for field testing of photovoltaic water pumps. GTZ Report Abt. 415, Eschborn (1992).
Hansen, J., Sato, M., Ruedy, R., Lacis, A., Oinas. V. Global warming in the twenty-first century: an alternative scenario. *PNAS* (USA) 97: 9875-9880 (2000).
Jackson, T. Principles of clean production. pp. 143-164. In Jackson, T. (Ed.). *Clean Production Strategies*. Lewis Publishers, Boca Raton (1993).
Jansen, H.D. Steffen, M.Ch. Abpressen von Öl - Technische u. logistische Verfahrensoptimierung - *BMEFL, Reihe A. Sonderheft*, Landwirtschaftsverlag Múnster-Hiltrup (1991).
Koppel, T. *Powering the Future: The Ballard Fuel Cell and the Race to Change the World*. Wiley, Chichester (1999).
Manne, A.S., Richels, R.G. A multi-gas approach to climate policy—with and without GWPs. Presented to the Energy Modeling Forum EMF-19, Washington, D.C. (March 22-23, 2000).Hart, S.L. Beyond greening: strategies for a sustainable world. *Harvard Business Review*. January-February (1997).
Mayer, O., Müller, J. Alternative sources of energy - solar energy photovoltaic pumps and their use in irrigation systems. *Natural Resources and Development* 47: 44-53 (1998).
NSTC. *Technology for a Sustainable Future*. National Sci. and Technol. Council, Science and Technology Policy, Washington, D.C. (1994).

OTA. *Enhancing Agriculture in Africa: A Role for US Development Assistance.* Office of Technology Assessment (OTA), Washington DC. (1998).
Ovshinsky, S.R. The road to decarbonized energy. *Nature* 406: 457-458 (2000).Krause, R. Plant oil : Local small-scale extraction and use in engines. *Natural Resources and Development* 47: 71-82 (1998).
Parikh, J.K. Role of science, engineering, economics and environment in energy system of the 21st century. *Current Science* 78: 122-124 (2000).
Posorski, R., Haars, K. Ökonomische Querschnittsanalyse photovoltaischer Pumpsysteme. Ein Kostenvergleich von PVP-Systemen und Dieselpumpen. Deutsche Gesellschaft für Technische Zusammenarbeit (GTZ) Eschoborn (1994).
Pye, M., McKane, A. Making a stronger case for industrial energy efficiency by quantifying non-energy benefits. *Resources, Conservation and Recycling* 28: 171-183 (2000).
UNEP. *UNEP Profile.* United Nations Env. Programme, Nairobi (1990).
WMO. *Meteorology and the Transfer of Technology.* World Meteorol. Organizn., Report No. 786, Geneva (1993).

Chapter 6
Rural Development

INTRODUCTION

In villages, land and human survival are intertwined. The use or misuse of land has altered soil profiles, threatening its role in serving the socio-economic need of its people. Land is a prime national resource in every country and water resources are finite, which need to be conserved, developed and properly managed.

Land records and reform should be up-to-date and transparent, with land ownership records computerized at different levels. Any sound community-based participatory approach calls for providing women with land rights, wherever applicable.

Part of the regulatory agenda lies in preventing the use of agricultural land for non-farm use and protection of topsoil. The latter's use in brick manufacture should be prevented by employing alternatives such as fly ash. Social afforestation programmes can aid re-vegetation in agriculturally marginal soils. Irrigation efficiency of disturbed lands such as mine spoils also needs to be restored. Monitoring and evaluation aspects include transparent information posted annually on extent of drinking water shortage in villages, reduction of rain-fed areas, achievement of gender and social equities. The concept of a 'virtual' college links farming communities with scientists via interactive mode, secure transfer of scientific information as inputs for farmers and a bio-industrial watershed model integrating the processing at the village level, ensuring value addition and quality. For implementation, the choice rests on village headmen, self-help groups and target-oriented village level groups (Sen, 2001).

A community-oriented food and water security system for land resources associates rural knowledge centres in partnership with local Institutions (PRI). Sustainable development involves integrated information management solutions, databases for support operations and decision-making. Sen (2001) listed the following recommendations with respect to the participation of the scientific community:

- Assess erosion impact on productivity (in economic terms) at farms, watersheds, eco-regions and at a national scale;
- Evaluate off-site impacts of soil erosion in terms of water quality, air quality, emission of greenhouse gases and sedimentation of reservoirs and waterways;
- Develop indigenous prediction models and conservation measures;
- Determine soil carbon sequestration potential to mitigate the greenhouse effect through desertification control, restoration of degraded soils and ecosystems;
- Identify soil/land quality indicators which can be quantified by scientists and which farmers can understand;
- Assess once every 5–10 years, the state of natural resources (land, water and vegetation) at regional and national scales, using remote sensing and other modern techniques.

In the past, water resources have usually been managed in a fragmented, ad hoc manner. In some places, new approaches to management now link water demand and water supply. One such approach is Integrated Water Resources Management (IWRM), which promotes the coordinated development and management of water, land, and related resources, in order to maximize the resultant economic and social welfare in an equitable manner without compromising the sustainability of vital ecosystems. IWRM is as much concerned with the management of water demand as with its supply. To address these concerns, integration has to occur within and between two basic systems.

- The natural system, with its critical importance for resource availability and quality;
- The human system, which determines the resource use, waste production, and pollution of the resource, and which also sets development priorities.

Tangible improvements in the sustainable use of freshwater resources can be achieved by protecting and enhancing local watersheds, reducing water pollution, improving the availability and efficiency of water and environmental sanitation services, and promoting public health.

Although much is being talked about sustainability, there are certain key obstacles as well as opportunities to local sustainability. One important issue is how to institutionalize sustainable development into municipal operations and planning.

Obstacles include:

- Inappropriately allocated financial and human resources.
- Difficulty in communicating the concept of sustainability.
- Jurisdictional and institutional fragmentation, and
- Psychological and institutional inertia.

Some opportunities for promoting local sustainability include green purchasing, revolving loan funds for sustainability projects, incentive programmes for local businesses, reviewing bylaws and land use planning policies in order to remove barriers to sustainable planning, and integrating sustainability into primary and high school education programs.

RURAL TRANSFORMATION THROUGH APPROPRIATE TECHNOLOGY

Rural technology policy in many developed countries needs to be changed so as to enable the poor to participate and share power in the process of technological transformation. Among the new technologies, organic agriculture (including production, processing and marketing) provides diverse opportunities. Organic agriculture offers the following holistic, multifunctional and sustainable solutions to rural problems:

1. The land available for cultivation in the villages is enriched with high fertility to generate food and commercial crop production needed for the population and the growing economies.
2. Employment and livelihood opportunities for the rural poor are greatly increased and absolute poverty is decreased.
3. Effective waste recycling occurs in the rural areas, opening up a vast potential for rural income generation and pollution control.
4. Nutrition and health status of rural communities has improved.
5. Biodiversity in rural areas and surrounding forest areas is sustained and enhanced.
6. Natural resources, especially in terms of land, water and biological wealth can be managed efficiently in environment-friendly manner.

Composting

Soil organic matter effectively regulates crop productivity. The mineralization of decomposing residues releases essential plant nutrients. The activities of micro-organisms and soil fauna help promote soil aggregation, which reduces erosion and promotes greater moisture infiltration. Maintenance of soil organic matter in low-input agroecosystems increases nutrient retention and storage, increasing buffering capacity and water holding capacity of the soil.

In conventional agriculture, little or no agricultural residues are returned to the soil. The resulting decline in soil organic matter frequently means lower crop yields or poorer plant biomass productivity. Composting effectively increases soil organic matter by recycling plant residues and animal wastes.

Spinning and Weaving

Since the 1950s and 1960s, decentralized spinning and weaving in villages has virtually disappeared. Millions of rural people have lost their livelihood, while billions of rupees flowed to a small number of textile mill owners, making them extremely wealthy. Today, the textile industry earns profits of around billions of rupees annually. Consequently, after five decades of independence, the notion that spinning in every house and weaving in every village is impracticable has gained popularity. According to Desai (1999), spinning in every house and weaving in every village by exploiting modern technology and small machines operated by electricity can still be made possible even today. In this context, the following points need to be considered.

1. As cotton is cultivated in almost 80% of India's villages, the raw material is readily available almost everywhere.
2. Spinning and weaving at the village level eliminates the need for baling of cotton and saves considerable expense otherwise required for this purpose. Also, the costs of transporting cotton bales from all over India to the few textile centres can be saved.
3. In a large textile mill, nearly half of the heavy and costly machinery is for breaking the bales, loosening up cotton and making the fibres of the pressed cotton parallel. Decentralized spinning means that loose cotton will directly go to spinning, eliminating the need for the complicated machinery. Heavy costs can be saved and better results achieved with simple and smaller machines for spinning of loose-ginned cotton. Cloth made from this cotton is 15 to 20% more durable than that made from the pressed bales. This means that village level spinning and weaving will yield more durable cloth than the mill cloth.

The few (centralized) mills have to procure cotton from half a million villages and, in return, market cloth to them. This entire system smacks of injustice, exploitation and profiteering. If spinning and weaving is decentralized to every village, this exploitative and unjust system will collapse.

Tiny Textile Mills

If we consider a village population of 2,000 and a per capita annual requirement of cloth of 25 m, the annual requirement of this village will be 50,000 m. So, this cloth needs to be made in a year. Assuming 250 working days in a year, the daily production could be 200 m. For this volume of production through a tiny textile mill, Desai (1999) calculated the requirement of simple, electrically-powered machinery using a 100 watt, 0.5 HP motor. Most of the machinery is available in India except for pre-processing machinery from lint cotton to roving stage which can be developed by utilising the most modern technology. Decentralized production does not imply that everything is under one roof. The various machines are small enough to be moved from one home to another. Spinning can be carried out according to the convenience of the family members, and once their own requirement of cotton yarn is met, the spinning machine can be shifted to another home. So, only 25 spinning machines can meet the needs of all the homes in the village by rotation. Activities are distributed among several locations. Pre-processing machines may be at one particular place in the village. Similarly, weaving looms may remain permanently installed in the weavers' homes. Other machines only can be taken from one home to another.

According to Desai, the cloth manufactured at the village level may not cost more than about Rs 16-20 per metre. By comparison, a cloth of the same quality in the urban market will cost Rs 35-40 per meter. So, in the village, the minimum benefit is Rs 20 per metre. For the whole village, the total benefit may amount to almost one million rupees per annum. By extrapolating to the level of the whole country, the economic benefits can well be imagined!

Treadle (Pedal) Pump

Agriculture provides livelihood to a high proportion of the total population. The availability of irrigation water is extremely important for rural farming communities. Most farmers are small and marginal in nature, having very little access to resources or technology. Marginal farmers are those who have less than one hectare (2.5 acres) of land. Small farmers have between one and two hectares of land; semi-medium farmers own 2-4 hectares; medium farmers own 4-10 hectares and large farmers possess more than ten hectares of land.

As rainfall and groundwater resources are unevenly distributed, access to irrigation water poses problems especially for small and marginal farmers.

As considerable investment is needed to install bore wells and purchase electric or diesel pumps, poor farmers cannot tap groundwater resources themselves. Since they depend on their land to eke out a living, they have to either rely on slow and inefficient methods of manual irrigation, or buy water from owners of bore wells, or hire diesel pumps. As these latter options are expensive, the farmer is caught up in a debt-and-poverty trap. Traditional means of manual irrigation are time consuming and labour intensive.

Quite often, poor farmers depend strongly on one rainy season crop for subsistence and have to struggle for the rest of the year, some times, migrating far to work as daily wage labourers.

For sometime now, International Development Enterprises (IDE) have developed and marketed a small foot-operated irrigation pump in the Indian sub-continent. This simple technological device, called the treadle pump, helps the poor to make their small-holdings productive and lucrative. With irrigation provided by the treadle pump, poor farmers become able to grow crops which not only feed their families but also provide some income, season after season. This organization has concentrated on developing appropriate, low-cost, environment-friendly irrigation systems. In areas with high water table, it promotes the treadle pump. In semi-arid areas, it has adapted, developed and introduced affordable micro-irrigation technologies in the form of kits such as bucket kit, drum kit, micro sprinkler kit and overhead sprinkler kit.

The treadle pump was originally developed in Bangladesh. It is a low-cost, manually-operated water-lifting device suitable for areas where the water table varies from 10 to 25 feet. It effectively irrigates small plots of land and is ideally suited to small and marginal farmers. It is easy to install, requires little maintenance and has minimal operating cost. In principle, the treadle pump is similar to the hand pump. However, while a hand pump consists of a single barrel or cylinder and water is pumped up with one's hands, the treadle pump has two cylinders and requires foot operation for lifting water (hence called a treadle or pedal pump).

One individual can operate the pump by manipulating his/her body weight on the treadles while holding on to a bamboo or wooden frame for support. The pump is usually installed on 1.5" tube-wells (made of GI, PVC or bamboo) but can also be fitted on 3" and 4" tube-wells (by using relevant reducer sockets) for installing electric and diesel pumps. The treadle pump may also be used for drawing surface water from ponds, canals, streams and dug wells as well as by connecting a suction pipe to the pump with a GI bend-pipe. The treadle pump has three unique design features: use of full body weight and leg muscles, of twin cylinders and alternating strokes and the water outlet at ground level. Depending on the model, water discharge from this pump can range between 4,500-6,000 litres per hour.

Wind Pumps

Windpumps also serve as an economical irrigation aid in many areas and, like treadle pumps, are relatively easy to operate.

Attempts have been made to produce low-cost windpumps on a large scale, choosing a light-weight second generation design developed in the Netherlands. The Dutch designers tried hard to make a sturdy and simple machine that could be manufactured in villages without the need for workshops with sophisticated equipment. Any small workshop with a lathe, drill press and welding equipment could produce this windmill from A to Z—rotor, head mechanism, tower, pump—and also take care of its installation and maintenance. A few hundred machines were actually manufactured and installed in villages. The programme was 90% government-subsidized, with the beneficiary only required to provide the tower foundation and a small tank. And yet, the programme failed (Truntz, 1999), because of poor manufacturing standards, bad quality and very low prices. The alignment of the tower with the well was not properly taken care of. In some cases, the well itself was curved or not gravitationally centred. Inferior quality water used caused rapid corrosion. The manufacturer was unable to honour his warranty obligations.

Today, this windpump is still being produced in small numbers by a few enduring manufacturers. More blades are put in the rotor to slow it down and get more torque, weak components have been reinforced, and it is being installed where the site conditions match the site criterion given by the designers for low lift pumping in areas with low to moderate wind. Subsidy has been reduced from 90 to 50%.

The government now prefers the time-tested traditional American multiblade geared windpumps which are also 50% subsidized. Technically, this also may be a retrograde step because it is based on an outdated design of the 1930s. Furthermore, these geared windpumps are not very cost effective, and with a rotor diameter of just three metres, output is only moderate. Over the last few years, these windpumps have been installed at a rate of about 100 to 200 a year (Truntz, 1999).

The Auroville Ashram in Pondicherry, India is developing its own windpumps. Lacking adequate electrical power supply, many communities depend entirely on the windpump for their water needs, and in order to pump the substantial amount of water required for greening the desert-like Auroville plateau, large rotor diameters were chosen. The first prototype was hammered, welded and wired together with bits of scrap steel. This structure, with a rotor diameter of six metres and canvas sails, was mounted on an 8-metre wooden mast. It turned out that wood and canvas were not the right materials since they require a lot of maintenance and have only a short life in the tropics.

Furthermore, the wind/mills have to be protected from storms by rolling in the sails. A second prototype windmill, all in metal with a fully automatic safety device, was designed and built as the forerunner of the AV55, and a sound windpump design has now evolved slowly. The AV55 is by far the most-cost effective windpump in India and possibly in the world, since it is cheaper than solar pumps or even diesel-powered ones.

Later, the matching valve developed by the Eindhoven University of Technology (P. Smulders) was tested on the Auroville windpump. In the field, an increase in water output of 50% could be achieved with the matching valve compared with a conventional pump design operated by the same windmill. The matching valve is a float valve built into the piston of the pump and replaces the disc valve of the standard piston pump. In low wind the float valve remains open at the beginning of the upstroke. The rotor starts turning without any load, and does not have even to lift the pumprods as their weight is counter-balanced by a counter-weight fixed in the rotor. Once the pump piston has gathered speed, the valve will close as soon as calculated critical piston velocity is exceeded and pumping starts.

APPROPRIATE ENERGY TECHNOLOGY AND PHOTOVOLTAICS

Slowly but steadily, a wave of privatization of the electricity supply industry (ESI) in Asia is sweeping the region with a substantial impact on appropriate energy technology (AET). In essence, it is effortlessly bringing to the market a wide and expanding range of AET. Two main aspects of this minor miracle were: firstly, by cutting the financial connection between ESI and government subsidies, privatization is forcing electricity suppliers to re-evaluate their options from a business—as opposed to a public service—perspective. The net effect of the consequent imperative to provide competitively priced power from a least-cost combination of energy source, energy conversion technology and delivery system is to strike a happy balance between all the various options to reveal where and under what conditions a specific technology is genuinely least cost. AETs are increasingly being selected on this basis. Although they still typically occupy only niche markets, this has started changing where fossil fuel markets are also being privatized and, hence, de-subsidized. Secondly, privatization is enabling the establishment of institutional structure, especially retail power markets and their associated regulatory authorities that can tilt the economic playing field even more strongly in favour of specific AETs. They, thus, enable still expensive renewables to find genuine market niches while their costs are being lowered (Sharp, 1999).

Today, small, unconventional and dispersed power stations, which were usually banned under earlier regimes, are being preferred and regulatory control over power stations is being loosened. An AET future seems much more feasible with privatization than before (Sharp, 1994).

The World Bank, in relation to photovoltaics (PV), has pointed out that in shedding the subsidies they formerly enjoyed, power utilities discover the true cost of grid extension to remote areas. They should now also consider PV as an alternative to grid extension or allow the private sector to compete to provide energy services to formerly unserved populations. Many entrepeneurs prefer PVs.

As other AETs join PV, many communities have begun to get their electricity at its most affordable price, and the entire energy resource base is beginning to be used more efficiently. This, in turn, has strong environmental implications which are greatly enhanced when the AET itself is renewable.

It is becoming possible to market expensive renewably generated electricity at its true cost, yet without coercing consumers in any way.

This positive aspect of privatization has not yet penetrated Asia. No country has yet established the full retail power markets that are required. But Singapore and China are on the verge of doing so. Much more important, the entire region is on the same path.

Today, even transmission systems are being redesigned, while in the foreseeable future, AET will be all there is, both massive and mainstream. Significantly, power utility monopoly is now seen to have blocked much cost-effective distribution of rural electricity. It has been demonstrated that dispersed power distribution costs can be reduced far below the central power authority norms typically used for planning purposes.

Australia is now building 2 and 5MW-scale demonstration solar thermal/coal hybrid power stations that use, respectively, conventional parabolic 'big dish' solar collectors and a much denser array of compact linear fresnel reflectors (CLFR) to produce steam for the boiler-house. Both systems expect to upgrade to upto 200MW installations in the coming years when solar power prices may fall to about 7.5 Australian cents/kWh, so close to competitive that Green Power-style markets will have no difficulty in absorbing them.

As both environmental pressures and the demand for energy increase, privatization has opened the door to a bright AET future. Indeed, the AET frontier may now have shifted from technological research, development and demonstration to regulatory reform and the political processes that enable it (Sharp, 1999).

Transnational Advocacy Networks

Technology and cultural changes in recent years have stimulated the growth and development of transnational advocacy networks, organized round shared values and discourses, designed to target international organizations or the policies of a particular country. These networks have had impact in human rights and influenced environmental politics as well. These networks are important in bringing transformative and mobilizing ideas into the international system. Very few studies have been done on the effects of transnational corporations (TNCs) on the poor people.

Madeley (1999) explored and examined these unelected, undemocratic and largely unaccountable corporations. Drawing on the experiences of worldwide NGOs, he documented the impact TNCs have on the economic sectors in which they operate, namely: agriculture, forestry, fishing, mining, oil extraction, manufacturing, tourism and medical drugs. He suggested some practical alternatives and solutions to enable consumers, governments, shareholders, NGOs and other interested parties to develop methods in which to influence TNCs, encouraging them to move from global production to small, local, co-operative and resource conscious strategies.

RURAL LABOUR

The peasant transition process has now reached a critical juncture in some countries. Peasant labour redundancy can undermine rural welfare and political stability. Academics and policymakers of the twenty-first century cannot ignore the world's disappearing peasantries without endangering sustainable development and international security. Bryceson et al., (2000) provide a new insight into peasant studies and the western biases that have permeated it. They have traced patterns of peasant formation and dissolution over time and explored whether today's rural producers in Africa, Asia and Latin America are peasants in either a theoretical or practical sense. Their rich case study material from these continents brings out the pressures and opportunities that have befallen peasants, leading them to 'diversify' into a number of occupations and non-agricultural income-earning avenues. The relationship of peasants to the land has changed; some factors that influence this include multi-occupational livelihoods, intensified labour mobility and flexibility, straddled urban and rural residences and flooded labour markets.

Edwards (1999) has provided an analytical perspective on future co-operation at international level, arguing that as part of an interdependent world, the only way to tackle problems of global poverty and violence is to forge a new path at global level that balances heavy-handed intervention and complete *laissez faire* that facilitates problem solving without imposing unfair costs on anyone else.

BIOGAS PRODUCTION TECHNOLOGY

Biogas technology provides an alternate source of energy in rural areas that can meet the basic need for cooking fuel in villages. Using local resources such as cattle waste and other organic wastes, energy and manure are generated. Biogas is produced from organic wastes by the concerted action of various groups of anaerobic bacteria.

Microbial conversion of organic matter to methane has attracted much interest as a method of waste treatment and resource recovery. It is an anaerobic process that is carried out by certain groups of anaerobic bacteria.

Three basic characteristics of this process are the following:

(i) Most of the important bacteria involved are anaerobes and slow growing;
(ii) A high metabolic specialization occurs in these anaerobes; and
(iii) Most of the free energy present in the substrate is found in the terminal product methane. Since less energy is available for the growth of bacteria, less microbial biomass is produced and consequently, disposal of sludge after the digestion does not pose a serious problem.

Complex polymers are broken down to soluble products by enzymes produced by fermentative bacteria (Fig. 6.1, Group 1), which ferment the substrate to short-chain fatty acids, hydrogen and carbon dioxide. Fatty acids longer than acetate are metabolized to acetate by obligate hydrogen-producing acetogenic bacteria (Fig. 6.1, Group 2). The major products of breakdown of the substrate by these two groups are hydrogen, carbon dioxide and acetate. Hydrogen and carbon dioxide can be converted to acetate by hydrogen-oxidizing acetogens (Fig. 6.1, Group 3) or to methane by carbon-dioxide-reducing, hydrogen-oxidizing methanogens (Fig. 6.1, Group 4). Acetate is also converted to methane by aceticlastic methanogens (Fig. 6.1, Group 5). Nearly 70% of methane from biogas digesters working on cattle dung comes from acetate (Nagamani and Ramasamy, 1999).

Microbial diversity in biogas digesters is as high as that of rumen, wherein seventeen fermentative bacteria species have been reported to play a role in biogas production. It is the nature of the substrate that determines the type and extent of the fermentative bacteria present in the digester. Whereas in rumen, *Ruminococcus* sp. alone accounts for

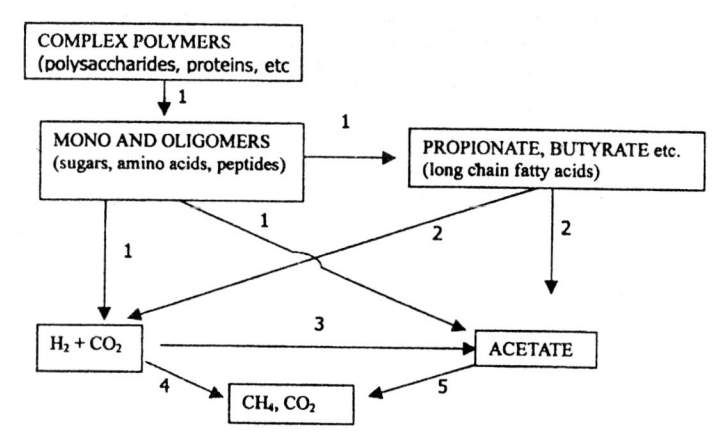

```
COMPLEX POLYMERS
(polysaccharides, proteins, etc)
        │ 1
        ▼                    1
MONO AND OLIGOMERS  ──────▶  PROPIONATE, BUTYRATE etc.
(sugars, amino acids, peptides)  (long chain fatty acids)
                    1
         1                2        2
         
         ▼      3              ▼
    H₂ + CO₂  ──────────▶  ACETATE
         4                5
         ▶  CH₄, CO₂  ◀
```

Fig. 6.1 Microbial groups involved in biogas production. Group 1, fermentative bacteria. Group 2, obligately hydrogen producing acetogenic bacteria. Group 3, hydrogen-consuming acetogenic bacteria. Group 4, carbon-dioxide-reducing methanogens. Group 5, aceticlastic methanogens (Nagamani and Ramasamy, 1999).

some 60% of the total population, in the biogas digester, the predominant species belong to the genera *Bacteroides* and *Clostridium* rather than the genus *Ruminococcus*. *Eubacterium cellulosolvens, Clostridium cellulosolvens, Clostridium cellulovorans, Clostridum thermocellum, Bacteroides cellulosolvens* and *Acetivibrio cellulolyticus* are some of the other predominant fermentative bacteria present in cattle dung-fed digesters.

Most of these bacteria adhere to the substrate prior to extensive hydrolysis. The particulate-bound bacteria show direct relation to the biogas yield from the digester.

Though several different products are formed by the action of fermentative bacteria, volatile fatty acids are the primary products of carbohydrate fermentation in biogas digesters, as they are in rumen. Studies on the interaction of cellulolytic bacteria, *Acetivibrio* sp. and methanogens, *Methanosarcina* sp. and *Methanobacterium* sp., using cellulose and cellobiose as substrate revealed that by using co-cultures, the growth of both *Acetivibrio* sp. and *Methanosarcina* sp. can be stimulated and that the methane content of biogas can be significantly enhanced (Nagamani and Ramasamy, 1999).

Most methanogens have a limited metabolic repertoire, using only acetate or C_1 compounds (H_2 and CO_2, formate, methanol, methylamines or CO), with methane being the end product of the reaction.

Methanosarcina sp. and *Methanosaeta* sp. form methane by aceticlastic reaction. Faster-growing *Methanosarcina* sp. predominate in high-rate, shorter-retention digesters where acetate concentration is higher; *Methanosaeta* spp. are predominant in low-rate, slow-turnover digesters. Pathway of methane formation by *Methanobacterium thermoautotrophicum* is shown in Fig. 6.2.

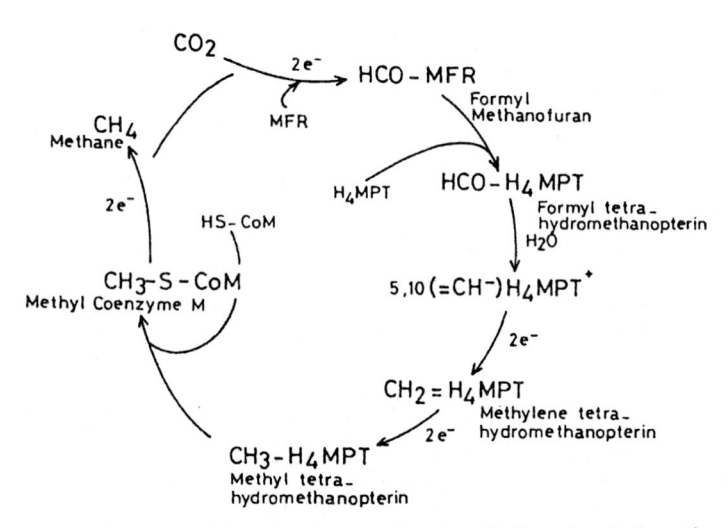

Fig. 6.2 Pathway of methane formation by *Methanobacterium thermoautotrophicum* (Nagamani and Ramasamy, 1999).

Both CO_2-reducing and aceticlastic-methanogens are important in maintaining stability of the digester. The failure in a biogas digester can occur if carbon dioxide-reducing methanogens fail to keep pace with hydrogen production. Sometime hydrogen in a biogas digester can build up rapidly to levels inhibitory to methanogenesis due to failure of the activity of hydrogen-scavenging organisms, shifting the fermentation products away from acetate. Moreover, failure of aceticlastic methanogens to keep up with acetic acid production leads to the accumulation of fatty acids, resulting in failure of the digester (Nagamani and Ramasamy, 1999).

Factors Affecting Biogas Production

Factors such as biogas potential of feedstock, design of digester, inoculum, nature of substrate, pH, temperature, loading rate, hydraulic retention time (HRT), C : N ratio and volatile fatty acids (VFA) affect the rate of biogas production (see Khandelwal and Mahdi, 1986). The performance of floating dome biogas plant seems to be better than the fixed dome biogas plant, showing an increase in biogas production by about 10%. Poultry droppings tend to show higher gas production as compared to other animal wastes. For increased gas yield, a pH between 7.0 and 7.2 seems best. The pH of the digester is a function of the concentration of VFAs produced, bicarbonate alkalinity of the system, and the amount of carbon dioxide produced.

Digester Design

Floating gas holder type and fixed dome type digester are being widely employed, but not much work has been done on the microbiological aspects of these digesters. In general, fixed dome digesters show higher microbial distribution of all trophic levels than floating drum type biogas digesters.

Alternate Feedstocks

Animal wastes are generally used as feedstock in biogas plants and their potential for biogas production is given in Table 6.1. But, scarcity of these substrates hinders the successful operation of biogas digesters. The availability of cattle waste can support only some 12-30 million family-size biogas plants against the requirement of over 100 million plants. A significant portion of 70-88 million biogas plants can be run with fresh/dry biomass residues. Of the available 1,150 billion tons of biomass, a fifth may suffice to meet this demand (Nagamani and Ramasamy, 1999).

Various substrates have been explored for biogas production. The two important parameters in the selection of particular plant feedstocks are economic considerations and the yield of methane for fermentation of that specific feedstock. Comparisons of the methane yield from freshwater aquatics, forage grasses, roots and tubers and marine plants showed that highest yield was obtained from root crops followed by forage grasses and freshwater aquatics. Marine species give the lowest yield of methane. Methane yield and kinetics are generally higher in leaves than in stems. Observations on biogas production from food industry wastes and different biomass sources have shown that pretreatment of feedstock improves the biogas yield and methane content from biomass-fed digesters.

Some alternate feedstocks and their potential for biogas production are given in Table 6.1. Several alternate feedstocks exist that have a potential for biogas production. It is time that substrate-specific biocatalysts are made available to reduce the lag period of biomethanation during the

Table 6.1 Selected alternate feestocks and their potential for biogas production (after Nagamani and Ramasamy, 1999).

Feedstock	Hydraulic retention time (days)	Organic loading rate (kg VS m^{-3} d^{-1})	Methane yield ($m^3 kg^{-1}$ VS)
Fruit wastes	8	3.8	0.030
	16	3.8	0.250
	16	9.5	0.420
Tomato processing waste	24	4.3	0.420
Banana peeling			
0.40 mm size	NA	NA	0.409
1.0 mm size	NA	NA	0.396
6.0 mm size	NA	NA	0.374
Cauliflower waste			
0.40 mm size	NA	NA	0.423
1.0 mm size	NA	NA	0.423
6.0 mm size	NA	NA	0.407
Mirabilis leaves			
0.40 mm size	NA	NA	0.341
1.0 mm size	NA	NA	0.329
6.0 mm size	NA	NA	0.327
Ipomoea sp.			
0.40 mm size	NA	NA	0.427
1.0 mm size	NA	NA	0.421
6.0 mm size	NA	NA	0.413
Wheat straw			
0.40 mm size	NA	NA	0.248
1.0 mm size	NA	NA	0.241
6.0 mm size	NA	NA	0.227
Paddy straw			
0.40 mm size	NA	NA	0.367
1.0 mm size	NA	NA	0.358
6.0 mm size	NA	NA	0.347
Gliricidia maculata leaves	5	4.95	0.034
Parthenium hysterophorus	10	2.48	0.117
	20	1.24	0.115
Lemna minor	NA	NA	0.106
Lantana camara	NA	NA	0.236
Night soil (NS)			
100% NS	NA	NA	0.500
50% NS + 50% cowdung	NA	NA	0.350
100% cowdung	NA	NA	0.130

start-up. Regular supply of inoculum and quality control on the marketable inoculum should help regulate the digester failures.

SOLAR COOKING

High-quality solar cookers can reach temperatures exceeding 200°C, which are high enough to cook food, bake bread and heat up an iron for ironing clothes. As rural people in many parts of the world heavily depend on wood for preparing meals, solar cooking can also make an important contribution to saving our forests. But even after three decades of implementation, solar cookers have not evolved into commercially-viable products that sell themselves.

In a solar cooker, the rays of the sun are converted to heat and conducted into the cooking pot. The solar cookers often make use of a box or parabolic reflector to concentrate the sun's rays. Solar power is inexhaustible, clean and free. By decreasing their dependence on conventional fuels, people not only save money but also non-renewable resources and the environment.

In rural areas, the available energy sources are mostly firewood, crop residues and animal dung. In urban and semi-urban areas, gas, kerosene oil and coal are being used for cooking purposes. The smoke from these fuels pollutes the environment. Unfortunately, contrary to the original intention, most solar cookers have been sold in cities and suburban areas. The cookers are still beyond the reach of most villagers.

In the past, solar cooking has often failed to make headway. The key to success is to have user-friendly, high-performance, durable and cost-efficient products. Previously, no cooker on the market completely satisfied all of these requirements, so projects often fell short of expectations.

If solar cookers are to be commercially sustainable, support for introducing them to the market is required to overcome market resistance, increase customer knowledge and ensure market penetration. There is need to develop low cost but durable models for wider use in rural areas.

PLANTS AS ENERGY SOURCE

Energy plants are those annual and perennial species which can be cultivated to yield solid or liquid energy sources. Organic residues and wastes of the most widely diverse production, also used for producing energy, though not energy plants, nevertheless have much potential for energy production.

Biomass for energy use includes roots, tubers, stems, or branches, leaves, fruits and seeds or even whole plants. Those plants and parts of plants are used preferentially, however, which have a high energy density, in order to achieve high yields.

All plants that store carbohydrates or oil are suitable for producing liquid energy sources. Cellulose, starch, sugar and inulin can be used to produce ethanol. Vegetable oils can be used as fuels (Fig. 6.3). Lignocellulosic parts can provide energy directly as solid fuels or indirectly after conversion (El Bassam, 1995).

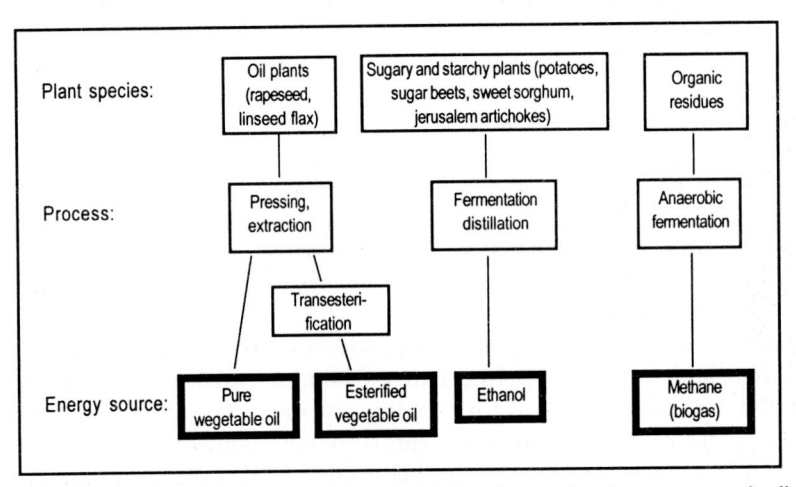

Fig. 6.3 Possibilities of energy production from starch, sugar and oil plants (after El Bassam, 1995).

By thermal and thermochemical processes (e.g. liquefaction or gasification), lignocellulosic raw materials can be used to produce fuels like methanol, biodiesel, synthesis gas and hydrogen and hydrolysis (saccharification) and by subsequent fermentation give rise to ethanol.

The chief aims of cultivating energy plants are as follows:

1. cultivation of sacchariferous and amyliferous (starch and sugar crop) plants to produce ethanol;
2. use of vegetable oils as fuels;
3. production of vegetable solid fuels to obtain heat and electricity directly or conversion for use as fuels; and
4. cultivation of biomass to produce biogas

There is considerable potential of agricultural and forestry waste and residue materials such as straw, industrial wood and forest wood residues for energy utilization.

For the production of solid fuels the following plant species deserve consideration:

1. Annual plants such as grain, rapeseed, mustard, sunflower and maize;
2. Perennial species utilized annually, such as *Miscanthus* and other reeds;
3. Fast-growing trees like poplar, aspen or willow with a perennial harvest rhythm (short rotation or cutting cycle); and
4. Tree species with long rotation (El Bassam, 1995).

Biomethanation is important for the disposal of organic residues and waste products generated in the processing of agricultural products and in animal husbandry. Here, environmentally relevant aspects are of primary concern.

A broad spectrum of green plants could be used for energy because, in contrast to combustion of the raw material, it might be used in its natural moist state.

The plant species most suitable for the production of biogas are those which are rich in easily degradable carbohydrates, such as sugar and protein.

Raw materials from lignocellulose-containing plant species are not suitable for biogas production.

CULTURE AND BIODIVERSITY

At the drawn of a new millennium, some people are looking back in history to search for security and to seek solace and guidance from their strong and resilient traditional cultures.

Creativity constitutes the core of all cultures. It is the identity of a people — how they sing their songs, dance to their music, design their ceremonies, festivals and rituals. In many traditional societies, these various cultural expressions are strongly linked to the surrounding nature, flora and fauna.

Many indigenous people and rural communities are reviving their old customs in relation to biodiversity — some of these suffered harm from the growing influence of modern globalization — with a view to be able to meet the challenges of the new millennium.

Some people consider biological and cultural diversity as an obstacle to improving production and economic growth in rural communities. This may be why many development efforts include the diffusion of monocultures and high-yielding agro-chemical dependent varieties. Such an approach, however, greatly changes traditional diversified agroecosystems with the long-term consequence of genetic erosion and the deterioration of natural resources. This can eventually lead to a reduction in opportunities for ensuring local self-sufficiency in food.

Several rural societies have recently suffered from the loss of biological diversity and cultures and traditions have become less diverse. But there are also experiences of rural people strengthening their knowledge, traditions and spirituality to meet the needs of the next millennium.

Most indigenous people have the greatest cultural diversity in the world and live in the areas of highest biological diversity. They have nurtured species variation for thousands of years.

Rural people's livelihoods and social structures and cultures are largely determined by nature and vice versa. This close interrelationship between nature and culture goes a step beyond the concept of 'natural biodiversity' towards 'biocultural diversity' (Dankelman and Ramprasad, 1999).

Recent decades have seen a great erosion of biodiversity both in natural environments and on agricultural landscapes. In the name of development the diverse agricultural landscapes have been transformed into a monotonous environment in many parts of Europe. All over the world, the Green Revolution has replaced numerous species and varieties of crops, plants and trees with monocultures that are vulnerable to pests and diseases.

The erosion of natural biodiversity has gone hand-in-hand with diminishing cultural diversity. With increasing globalization, cultural expressions like songs, dance, and rituals are being forgotten or considered outdated by the younger generations.

The market-economy and globalization of material needs tended to increase homogeneity in culture and values, and so the linkages between culture and nature get blurred.

A diverse and balanced ecosystem tends to be highly flexible and resilient because there are many species with overlapping functions that can substitute each other. In a more diverse network, the patterns and relationships are more complex. A diverse community is flexible and can adapt more easily to changing circumstances. This requires awareness of and respect for different functions and perspectives, based on communication. Isolation of groups and individuals in a society leads to fragmentation and can be a source of conflict. Similarly, isolation and fragmentation in ecological systems can threaten the survival of species and ecosystems.

Biocultural Diversity

In general, most indigenous people have a holistic and sustainable life style — what is taken from the earth is given back in some way.

Conservation and preservation are central to these regulations. Some groups are now working to revive indigenous knowledge and trying to understand and challenge them by means of experiments and innovations. Many farmer groups and indigenous people are engaged in endogenous development. There is also enthusiasm among farmers, spiritual leaders and some policy bodies to experiment with ancient knowledge in the modern context. Of course, there are some constraints in the process of strengthening biocultural diversity. Some negative aspects often exist within the traditional cultures themselves (e.g. restrictions on women, black magic or the abuse of power related to secret knowledge). In extreme cases, some people have fundamentalist tendencies (Dankelman and Ramprasad, 1999).

Table 6.2 lists some specific examples of the relationship between culture and biodiversity.

Table 6.2 A few examples of relationship between culture and biodiversity (after Dankelman and Ramprasad, 1999)

Country	Cultural aspect	Related biodiversity	Development potential
Bolivia	Potatoes have souls; Mother Earth is sacred	Many varieties related to altitude, microclimate and traditions	Search for and promote traditional potato varieties
India	Mango is used in rituals; its leaves protect the house	Many varieties of tender mango	Rural competition for revival of tender mango varieties and related traditions
	Tribals have strong relations with plants and animals	Plants, trees and animals are protected	Changing hunting ritual into conservation ceremony
	Sacred seeds and ceremonial germination tests	Diversity of food crops is maintained by women	Conservation and dissemination of traditional food crops
Indonesia	Adat house is place for worship and teaching	Different plants are used in pest control	Revival of adat house and traditional pest control in crops
Zimbabwe	Spirits live in nature (e.g. in a baobab tree)	Wetlands are needed as habitats for spirits	Rehabilitation of sacred wetlands

A vast web makes up our cultural, spiritual and social diversity in which cultural and spiritual values have central importance for the appreciation and preservation of all life. It is value that gives us a true reflection of worth.

Language cognition and speech encode indigenous knowledge systems and are critical for preservation of diversity. The complex issue of indigenous people and the problems of preserving their relationships both with and within their societies are equally important in this context and so also voices of the world — expressions of concern and disquiet over the declining world diversity; holistic health practices where environment and diet are integrated into indigenous medical health systems; the importance of developing effective intellectual property rights, territorial and land rights to enhance and maintain local control (see Posey, 2000).

An effective integration of environmental, social and economic policies is the cornerstone of sustainable and equitable development. Schnurr (1998) provided concrete examples of the relationships among these three imperatives and examples of meaningful policy integration, from both Northern and Southern perspectives. The complexity of local-global interaction in environmental management can be partly understood by focussing on understanding resource management as a socio-cultural concept. There is an urgent need to combine different perspectives in environmental management, both of different levels and of separate traditions and disciplines, as well as cultural and organizational experiences.

There are also specific examples of how local people have learned to conserve biodiversity in an extraordinary range of environments and social conditions, and how many world religions are re-assessing their roles as stewards in the light of environmental impoverishment.

Two of the most important currents in today's thinking are community participation and sustainable development.

Some issues that have become more and more prominent in the concerns of communities throughout the world are the environment; gender and development; ethnic and racial conflict; intercultural understanding; and building participatory governance.

Influential thinkers on community and society need to address the fundamental questions about the needs of the future—the need to participate and belong; the need to conserve resources and to get more from less—and the kind of society that will successfully create a community and achieve sustainable development.

The objectives of sustainable people-centred development and the processes required to achieve it should focus on the five factors which determine effectiveness: suitable organizational design; competent leadership and human resources; appropriate external relationships; mobilization of high quality finance; and the measurement of performance coupled to 'learning for leverage'. Fowler (1997) has explained the capacities needed and how they can be assessed and improved.

AGROBIODIVERSITY

Agrobiodiversity has been decreasing in many industrialized countries. Until recently, only a few organizations perceived agrobiodiversity as crucial to sustainable development.

In the Netherlands, for instance, agrobiodiversity has seriously decreased since 1950. Three major crops in the Netherlands are potato, sugarbeet and wheat. In 1989, the top three varieties of sugarbeet and wheat covered 69% and 79%, respectively, of the total planted area with these crops. In barley and rye, this was even higher—89% and 95%, respectively. In livestock production, by 1996, over some 900,000 inseminations had been carried out in the Netherlands with semen from one single Holstein Friesian bull, named Sunny Boy (Kieft, 1999). The survival of many Dutch traditional breeds of cattle, goats, sheep and horses is being increasingly threatened.

Dutch landscapes have been much simplified and in the previous century, seven native bird species have become extinct, especially birds that thrive in specialized habitats, because of change of specific habitats and the upscaling of agriculture. Homogenized agriculture has also influenced the diversity of other wild organisms. Modern farming trends have turned out to be neutral for mosses, higher plants and mammals, and moderately positive for birds (see Kieft, 1999).

A potential source of support for biodiversity within conventional agriculture appears to be the consumers' desire for having higher quality and more assorted foods of a regional character.

Biodiversity in organic agriculture is generally greater than in conventional farming, particularly for genetic diversity, life-support functions and landscape diversity. In terms of acreage, however, eco-farming is still very marginal and occupies between 1% and 10% of Europe's agricultural area.

Organic farming gives considerable importance to biodiversity because it is both a production factor and important against pests and diseases. In the European Union, genetic manipulation is not accepted in organic farming.

The four dimensions of biodiversity are:

Genetic level: the number of different genes in (wild) species, breeds and varieties of plants, animals and other organisms.

Life support level: the number of different organisms with life support functions for agricultural production, like pollination, natural enemies of diseases and plagues, and soil-organisms for improving soil fertility and structure.

Nature and landscape level: the number of different elements without direct agricultural production function like meadow birds, flora and fauna in hedges, field borders and along ditches.

System level: the number of different agro-ecosystems, characterized by a certain combination of crops or animals and technology.

RURAL DEVELOPMENT IN THIRD WORLD

In 1992, the total world population was about 5.3 billion people. Of this number, 77% were in low and middle income economies, 51% in Asia and the Pacific, 15% in Africa and the Middle East, 7% in Latin America and the Caribbean, and 4% in Europe. An estimated 80% of the population in the Third World countries lived in rural areas, implying that nearly two-thirds of the entire world population lived in these rural areas—nearly all at a poverty level or below—with inadequate diet, housing and health conditions. According to the World Bank, more than one billion people were in 'acute' poverty conditions and barely existing on $1 per day or less (Brown, 1992; World Bank, 1992; UNDP, 1992).

The 3.3 billion villagers in the Third World lived in about two million villages, having an average population of about 1500.

The major underlying cause of poverty is a lack of access of the poor to the resources that they require for survival and development. So, how can Third World villagers gain access to the resources essential for their full participation in the twenty-first century? Appropriate methods should be designed to build an effective bridge between global resources and the rural sector. The technical, social, scientific, financial and knowledge resources of our global society should be made accessible to two-thirds of the world's people residing in villages, who lack the practical resources to do their own self-development.

The local people should be involved actively in both the resources and environmental concerns of the broader society. The following are some important tasks that can be the basis of a design for large-scale, sustainable village-based development that can create this necessary bridge of access to essential resources.

1. Expand and empower human resources through local participation in planning, designing and implementing village-based programmes; an external activator force skilled in helping villagers by building resource linkages, and basic mutual agreements;
2. Establish a viable development framework through a Basic Development Unit (BDU) of 30-40 contiguous villages with a population of approximately 50,000 and a multi-sector strategy to deal with the varied needs identified by the villagers in village planning events;
3. Build a resource access mechanism through a BDU service center training village-appointed trainers in appropriate hard and soft

technologies (see Albertson, 1991) and building linkages to resources and a residue of locally managed organizations operating sustainable economic and social enterprises;
4. Create sustainable development patterns through an internally sustainable process that builds local technical capacity and management skills and insures continuing socio-economic improvement and an environmentally sustainable programme of land and natural resource use to insure long term viability; and
5. Maintain a dynamic learning process through rigorous monitoring and evaluation of project process, key variables and structure through baseline data study and ongoing data collection and documentation of lessons learned to distill practical wisdom for programme development and extension.

There is no doubt that mobilization of human energy, skills and vital interests into a force for locally-determined change can be the engine that drives effective development efforts. However, because local village people have been isolated from many resources and options for their future, an external activator is usually needed to act as a catalytic agent to help identify and focus clear, achievable developmental aims.

Local Participation

The crucial element of development action is that villagers should identify their basic needs and problems and then find solutions. There are numerous cases where well intentioned efforts to provide technology failed because recipients were inadequately involved in the planning, design and implementation process. The development and use of participatory skills is essential both for planning and for adaptation and use of technologies needed to achieve community determined objectives. It must be ensured that all villagers be involved. Towards this goal, one model for village development, as proposed by Faulkner and Albertson (1986), is shown as a 'Development Wheel' in Fig. 6.4. It shows that the *process or development begins with the people.*

Internally Sustainable Development

The development of human resources capable of sustaining the speed of development is a result of the accumulation of local knowledge, management skills and organizational abilities. These skills are maintained by establishing patterns of daily application that make the technologies that have been introduced actually meet the needs and desires of the people.

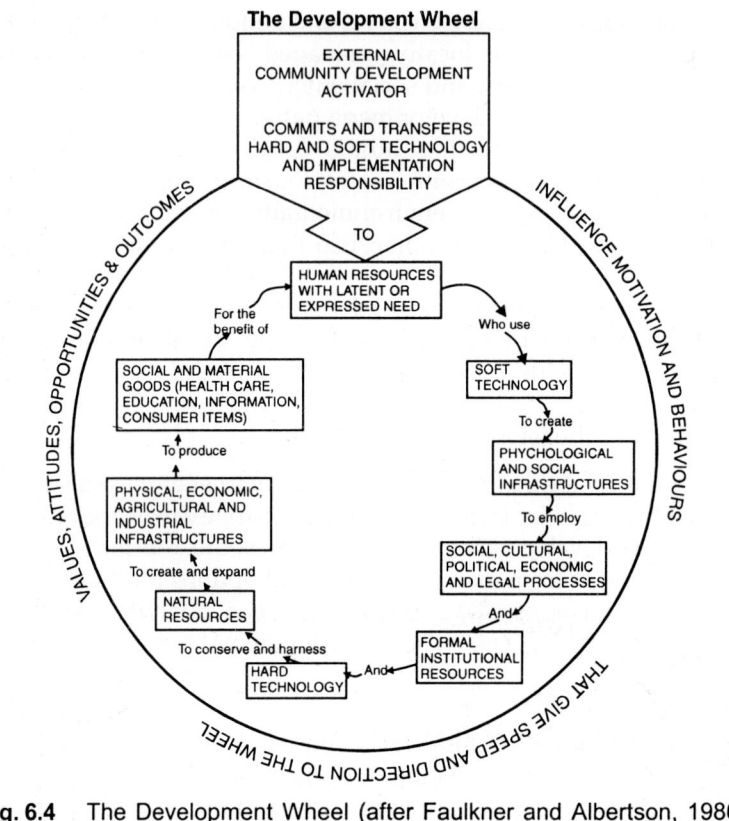

Fig. 6.4 The Development Wheel (after Faulkner and Albertson, 1986).

Environmentally Sustainable Development

Environmentally-sustainable development is that which meets the needs of the present generation without compromising the needs of future generations. This is quite a difficult challenge to apply in meeting the needs of Third World villagers. Possibly, a service center approach to the development of 20 to 40 villages may provide the resources, training and organizational development to enable the people of these villages to use environmentally-sound practices in building infrastructure and industry and improving agriculture. When villagers seek solutions to their needs in the context of a larger area, they understand how their efforts can either complement or diminish the well being of other villages. As a part of the body larger than a single village, it is possible for villages of the BDU to share a broad spectrum of environmentally sound resources for infrastructure, agriculture and industry that would generally not be affordable on an individual village basis. By negotiating the delivery of basic services through a Service Centre, governments can promote the use of technologies that fit within an environmentally sustainable, long-term process of development.

INTEGRATED RURAL DEVELOPMENT

Integrated rural development (IRD) became a buzzword in the late 1960s and 1970s. It involves the development of strategies that produce qualitative and quantitative changes within a rural population through improved living conditions, increased production capacity and the creation of social infrastructure necessary for increased production. Its work plans include mechanization of agriculture, provision of electricity, clean potable water, decent housing, marketing and storage facilities for farm products, improvement of networks of feeder and access roads and the organization or re-organization of human settlements.

Socialist Ideology and Development from Below

The socialist approach to development involves commitment to the socialist ideology as a vehicle for bringing about rapid rural transformation and social development. The approach differs from the conventional one because of the professed aim of reaching and involving all people in rural areas and the associated self-help ideology.

Any sound regional development planning and management in the future must address the following three critical challenges:

(i) The world has become a global village linked by a continuous flow of ideas through the best communication systems the earth has ever had.
(ii) The world is rapidly urbanizing. Unlike in developed countries where the largest mega-cities are showing signs of slowing down or even decline, such cities in the developing world are likely to grow in size and functions. And yet, a great majority of people in several developing countries will continue to live in rural areas and will continue to be faced with problems of rural poverty and lack of access to basic amenities and infrastructure (Ayeni, 1999). Previous attempts at planning rural areas as appendages of urban based strategies have not been successful. What is needed is a policy of settlement management that will take care of all types of settlement, rural and urban.
(iii) Sustainable human settlement development strategy needs to be created in the face of previous misuse or overuse of land, escalation of deforestation, over-consumption of energy and the attendant problem of environmental pollution, and the problems of water supply, sanitation and waste management (Ayeni, 1999).

REFERENCES

Albertson, M.L. *Appropriate Technology.* Hydraulics Division ASCE, Nashville, Tenn. (1991).
Ayeni, B. Socialist ideology and development from below. *Habitat Debate (UNCHS)* 5 (1) : 15 (1999).
Brown, L.R. *State of the World.* Worldwatch Institute, Washington, D.C. (1992).
Bryceson, D., Kay, C., Mooij, J. (eds.) *Disappearing Peasantries ? Rural Labour in Africa, Asia and Latin America.* Intermed. Tech. Publ., London (2000).
Dankelman, I., Ramprasad, V. Biodiversity in a cultural perspective. *Compas Newsletter* (2): 4-6 (1999).
Desai, V. Tiny mills with a great future. *Gate* 4/99: 9-13 (1999).
Edwards, M. *Future Positive: International Co-operation in the 21st Century.* Earthscan, London (1999).
El Bassam, N. Possibilities and limitations of energy supply from biomass. *Natur. Resources and Develop.* 41: 7-21 (1995).
Faulkner, A.O., Albertson, M.L. Tandem use of hard and soft technology: An evolving model for third world village development. *Intern. J. Applied Engineer. Education* 2: 2 (1986).
Fowler, A. *Striking a Balance: A Guide to Enhancing the Effectiveness of Non-governmental Organizations in International Development.* Earthscan, London (1997).
Khandelwal, K.C., Mahdi, S.S. *In Biogas Technology : A Practical Technology.* Tata McGraw-Hill Publishing Company Limited, New Delhi, p. 128 (1986).
Kieft, H. Biodiversity in Dutch Agriculture. *Compas Newsletter* (2): 26-27 (1999).
Madeley, J. *Big Business, Poor People — The Impact of Transnational Corporations on the World's Poor.* Zed Books, London (1999).
Nagamani, B., Ramasamy, K. Biogas production technology: An Indian perspective. *Current Science* 77: 44-55 (1999).
Posey, D.A. (Ed.). *Cultural and Spiritual Values of Biodiversity.* UNEP, Nairobi (2000).
Schnurr, J. *The Cornerstone of Development — Integrating Environmental, Social and Economic Policies.* IDRC, London (1998).
Sen, N. The community caring for its own food and water security. *Current Science* 81: 236 (2001).
Sharp, T. Privatisation brings energy AT to the market. *Gate* 4/99: 47-50 (1999).
Truntz, R. Full sail ahead. *Gate* 4/99: 21-25 (1999).
UNDP. *Human Development Report, 1992.* Oxford Univ. Press, New York (1992).
World Bank. *World Development Report 1992: Development and the Environment.* Oxford Univ. Press, New York (1992).

Chapter 7
Wastes and Pollution

INTRODUCTION

The throw-away century has ended, leaving a legacy of changing climate, polluted seas, dying forests and toxic air. The twenty-first century requires a new approach to production, based on clean technology and recycling. Companies which have adopted these methods have already registered enormous financial savings.

In addition to gaseous and liquid wastes released into the environment, humanity piles up mountains of solid refuse. The countries of the European Union (EU), for instance, throw away an estimated 2 billion tonnes of waste every year—ranging from the contents of domestic dustbins to sewage sludge, mining residues and ash from power stations (UNEP, 1990).

Only a small proportion of this waste is 'hazardous' (needing special handling or disposal to avoid harming health, the environment or both). In the mid 1980s, the countries of the Organization for Economic Cooperation and Development (OECD) generated about 300 million tonnes of hazardous wastes a year (as against 370 million tonnes of municipal wastes). Industrialized countries probably generate over 90% of the world's annual total of some 325-375 million tonnes. Most of this comes from chemical and petrochemical industries.

Faulty disposal can make even a relatively harmless waste troublesome. In the 1950s and 1960s, some 2,000 people at Minamata and Niigata (Japan) suffered crippling neurological diseases after consuming fish poisoned by mercury wastes discharged into the sea. Wells have often been contaminated by leaks from chemical dumps. Land devastated by defoliants and pesticide wastes near Denver, Colorado, 5 decades, ago has still not been fully decontaminated.

It is difficult to overemphasize the need for all countries to cut back on the quantity and toxicity of the wastes they generate; to manage them in an environmentally-sound way; and to dispose them of safely and as near to the source of their generation as possible. Countries should only export wastes if they cannot dispose off them safely at home.

Waste management in rapidly urbanizing cities has become a challenging task, particularly in developing countries.

Cooperation from the private sector and civil society groups is required to deal with the problem effectively by means of sound practices for sustainable waste management. Urban wastes strongly impact on freshwater resources. Use of environmental technologies in waste management can help minimize this impact.

A set of indicators for solid waste management is badly needed to allow monitoring of country/city performance and to help establish a set of common principles. These could also provide a basis for target-setting action plans, as well as performance monitoring in participating nations. Policies based on the four Rs (reduce, reuse, recycle, recovery) must be developed and basic principles agreed upon, such as the responsibilities that waste managers impose on waste producers. The problem of used oils and their environmental impact needs to be considered in all urban waste management initiatives.

The problem of eutrophication of lakes and reservoirs needs to be considered from several different perspectives, including economic, social and cultural, educational, technical, legal, management and institutional.

Eastern and Central European countries are currently in a phase of transition from centrally-planned to market-oriented economies. Although the region has relatively abundant water with many rivers and lakes, it is facing several instances of environmental pollution, water resource depletion and threats to development. A lack of efficient technologies has contributed to high rates of water consumption, especially in power generation, mining and the steel industry. Consequently, water tables have fallen and some wells have dried up. To tackle these problems, several technologies are being applied for freshwater augmentation, water quality improvement, waste-water treatment and reuse, and water conservation. There are also technologies oriented to the power generation industry and mining industries, such as recycling/reuse of used water, and the use of mine water. Technologies related to water supply systems, such as enlarging/lining pipes, water-saving fixtures and water meters, can provide good sources of 'new' water through enhanced water-pipe network efficiency. In wastewater treatment, the use of cost effective, renewable natural processes needs to be encouraged. These processes include *Lemna* technology using duckweed, and hydrobotanical or wetland treatment. Some remedial technologies for polluted sites and lakes are also being explored.

ENVIRONMENTAL ENGINEERING, WASTE DISPOSAL AND RESOURCE RECOVERY

Environmental engineering has been mainly concerned with developing technologies for waste disposal. The major concern has been to dispose of wastes in the most convenient and least expensive way, not to convert them into useful resources. There has been little interaction between disposal agencies on one hand and populations that produce the wastes, on the other.

Neither wastes nor resources are rigidly defined concepts, but rather, both depend on customs or values (real or perceived) and available knowledge and technology to determine whether a substance is considered to be a potentially useful resource or a useless waste. It is sometimes possible to process useful components from what had previously been considered entirely useless wastes (Iranpour et al., 1999).

Waste reclamation can have the following potential benefits: (1) to reduce consumption of natural resources, such as ore minerals and fresh water; (2) to reduce pollution produced by discharging untreated waste; and (3) the energy conservation effect of reclaiming wastes. Reclamation usually consumes less energy than producing new materials. Therefore, increasing reclamation not only reduces pollution but saves energy.

However, in existing reclamation industries, producing useful materials from natural resources usually requires large amounts of energy and imposes additional costs for transportation from where the resources are available to where the product is used. Energy may also be needed to treat wastes in order to render them safe and to transport them from treatment centers to disposal sites. All this extra energy is lost when the waste is discarded. On the other hand, there are many materials, e.g. the steel and aluminium scrap industries, for which the energy costs of reclamation are a small fraction of the costs of new production.

Efforts are being made to use waste heat from high-temperature industrial processes like power generation and metal smelting. Indeed, waste reclamation should be a part of the overall energy picture and, in some cases such as reclaiming wastewater instead of desalinating ocean water, the total expected savings can be locally or regionally important.

Environmental engineers face a challenge to seek new methods of separation, conversion and reclamation.

Emerging Technologies

Some new technologies have already been commercialized and are being used in waste water reclamation. A combination of microfiltration and ultra-violet disinfection has proved very helpful in wastewater reclamation.

Ultraviolet (UV) disinfection has emerged as an alternative to disinfection with chlorinated compounds in reclamation. Preceding UV disinfection with microfiltration provides several benefits: (i) it reduces the turbidity of the effluent; (ii) it removes bacteria and larger organisms; and (iii) it reduces introduction of chlorinated byproducts into the environment (Jolis and Hirano, 1993; Iranpour et al., 1999).

Microfiltration also greatly reduces the UV dose needed to achieve mandated levels of virus inactivation. A low pressure UV disinfection system has been developed that is highly effective in virus inactivation (Iranpour et al., 1999).

Another novel technology potentially applicable to wastewater reclamation is online monitoring of biological oxygen demand (BOD) for improved control of the primary and secondary treatment in a reclamation plant. Automatic instruments containing a supply of microbes and an oxygen concentration monitor are now available so that the nutrients in a sample of wastewater are consumed in a few minutes (rather than in five days, as in standard BOD tests), with determination of the corresponding oxygen uptake (e.g. Anonymous, 1994).

Controlled Biodegradation

Secondary wastewater treatment as well as sludge digestion exemplify a strategy of replacing uncontrolled biodegradation in lakes, rivers and the ocean with controlled biodegradation under conditions that minimize harm to people. Other examples are composting and landfills designed for gas production. These methods of waste disposal are well established but some additional possibilities for applying this strategy are outlined below.

1. Termites for Wood Fibre Degradation. Woody materials make up a large fraction of municipal solid waste stream in some countries. Around 40% of all solid waste can be paper, and there is also waste wood, much of which is not suitable for processing into paper. Moreover, paper fibres can only be recycled a few times and unlike aluminium beverage cans, they cannot be reused in material of constant quality. Hence, there is a large supply of useless wood fibre material. Controlled termite colonies might provide an alternative, by analogy with the biodegradation tanks in wastewater treatment. Termite feeding fragments the fibres needed for rapid bacterial decomposition.

2. Acclimation of Bacterial Communities to Degrade Toxic Organic Chemicals. Suitable strains of certain bacteria, e.g. acidogenic bacteria, antibiotic-resistant bacteria, toxin-degrading bacteria, etc., can help in degrading toxic chemicals such as polychlorinated biphenyls and volatile organic chemicals.

Wastewater Aeration

Since the surface-to-volume ratio of bubbles increases with decreasing bubble diameter, the best ceramic fine-pore diffusers for conventional large wastewater aeration basins achieve much higher efficiencies of oxygen transfer into the water than the oxygen transfer efficiencies (OTEs) of coarse-bubble devices such as spiral-roll diffusers. The use of fine-pore diffusers has proved highly useful in this context (Iranpour et al., 1999).

Another possibility being explored is to use microfiltration fibres below the bubble formation pressure which allows surface tension to maintain an air-water interface across the pores. Wastewater can then be aerated by letting it flow over such fibres in devices similar to those used to aerate blood in heart-lung machines. Alternative use of either microporous membranes or polymers that are permeable to oxygen may allow bacterial films to grow on the surfaces of other arrangements or submerged tubes carrying air or oxygen in tanks similar to those used for aeration now. The idea is to supply bacteria with oxygen by diffusion through the pores or tubes instead of relying on transfer through bubble surfaces in the short period that the bubbles rise through the water. This technology has not yet come into use in full-scale wastewater systems.

Geothermal Pyrolysis

The possibility of converting organic wastes into petroleum-like materials by geothermal pyrolysis is attracting attention.

Heating organic materials with little or no oxygen pyrolyzes them into oily or tarry substances, with release of water, CO_2, CH_4, and other simple gases, and production of a charred residue if the temperature is high enough. There is a great sensitivity to temperature of both the rate of the reaction and the composition of the products. Finally, there is the availability, underground of temperatures of 150°C or higher at depths accessible to drilling. All these facts raise the hope that geological heat may be sufficient in some places to cause pyrolitic reaction that would produce useful chemicals from organic wastes, such as sewage sludge, if the reaction were allowed to proceed for weeks or months (Iranpour et al., 1999).

Glass furnaces and many other high-temperature industrial processes such as petroleum refining, metal smelting and power generation produce waste heat at temperatures that may be used for bread baking, milk pasteurization or sterilizing canned products.

According to Iranpour et al., (1999) certain improvements/opportunistic innovations and strategies such as controlled biodegradation can give a new orientation to environmental engineering for resource recovery from wastes. They would also change the pace of environmental protection efforts, in which activities providing food, shelter and transportation etc. would seek to accommodate themselves within certain environmental limits.

WASTEWATER TECHNOLOGY

The ever-escalating costs of upgrading, maintaining and operating the existing municipal wastewater treatment infrastructure, expanding it to support residential and industrial growth, and providing a new infrastructure where it does not exist poses a serious problem for municipalities. The requirement for the provision of a treatment facility that will reduce or eliminate toxic chemicals, in addition to the conventional pollutants, further complicates this issue. Toxic chemicals control also requires that the treatment, handling and disposal of treatment process sludges or residues receive due attention.

The solutions to these problems are highly complex. Advances in reactor design, computer control technology and analytical capability have spurred scientists and engineers to develop cost-effective solutions in many of these cases. In general, emphasis has been on the optimization of the design and operation of existing wastewater treatment systems. Although this approach has resulted in the development of some new technology, most of the advancements have taken place through upgrading and modifying existing unit operations.

For industries discharging to a municipal sewer system, product reuse, generation of alternative products and modifications to existing processes constitute the essential initial components of an effective municipal waste management program. If source control is essential, metal recovery systems or energy efficient high-rate anaerobic systems need to be considered. If hazardous or toxic wastes which cannot be eliminated, recovered or treated at an on-site facility, are present, then these wastes should be passed on to some centralized treatment, handling and disposal facility. The municipality should ensure that these source control programmes are operative and effectively controlled. The impact of industrial waste source control on the hydraulic and organic loading capacity and operating complexity of the municipal wastewater treatment facility can be very significant, eventually resulting in substantial savings in capital and operating costs (Jank, 1988).

A very good example of cost-effective industrial waste source control is the application of high rate of anaerobic technology for the pretreatment of high organic strength, high temperature wastes. The methane generated from the anaerobic process located at the industrial plant can be used by the industry to offset the purchase of natural gas or fuel oil. The percentage reduction of organics usually suffices to provide an effluent that meets municipal industrial pretreatment requirements, and also eliminates or significantly reduces industrial waste discharge surcharges. Significant energy savings accrue at the central treatment facility as a result of reduced aeration requirements and corresponding reductions in sludge production. Dairy, brewery, distillery, slaughter house and fruit and vegetable processing wastewaters represent excellent examples where high-rate of anaerobic technology may be used at source to reduce the total cost of wastewater treatment for both the industry and the municipality (Jank, 1988).

Notable advances in optimizing, designing and operating municipal wastewater treatment plants have come from automated process control. These include techniques for improving the efficiency of the aeration equipment and the design and operation of the secondary clarifier, both of which can potentially yield capital, operation and maintenance savings. On-line instrumentation, coupled to computers programmed with appropriate automated control strategies, improves the operation of the activated sludge process and many other unit operations within the treatment system. The computer control procedures are used to develop a process audit to assess the operational performance of full scale wastewater treatment plants. This audit helps identify design deficiencies, the anticipated life of the facility (following the correction of design deficiencies), the specifications for the computer control system and the potential energy savings following incorporation of computer control.

In Canada, properly designed and operated municipal treatment facilities have been shown to produce an effluent having an acceptable level of trace organics and heavy metals. Computer controlled facilities help provide consistent effluent quality, an important factor for those plants designed to provide toxic chemicals control. Appropriate sludge treatment, handling and disposal facilities need to be included in the treatment facility.

On economic and technical grounds, whenever possible, land application of sludge should be the preferred disposal option, because the environmental risk associated with land application is minimal. Anaerobic sludge digestion is frequently an integral component of the land application option. Improvement of digestion mixing by optimizing the number of mixing devices and the mixing energy can help extend the life of the existing digester facilities at a fraction of the capital cost of providing new digesters (Jank, 1988).

In recent years, there has been a growing trend in several developed countries towards realizing the importance of the sludge train in both the design of wastewater treatment plants and the evaluation of capital and operating and maintenance costs. Sludge treatment costs amount to roughly 50% of the total wastewater treatment costs. Sludge treatment costs at four integrated sludge management systems in Canada, all utilizing incineration, were quite high and ranged from $350 to $1042 per metric ton of dry sludge (see Campbell and Bridle, 1988). This highlighted the need of improving the efficiency of sludge management either by optimizing and upgrading existing facilities, or by introducing new technology. New processes such as starved air incineration, gasification and liquefaction have been developed. The most promising new technology appears to be low temperature conversion of sludge to fuel products (Campbell and Bridle, 1988). The basic idea of low temperature conversion of sewage sludge to produce fuel products has been known since long. Low temperature conversion of sludge to oil represents a promising alternative to sludge incineration. According to Campbell and Bridle, a 1 kg/h bench-scale reactor gives oil yields ranging from a low of 13% for an anaerobically digested sludge to a high of 46% for a mixed raw sludge. Char yields range from 40 to 73% at the optimum operating temperatures.

Thermal conversion of sludge to fuel offers the wastewater industry three potential advantages: (1) reduced cost for municipal sludge treatment; (2) an increase in the energy efficiency of sludge treatment systems, not only in the generation of that energy but also its utilization; and (3) the improved operation of treatment plant.

ACTIVATED SLUDGE PLANTS FOR ENHANCED SUSPENDED SOLIDS REMOVAL

Activated sludge treatment plants yield a purified effluent by efficiently converting the soluble organics in wastewater to biomass, entrapping colloidal particles in the biomass, and separating the latter from the liquid stream. Most activated sludge plants can produce effluents with a soluble BOD_5 of less than 5 mg/L. But the escape of suspended solids from the final settler greatly limits the quality of the effluent achieved by an activated sludge plant. Most organic matter in the effluent exists in the form of suspended solids. Depending on the hydraulic and solids retention time of an activated sludge plant, suspended solids which escape from the final settler can account for 50% to 90% of the effluent BOD_5 (Chapman, 1988).

The efficiency of phosphorus removal is also decreased by poor suspended solids removal. As much as 80% of effluent phosphorus can be associated with the effluent solids.

Enhanced removal of suspended solids can be achieved by giving careful attention to the design of three components of the activated sludge system—the pumps and controls governing the rate of inflow and recycle; the aeration tank and oxygen transfer system; and the final settlers. Pumps are selected and controlled to prevent the generation of hydraulic transients which can degrade effluent quality. The oxygen transfer system and the aeration tank are designed to provide a mixed liquor with good settling and clarification properties (Chapman, 1988).

Initially, the final settler is designed to prevent a thickening failure by specifying the correct combination of final settler area and recycle rate for the particular settling characteristics of the mixed liquor. Thickening failure can occur if the rate at which solids enter the final settler exceeds the rate at which they are removed. The persistence of such an imbalance can cause solids to be added to the sludge blanket, eventually resulting in a great loss of solids when the blanket reaches the weirs of the settler. Techniques that enable the designer to assess the thickening capacity of a final settler, have been designed (see Chapman, 1988).

When it has been ensured that the final settler has adequate thickening capacity, optimization of the clarification function of the unit is attempted. Clarification is controlled by overflow rate, tank depth and the mixed liquor suspended solids (MLSS) concentration. Maximum clarification may be achieved by the use of flocculating centrewells and effluent weirs with horizontal baffles to deflect density currents.

Suspended solids account for most of BOD_5 in the effluent from an activated sludge plant. Similarly, the efficiency of phosphorus removal by precipitation is reduced by the escape of suspended solids in the final effluent.

Poor suspended solids removal may be due to several factors. Hydraulic transients which are generated by influent or recycle pumps degrade effluent quality. In the aeration tank, inadequate attention to the selection of SRT (solids retention time), D.O. levels and tank configuration can result in a floc which settles and clarifies poorly. A circular flow pattern in the final settler is generated by the momentum of the incoming mixed liquor and density differences between the mixed liquor and the clarified supernatant in the settler. This current creates both mixing and scour within the settler (Chapman, 1988).

According to Chapman (1988), it is possible to enhance the removal of suspended solids by the activated sludge process by:

1. Selecting and controlling influent and recycle pumps to avoid the creation of large hydraulic transients,
2. Increasing the freeboard and the hydraulic resistance of the outlet weirs of the aeration tank to maximize dampening,
3. Selecting an SRT with due regard to its effect on the settling and clarification properties of the mixed liquor,
4. Preventing sludge bulking by providing adequate levels of dissolved oxygen in the aeration basin and selecting process configurations (such as tapered aeration) which create a substrate concentration gradient along the length of the tank,
5. Reducing levels of turbulence within the aeration tank by selecting efficient oxygen transfer equipment and controlling the level of dissolved oxygen,
6. Reducing floating solids in the final settler using a small anoxic zone in the aeration tank to provide denitrification prior to final settling or reducing the SRT to minimize nitrification,
7. Reducing the mixed liquor suspended solids (MLSS) concentrations by reducing the SRT, designing larger aeration tanks, or making provision for step feed,
8. Designing final settlers with adequate tank depth, and
9. Incorporating special features into the design of final settlers such as flocculating centrewells or effluent weirs with a short horizontal baffle to deflect the circular current created by the incoming flow.

Sewage sludge is a nutrient rich, largely organic by-product of municipal wastewater treatment that must be removed from the treatment facility. The disposal options include ocean dumping, incineration, landfilling and utilization on agricultural land. Recent high energy costs and environmental awareness have increased interest in sludge disposal by utilization on agricultural land. When managed properly, this practice minimizes environmental risks, takes advantage of the fertilizer and soil conditioning value of sludge and is frequently the least expensive disposal option (Webber, 1988).

Two types of environmental risk are associated with sludge utilization on agricultural land. Temporary risks disappear within a few years following sludge application. These include malodour, pathogens, groundwater contamination with nitrate-nitrogen and phytotoxicity due to soluble salts or toxic biodegradation products from inadequately stabilized sludge. Persistent risks remain much longer and include increased concentrations of industrially produced organic compounds such as polychlorinated biphenyls (PCBs) and waste metals in soil. The half-life of persistent organics such as PCBs in soil is about ten years while that of most waste metals is about a thousand years.

Sewage sludge utilization on agricultural land needs to be properly regulated to minimize the risks associated with heavy metal buildup in soil. Consequently, guidelines for this purpose have been developed in many countries e.g. Canada.

In Canada, agricultural utilization of sludge has been practised since long and has usually been found to be much cheaper than disposal where there is relatively easy access to land. It is an important sludge management option.

The risks associated with waste metal application to agricultural land are well recognized. Guidelines to limit loadings to soil have been developed and they generally correspond with the mid-range of values adopted by many other countries (Table 7.1). Canadian research on waste metal uptake by plants from sludge treated soils indicates that even the maximum suggested values are not likely to cause significant crop production or animal and human health problems. Indeed, some relaxation of limits may be warranted in the future.

OPTIMIZATION IN WASTEWATER TREATMENT TECHNOLOGY

Municipal wastewater treatment technology was first developed over a century ago. Wastewater treatment technology implementation involving intermittent sand filtration began in the mid-1850s. Some new and improved treatment technologies were developed in the period 1850 through 1900, when there was the progression from intermittent sand filtration to sewage farms to trickling filters. Almost immediately, following the turn of the nineteenth century, industry welcomed the introduction of the Imhoff tank and the more celebrated activated-sludge process. With such a vast variety of technologies available and being implemented, a general feeling emerged that water pollution problems could be solved simply by building a few more systems or expanding the existing ones. The possible impact of the industrial revolution or the manufacturing demands necessary to support World War I were not considered.

The driving force behind the development and implementation of new technologies at the turn of the century was control of waterborne diseases. The trend today is more toward regulatory compliance, namely, higher effluent standards and enhanced environmental awareness. However, regardless of the nature of the driving force, the result for the wastewater industry has been the need to produce higher quality effluents, with the accompanying higher costs (Lue-Hing, 1998).

In certain circumstances, systems capable of meeting effluent requirements at peak performance produce poor effluent quality during a gross

Table 7.1 Waste metal guidelines for agricultural utilization of sludge (after Webber, 1988).

Maximum Acceptable Loadings to Soil (kg/ha)

	Canada	Denmark	Finland	France	Germany	Netherlands	Norway	Sweden	United Kingdom	United States	All Countries Range	Median
As	15					2			10	2-15	0.1-20	5
Cd	4	0.2	0.1	5.4	8.4	2	0.2	0.075	5	5, 10, 20	0.4-30	
Co	30						0.4	0.25				
Cr				360	210	100	4	5	1000	125, 250, 500	4-1000	210
Cu				210	210	120	30	15	280		30-500	210
F									600			
Hg	1			2.7	5.7	2	0.14	0.04	2		0.14-5.7	2
Mn							10					
Mo	4								4			
Ni	36			60	60	20	2	2.5	70	50, 100, 200	2-200	60
Pb	100			210	210	100	6	1.5	1000	500, 1000, 2000	6-2000	210
Se	2.8							5			2.8-5	
Zn	370			750	750	400	60	50	560	250, 500, 1000	60-1000	500

overload. In these cases, a reasonable solution could be to expand or install new technology, or to get more from an existing one.

No single factor drives optimization. Economics is an important driver but so also the influence of the academia, the desire of front-line practitioners to increase productivity, more stringent regulatory requirements, public demand for more efficient operations, enhanced environmental awareness, and serendipitous discoveries needs to be appreciated. Among all these it appears that economics, regulatory requirements and environmental awareness will continue to be critically important in the foreseeable future.

Optimization can involve complex modifications, innovative alterations or simple adjustments. Some areas in which it is quite effective include energy conservation, electrification, automation, computer-aided operations, strategic monitoring, use of surrogate parameters and a wiser use of sensors. Some optimization efforts have a short shelf life, are designed specifically for an emergency, are not widely adopted or fail to perform as claimed or otherwise quickly outlive their usefulness. Examples of such efforts include rotating biological contactor-enhanced activated-sludge systems, physical-chemical treatment for municipal wastewater, and breakpoint chlorination (Lue-Hing, 1998).

Among the more lasting technologies—tapered aeration, oxygen activated sludge, high-rate anaerobic digestion, two-stage anaerobic digestion, thermophilic anaerobic digestion, sidestream dewatering to increase digester feed concentration—permanent status may justifiably be assigned to tapered aeration, oxygen activated sludge, and high-rate anaerobic digestion.

The novel concept of optimizing individual unit processes and mating them sequentially offers a potentially promising scope for expansion. This concept may be extended successfully to treatment schemes with multiple unit processes operating sequentially, such as systemic optimization, or to unit processes not traditionally mated sequentially or otherwise. This approach permits effective exploitation of optimization improvements in each unit, so that the advantages start piling up.

The systemic/sequential concept is already being applied with some regularity to optimization projects. The trickling filter-activated solids process and Chicago's anaerobic digestion-lagooning-air-drying process to produce Class A biosolids are two specific examples. The former incorporates, in sequence, unit processes not traditionally mated, whereas the latter incorporates, also in sequence, three optimized unit processes, two with a limited history of traditional mating and the third with no such history. In both of these process schemes, the sequential approach exploits the individual unit optimization effects to produce a net optimization outcome that is cumulative.

Optimization has proved to be eminently successful. It has solved many short-term and long-term problems and is fairly inexpensive.

URBAN WASTEWATER USE FOR PLANT BIOMASS PRODUCTION

Irrigation of crop fields with wastewater (sewage) effluent is a good alternative to discharging the effluent into river/ocean systems. In some countries, there is a growing interest in crops, pastures and trees being irrigated with recycled water (Edgar and Stewart, 1979).

Artichokes (*Helianthus tuberosus*) can grow and produce under high fertilizer levels, especially under irrigation using piggery effluent. Therefore, artichoke can play a role in wastewater/effluent disposal strategies.

When grown under irrigation using wastewater, artichoke crop does not show any signs of visible damage even in comparatively high nutrient load of the water. The crop establishes good canopy cover within 2 months after planting, showing no signs of toxicity during even early stages of growth. It uses about 6 ML/ha. of wastewater for completing its life cycle and successful production of tubers.

Seasonal differences in climate result in monthly water use ranging from 50-180 mm; this is comparable to the water use of a perennial pasture with a monthly water use of about 150-200 mm for the same period. The water use of this crop points to the usefulness of the crop in land disposal of municipal wastes in areas where evaporative demands are high (Parameswaran, 1999).

The water may be applied by furrow irrigation. A dense canopy cover from crop growth helps in reducing the evaporation of water from the surface soil and can change the microclimate. Wide plant spacing or enhanced air flow within and under the canopy is beneficial in growing crops like artichoke under furrow irrigation using wastewater. The ability of the crop to grow and give high yields under wastewater irrigation should be exploited in land disposal of municipal/farm wastewater strategies, along with pasture and tree systems.

Artichoke is tolerant of comparatively high nutrient loads, which is another added advantage of the crop to sustain growth and production under wastewater irrigation.

A great potential exists for production of 'value added' products from artichokes which should be tapped; e.g. fuel for transport (ethanol), inulin, citric acid and lysine production.

ASSESSMENT AND MANAGEMENT OF HAZARDOUS WASTES

Biodegradation studies and physicochemical characterization are crucial to the overall assessment of hazardous wastes (see Hess et al., 1998).

Lesage et al., (1997) used multivariate plotting techniques and principal component analysis to distinguish between naturally occurring and anthropogenic hydrocarbons in groundwater. They found that propane and pentene are the most useful chemical parameters in evaluating the difference between natural and anthropogenic sources. Powers et al. (1997) showed that multivariate analysis can provide considerable insight into contaminant source characteristics by elucidating correlations in groundwater related to type of nonaqueous phase liquid (NAPL) and proximity to the contaminant source. According to them, current site assessment may fail to generate adequate information regarding the type or distribution of NAPLs. The analyses provided an additional level of interpretation regarding the distribution of NAPL not possible with standard evaluation techniques used during remedial investigation.

Soil quality assessment is very important in terms of intended land use and the important role of soil scientists.

HAZARDOUS WASTE TREATMENT

Munaf and Zein (1997) have used rice husk to remove chromium, zinc, copper, and cadmium from wastewater. Metal uptake efficiencies by rice husk were significantly affected by particle size, pH and temperature. Metal removal efficiencies of 79%, 85%, 80%, and 85% for chromium, zinc, copper, and cadmium, respectively, were supported at the optimal conditions. Certain barks, after grinding and treatment with formaldehyde in acid media, can also be used as a biosorbent for removal of various metals.

Watanabe (1997) has reviewed current developments in phytoremediation. Indian mustard seedlings grown in aerated water can concentrate divalent cations (lead, strontium, cadmium, and nickel) 500 to 2000 times over a range of metal concentrations and in the presence of competing ions. Intracellular cadmium accumulation was inhibited in shoots (competitively) and roots (noncompetitively) by calcium, zinc and manganese. Cadmium can bind to phytochelatins. The addition of chelates to lead-contaminated soil (2500 mg/kg) increases shoot lead concentrations in corn and peas by a factor of more than 20 (Huang et al., 1997).

Taylor et al., (1997) isolated (in order of descending number) sulfate reducers, anaerobic heterotrophs, aerobic heterotrophs, iron oxidizers, iron reducers, sulfur oxidizers and fungi from certain contaminated groundwater recovery wells prone to biofouling. They reported that aerobic and anaerobic heterotrophic bacteria can use the principal contaminants as sole carbon and energy sources and can re-establish a population in well water in 6 weeks.

Both synthetic surfactants and biosurfactants produced from *Pseudomonas aeruginosa* increase the solubility and biodegradation of phenanthrene. Experiments with two types of biosurfactants—monorhamnolipid and dirhamnolipid—showed that the effect of biosurfactant on phenanthrene degradation depends on the solubilization capacity of the surfactant (see Cha et al., 1998).

NONPOINT SOURCES OF POLLUTION

Nonpoint source (NPS) pollution originates from diffuse land areas that intermittently contribute pollutants to surface and ground water. Agriculture and forestry are two very important sources of diffuse pollution.

Implementation of best management practices (BMPs) usually can lead to improved water quality.

There is a strong need of regulating nonpoint sources of water pollution from agriculture, particularly from animal-feeding operations.

Quality of Water Resources

Surface water

All over the world, nonpoint sources are being recognized as a chief source of pollution to surface waters. Unpaved roads resulting from rapid development during the past few decades were determined to be the primary cause of erosion and increased sedimentation of the marine ecosystem in some areas. Cattle grazing has affected infiltration, runoff and soil loss during rainy season on a natural pasture in the Ethiopian highlands. Soil trampling and reduced vegetative cover increased surface runoff and soil loss (particularly on upper slopes) and reduced infiltration (particularly on tilled soils and soils with high silt content).

France (1997) documented a 60% reduction in litterfall as a result of riparian clear-cutting of forest stands surrounding ten Canadian lakes, decreasing the protective ground surface cover and retention of organic duff. A rainfall simulation experiment further suggested that the erosion of sandy loam can be twice as great under litterfall conditions, representative of clear-cut compared with forested shorelines.

Several studies have reported on the fate of pesticides in the aquatic environment. Nine alachlor metabolites could be identified in river water after 4 weeks of incubation with the parent compound (Mangiapan et al., 1997). Smaller water bodies seem to be more effective in degrading the pesticides than larger ones, and previous exposures through past applications also tend to enhance degradation. The relative effect of hydroxy radicals and direct photolysis of atrazine is altered by the presence of dissolved organic matter with more direct photolysis.

The mechanism of removal of 2,4-D in slow sand filters has been determined to be microbiological instead of sorption and tended to be complete once the system matured (Woudneh et al., 1997). Reducing application rates is the primary tool used to prevent groundwater contamination in many areas. Hanson et al. (1997) found that varying rates of 0.56-4.48 kg atrazine/ha. did not change the relative movement of either the parent compound or two of its metabolites and that absolute differences were directly proportional to the rate of application.

Urban land uses contribute varying levels of metals, bacteria, sediment, nutrients, and organic chemicals to surface waters. Concentrations of metals in runoff from industrial sites often vary significantly for various categories of businesses.

Groundwater

Pesticide contamination of groundwater is a matter of serious concern. It has been found that a vast majority of municipal wells sampled in the USA were contaminated with pesticides and that the different types of aquifers affect pesticide leaching differently. Wells in alluvial deposits had the highest number of pesticide detections and bedrock land forms had the least (Kolpin, Kalkhoff et al., 1997). Kolpin et al., (1997a) used groundwater pesticide data from 1982 and compared it with current data to determine changes in pesticide concentrations. The data indicated that atrazine concentrations had declined, whereas alachlor concentrations had increased.

Processes affecting the movement of pesticides into groundwater have been studied by several researchers. Sorption of pesticides on different types of colloids and subsequent movement of these colloids through the soil profile has been studied, and almost no increase in the transport of atrazine could be found (Seta and Karathanasis, 1997).

It appears that the concentrations of atrazine and nitrate are positively correlated to the amount of irrigated land area and that alachlor is inversely related to the amount of highly erodible land.

Concentrations of nitrogen in groundwater are an important concern because many people use groundwater as a drinking water source. Much of the nitrogen entering our streams and lakes comes from nitrate in groundwater. When urea was applied as a deicing agent during the winter, large concentrations of ammonium were detected at all soil depths sampled; however, over time, the ammonium became oxidized to nitrate, which leached into the groundwater (Swensen and Singh, 1997).

Using ^{18}O, nitrate, and potassium concentrations in an assessment of groundwater quality in Delhi, Datta et al. (1997) showed that fertilizer caused high nitrate concentrations.

Best Management Practices for Nonpoint Source Control

Skinner et al. (1997) examined the extent of current knowledge of management systems in the UK for mitigating the detrimental effects of agriculture. Site-specific recommendations for the use of the riparian buffer, controlled drainage, and in-stream wetland management systems in the Neuse River Basin of North Carolina were presented by Gilliam et al. (1997).

Conservation tillage practices have been reported to exert a beneficial effect on the quality of surface runoff (see Line et al., 1998).

Nitrogen and phosphorus runoff losses from plots of silt loam soil with surface swine manure application were lower when the manure was disked in, than for no-till practices, indicating that disked till should be preferred when swine manure is used as a fertilizer (Gupta et al., 1997).

Sojka and Lentz (1997) claimed that polyacrylamide added to irrigation water is an effective practice for controlling furrow erosion and, therefore, can be an attractive alternative to more difficult conservation practices.

Constructed wetlands consisting of gravel, cattail, common reed and spike rush have been reported to remove about 40% of the orthophosphate, 45% of the suspended solids, and 50% of the dissolved copper in effluent from a stormwater detention pond (Rochfort et al., 1997).

POLLUTION, SANITATION AND HUMAN HEALTH

The prevalence of pollution is an inherent feature of our civilization. To control this scourge, for which people alone are responsible, people themselves will have to come to terms with humankind's growth and industrialization while strengthening the capability of neutralizing its ecological consequences. Pollution does exist, but humans must take control of it. This obligation makes it a problem of organization of the consumer society in a context of overpopulation. It is more a political concern than a scientific one. Harmonious cooperation between the politicians and the technologists is essential.

Even today, many people in some developing countries dump their rubbish right beside their homes; think that rivers, lakes and seas make the best receptacles; or that stray cattle, dogs and other animals will act as scavengers. As soon as any shore in inhabited or frequented, it becomes polluted. This observation is too commonplace to require any proof.

When human populations evolve and nomadic civilizations give place to sedentary ones along rivers, lakes and seas, pollution becomes chronic and increases commensurately. It increases with the rising curve of population. Progress in technology leads to industrialization. Consequently, domestic sources of pollution are supplemented by discharge of industrial and urban wastes.

In olden times, the natural media could, within reasonable limits, cope with the discharges through degradation processes that neutralized or transformed the wastes. Today, the situation is very different. As a result of population growth, intensive industrialization, agricultural development, and the spread of built-up areas and seaside resorts on coasts and estuaries, the critical point has not only been reached but passed.

MAJOR SOURCES OF POLLUTION

Some important sources of pollution are urban pollutants, industrial pollutants, pollutants of animal origin, atmospheric pollutants, natural pollutants of terrestrial origin in the open sea, and pelagic pollutants.

Urban pollutants (domestic wastes) include sewage, garbage, trash, slaughterhouse wastes, hospital wastes, etc.

The main examples of industrial pollutants are wastes from:

(a) food processing (packing plants);
(b) paper mills;
(c) the tanning and textile industries;
(d) dairies and milk processing;
(e) all types of chemical industry;
(f) physical industries (atomic plants, electrical generating stations, thermal power stations);
(g) agricultural industries, industrialized farming and stock raising; and
(h) the petroleum industry, refineries, the petrochemical industry.

Harmful substances are transported by air currents. Winds blowing from the continents towards the sea carry bacteria, viruses, parasites and toxic substances, notably pesticides and heavy metals. Rain brings down these pollutants into rivers and oceans. The runoff reaches estuaries and the seashore laden with everything it has picked up on its way. Water which has traversed agricultural regions regularly treated with nondegradable pesticides, viz., organochlorines, organometallics, fungicides or rodenticides, becomes highly polluted and poses serious hazards.

Pelagic Pollution

Alongside contamination of the ocean by objectionable substances from the land resulting from inland or coastal human activities, there are forms of pollution that take place in the open sea. The establishment of certain industries far out to sea makes them a potential threat. This applies in particular to drilling for oil and the working of the deposits found under the open sea. In such areas, pollution is liable to occur continuously, if protective measures are not vigorously applied, or accidentally in case of serious damage at a specific point in the chain of operations.

Ships carrying petroleum also constitute a threat of offshore pollution of the oceans.

There is strong evidence that pathogenic bacteria and a good many parasites do survive very well in natural seawater.

Studies on marine microbiocenosis all confirm that the oceans are populated mainly by bacteria of terrestrial origin, or by ubiquitous bacteria which are found not only in the sea but also in fresh water and in every kind of soil. These universal-type bacteria are unaffected by the degree of salinity and are described as euryhaline.

Alongside these microorganisms, there exist others which are very fastidious—the strictly halophilic bacteria are restricted to a narrow range of salinity values.

The transport and fate of microorganisms in the marine environment is diagrammatically illustrated in Fig. 7.1.

COMBATING POLLUTION

The great diversity of pollutants, their undeniable dangers and the uncertainties of their fate even when they are discharged relatively far from the shore or in deep water, are all very well known. Appropriate laws have been developed in response to all sorts of incidents caused by discharge of wastes into the sea and estuaries and these laws need to be enforced and respected.

The basic principle of environmental health therefore consists in:

(a) respect for the laws prohibiting discharge of wastes; and
(b) refusal to grant any waivers.

To uphold this principle, the wise course is to give possible offenders (the potential polluters) the means of not contaminating the environment. A new technology (termed 'rupology' by Brisou, 1976) of waste products has been developed which deals with the study of waste (rupos = rubbish, wastes).

Wastes can be eliminated by burying or burning; directly utilized: remelting of metals, utilization of automobile bodies for fish to breed in, conversion into building materials, utilization of woody debris, fertilization of red muds, etc.; processed by biodegradation and marketing of the finished products and residues, manufacture of animal feeds, fertilizers or composts.

Many studies have been devoted to the treatment of wastes from papermaking, stock-raising, tanning and brewing, to the bacterial treatment of alkanes, collagen residues (from glue factories), terpenes, hydrocarbons and fish wastes.

Wastes and Pollution 309

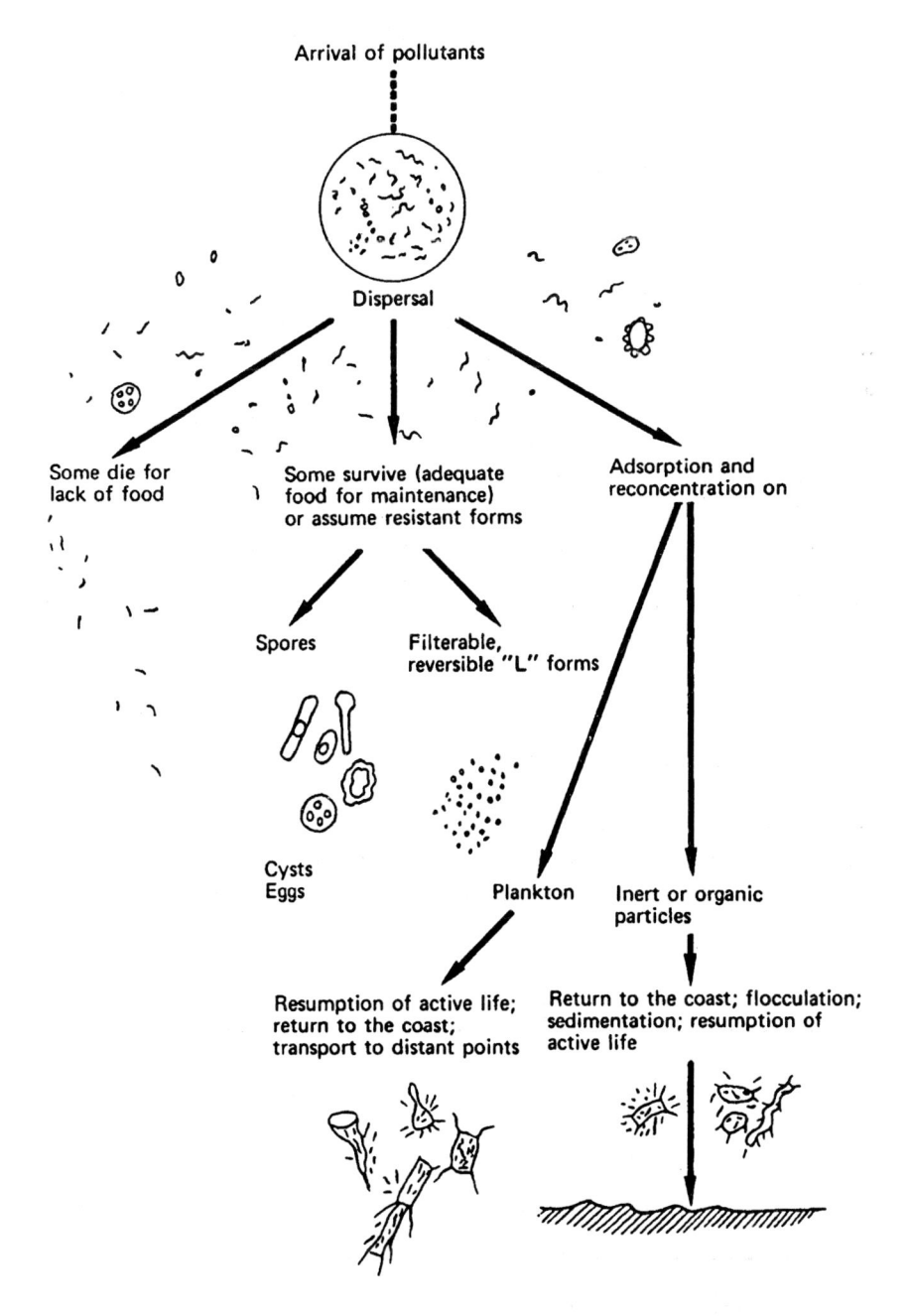

Fig. 7.1 Fate of micro-organisms in the marine environment (after Brisou, 1976).

Stable substrates include toxic, neutral and offensive or 'dirtying' substances. The toxic ones, depending on their affinities, are concentrated by certain tissues such as the liver, the kidney, the brain and the genital glands. It is, therefore, important to eliminate them and prevent their being discharged into water bodies.

The degradable substances are very interesting as they give many by-products when they serve as substrates for microorganisms.

Millions of substances are being discharged into the environment. Many of them act as pollutants.

Food-processing wastes exposed to the action of lower fungi can be made to yield marketable products such as steroid hormones, proteins and fumaric acid.

Sewage can be made to yield methane, an excellent source of energy. Liquid manure and waste water from slaughterhouses are good sources of methane. Bacteria can synthesize hydrocarbons from methane. Sewage treated with algae, bacilli, clostridia, lactobacilli and propionobacteria can be made to yield proteins, animal feeds, enzymes, sulfur, acetone, butanol, recoverable lactic, propionic and acetic acids, vitamin B12 and, of course, methane (Brisou, 1976).

A sound environmental health plan should provide for the recovery of all the sewage of a region in order ot derive the maximum yield from it, through controlled fermentation under the supervision of competent bacteriologists.

In areas undergoing intensive urbanization and industrialization, it is a good move to establish a complete inventory of pollutants. Some can be immediately neutralized or recovered by the industry concerned, others collected and utilized for the benefit of the community.

Table 7.2 and Figures 7.2 to 7.4 illustrate some of the above points.

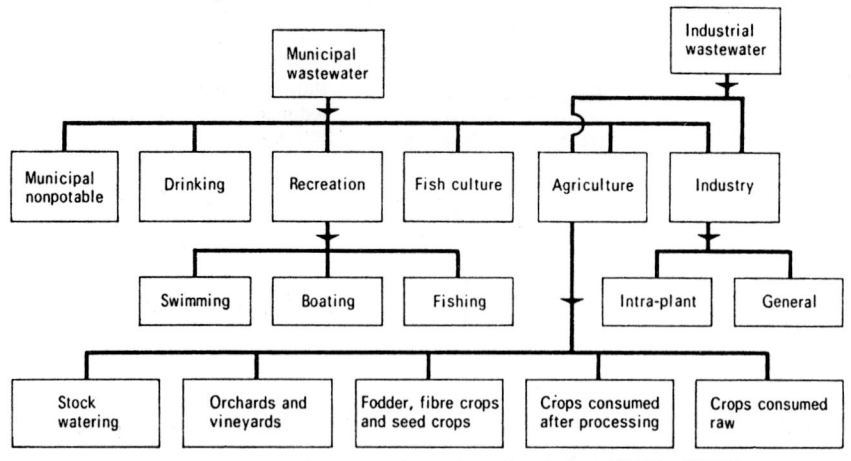

Fig. 7.2 Intentional reuse of wastewater (after WHO, 1973).

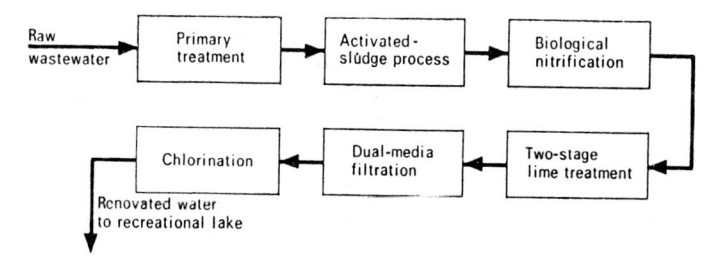

Fig. 7.3 Treatment system for producing recreational lake water (after WHO, 1973).

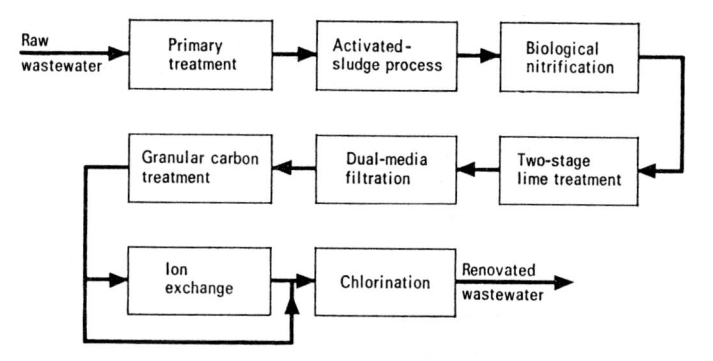

Fig. 7.4 Treatment system for producing potable water (after WHO, 1973).

RECYCLING HUMAN WASTE

In recent years, there has been a growing trend in some countries to fertilize farms with human waste. As greater quantities of human waste are produced, and as traditional dumping areas become scarce or increasingly costly, policymakers have resorted to this ancient strategy to dispose of the nutrient-rich material. But because modern waste flows are dirtier than those of centuries past, capturing the benefits of reuse with minimal risk is a daunting task.

Today's economies are consuming about one third-more resources and eco-services than nature can deliver sustainably. Reducing the human footprint will therefore require a much higher level of material reuse. As recycling goes beyond newspapers and aluminium cans, we should distinguish between beneficial recycling—the return of materials to advantageous and environmentally benign uses—and careless reuse, which is often no more than dumping under a green label. This is all the more important for industrialized nations active in recycling human

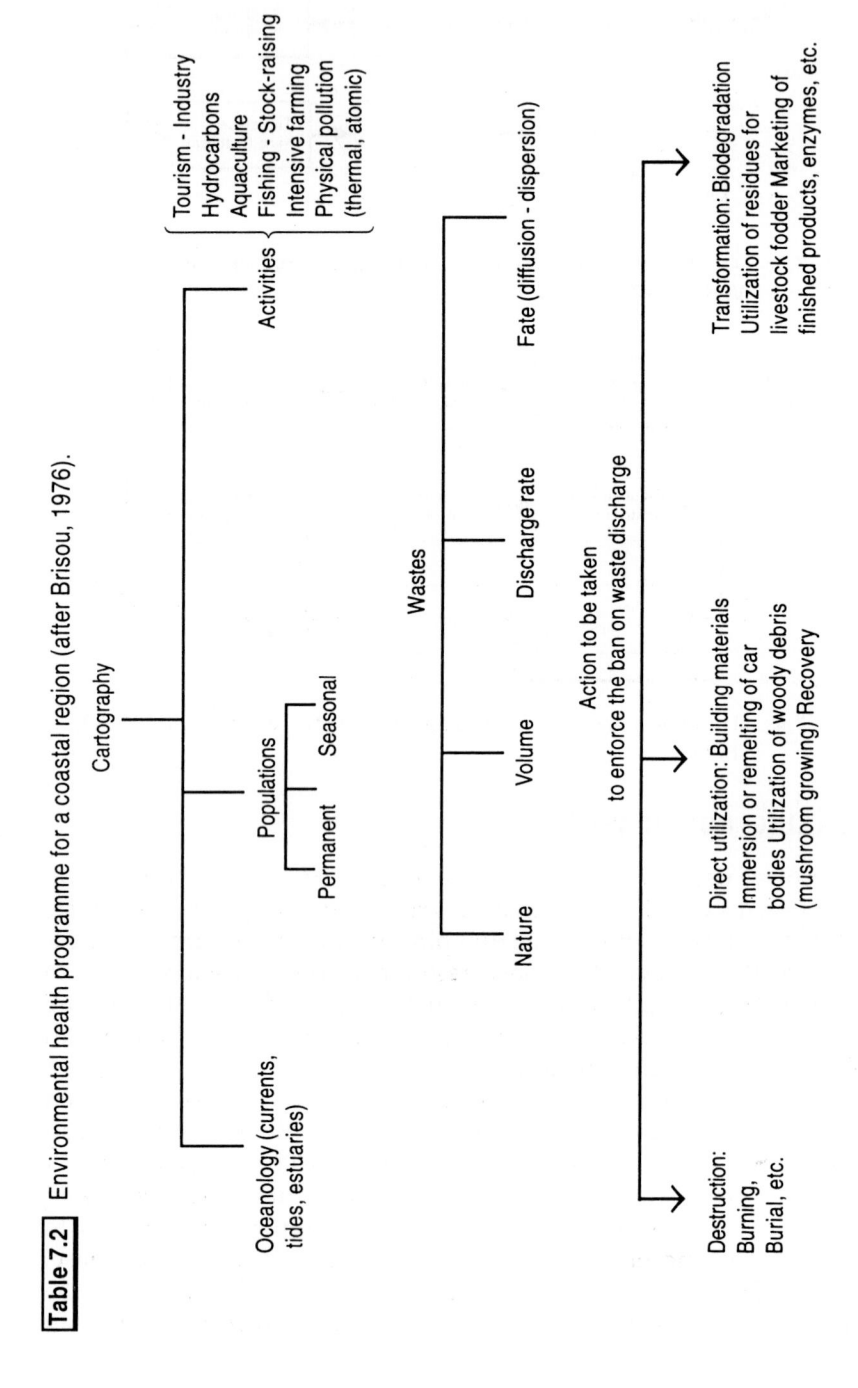

312 Environmental Technology and Biosphere Management

Table 7.2 Environmental health programme for a coastal region (after Brisou, 1976).

Cartography
- Oceanology (currents, tides, estuaries)
- Populations
  - Permanent
  - Seasonal
- Activities
  - Tourism - Industry
  - Hydrocarbons
  - Aquaculture
  - Fishing - Stock-raising
  - Intensive farming
  - Physical pollution (thermal, atomic)

Wastes
- Nature
- Volume
- Discharge rate
- Fate (diffusion - dispersion)

Action to be taken to enforce the ban on waste discharge

Destruction: Burning, Burial, etc.

Direct utilization: Building materials Immersion or remelting of car bodies Utilization of woody debris (mushroom growing) Recovery

Transformation: Biodegradation Utilization of residues for livestock fodder Marketing of finished products, enzymes, etc.

waste and is even more timely for the many developing country cities that are busy planning and designing sanitation systems for the twenty-first century. Recycling human waste isn't like recycling newspapers. The use of 'night soil' on crop fields is ancient, but modern sewage and farming systems have greatly complicated the risk of using our most obvious fertilizer (Gardner, 1998).

Organic recycling has been commonly practised in Asia since times immemorial. Eventually, it also became more widely practised outside Asia as burgeoning cities scrambled to get rid of their waste. By the mid-nineteenth century, in many European cities sewage was collected by scavengers and delivered to nearby 'sewage farms' for use as fertilizer. The practice soon spread to many cities in some other countries. Indeed, by the early twentieth century, with sewers commonly in use in more developed countries, land application of sewage constituted the only method of disposal in many metropolitan areas. In this pre-modern era and many developing countries even today, the major health risk from sewage came from its pathogen content. Untreated sewage is rich in bacteria, viruses and parasites, which can spread to people through water supplies, food fertilized with waste, or direct contact. Outbreaks of cholera and other infectious diseases have occurred in developed countries in the past, and continue to occur in developing countries today.

In the twentieth century, underground sewers became increasingly common and some were connected to treatment plants. These two technologies reduced the pathogen menace considerably. However, new contaminants emerged in many cities. Wastewaters from industries, often containing toxic chemicals and heavy metals, mixed with human waste when discharged into public sewers. Sewage treatment processes were somewhat effective in killing pathogens but did not eliminate these other contaminants. Instead, the pollutants simply accumulated in the sludge. But because most sludge was destined for disposal at a landfill, in an incinerator, or even on the ocean bottom, there was not much reason to worry about these substances.

Over the past few decades, however, several developments re-focussed interest in recycling human waste to land. Urban growth and an increasingly sewered population led to concentrating more and more human waste in urban areas. Waste disposal sites became less available or more expensive. Ocean dumping of sewage sludge has been outlawed both in the United States and in Europe. Incineration is costly and landfills leak greenhouse gases—globally, landfills account for about 10% of the world's human-origin emissions of methane.

Meanwhile, many countries located in arid or semi-arid zones and faced with increasing water scarcity have started tapping wastewater for irrigation. The use of manufactured fertilizers—the technology that has largely undercut the practice of reapplying wastes to soils—has been blamed for a variety of pollution ills, from high nitrate levels in drinking water to nuisance algal blooms that rob fish and other aquatic animals of oxygen.

All of these problems are related to the disruption of the organic loop, which broke the circular flow of human waste from people to farms, then back to people in the form of food. The natural response to resolving these problems could be addressed by rejoining the loop by once again recycling organic waste.

A materials loop can be reconnected in several ways. Ideally, recycling should imitate nature's cycling process, which is much more efficient and benign than any human designed process. But recycling becomes quite difficult in modern industrial economies, which mix materials in combinations and concentrations.

Even for such an organic material as human waste, industrial economies have greatly complicated the prospects for reuse. On its movement from toilet to treatment plant, excreta mixes with toxic chemicals and heavy metals, which are poured down household drains and leached from household plumbing, or dumped into sewers by industry. Virtually, all of the thousands of chemicals and metals flowing through modern economies—including PCBs, pesticides, dioxins, heavy metals, asbestos, petroleum products and industrial solvents—potentially become a part of sewage flows (Gardner, 1998). Some of these materials degrade quickly without harming the environment but others are very long-living and persist. Natural cycling keeps potentially polluting materials spread thin. In contrast, human economies tend to concentrate harmful pollutants in our waste streams. Returning of these wastes to farmland is usually more dangerous than beneficial.

It is also quite important to understand the hazards posed by contaminants in recycled material. The full effects of sludge on the environment and humans are not known. We know very little about the long-term behaviour of metals in sludge-applied soils, for example. Conceivably, heavy metals could eventually be freed up and absorbed by crops. As organic matter in sludge breaks down over time, the bonds that keep metals from migrating (either down into groundwater or up into crops) are weakened. Metals that are largely immobile in the short run may well surface in our food and drinking water in the long run.

Similarly, our understanding of the threat posed by many toxic chemicals is not complete. Persistent chemicals, e.g. the PCBs and dioxins found in sewage sludge are now suspected of mimicking hormones and causing reproductive abnormalities in humans and wildlife, even when present in traces (Gardner, 1998). Wisdom warrants counsel against allowing such chemicals to become part of our drinking water or food supply.

Stricter pretreatment programs need to be instituted and effectively enforced to cover a broader range of polluters, and to enforce a lower level of dumping. The cleaner the sludge, the more likely its organic richness that could eventually be applied to cropland (Gardner, 1998).

Industrial nations have been depending on technologies designed for *disposal*, rather than *recycling*. In many ways today's flush toilets, sewers and treatment plants are inferior technologies for recycling human waste. Sewers commonly serve residences and industry together, a practice that contaminates organic matter with heavy metals or toxic chemicals. From a recycling perspective, it is much better to prevent human waste from mixing with other wastes in the first place. Waste streams can be segregated by using separate sewers for human and other wastes, by treatment of industrial waste at the factory, or by treatment of human waste in residences or office buildings. Further, treatment plants are designed to recycle water but not sludge. They often remove some of the material that should be recycled.

Some technologies modify conventional sewage treatment to produce a product that gets closer to a safe recyclable. One such process is called advance alkaline stabilization and is used by an Ohio-based American company, N-VIRO. This company mixes roughly equal parts of sewage sludge and alkaline cement kiln dust to produce a sludge product called 'N-VIRO soil'. The alkaline dust raises the pH level, which prevents most of the sludge's metals from leaching or being taken up by plants. The dust pasteurizes the sludge, killing pathogens. The N-VIRO process reconnects the organic loop safely but it is still used in conjunction with some disposal technologies, such as sewers that mix domestic and industrial wastes. In this contex, 'N-VIRO soil' is riskier than waste processed entirely with recycling technologies. The N-VIRO process does not eliminate or neutralize toxic chemicals. The process is most promising, therefore, where contaminant levels are very low, both in the sludge and in the kiln dust that is mixed with it (Gardner, 1998).

It appears that the more the reuse of human waste depends on conventional *disposal* technologies, the less likely that such reuse will amount to benign recycling.

This is good news for many cities in developing countries that have not yet committed to a particular system of sanitation, also for those cities that plan to overhaul or rehabilitate their old systems. These cities can leapfrog ahead to alternative technologies designed for recycling.

Many alternative systems require the separation of human waste from other contaminants. This may be achieved by treating industrial wastes at the source, so that they never enter the public sewer system. Nature's cleansing processes can be imitated by using plants, microorganisms, and fish, in combination with solar energy, to progressively treat industrial wastes in a series of pools and constructed wetlands. There exists a good market for these facilities.

Household treatment has the merit of isolating human waste from home-grown chemicals and metals—detergents, soaps and cleaning solvents, for example, and copper and other metals that leach from plumbing. Composting toilets are one viable on-site recycling technology. They look like standard flush models (without the water tank, as most use no water) and can hold up to several years' worth of excreta. These systems create a fertilizing product that can be applied to home gardens or is collected and sold to farmers. The concept of *neighbourhood sewage wall* channels sewage through a series of terraced planters that progressively filter and purify the waste. Each terrace would contain the plants and bacteria best suited for the various stages of treatment. The resulting effluent could be used on local gardens and the plants could be harvested and composted (Gardner, 1998).

Flush toilets have been shown to account for 20-40% of the residential water use in industrial countries; the proportion is higher in developing countries. With the population of 'water-stressed' countries expected to more than triple by 2025; any technology that can reduce water demand deserves serious consideration.

Nature-centered recycling is a good option for the reuse of various materials, e.g. human waste and sawdust as societies struggle to eliminate the flow of resources straight to the dump (Gardner, 1998).

TREATMENT OF SANITARY LANDFILL LEACHATES

Leaching occurs from every landfill that has not been sealed with some impermeable surface. When, as a result of precipitation and/or groundwater which the landfill body cannot fully absorb, water percolates through the landfill, soluble materials from the refuse as well as from biochemical transformation processes are taken up and are washed out. Deposition of refuse, such as slurry, with a high water content, also contributes to the leachate.

The pollution of seepage water depends on the composition of the refuse, on its age, on the management of the landfill, on the hydraulic and hydrogeological conditions, and on the frequency and duration of precipitation events (Biehler and Hägele, 1995). When assessing the quality of leachates from domestic refuse landfills, two phases are pertinent: the acid fermentation phase, and methane formation. Acid fermentation denotes insufficient anaerobic biological breakdown to the lower fatty acids in the early years of a landfill's management. This leachate has a pH-value between 4.5 and 7.5, high pollution, and traits that favour biological sewage purification. With the increasing age of a landfill, a virtually complete anaerobic biological decomposition into methane and carbon dioxide takes place.

Leachates from the methane phase typically have slightly alkaline pH value (between 7.5 and 9.0) and significantly lower pollution loads; these leachates are not suitable for biological treatment.

In several advanced countries, a great deal has been spent on the cleanup of old waste dumps but too little on the prevention of future contamination. This has happened in part because of poor decision-making in the management of hazardous waste and in part because, for better or worse, the debate over environmental policy has been dominated by lawyers and economists, rather than by scientists (see Freeze, 2000). There is need for thinking people to reject the polarization of environmental policy, which produces strong fluctuations in our answers to such questions as how stringently to regulate, how much risk to tolerate, and what priorities to set.

Landfills are a necessary evil that inevitably leak, and the best way to limit the damage they cause to the environment is to place them in geologically desirable locations. To avoid wasting money on remediation, we should make the siting process more rational, while respecting the ethical principle of 'intergenerational equity' as a factor that should have greater influence on environmental decision making.

It is now generally agreed that in many countries, environmental regulation has become highly complex, inconsistent and has several shortcomings. This has made it a challenging task to simplify prospective regulation. Such simplification is very different from the task of cleaning up the legacy of past mistakes.

General Overview of Available Leachate Treatment Processes

The various treatment processes of landfill leachates can be classified as follows:

A. Biological: aerobic (revitalization, oxidation); anaerobic (anaerobic reactor); sessile organisms (rotating disc filter, gas dome); and plant sewage bed.
B. Chemical oxidation: hydrogen peroxide; and ozone.
C. Chemical/physical : absorptive (such as flocculation); membrane (reverse osmosis, ultra-filtration); and stripping.
D. Thermal: evaporation, vaporization; and incineration

Of the above biological processes, reverse osmosis and vaporization are among those applied most frequently. Also combinations of several processes, e.g. biology-flocculation/precipitation-adsorption;

biomembrane; biology-reverse osmosis; biology-chemical oxidation-biology; flocculation/precipitation + adsorption + reverse osmosis; and ultrafiltration + reverse osmosis (Biehler and Hägele, 1995). Some of these processes are diagrammatically illustrated in Figs. 7.5 to 7.7.

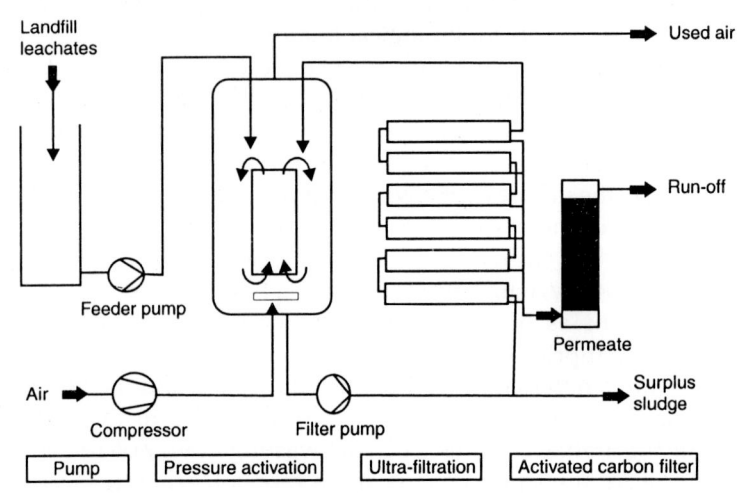

Fig. 7.5 Biomembrane process for the treatment of leachates (after Wagner and Wienands, 1989).

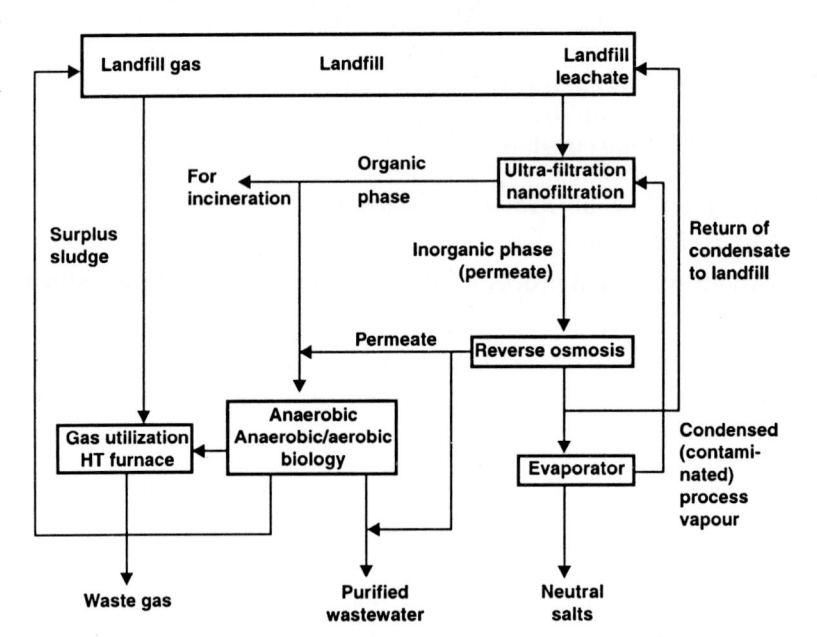

Fig. 7.6 Combined physio-biological treatment of landfill leachates (after Biehler and Hägele, 1995).

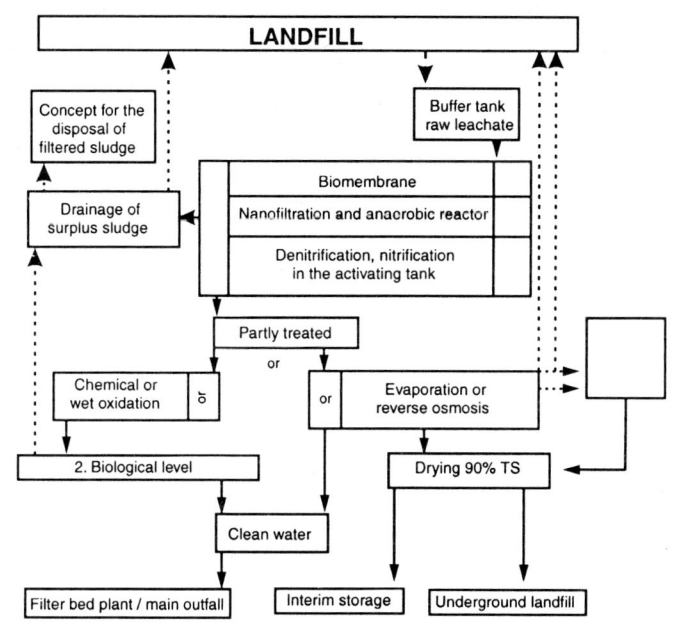

Fig. 7.7 Available processes for the treatment and disposal of leachates and associated residue (after Biehler and Hägele, 1995).

BIOREMEDIATION

During the last few decades of industrial development, the amount and variety of hazardous substances has increased drastically. An estimated 100,000 human–made chemicals are in use, and hundreds of new chemicals are produced each year. Due to the increase in industrial and agricultural activities and exports of wastes, not only the traditional industrialized nations, but all countries are faced with widespread soil pollution. Many synthetic compounds that are not related to natural ones, persist in the environment. Essentially, there are three major categories of sites with polluted soils: (1) sites polluted by either spillage or leakage during production, handling or use of industrial material. This includes activities to harvest raw materials, such as mining and oil drilling; (2) locations used as disposal sites for diverse waste; and (3) farmlands excessively exposed to pesticides. Contaminated land sites are health hazards for human beings and so are unsuitable both for housing and agriculture. The downward migration of pollutants from the soil into the groundwater is of serious concern in those developing countries where groundwater is directly used for drinking without any prior treatment.

Most organic chemicals are subject to enzymatic attack of living organisms. These activities come under the term *biodegradation*. However, the end products of these enzymatic processes might differ greatly. For instance, an organic substance might be mineralized (transformed to carbon dioxide and water). It might also be converted to a product that binds to natural materials in the soil, or to a toxic substance (Lehmann, 1998).

Bioremediation involves the use of microorganisms to remove or detoxify pollutants, usually as contaminants of soils, water or sediments that otherwise threaten public health (Crawford and Crawford, 1996). Microorganisms are sometimes the only means—biological or non-biological— to convert synthetic chemicals into inorganic compounds. Bioremediation has now emerged as an industry that is driven by its particular usefulness for sites contaminated with petroleum hydrocarbons.

Bioremediation is the technological process that harnesses biological systems to clean-up environmental pollutants. Microbial systems are generally employed in bioremediation programmes, commonly for the treatment of soils and waters contaminated with organic pollutants. Micro-organisms can degrade a variety of organic pollutants.

The merit of a bioremediation approach lies in its potential to treat contaminants on site with relatively little disturbance to the contaminated matrix. Micro-organisms make it possible that organic pollutants are completely mineralized to inorganic materials. In contrast, removal of contaminated material to landfill sites, or extraction of contaminants using physical processes such as soil washing, fail to destroy the contaminants present: they merely concentrate the contaminated material in a different location. However, physical treatments are rapid and their outcome is generally predictable in the short term; they are also relatively inexpensive. Bioremediation, too, can be fairly cheap. The unpredictability of bioremediation arises from a lack of knowledge about the behaviour of microbial populations in natural environments and about how physical, biological and chemical factors might interact to control microbial activity against environmental pollutants. This realization has shifted the focus of bioremediation from isolation and construction of 'surperbugs' to determining the factors that limit pollutant transformations and mineralization in natural environments (Head, 1998). Access of microorganisms to pollutants in situ critically determines the success of bioremediation. Hence, methods are being developed to enable predictions regarding the feasibility of bioremediation based on pollutant bioavailability *and* biodegradation.

It is easy to show biodegradation of specific compounds in a contaminated environmental sample. Spiking a sample of contaminated soil with some pollutant chemical and monitoring its loss and the appearance of degradation end products in comparison with sterilized control samples usually demonstrates rapid biodegradation. Reduced bioavailability results from the interaction of pollutants with both organic and inorganic components of the soil matrix. Access of microorganisms or their enzymes to the pollutant molecules is constrained and with increasing contact time (ageing) the proportion of the pollutant that becomes biologically unavailable increases.

Bioavailability is undoubtedly a key factor in determining the feasibility of bioremediation. According to Bosma et al., (1997) the intrinsic microbial activities limit bioremediation in very few cases; in most cases, it is mass transfer limitation that prevents the full exploitation of the microbial degradative potential. While reduced bioavailability may lead to failure of bioremediation, it can also be responsible for reduced toxicity of pollutant residues (Head, 1998). Bioavailability can be estimated using microbial bioassays, differential solvent extraction techniques, and a combination of desorption, transport and biodegradation kinetics (Bosma et al., 1997).

The biodegradation of a contaminant in situ is a function of both the catabolic activity of the microbes present and transport of the contaminant to microbial cells with the ability to degrade the contaminant. The latter may be the most important factor in determining the success of a bioremediation programme.

Concentration of contaminant compounds and residue toxicity are the commonest criteria specified by legislative bodies to define successful bioremediation. Chemical analyses (e.g. GC-MS) and toxicological assays (e.g. Microtox) can be applied to accurately identify and quantify organic contaminants or assess residual toxicity following treatment. Nevertheless, the task is highly complicated because of the heterogeneous distribution of pollutants. Therefore, alternative strategies have been devised that either alone or in combination may provide the proof to show that bioremediation has been actually effective (Head, 1998).

The biodegradation of contaminants containing a complex mixture of compounds with different susceptibilities to biodegradation may be assessed by measuring the ratio of the degradable components to a poorly degradable component in the mixture.

Whilst assessment of crude oil biodegradation has been the main application of this approach it can also be applied to other complex mixtures. PCBs, for example, are normally found as a mixture of congeners with similar physico-chemical properties but different biological fates and therefore lend themselves to this approach (Cerniglia, 1992; Knackmuss, 1996). As 3,4-3',4'-tetrachlorobiphenyl is more readily biodegradable than 2,3,6-3',4'-pentachlorobiphenyl, the ratio of the two congeners is used to monitor bioremediation of PCBs in river sediments.

Except in fermentation and disproportion reactions, the oxidation of organic compounds is coupled to the reduction of exogenous electron acceptors. In soil and sediment environments these are usually O_2, NO_3^-, Fe^{3+}, Mn^{4+}, SO_4^{2-} or CO_2. Depletion of these species or accumulation of their reduced products in contaminated relative to uncontaminated material can indicate organic pollutant metabolism. This becomes possible if the concentrations are compared with levels of a conservative tracer (e.g. chloride or bromide) having physico-chemical properties similar to the analyte of interest, but is not susceptible to biological reduction.

The ability to isolate and enumerate bacteria from contaminated sites capable of degrading a particular pollutant is a line of evidence frequently used to support the feasibility of bioremediation. This applies if the population of degradative bacteria increases following implementation of a treatment to stimulate biodegradation. Culture-based methods underestimate both qualitative and quantitative measures of microbial populations by orders of magnitude. This has prompted the development and application of nucleic acid-based techniques to study the ecology and diversity of microorganisms in nature. Molecular biological methods are now being exploited to study bioremediation. This is analogous to the isolation of bacteria with appropriate catabolic properties from a polluted site, as a pointer that competent degradative populations are present at the site, and that bioremediation potential exists (Head, 1998).

Indeed, molecular biological techniques are excellent tools that increase our understanding of the distribution and expression of important catabolic genes and microbial population dynamics during biotreatment (Massol-Deya et al., 1997) but they cannot yet be applied for routine monitoring applications.

It is now being realized that it is not our knowledge of pollutant catabolism that limits the success of bioremediation, but rather, an insufficient understanding of the interplay between the biotic and abiotic factors that determine the outcome of any particular remedial strategy. Bioremediation ought to be considered as a natural bioengineering process that takes account of these interactions. This can ensure that bioremediation is a successful technology (Head, 1998).

Soil Bioremediation

In soil bioremediation, a general distinction is made between in situ treatments, i.e. on the contaminated site itself, and ex situ treatments, where the soil is excavated and processed elsewhere. The latter covers a wide range of technologies, from relatively simple land farming to costly bioreactor treatments. Bioreactors allow a more rigid control of the whole process and can accelerate degradation. However, because the soil has to be removed, ex situ treatment is generally more costly.

Under field circumstances, microbial activity is often restricted by nutrients and oxygen. In situ bioremediation stimulates the indigenous microflora by supplementing the limiting factors, e.g. by aeration and adding nitrogen and phosphate. Some hazardous substances are preferably degraded anaerobically. The chlorinated organic solvent *tetrachloroethylene* (PCE), for example, is degradable in a two-step approach. Firstly, oxygen has to be removed so that PCE is transformed to dichloroethylene by anaerobic bacteria. Then, the soil is aerated and further degradation takes places by aerobic microbes. Whereas most degradable substances serve as a carbon and energy source for microbial growth, others do not. In this case, another substance needs to be added as external energy source. For the degradation of PCE, the anaerobic degradation stage is fuelled by methanol, whereas in the subsequent step aerobic bacteria get their energy from additional phenol.

Toxic metals cannot be degraded bacterially, so bioremediation processes aim at sequestering the metals. Here, metals become unavailable to biological processes and so are no longer toxic.

Today, petroleum and petroleum derived products cause the most pervasive environmental contamination. They are generally susceptible to naturally occurring microbial activity, and have become a principal target of bioremediation. Hydrocarbon-degrading microbes occur everywhere. Several different genera of oil-degrading bacteria and fungi have been identified for both in situ and ex situ bioremediation purposes (see Table 7.3).

Such geomorphological features of the site as soil type, pH and organic matter content influence the clean-up. The future use of a site is also important because it determines the tolerance level of the pollution that may remain in the soil.

Bioremediation broadly includes five approaches:

A. *Ex situ treatments:* Contaminated soil is excavated and treated at another site
1. *Bioreactors.* Liquids, vapours, or solids in a slurry phase are treated in a reactor. Microbes can be natural, cultivated, or genetically engineered. Processes can be monitored, regulated and modelled mathematically very precisely.
2. *Solid-phase technologies.* Contaminated soils are excavated, placed in a containment system through which water and nutrients percolate. Particularly useful for petroleum-contaminated soils.
3. *Composting.* This is a variation of solid-phase treatment. Large amounts of degradable organic matter are added to a contaminated

Table 7.3 Soil bioremediation for various substances

Contamination	Volatility	Biodegredability Aerobic	Biodegredability Anaerobic	Solubility in water	In situ possibilities
Hydrocarbons					
Gasoline	+	+	–	+	yes
Kerosene	±	+	–	+	yes
Gasoil	–	±	–	–*	yes
Domestic fuel	–	±	–	–*	yes
Lubricants	–	–	–	–	no
PAH					
Light (2–3 rings)	±	+	–	±*	yes
Heavy (4–5 rings)	–	–	–	–*	no
Chlorinated Hydrocarbons					
Aliphatic (per, tri)	+	–	+	+	yes
Chlorobenzene	+	+	–	+	yes
Pesticides	–	±	–	–	no
PCB	–	–	–	–	no
Heavy metals	–	–	–	±*	yes
Aromatics (BTEX)	+	+	±	+	yes

PAH : Polyaromatic Hydrocarbon; Pesticides: Organochloro–pesticides (e.g. DDT); PCB: Polychlorinated Biphenyl;
BTEX : Benzene, Toluene, Ethylbenzene, Xylene
Source: J.P. Okx, L. Hordijk and A. Stein. Managing Soil Remediation Problems." *Environmental Science and Pollution Research*, 3 (4): pp. 229–235 (1996).
* Solubility can be enhanced by detergents (for hydrocarbons) or by acidification (heavy metals)

material. The process consists of anaerobic incubation for several weeks or months.

4. *Land farming.* Contaminated sludge, soils or sediments are spread on fields and cultivated in the same way as a farmer might plough and fertilize agricultural land. Petroleum-contaminated soil can be cheaply cleaned up by microbial activity. Although its application is restricted to readily degradable material, there can be some leaching into groundwater.

B. *In situ treatments:* The treatment of the contaminated soil takes place at the site of the contamination.

5. In situ *bioremediation.* Involves the stimulation of indigenous microbial populations (e.g. by adding nutrients or aeration). As the soil is not removed this method is quite cost effective. But precise control of the biological processes cannot be made.

Non-biological treatments of waste such as landfill, chemical extraction, electroreclamation and incineration are still the techniques in common use. With the exception of landfill, the physical-chemical processes on which they are based are fast and controllable, but their demand for energy in high. In contrast, bioremediation approaches are less energy demanding. The degradation of a contaminant in the soil is not a linear process over time. Degradation processes slow down with the falling concentration of pollutants in the soil. Consequently, bioremediation is time–consuming and unable to clean the soil completely. Synthetic compounds such as *polychlorinated biphenyls* (PCBs), highly substituted nitro compounds and *polyaromatic hydrocarbons* (PAHs) are very recalcitrant to microbial attack. For soils contaminanted with these substances, incineration or chemical treatments have an edge over bioremediation (Alexander, 1994).

Genetic engineering may allow to construct tailor-made microorganisms having improved degrading capabilities for toxic substances. but little actual progress has been made in developing robust strains of organisms for in situ use. Of about 30 approved field tests of recombinant bacteria carried out world-wide since 1986, only one was for bioremediation purposes! GMOs are usually outcompeted by naturally -occurring organisms.

On the other hand, when exposed to the contaminating substances, they are supposed to degrade, GMOs survive better than naturally occurring bacteria. But there is concern about their potential effect outside the treatment area. While recombinant strains may appear harmless in the laboratory, it is impossible to assess their impact in the field.

When restricted to contained use in bioreactors, GMOs can be used for the treatment of industrial discharges which are reasonably well-defined and selected. However, under field conditions, in situ bioremediation techniques continue to be more promising than the application of GMOs.

Bioremediation in Developing Countries

Currently, bioremediation is mostly applied in developed countries. In many developing countries, a warm climate and high humidity naturally enhance microbial remediation processes.

Several attempts in bioremediation are underway in the oil producing developing countries, e.g. Brazil.

PHYTOREMEDIATION

Phytoremediation refers to the use of plants for cleaning up contaminants in soil, groundwater, surface water and air. It encompasses: (1) phytoextraction or phytoconcentration, where the contaminant is concentrated in the roots, stem and foliage of the plant; (2) phytodegradation, where plant enzymes help catalyze breakdown of the contaminant molecule; (3) rhizosphere biodegradation, where plant roots release nutrients to microorganisms which are active in biodegradation of the contaminant molecule; (4) volatilization, where organics are transpired through plant leaves; and (5) stabilization, where the plant converts the contaminant into a form which is not bioavailable, or the plant prevents the spreading of a contaminant plume (Neare, 1999). Phytoremediation is applied for lightly contaminated soils, sludges and waters where the material to be treated is at a shallow or medium depth and the area to be treated is large, so that agronomic techniques are economical and applicable for both planting and harvesting. The site owner should be prepared to accept a longer remediation period.

Some advantages of phytoremediation are: (1) low cost compared mechanical methods for soil remediation; (2) passive and solar driven; (3) faster than natural attenuation; (4) reduction in amount of contaminated material going to landfills; (5) energy recovery from controlled combustion of harvested biomass; and (6) high public acceptance (Neare, 1999).

Some limitations of phytoremediation are:

1. Generally slower than other treatment methods, and is climate dependent;
2. The contamination system to be treated must be shallow;
3. Usually requires nutrient addition, and mass transfer is limited;
4. High metal and other contaminant concentrations can be toxic to some plants;
5. Access to the site needs to be controlled, as contaminants being treated by phytoremediation may enter ground-water or may bioaccumulate in animals;
6. For mixed contaminant sites (i.e. organic and inorganic), more than one phytoremediation method is usually required; and
7. The site must be large enough to utilize agricultural machinery for planting and harvesting.

Table 7.4 compares other remediation techniques with phytoremediation.

Certain production systems have attempted to use willows (*Salix*) as a vegetation filter for treatment of sewage water and examined growth responses with municipal sludge and ash applications. The success of willow filters varies considerably with the development stage of the plantation and also with site conditions, but some European municipalities already have planted willow filtration stands. leaching of nitrate limits the size of the sewage water dose. Additions of sludge and ash resulted in *Salix* stem mass equal to that produced by commercial fertilizers. As a result of increased pH in the lime and ash treatments,

Table 7.4 A comparison of phytoremediation and other remediation techniques (after Neare, 1999)

Treatment	Advantages Compared to Phytoremediation	Disadvantages Compared to Phytoremediation
Solidification/ Stabilization	Not seasonally dependent; well established; rapid; applicable to most metals and organics; simple process.	Site not restored to original form; leaching of the contaminant is ricky; can result in significant volume increase.
Soil Flushing/ Soil Washing	Not seasonally dependent, except in cold climates; methods well established for several types of sites and contamination.	Metal removal using water requires pH change; additional treatment steps and chemicals add to costs.
Bioremediation	Established and accepted; a bioreactor can be utilized for *ex situ* work; often faster than phytoremediation.	Requires more nutrient addition than phytoremediation; applicable to organics only.
Electrokinetics	Not seasonally dependent; can be used in conjunction with phytoremediation to enhance rhizosphere biodegradation.	Useful for soil only, not wetlands; uniformity of soil conditions required.
Chemical Reduction/ Oxidation	Not seasonally dependent; relatively short treatment time frame; usually off-site.	Requires excavation; uses chemical additives; fertility of the soil can be affected.
Excavation/Disposal	Rapid, immediate solution	Transfers contaminants to landfills; does not treat.

manganese and cadmium content in stems decreased. Net accumulation in the topsoil of Cu, Mn and Ni occurred in all ash treatments; Zn and Cd accumulation was dependent on ash source and type.

Willow can also prove useful in other ways, e.g. in coupling water quality improvement with producing feedstocks for central heat and power production. Small farmers can plant willows strategically between food production fields and small streams which drain into some lake used for recreational purposes by local residents. Water quality in the lake gets improved; so, residents are willing to pay a higher price for the heat and power produced by combustion of the harvested willows.

Some attempts are underway to frame environmental guidelines for development of sustainable energy from biomass, e.g. biomass harvesting and sustainable forest management. Best management practices and certification schemes for environmental sustainability of conventional forestry bioenergy production systems are also being reviewed. There is considerable interest in some quarters in the environmental sustainability of short-rotation forestry and in designing a template for sustainable forest bioenergy production systems. This template needs to be based on essential concepts incorporated in existing guidelines and should be generic enough to meet the requirements of most individual state, company, or international regulatory structures.

PLASTICS RECYCLING AND WASTE MANAGEMENT

Increasing environmental awareness of the environment has stimulated concerns related to current life styles and indiscriminate disposal of wastes. In the USA, the municipal solid waste (MSW) produced annually has declined slightly from 211.5 million tons in 1995 to 209.7 million tons in 1996 (Subramanian, 2000). Recycling rates and composting rates are increasing. Disposal in landfills is decreasing. Waste disposal by combustion is also increasing, primarily due to the improved efficiencies of the new incinerators and their ability to remove particulates and noxious gases. Plastics form a small but significant part of the waste stream. The amount of plastics being recycled is showing an increasing trend.

The goal of any sustainable growth should be that the efficiency of energy utilization at every stage of the system, from the production of the goods to the disposal of the wastes is maximized.

Contrary to popular belief, plastics may not be the most prevalent material in landfills—paper and paper products often make up the largest percentage of a landfill's contents. Food items, and yard wastes and plastics constitute some other large fractions. Most of the waste products are generally disposed in landfills.

Modern landfills are designed to safely house wastes so that their uncontrolled degradation does not endanger groundwater with pollutants. Such landfills could, in many cases, be used after they are capped, to construct parks, golf courses or even airports (Subramanian, 2000).

Plastics now form an integral part of our lives. The amounts of plastics consumed annually have been growing steadily over the past three decades.

The waste plastic collected from the solid wastes stream is a contaminated, assorted mixture of a variety of plastics. This makes their identification, separation and purification, a challenging task. In the plastics waste stream, polyethylene forms the largest fraction, which is followed by PET (polyethyleneterephthalate). Lesser amounts of several other plastics can also be found in the plastics waste stream. Any attempt to manage such large quantities of a diverse, contaminated mixture of plastics in an energy efficient and environmentally benign manner needs to be addressed by using an integrated approach, comprising source reduction, reuse, recycling, landfill and waste-to-energy conversion.

Of course, an important aspect of the integrated waste management approach is to minimize the amount of plastics used.

Plastic recycling has grown considerably in some industrialized countries during the last few years. Recycling of rigid plastic containers, waste HDPE (high density polyethylene) bottles and waste PET bottles is also increasing. Hundreds of business houses that handle and reclaim post-consumer plastics have come up. A wide variety of new products, such as single-use cameras, park benches, sweaters, jeans, videocassettes, detergent bottles and toys are being made with or packaged in post-consumer recycled plastics (Subramanian, 2000).

Durable plastics, as opposed to most packaging and convenience goods which are discarded after a single use, tend to have a life span of 3 or more years. Automobiles, computers, household appliances, carpets and fabrics come into this durable category. The use of plastics in durable appli-cations is growing as design engineers, manufacturers and consumers continue to rely on its performance, low cost and design benefits. It is quite difficult to recover plastics from such durable goods because quite often they are integrated with several other plastic and non-plastic components. Their separation, recovery and purification is done in several stages and generally, the volumes of such materials available for recovery are small. Business equipment and computer manufacturers, who are currently recovering precious metals from such products, are testing the recovery of plastic housings and other components from them. Automotive companies are interested in effective recycling of plastic components so as to increase the use of materials having recycled plastics content.

One way to manage solid waste is to recover the energy value of products after their useful life. One such method involves combustion of municipal solid waste (MSW) or garbage in waste-to-energy (WTE) facilities.

As plastics are generally made from petroleum or natural gas, they have stored energy values greater than any other material commonly found in the waste stream. Polyolefins commonly used in packaging can generate twice as much energy as some coals and almost as much energy as fuel oil. Processing of plastics in modern WTE facilities helps other wastes combust more completely, leaving less ash for disposal.

REFERENCES

Alexander, M. *Biodegradation and Bioremediation*. Academic Press, San Diego (1994).

Anonymous. Nissin BOD Rapid Measuring Instruments: BOD-2000 and BOD-2200. Instruction Manuals, Central Kagaku Corp., Tokyo (1994).

Biehler, M.J., Hägele, S. Treatment processes of sanitary landfill leachates. *Natur. Resources and Develop.* 41: 64-84 (1995).

Bosma, T.N.P., Middeldrop, P.J.M., Schraa, G., Zehnder, A.J.B. Mass transfer limitation of biotransformation: quantifying bioavailability. *Environ. Sci. Technol.* 31: 248–252 (1997).

Brisou, J. *An Environmental Sanitation Plan for the Mediterranean Region.* World Health Organization, Geneva (1976).

Campbell, H.W., Bridle, T.R. Conversion of sludge to oil: A novel approach to sludge management. pp. 41-50 In *What's New in Wastewater Technology?* Environ. Canada, Ottawa (1988).

Cerniglia, C.E. Biodegradation of polycyclic aromatic hydrocarbons. *Biodegradation* 3: 351–368 (1992).

Cha, D.K., Sarr, D., Chiu, P.C., Kim, D.W. Hazardous waste treatment technologies. *Water Environment Research* 70: 705-721 (1998).

Chapman, D.T. Designing activated sludge plants for enhanced suspended solids removal. pp. 93-105. Environ. Canada, Ottawa (1988).

Crawford, R.L., Crawford, D.L. *Bioremediation: Principles and Applications.* Cambridge University Press, Cambridge (1996).

Datta, P.S., Deb, D.L., Tyagi, S.K. Assessment of groundwater contamination from fertilizers in the Delhi area based on ^{18}O, NO^{3-} and K^+ composition. *J. Contam. Hydrol.* 27: 249-256 (1997).

Edgar, J.G., Stewart, H.T.L. Wastewater disposal and reclamation using eucalyptus and other trees. *Prog. Water Tech.* 11: 163-173 (1979).

France, R.L. Potential for soil erosion from decreased litterfall due to riparian clearcutting: Implications for boreal forestry and warm- and cool-water fisheries. *J. Soil Water Conserv.* 52: 452 (1997).

Freeze, R.A. *The Environmental Pendulum: A Quest for the Truth about Toxic Chemicals, Human Health and Environmental Protection.* Univ. of California Press, Berkeley, CA (2000).

Gardner, G. Recycling human waste: Fertile ground or toxic legacy? *World Watch* 28-34 (Jan. Feb., 1998).

Gilliam, J.W., Osmond, D.L., Evans, R.O. Selected agricultural best management practices to control nitrogen in the neuse river basin. N.C. Agric. Res. Serv. Bull. 311, N.C. State Univ., Raleigh (1997).

Gupta, R.K., Rudra, R.P., Dickinson, W.T., Wall, G.J. Surface water quality impacts of tillage practices under liquid swine manure application. *J. Am. Water Resour. Assoc.* 33: 681-690 (1997).

Hanson, J.E., Stoltenberg, D.E., Lowery, B., Binning, L.K. Influence of application rate on atrazine fate in a silt loam soil. *J. Environ. Qual.* 26: 829-835 (1997).

Head, I.M. Bioremediation: towards a credible technology. *Microbiology* 144: 599–608 (1998).

Hess, T.F., Büyüksönmez, F., Watts, R.J., Teel A.L. Assessment, management, and minimization. *Literature Review* 70: 699-705 (1998).

Huang, J.W., Chen, J., Berti, W.R., Cunningham, S.D. Phytoremediation of lead-contaminated soils : Role of synthetic chelates in lead extraction. *Environ. Sci. Technol.* 31: 800-810 (1997).

Iranpour, R., Stenstrom, M., Tchobanoglous, G., Miller, D., Wright, J., Vossoughi, M. Environmental engineering: Energy value of replacing waste disposal with resource recovery. *Science* 285: 706-710 (1999).

Jank, B.E. Introduction. pp. 1-3. In *What New in Wastewater Technology?* Environment Canada, Ottawa (1988).

Jolis, D., Hirano, R. Microfiltration and ultraviolet light disinfection for water reclamation, Report for Environmental Engineering Section, Bureau of Engineering, City and Country of San Francisco, CA (1993).

Knackmuss, H.J. Basic knowledge and perspectives of bioelimination of xenobiotic compounds. *J. Biotechnol.* 51: 287–295 (1996).

Kolpin, D.W., Kalkhoff, S.J., Goolsby, D.A., Sneck-Fahrer, D.A., Thurman, E.M. Occurrence of selected herbicides and herbicide degradation products in Iowa's ground water, 1995. *Ground Water* 35: 679-695 (1997).

Lehmann, V. *Bioremediation: A solution for polluted soils in the South? Biotechnol. Develop. Monitor* No. 34: 13–17 (1998).

Lesage, S., Xy, H., Novakowski, K.S. Distinguishing natural hydrocarbons from anthropogenic contamination in ground water. *Ground Water* 37: 149-156 (1997).

Line, D.E., McLaughlin, R.A., Osmond, D.L., Jennings, G.D., Harmna, W.A., Lombardo, L.A., Spooner, J. Nonpoint sources. *Water Environ. Res.* 70: 895-912 (1998).

Lue-Hing, C. (Ed.). Optimization in wastewater treatment technology: an option worth remembering. *Water Environ. Res.* 70: 259-260 (1998).

Mangiapan, S., Benfenati, E., Grasso, P., Terreni, M., Pregnolato, M., Pagani, G., Barcelo, D. Metabolites of alachlor in water : Identification by mass spectrometry and chemical synthesis. *Environ. Sci. Technol.* 31: 3637-3640 (1997).

Massol–Deya, A., Weller, R., Rios–Hernandez, L., Zhou, J.Z., Hickey, R.F., Tiedje, J.M. Succession and convergence of biofilm communities in fixed–film reactors treating aromatic hydrocarbons in groundwater. *Appl. Environ. Microbiol.* 63: 270–276 (1997).

Munaf, E., Zein, R. The use of rice husk for removal of toxic metal from waste water. *Environ. Technol.* 18: 359-365 (1997).

Neare, J. Phytoremediation. *UNEP IETC Newsletter.* p. 8 (Sept, 1999).

Parameswaran, M. Urban wastewater use in plant biomass production. *Resources, Conservation and Recycling* 27: 39-56 (1999).

Powers, S.E., Villaume, J.F., Ripp, J.A. Multivariate analyses to improve understanding of NAPL pollutant sources. *Ground Water Monit. Remed.* 17: 130-140 (1997).

Rochfort, Q.J., Anderson, B.C., Crowder, A.A., Marsalek, J., Watt, W.E. Field-scale studies of subsurface flow constructed wetlands for stormwater quality enhancement. *Water Qual. Res. J. Can.* 32: 101-110 (1997).

Seta, A.K., Karathanasis, A.D. Atrazine adsorption by soil colloids and cotransport through subsurface environments. *Soil Sci. Soc. Am. J.* 61: 612-618 (1997).

Skinner, J.A., Lewis, K.A., Bardon, K.S., Tucker, P., Catt, J.A., Chambers, B.J. An overview of the environmental impact of agriculture in the UK. *J. Environ. Manage.* 50: 111-1118 (1997).

Sojka, R.E., Lentz, R.D. Reducing furrow irrigation erosion with polyacrylamide (PAM). *J. Prod. Agric.* 10: 47-55 (1997).

Subramanian, P.M. Plastics recycling and waste management in the US. *Resources, Conservation and Recycling* 28: 253-263 (2000).

Swensen, B., Singh, B.R. Transport and transformation of urea and its derivatives through a mineral soil. *J. Environ. Qual.* 26: 1516-1520 (1997).

Taylor, S.W., Lange, C.R., Lesold, E.A. Biofouling of contaminated ground-water recovery wells: Characterization of microorganisms. *Ground Water* 35: 973-1000 (1997).

Wagner, F., Wienands, H. Das Biomembran Verfahren. *Müll und Abfall* 10/89, 528-533 (1989).

Watanabe, M.E. Phytormediation on the brink of commercialization. *Environ. Sci. Technol.* 31: 182A-189A (1997).

Webber, M.D. Waste metals—the Canadian approach to limiting metals on land from municipal sludges. pp. 107-116. In Environ. Canada, Ottawa (1988).

WHO. *Reuse of Effluents : Methods of Wastewater Treatment and Health Safeguards.* World Health Organization Technical Report Ser. No. 517, WHO, Geneva (1973).

Woudneh, M.B., Lloyd, B., Stevenson, D. The behaviour of 2,4-D as It filters through slow sand filters. *AQUA* 46: 144-150 (1997).

Chapter 8
Animal Husbandry and Wildlife Management

INTRODUCTION

Humans interact with domestic animals and some species of wild animals in various ways. Sometimes, there is a clash between rational attitudes toward near-wild animals based on their usefulness and emotional reactions. Sometimes, animal welfare officials and the general public do not see eye-to-eye. The usefulness of domesticated animals to humans has a strong influence on their relationship and has led to a conflict between apparently purely rational developments in breeding and rearing on the one hand and emotional aspects on the other. Often, farm animals are treated rather cruelly. The basic emotional attitudes of humans need to be reconciled with rational considerations along with better knowledge and an understanding about animal welfare, from the animal viewpoint, needs to be gained.

Animal welfare is a special aspect of the relationship between humans and animals. It is distinct from species conservation (Erz, 1991). There is a wide spectrum of different relationships between animals and humans. For humans, animals represent quarry, competitors for food, commensals (water) and symbionts; for numerous animal species, humans represent enemies and competitors for territory and food or potential quarry. On the human side also, there are big emotional differences (Gärtner, 1986). Animal welfare relates to individuals from certain groups of species which are emotionally highly valued by humans and which have special relationships with humans.

Domestication and taming have thrown light on novel features of the symbiosis between humans and some animal species. The active influence of humans on genome, characteristics and behaviour with regard to the use expected of an animal species or population, distinguishes domestication from purely biological symbioses (Herre and Röhrs, 1973).

Mallinson (1995) and Grauvog (1996) suggested that the relationship between humans and animals should be based on a human ethic, which should determine the attitude of humans towards the animal species and their living conditions, irrespective of whether they are wild animals in anthropogenic habitats, zoos, farm animals, laboratory animals or pets. There should be appropriate husbandry for each animal species, which should be based on an in-depth understanding of the behaviour of the species concerned. Any management method should be based on an animal-human relationship, which is the product of the conflicting interests of the various elements and demands of the biological nature of the animal on the one hand and the state of development of human culture and society on the other.

Animals, like humans, also have emotions. Their emotional symp-toms are the outcome of evolution, developed in the evolutionary battle with the environment as an important element in the preservation of individuals, populations and species. Therefore, they are strongly related to environmental factors. Emotions represent an appraisal of external phenomena in the light of the body's own resources, but the subjective component is inaccessible to humans. We can only point out some analogies (Sambraus, 1994). However, this does not mean that functional patterns which have turned out to be beneficial from the evolutionary point of view are fundamentally different at the subjective level.

The complex human concepts of various emotions (Immelmann et al., 1988) are rather unsuitable for animals. Only those concepts which are not dependent on subjective human attitudes can allow a proper description of emotions, irrespective of the species.

Human behaviour should be based on the principle of might giving way to right, especially for a weaker partner. We humans should shoulder a greater responsibility because of our greater freedom of action than other species. Animals need to be fully protected from damaging intervention which causes suffering as a result of the dominance of the stronger party. Basic information is needed about individual animal species as well as animal requirements for the designing of good management methods. Effective dissemination of ecological and ethological information is a must for a better understanding of the relationship between humans and animals.

Livestock farming is based on a strong interaction between man, animals and the environment. Humans have different attitudes to animals and there are different consequences for livestock production. The objectives of livestock farming go beyond the production of material products such as milk, meat, eggs, wood and leather (see Fig. 8.1)

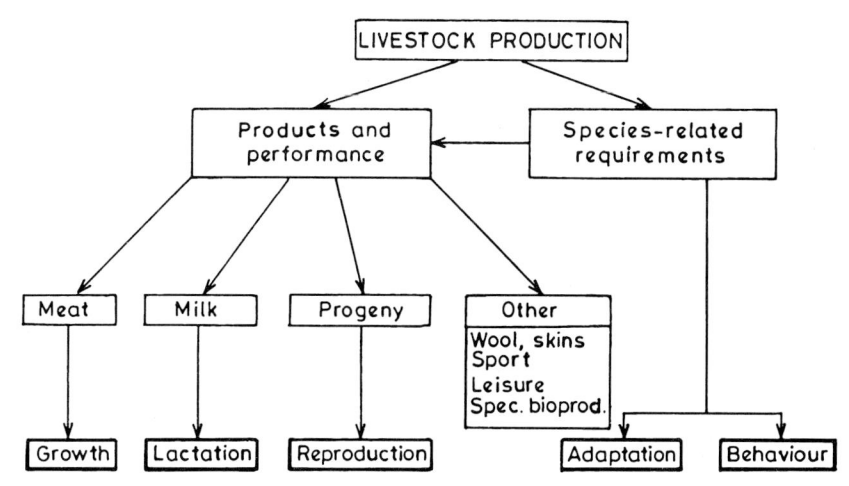

Fig. 8.1 Livestock products, uses and management.

Figure 8.2 summarizes the products and functions of livestock farming as outputs together with the inputs, i.e. the production criteria and sevices which animals need in order to perform on behalf of man.

Livestock-Crop Interactions and Ecological Sustainability

Since times immemorial, agriculture has been a crucial test of the endurance of nature. Natural ecosystems have been replaced by human-engineered artificial, low diversity and highly productive systems, at the expense of significant energy and resource inputs.

Modern industrial agriculture has triumphed over nature and attained a high level of production and labour efficiency, but in the process, it has become unsustainable (Naegel, 1996). The success of modern agriculture has resulted in high costs to the larger environmental, social and economic systems. The management of renewable resources is a crucial factor in achieving sustainable agricultural development in Third World countries. There is no doubt that sustainable agricultural production systems should involve the successful management of resources for agriculture to satisfy changing human needs while maintaining or enhancing the natural resource base and avoiding environmental degradation.

The interaction of crops and livestock can play a key role in achieving ecological sustainability by intensifying nutrient and energy cycles. Crop

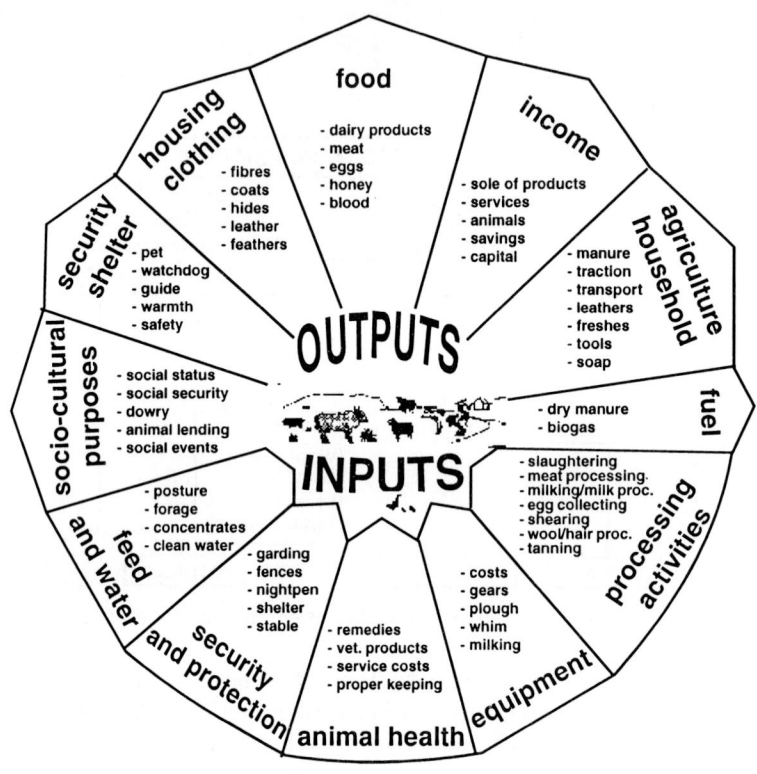

336 Environmental Technology and Biosphere Management

Fig. 8.2 Output/input correlations between humans and animals (after Neidhardt et al., 1996).

residues are an important source for livestock feed in small-scale farming systems. Livestock herdsmen and small-scale farmers can mutually benefit from a close cooperation.

Livestock serves as a kind of savings account, with the offspring as interest. In industrialized countries, animals play a major role in cushioning trade and market disruptions. When there is a grain surplus, declining prices encourage the use of grain for animal feed. Conversely, when supplies of grain are short and prices rise, livestock numbers can be reduced and/or alternative feeds and feeding systems can be used (Naegel, 1996).

Integrating fish production with agriculture (Fig. 8.3) is a traditional practice in many developing countries. The integration of aquaculture into existing agricultural production systems allows reducing feed costs by recycling animal and crop wastes into fish feed. Integrated agricultural/aquacultural production systems can allow small-scale farmers to achieve high fish yields with fairly low input costs.

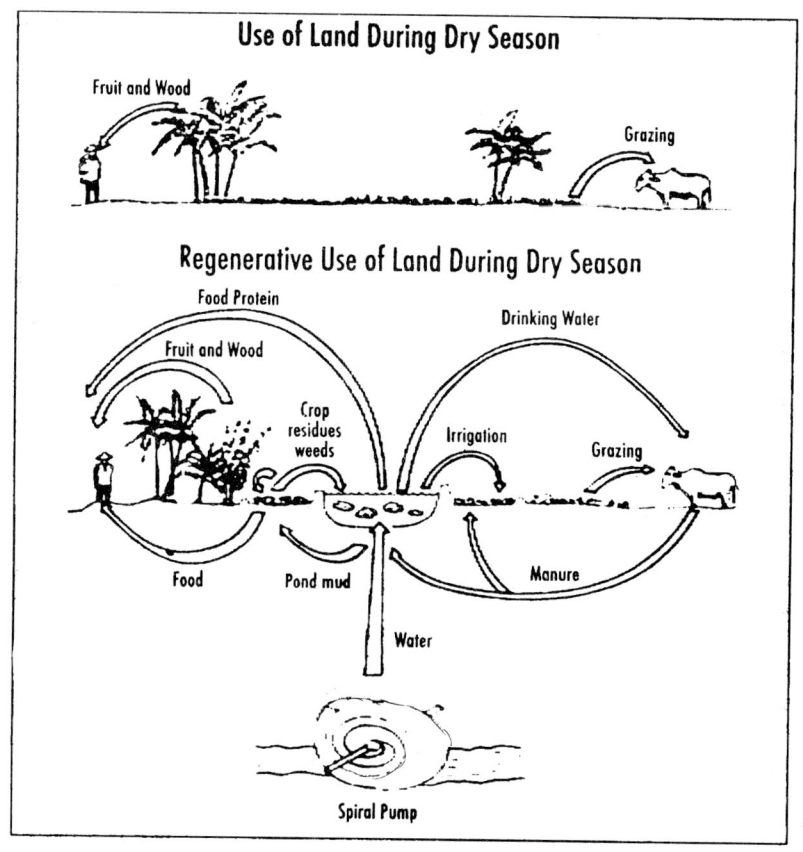

Fig. 8.3 Integration of agricultural/aquacultural systems (after Naegel, 1996).

LIVESTOCK DIVERSITY

After decades of development, setbacks with alien breeds, scientists and developers are now realizing the vast animal genetic resources that ordinary farmers and herders have developed through the ages, especially in the south. Today, all over the world, rural people keep about 4,500 breeds of domestic animals of more than 40 species. As many as 150 reported varieties of cattle, 60 of sheep, 50 of goats and considerable (but less well documented) biodiversity in horses, donkeys, mules, chickens, pigs and dromedary camels are currently found in Africa.

African stockraisers very often have a rich knowledge of animal husbandry. Many pastoral and agropastoral peoples keep detailed mental or oral livestock stock records.

About one-third of the world's livestock breeds are currently at risk of disappearing and the extinction rate now stands at about six breeds per month. Among Africa's nine traditional cattle breeds showing resistance to blood parasites, all but three are endangered. A number of African breeds of sheep, horses, donkeys and poultry as well as cattle have already become extinct.

Western society has successfully developed powerful tools through its reductionist world view, but it has not yet evolved the maturity to use these tools with discrimination and compassion. The western world has been creating uniformity, standardization and monocultures in nature and in society. However, nature abhors uniformity—it produces not only species diversity, but also individual diversity. Throughout their long history, human beings have contributed to this through the free exchange of knowledge and biological specimens.

The western industrial worldview usually sees nature as a resource to be dominated and used to satisfy human needs (see Fig. 8.3). In recent decades, global corporations have consolidated their control over nature and society to the point where they are more powerful than most nation states. They have forced governments to participate in the World Trade Organization, also called 'the new Global Government' (Hosken, 1999). Through the WTO the corporations can further consolidate their control over minerals, fossil fuels, genetic material of living organisms, scientific research, public institutions and indigenous knowledge.

There are two major schools of thought about global trade interests in biodiversity and local knowledge. One believes in the commercialization of living organisms and knowledge being an inevitable trend. This viewpoint aims at extending intellectual property rights to communities and giving them a financial deal in exchange for benefit sharing. This may be a just and fair demand under the circumstances but is proving difficult to achieve. The second school views this trend of commercialization of living organisms and knowledge as a self-destructive, one-dimensional reality. It feels that any further dissection of nature, knowledge and culture will lead to greater disintegration of biological and cultural systems and human potential (Hosken, 1999).

The WTO intends to create global rules for all countries. In the area of biodiversity and knowledge, the Trade Related Intellectual Property Rights Agreement (TRIPs) compels countries to allow commercial monopoly control of plant varieties.

Objections to the injustice of excluding communities and countries from developing their own biodiversity resulted in the World Intellectual Property Organization (WIPO) which desires to extend intellectual property rights to communities.

ANIMAL HUSBANDRY AND SUSTAINABLE TROPICAL FARMING SYSTEMS

It is quite difficult, if not impossible, to achieve a locally and environmentally compatible and sustainable, diversified, indigenous farming systems in the tropics, without appropriate methods of utilizing, keeping and breeding of suitably adapted, high performing farm animal populations. Such systems need to be geared chiefly to supplying local markets, not just raising export crops. In tropical crop-growing systems, whether rainfed or irrigated, farm animals provide valuable inputs at reasonable costs—e.g. draught potential for tilling the soil, tending the crops and transporting the farm produce, or the manure needed to improve and enrich the soil. Using farm animals on the fields minimizes the physical damage to soil parameters, while growing fodder crops improves the rotation system and helps minimize erosion. In marginally-arable areas, fodder crops are also used for barter trade with nomadic herdsmen and, thereby, improve the livelihood of the parties engaged in this exchange of animals or animal products in return for feedstuffs and residual products which humans cannot use.

Occasionally, such factors as shortage of land, lack of capital to buy draught animals and implements, insufficient work for a family farm, or lack of experience in dealing with cooperative draught animal husbandry and of loans to promote small agricultural market production, present harmonious integration of arable and stock farming. According to Poetschke (1997), effective integration of animal production on an arable and crop farm is based on the following four premises:

1. Usually, priority must be given to crop growing with animal production as a secondary enterprise. Animal production, regardless of whether it relies on external inputs, should only complement the enterprise.

    Animal production on an arable farm should utilize and convert the non-marketable or waste products of crop growing; utilize non-marketably, soil-improving catch crops (e.g. legumes), rotational fodder crops or the natural vegetation on fallow land to produce marketable animal products; provide soil-improving natural organic growth promoters that directly benefit the crops before tilling in the form of nutrients (fertilizing effect) and humus to improve soil structure and by enriching and activating the soil fauna (that can have a beneficial impact on cropping densities).
2. Integrating animal production with plant production should result in a beneficial, harmonious merger. From among diverse farm animal species, only those should be selected which can contribute to the utilization or processing of various products of plants or

foods; i.e. they should match the crop growing profile with a long term crop rotation as well as the marketing potential.

When feed supplies are temporarily uncertain or doubtful preference should be given to less demanding, less specialized, more multi-purpose animal species that are not very sensitive to deficiencies in feed supplies. Autochthonous breeds and local types usually prove more hardy and tend to better survive adverse conditions and acute feed shortages.

3. Animal production in tropical areas should be reduced to the level of supplying their own needs, to a so-called 'subsistence level'. The arable functions mentioned under Premise I above should be replaced with green manuring, composting and biomass production.
4. Animal production in tropical zones can satisfy the ecological standards required of agricultural production only if its exploitation of resources does not lead to their devastation and the waste products arising from animal production do not cause degradation and contamination, for example, of the fertile soil. Environmentally compatible tropical animal production should be limited to a stocking level which is sustainable in the long term, a herd size which depends on the production location and an economically justifiable, possibly limited utilization intensity (Poetschke, 1997).

Prudence warrants that higher priority be given to sustainable tropical animal production at a rather low performance level, rather than making unrealistically high demands on environmentally sensitive high performance animal material. This may be accomplished by merging arable farming and animal production harmoniously at some common limited agricultural tropical location that is clearly defined as regards its productivity and economic use in a long term perspective.

Production Conditions and Animal Breeding

An effective integration of production conditions and animal breeding in the tropics may be achieved if the following objectives are kept in view:

1. In the predominantly subsistence farming sector, the physiological adaptability, function, existing distribution and minimum quality requirements (as regards constitution and reproduction) of autochthonous animal material need to be redefined. Basic management standards should be recommended and enforced.
2. The specific quality of animals and livestock products produced should be judged, evaluated and marketed as an alternative to import products.
3. Location specific environment/farm animal correlations should be examined so as to effectively improve actual production conditions.

4. Autochthonous animal populations should, if necessary, be improved, diversified (split up into lines) or graded up provided the particular conditions reinforce such a breeding aim.

The integration of animal breeding and animal husbandry into a sustainable, tropical farming system geared to the location and environment should be an all-embracing, harmonious development process. The aims of the animal production enterprise should be geared to the capacities of an equally sustainable crop production enterprise in the long term, unless the value of the animal production enterprise can be increased independently of the farm by using assured supplies of external resources such as bought-in compounds or complete diets. Sometime it is not possible temporarily to fulfil both the supply and waste disposal functions of animal production as well as its aims; in the case of a breakdown in infrastructure during the rainy season, a production strategy is needed which is safeguarded by adequate storage facilities (Poetschke, 1997).

Farm management is an important factor involved in shaping of the integration process. It might prove useful sometime to break the farming pattern—depending on the type of stock and taking into account seasonal influences; e.g.

1. Reduction of stock numbers as much as possible during long periods of fodder shortage;
2. Planned fluctuations in the production of and use of, mono-gastrids according to the season (by keeping only breeding animals in periods of feed shortage);
3. Curtailing total expenditure by introducing more draught animal power into the processes of animal and crop production, including the wider availability and use of riding animals.

Animal production has a crucial role in fulfilling the criteria for socio-economic sustainability and in relevant projects in tropical agriculture. But its real economic success will depend in the final analysis on the impact of its performance in the market phase transactions and economic potential.

ANIMAL HUSBANDRY AND GLOBAL WARMING

Intensification of animal husbandry often increases environmental pollution problems. Increased emissions of ammonia, methane and nitrous oxide (N_2O) influence the global climate as well as regional soil quality. Climate change threatens agriculture in developing countries because adaptation measures such as increased irrigation are beyond the means of poor farmers. However, by taking suitable precautions, the effects of intensive animal husbandry on climate and soil can be kept to a minimum.

The global climate is governed by many factors including (besides CO_2), methane and nitrous oxide both of which are released in substantial qunatities by agricultural activities. Ammonia (NH_3) is a gas mostly emitted through animal husbandry and it exerts an indirect influence on the atmospheric greenhouse effect (Fig. 8.4) due to physical and chemical processes in the atmosphere and soil (Seidl, 1999).

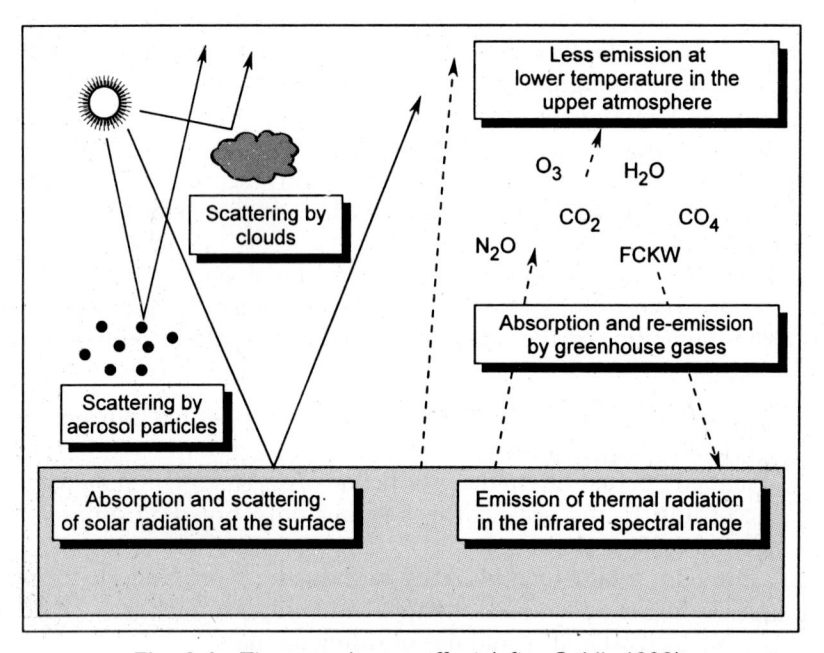

Fig. 8.4 The greenhouse effect (after Seidl, 1999).

Chemical reactions of the trace gases released from industrial and agricultural sources into the atmosphere form aerosol particles (diameters ranging from 0.01 to 10 µm), which have an influence on the greenhouse effect. Furthermore, the reaction products alter the oxidation capacity of the troposphere and result in the formation of acids. NH_3 plays an important role in these processes. Whereas in the atmosphere, NH_3 neutralizes acid aerosol particles and acid rain, in the soil it becomes nitrified to nitric acid.

Animal husbandry produces two important GHGs, namely; CH_4 and N_2O whose concentrations have increased considerably over the past century. CH_4 is formed by anaerobic methanogenic bacteria and archaea. It is produced throughout the biosphere wherever organic matter is decomposed anoxically. In animal husbandry it arises by the fermentation of feed in the stomach of ruminants and nonruminants. Their ability to digest cellulose enables ruminants to produce much methane depending, e.g. on the quantity and quality of the feed.

Another source of CH_4 associated with animal husbandry is the decomposition of animal wastes. These sources mainly consist of organic material, which produces CH_4 when decomposed anaerobically. (Under aerobic conditions, however, these wastes produce CO_2). As CH_4 is a more effective greenhouse gas than CO_2, it is preferable to have this material decomposed in the presence of oxygen rather than in a closed system (Seidl, 1999)

Micro-organisms are also involved in the production of N_2O, mainly via nitrogen reactions in the soil. Under anaerobic soil conditions, redox reactions take place between nitrate (NO^-_3) and oxidized hydrocarbons and nitrate is reduced to N_2O or molecular nitrogen N_2 (denitrification). N_2O also arises from the microbial oxidation of ammonium (NH_4^+) to nitrate (NO_3^-) (nitrification).

In animal husbandry, N_2O comes from the decomposition of animal wastes and fertilized cattle ranges, which together, contribute about 13% to global N_2O emissions. Any further intensification of animal husbandry will increase the amount of animal waste, further enhancing N_2O emissions.

Ammonia is emitted substantially through animal husbandry. It is not a greenhouse gas itself but its interactions with anthropogenically emitted trace gases and its microbial reactions in the soil can potentially influence the greenhouse effect as well as the soil quality. But NH_3 also has positive effects on the environment. Chemical reactions in the atmosphere transform the gases sulphur dioxide (SO_2), nitrogen monoxide (NO) and nitrogen dioxide (NO_2), which largely originate from industrial sources and road traffic, into sulphuric acid (H_2SO_4) and nitric acid (HNO_3). NH_3, on the other hand, is the only alkaline trace gas in the atmosphere and so at least partly neutralizes acid aerosol particles and acid rain. In regions where NH_3 levels are too low to neutralize sulphuric acid, acid aerosol particles can become a grave environmental threat.

Where there is enough NH_3 to nuetralize sulphuric acid, it can also react with nitric acid to ammonium nitrate (NH_4NO_3), which forms aerosol particles. Aerosol particles tend to scatter solar radiation back into space, so decreasing the amount of energy available for the warming of the earth. In this way, aerosol particles can cause a global cooling which counteracts the warming caused by GHGs. At the global level, this effect is probably largely attributable to particulate sulphate and organic compounds. Nevertheless, in regions with intensive animal husbandry as

344 Environmental Technology and Biosphere Management

well as heavy traffic emissions, a substantial fraction of the aerosol mass may consist of ammonium nitrate (Seidl, 1999).

Ammonia, nitric acid and ammonium nitrate dissolve in cloud and rainwater and get deposited on the ground by precipitation. The nitrogen input to the soil leads to the eutrophication of natural ecosystems and the various problems associated with this. This additional input enhances microbial nitrogen reactions in the soil, thus increasing the release of N_2O into the atmosphere.

Nitrification involves oxidation of ammonia first to nitrite (NO_2^-) and then to nitrate (NO_3^-). As nitrate forms nitric acid in aqueous solution, this eventually leads to soil acidification. Under certain conditions, these acids are neutralized by calcium or magnesium. However, once the soil's neutralization capacity is exhausted, the pH value falls. A decrease in pH below 4.5 leads to the destruction of mineral clays and the release of potentially toxic metal ions of aluminium, manganese and iron. When the concentration of aluminium ions exceeds that of calcium ions, there can be some forest damage, for example. The critical level is considered to be a ratio of 1:1.

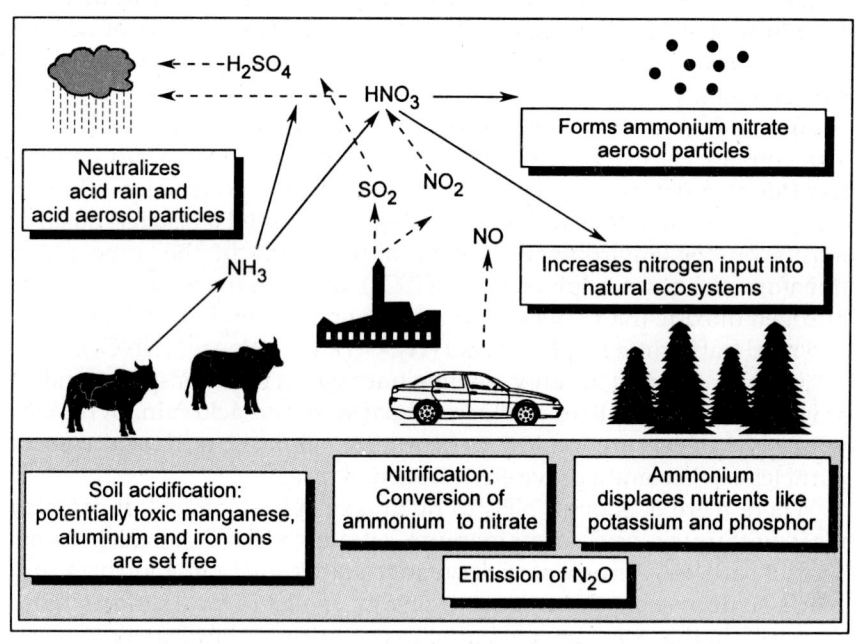

Fig. 8.5 Effects of ammonia emissions (after Seidl, 1999).

Forest damage also results directly from high ammonia concentrations. Trees take up ammonia from the air through their stomata. This acidifies the leaves. There is a reduction of the trees' uptake of such important nutrients as potassium, calcium and phosphorus ions and consequently to deficiency symptoms. The critical concentration ratio between ammonia and potassium in soil is considered to be 5:1.

In view of its high solubility, ammonia resides in the atmosphere for a shorter time compared with CH_4 and N_2O, so its effects on the environment are more regional. The above problems are therefore likely to be more prevalent in regions with intensive animal husbandry.

Nitrogen Leakage

According to Socolow (1999), less than half of the fixed nitrogen added to agriculture ends up in harvested crops and less than half of the fixed nitrogen in these crops end up as human food. Thus, there is considerable leakage of fixed nitrogen which needs to be reduced. J.N. Galloway (*J. Agroecosystems*, 1999) has suggested that all leakage be assigned to either crop wastes, animals, or people. Ideally, all crop wastes, manure and food wastes should return to the field so that no nitrogen is lost through volatilization runoff or denitrification. In such a system, maintaining constant food production would require no external inputs of nitrogen. The required inputs of nutrients and energy will come from human, animal and industrial/agricultural wastes rather than from chemical fertilizers and fossil fuels. But this kind of lofty scenario appears to be Utopian, at least for the next few decades.

ANIMAL HUSBANDRY AND THE ENVIRONMENT

Humans use domestic animals for transport and haulage work (as an energy source). They use diverse animal products as food or as raw material for clothing. Ownership of animals serves man as a means of building up reserves and making provisions for times of need. When millennia ago, humans began to domesticate useful animals and evolved from hunter-gatherers to pastoralists or herders, they satisfied their needs far more efficiently than they were able to before. The integration of animal husbandry and arable farming contributed substantially to food production and laid the foundation not only for unprecedented economic development but for a sustainable growth potential in food production. Animal husbandry is an integral component of agricultural land use systems all over the world. Using only a few domesticated animal species (cattle, sheep, goat, pig and poultry everywhere and camel, buffalo, lama and yak in some parts of the world), humans have been able to satisfy their needs under almost all ecological conditions by selecting appropriate animal species and types of use. The relative significance of animal husbandry in meeting human needs depends greatly on the natural conditions of the particular place and also on economic conditions. In marginal areas that preclude arable land use, animal husbandry offers the sole basis for colonization. In more favourable conditions, a wider range of animals and arable farming systems are available out of which the best choice can be made in the light of economic and market-related factors (Peters, 1999).

Ecological factors determine the type of system in which animal husbandry is practised and influence one's choice of a land use system, but the concrete perspective of animal husbandry is determined largely by superposed anthropogenic influences and socio-economic and locational factors. Various countries and regions differ greatly in their social and economic status. This becomes explicit when one compares their per capita gross domestic product (GDP), or perspectives for economic development. Per capita GDP ranges from about 21390 US dollars in OECD countries to about 320 dollars in Southeast Asia. In Africa it is currently declining at 1.3% a year, whereas in South Asia, it is rising by an annual 9%. With this divergence of living conditions, efforts made to secure the food supply and demands placed on animal husbandry in securing people's subsistence, must differ considerably from region to region.

Food security implies the following aspects: equilibrium between food demand and availability; overcoming temporary food shortage; sufficient supply with micronutrients (iron, iodine and vitamin A) and essential amino acids; and absence of nutrition-related deficiency diseases (Peters, 1999). It can also imply the sustenance of one's capacity to work and earn a living and the development of economic sustainability. Another important element of the security of subsistence is the sustainability of the use of land and other natural resources. Even in the early nineteenth century, the concept of sustainability used to be a focal point in some discussions on the intensification of arable and animal farming. Today, when global environmental protection has attained great importance, it has again come into sharp focus all over the world. The idea of sustainability includes elements such as the conservation of soil fertility, protection of biodiversity and avoidance of adverse impacts on the environment. The role of animal husbandry in meeting these needs essentially comprises three functions—food production; insurance, capital accumulation, income generation; and integrated resource use (Fig. 8.6).

Except for sheep, the number of farm animals worldwide is currently increasing (Table 8.1). Cattle and sheep stocks in developing countries are growing more slowly than the population and only goat stocks are increasing in all regions, particularly in Asia. Intensive animals such as pig and poultry are becoming more and more important with high

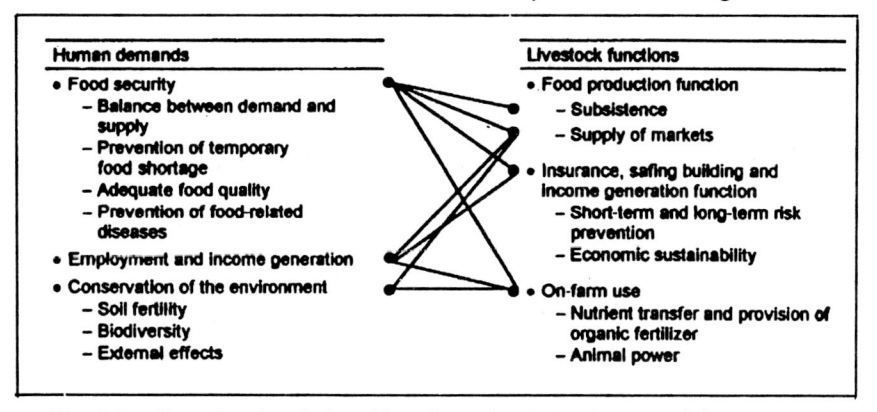

Human demands	Livestock functions

• Food security
  – Balance between demand and supply
  – Prevention of temporary food shortage
  – Adequate food quality
  – Prevention of food-related diseases
• Employment and income generation
• Conservation of the environment
  – Soil fertility
  – Biodiversity
  – External effects
	• Food production function
	  – Subsistence
	  – Supply of markets
	• Insurance, safing building and income generation function
	  – Short-term and long-term risk prevention
	  – Economic sustainability
	• On-farm use
	  – Nutrient transfer and provision of organic fertilizer
	  – Animal power

Fig. 8.6 Functional relationships in animal husbandry (after Peters, 1999).

annual growth rates in developing countries and slight reductions in stock in industrialized countries. In Asia and Latin America, animal stocks are growing proportionately faster than the population, whereas in Africa, they are increasing at the same rate while in Europe, more slowly than the human population. In North America also, animal stocks show decreasing trends (Peters, 1999).

In underdeveloped countries, animal products only contribute upto about 20% to the total food supply, which is much less than in Europe and North America (Table 8.2). Asia and South America show particularly high growth rates in per capita demand, whereas in Africa, the growth rate of animal production is only just keeping pace with population growth, with milk demand falling.

Growing demand for poultry products and pork in south Asia has led to a rapid intensification of poultry and pig farming in the proximity of distribution centres and to an increased use of animal feedstuffs. Intensification of milk and dairy production, by contrast, is far more complex, requiring extensive changes in equipment, consultancy, breeding and marketing. Thus, imports of dairy products by developing countries have increased linearly during the past few years.

In developing countries, the only way to better meet the demand for animal products is to increase stocks, as individual animal productivity is highly constrained by difficulties to intensify the production process. In these countries, approximately one-half of arable land area is cultivated with the aid of draught animals. Mechanized arable farming is often too capital-intensive and limited to certain better-developed sites (22% of total land). Therefore, animal draught power will continue to play an important role, particularly in the African continent. In Latin America and the Near East, by contrast, the use of tractors is on the

Table 8.1 Stock development (source: FAO production yearbook, 1997).

	Humans		Cattle/Buffalo		Population Sheep		Goat		Pig		Poultry	
	pop.	change	pop.	change	pop.	change	pop.	change	pop.	change	pop.	change
World	5.76	+1.5	1.320	+0.5	1.04	-1.8	-67	+2.9	0.92	+1.4	12.95	+3.5
Africa		+2.9		+1.1		+0.8		+0.7		+4.8		+2.7
Asia		+1.6		+1.8		+0.8		+4.5		+3.9		+6.3
Latin America		+1.8		+1.2		-2.6		+0.7		+1.6		+8.0
Europe		+0.2		-0.1		-2.2		+0.5		-1.5		-1.2
North America		+1.0		+1.4		-3.5		+0.4		+1.4		+2.9

(N.B: pop. = population in billions; change = % annual change from 1990 to 1996).

Table 8.2 Per capita supply of animal products in different parts of the world (Source: FAO Production yearbook, 1997).

	Meat						Eggs		Milk	
	Ruminant		Pig		Poultry					
	kg	PC	kg	PC	kg	PC	kg	PC	kg	PC
World	11.2	-0.8	14.9	+2.1	10.1	+5.2	8.2	+2.3	92.6	-1.4
Africa	6.8	-1.9	1.0	+1.9	3.1	+0.7	2.3	-0.7	28.9	-1.4
Asia	4.3	+8.9	13.8	+7.8	5.8	+12.6	7.8	+8.3	38.6	+10.9
Latin America	25.1	-0.4	7.8	+2.4	18.9	+9.2	8.8	+0.4	111.9	+2.7
Europe	21.1	-2.7	42.6	-0.1	18.4	+1.8	13.3	-1.1	315.9	-1.4
North America	43.1	+0.8	30.1	+0.5	52.0	+4.7	16.2	+1.0	260.7	-0.2

(N.B. PC = % change per annum, 1990-96).

increase. In Asian countries, working animals still play a very important role, but this may slowly decrease in the coming years.

Nutrient transfer is an essential element in any strategy for sustainable agriculture. Integration of livestock and arable farming makes possible both an efficient utilization of residues and by-products of arable farming and also an additional uptake of nutrients in distant, extensively used pastureland and the transfer of nutrients contained in farm manure to intensively-used crop fields. Dung directly serves the soil with nitrogen, phosphorus and potassium. The humus in the dung manure improves soil pH and, consequently, phosphorus availability. Dung improves soil structure and nutrient availability, increases water infiltration and retention capacity and stimulates nitrogen-fixing soil bacteria (Turner, 1995).

There is paucity of mineral fertilizers in developing countries. This points to the great importance of nutrient transfer and dung management. Certain enterprises with substantial livestock and comparatively little arable farming can use dung to compensate for the losses incurred by annual cropping over the long term and so improve the sustainability of arable farming. Another element of the beneficial interaction between animal husbandry and arable farming is seen in the use of forage cropping. The following requirements must be fulfilled for the integration of forage cropping in an agricultural enterprise: year-round control of the farm area; high yield potential of animal husbandry and market-oriented production; and sufficient land that can be used without posing a risk to subsistence cropping (Peters, 1999).

Population growth, economic development, increasing purchasing power and growing markets for animal products allow livestock farmers to enhance the economic efficiency of their operations and secure a substantial part of their subsistence but this dynamic adaptation process is sometimes associated with drastic changes in inputs and the overexploitation of natural resources (see Table 8.3).

ANIMAL NUTRITION

The increase in world population to over 6 billion today has gone hand in hand with a sharp decline in the amount of agricultural land available per inhabitant (Fig. 8.7). This trend will also continue into the coming few decades so that less and less land will be available per person for the production of food. However, increased productivity has made it possible to keep the amount of food available globally per person at an almost constant level since 1985 (FAO, 1996). In the context of food supplies in the coming few decades, Flachowsky (1999) raised the issue of whether it is necessary to produce foodstuffs of animal origin and whether livestock production is justified. This is relevant because in some cases, feedstuffs

Table 8.3 Effects of population growth and economic development on the status of animal husbandry (after Peters, 1999)

Cause	Effects on animal husbandry	Change in the function of animal husbandry
Population density		
Rural	(a) increase in demand for vegetable food; (b) decrease in per capita land availability	(a) subsistence, (b) increasing importance of draught power and dung supply for arable farming
Early urban	increase in land use intensity	beginning market-oriented production
Increasing purchasing power and demand		
Increasing urbanization	increasing demand for animal products	intensification of integrated animal husbandry
Improvement of infrastructure	better cost-price relations in animal husbandry	dung, fodder cropping, use of draught animals
Increasing income	better operational integration and competitiveness of animal husbandry	increasing sale of products (income)
	increasing differentiation and intensification of animal husbandry (milk production, increasing importance of hen and pig)	specialization in market production

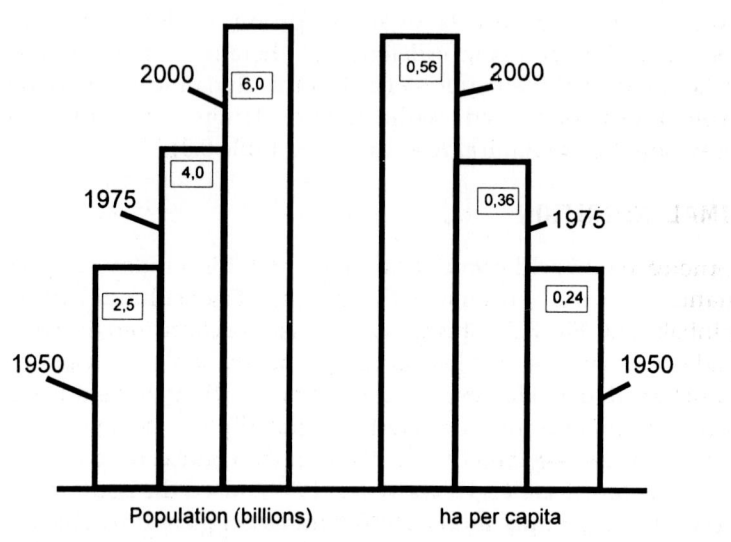

Fig. 8.7 Agricultural land and population (after FAO, 1996).

Population (billions): 1950 = 2.5; 1975 = 4.0; 2000 = 6.0
ha per capita: 1950 = 0.56; 1975 = 0.36; 2000 = 0.24

include materials which can be utilized more efficiently for direct human consumption (e.g. cereals, legumes and root crops). Many people now question whether it is necessary for humans to consume foodstuffs of animal origin (see Flachowsky, 1999). There is much evidence to show that a vegetarian diet is feasible for adults if supplemented with certain major elements (e.g. Ca) and trace elements (e.g. Fe, Zn, I), vitamins (e.g. A,E,B$_{12}$) and amino acids (e.g. lysine).

However, this vegetarian diet is not suitable for infants and growing children. For children and even for adults, it appears that nutritional requirements can be balanced more easily and nutritional deficiencies largely avoided if a certain percentage of the daily diet comes from animal products. Twenty grams of protein from animal products per capita per day (_ 1/3 of the recommended protein supply) means an annual requirement of about 7 kg per capita. With 6 billion people in the world, this amounts to a protein requirement of 42 million tons per year. The amount of foodstuffs of animal origin available varies considerably between different parts of the world (Table 8.4). On an average, about 25 g protein of animal origin is available per person per day and ranges between 9 g in Africa and 65 g in North America (FAO, 1996).

Table 8.4 Availability of foodstuffs of animal origin per person per year (kg) in different parts of the world (after FAO, 1996).

Region	Meat	Milk	Eggs
World	37.3	95.2	7.5
North America	93.2	195.2	14.1
Europe	58.5	219.6	9.2
Asia	24.4	37.9	6.5
Africa	12.6	30.4	2.3

The recommended requirement of 20 g of animal protein per capita per day puts heavy demands on world protein supplies from farm and wild animals and fish. These requirements may be met at least partly by improving feed production, storage and preparation as well as by animal breeding, health improvement, management and nutrition.

If natural resources remain unchanged or decrease but the world population increases, every inhabitant will have less energy, food and water. It is, therefore, important to use resources economically when producing foodstuffs of animal origin. This applies to the production, storage and preparation of feedstuffs as well as to their efficient utilization in animal feeding.

For animal nutrition, the following conditions are required for efficient feed production: high and stable yields, a high energy and nutrient content, high energy and nutrient digestibility and availability, low content of antinutritional ingredients, high feed intake by farm animals (high energy consumption) and high breakdown rate of potentially utilizable sclerocarbohydrates (Flachowsky, 1999).

Transgenic plants that have been modified can be good fodder feed. Increased content and availability (digestibility and absoprtion) of the main plant ingredients such as starch and other highly digestible carbohydrates, protein and selected amino acids (e.g. lysine, methionine), fats and selected fatty acids (e.g. certain unsaturated fatty acids); An increased resistance to fungi (mycotoxins) and pests. Plants may be bred with high concentrations of special ingredients in various plant parts, such as certain enzymes (phytase, non-starch-polysaccharide (NSP) degrading enzymes, major and trace elements and vitamins. Fodder/feed with improved storage and preservation properties would be useful, also plants showing high microbial breakdown of potentially utilizable substances in order to obtain a high roughage intake (Flachowsky, 1999).

Utilization of by-products of Agriculture and Food Industry

Processing of vegetable crops foodstuffs and industrial raw materials generates large amounts of by-products such as cereal straw, sugar beet leaves, etc., which at present, are not being utilized optimally for various reasons.

In the past several decades, sustained efforts have been made to increase the feed value and to improve the utilization in animal nutrition of fibrous feedstuffs (Sundstol and Owen, 1984) (Table 8.5).

Feed Supplements

In order to protect the environment and prevent wastage of resources when producing foodstuffs of animal origin, it is necessary to match energy and essential nutrient supplies as closely as possible to the requirements of the particular animal species and its performance level. Individual feeds and feed mixtures can rarely meet all requirements, so they have to be supplemented in some cases.

Table 8.6 shows one system of classifying feed supplements.

Figure 8.8 shows the various resources for using feed supplements. Increased amounts of certain additives (e.g. I, Se, vitamins A and E) can enrich foodstuffs of animal origin. The effectiveness of non-essential additives depends on various factors, as exemplified by using antibiotics in the feed (Fig. 8.9).

Feed additives can greatly help to convert feed ingredients into useful animal products efficiently and hence to reduce excretion levels of nitrogen, phosphorus, methane and other substances. For high

Table 8.5 Methods for improving the feed value of cereal straw and other low quality roughage (after Flachowsky, 1999)

Treatment method	Objective
PHYSICAL	
Mechanical treatment - crushing - compacting - separation into high value and low value fractions - high-energy irradiation	To increase straw intake
Other methods -pressure -steam -electricity	To increase digestibility and/or straw intake
CHEMICAL	
Breaking down with NaOH, KOH, Ca(OH)$_2$ or NH$_3$ compounds -wet methods -damp methods -dry methods	To increase digestibility and feed intake
Delignification using peroxides Treatment with acids or other chemicals	To increase digestibility to "partially sweeten" straw, to utilize it in non-ruminant feeding and as a raw material for fermentation processes
BIOLOGICAL	
Predigestion by enzymes and microbes	To increase digestibility
Production of edible fungi/white rot fungi	To produce these fungi, residues can be used as feedstuffs
Microbial fermentation for single-cell or alcohol production	To develop straw as a feedstuff (protein) for non-ruminants and as an energy source

performance, the feed must meet all an animal's requirements. With a higher performance, the maintenance requirement is relatively lower and excretion levels per product unit are reduced.

Meeting amino acid requirements exactly, by adding free amino acids and reducing the protein content, reduces N-excretions in poultry and pigs substantially (Schulz, 1992).

According to Flachowsky and Schulz, the following developments may be expected in the area of feed additives in the course of the next two decades:

Table 8.6 Classification of feed supplements (after Flachowsky and Schulz, 1997).

Supplements with essential ingredients	Supplements with non-essential ingredients	
	Present in feedstuffs or in the body	Not present in feedstuffs or in the body
Amino acids	**Enzymes**	**Antibiotics**
(Individual feedstuffs acc. to feed legislation) Lysine, methionine, threonine, tryptophan, etc.	Phytase Non-starch poly-saccharide (NSP)-splitting enzymes	Avilamycin, flavomycin, monensin, salinomycin, spiramycin, tylosin, virginia-mycine, zinc bacitracin
Minerals		
Major elements	**Organic acids and their salts**	**Chemobiotics**
(Individual feedstuffs acc. to feed legislation) Ca, P, Mg, Na	Formic acid, propionic acid fumaric acid, malic acid, calcium formate	Olaquindox, carbadox
Trace elements	**Probiotics**	**Probiotics**
e.g. Mn; Zn; Cu; Fe, Se, I	Lactic acid bacteria	Various bacteria, yeasts
Vitamins and provitamins		
Fat soluble vitamins	**Buffer substances**	**Adjuvants**
A; D; E; K; b-carotene	$NaHCO_3$, MgO, $CaCO_3$	Antioxidants, flavourings, colourings, emulsifiers, pelleting adjuvants, etc.
Water-soluble vitamins		
B_1, B_2, B_6, B_{12}, niacin, pantothenic acid, folic acid, biotin, choline	**Antibodies**	**Prophylactic substances**

- provision of more single amino acids for optimal feed improvement;
- manufacture and use of such organo trace elements (as mineral proteinates);
- provision of rumen-resistant substances (e.g. amino acids, B vitamins);
- extension of the range of antioxidative substances;
- new ways of producing different enzymes (e.g. by means of genetically altered crop plants) or extension of the host spectrum;
- production of antibodies that act in the digestive tract; and
- genetically altered microorganisms with specific functions in feed preparation and in the digestive tract (e.g. improved utilisation of skeletal carbohydrates, lignin degradation, degradation of antinutritive substances, reduction of energy loss in the rumen, selective inhibition of various microbes, etc.).

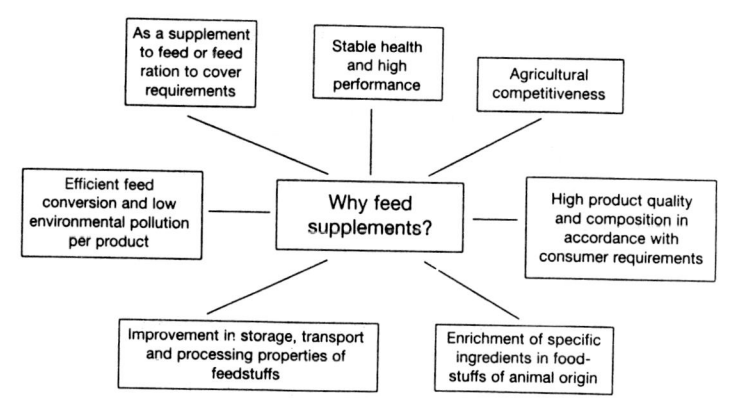

Fig. 8.8 Reasons for using feed supplement (after Flachowsky and Schulz, 1997).

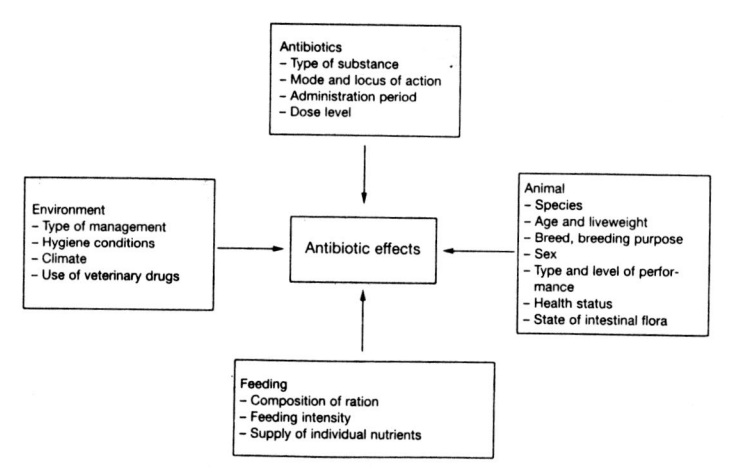

Fig. 8.9 Important factors affecting the action of non-essential feed supplements (after Flachowsky and Schulz, 1997).

IMPACT OF ANIMAL BREEDING ON CONSERVATION OF NATURAL RESOURCES

Animal breeding serves several biological, economic and sociological functions. In the context of the nutrient composition and taste, the main aim of animal breeding is to utilize and improve natural resources in order to produce high quality food.

The need to meet the growing demand for animal food products has led to a continuous increase in stock of the more important domestic animal species, especially in developing countries, where a doubling of the population from 1960 to 1990 was accompanied by an approximate 50% stock increase among ruminants and a three to fourfold increase among monogastric animals such as pigs and poultry (Horst, 1997).

Some potential hazards of intensified animal production are overgrazing of pastures, especially in arid regions; microbial or gaseous emissions from animal husbandry; trace gas emissions, with consequences for the climate; and excessive nutrient levels in animal excretions and the resulting impacts on soil and water quality.

Table 8.7 compares the consumption and wastage of energy and resulting CO_2 emissions from animal food production.

Table 8.7 Product-related cost of agricultural food production expressed in terms of fossil primary energy and CO_2 emissions (after Abel, 1996).

Product	Energy expenditure (MJ/kg)	Emissions (kg CO_2/kg)
Wheat		
conventional	20.8	1.63
integrated	20.2	1.59
organic farming	24.4	1.89
grain legumes	13.0	1.01
rape	27.3	2.15
beef	405	31.1
pork	334	26.5
poultry meat	292	23.1
eggs		
slurry system	222	17.7
drying of faeces	249	19.7

Two approaches that may help limit the environmental impacts of the growing demand for animal products are to modify present animal husbandry and feeding methods; and to reduce maintenance feed intake by increasing the productivity of the individual animal (Horst, 1997).

Stress tolerance and disease resistance are two interesting breeding aims from the viewpoint of ecology and economy. They serve either to maintain biological potentials that have been reproduced at great cost, or to lower production risks by reducing mortality and curb the requirement for expensive drugs.

One important goal of long-term breeding strategies in non-commercial production systems is to advance from single-purpose (e.g. dairy products, eggs, or meat) to multi-purpose livestock. Such versatile breeds are particularly valuable to the small holder or marginal farmer because they allow him to capitalize on local operating conditions and adapt flexibly to changes. This underscores the importance of ensuring that local and indigenous animal resources are conserved and allowed to develop dynamically, especially in adverse environments such as those prevailing at marginal sites. In some cases, this may require crossing with populations endowed with additional useful traits. European (e.g. German) livestock have proved useful in this context (Table 8.8).

Table 8.8 Examples of genetic resources from Germany being used in tropical regions (after Horst, 1997).

Breed	Country where used	Function	Breeding system*
Brown Swiss cattle	South Africa	Meat, suckling cow	PB, XB (Lo)
Pinzgau	Namibia, Cameroon	Meat, suckling cow	PB, XB (Lo)
Hinterwälder	Bolivia	Milk (meat)	XB (E; Lo)
Merino sheep	South Africa, Turkey, Australia	Fine wool (meat)	PB
Milk sheep	Turkey, Israel	Milk (meat)	XB (Lo)
German Cameroon sheep	Malaysia	Meat	XB (Lo)
German Improved Fawn	Malaysia, Burundi, Sri Lanka	Milk and meat	XB (Lo)
Transylvanian naked neck hen	Worldwide	Eggs, meat	PB, XB (E)

*PB = pure breed
 XB = crossbreed
 Lo = local breed
 E = exotic breed of Euro-American origin

PEOPLE-WILDLIFE INTERACTIONS AND WILDLIFE MANAGEMENT

The outcome of wildlife management in the last century has been that some scarce species have become quite abundant or even overabundant. Many situations have arisen that require an urgent, concerted action to reduce conflicts between people and species of wildlife that were scarce just a few decades ago. Wildlife managers must now try to operate within a complex interface of biological and sociological forces. They will be increasingly required to deal with the difficulties of managing wildlife so as to optimize benefits to a society that is living with wildlife. Wildlife managers in some countries have already incorporated human dimensions into decision making in their search for solutions to people-wildlife problems (Decker and Chase, 1997).

Some of the difficult but crucial questions faced by conservation professionals are: Should conservation efforts be directed toward rare individuals, remnant populations, endangered species or threatened ecosystems as a whole? In there justification in focusing attention on 'charismatic megavertebrates' (whales, tigers, pandas) at the expense of anonymous but ecologically-significant species? Should we manage for exotics or for local, native species? And to what extent may we legitimately disrupt the lives of individual animals in the name of species preservation and conservation education? (see Norton et al., 1995). Norton et al., (1995) considered the ethical dilemmas involved in the captive breeding of species threatened by habitat destruction, dilemmas that arise when choices must be made regarding the welfare of individual animals and when steps must be taken to ensure the future of threatened populations in the wild.

Arguments are being advanced for abolishing zoos and increasing their role in both in situ and ex situ conservation efforts, for redirecting their efforts toward local fauna and habitats and for expanding their contribution to international conservation and habitat-preservation efforts. The recreational and educational values of zoos and the unrealistic nature of many reintroduction strategies need to be highlighted. There are debates over the procurement of individuals from the wild, the fate of surplus animals in captive-breeding programs, the care of captive animals in more-or-less naturalistic designed habitats and the use of captive animals in research. Zoos and captive-breeding programs must navigate rough waters as they search for workable solutions to the dilemmas posed by their four-pronged mission, viz., conservation, education, research and recreation.

Human Dimensions of Wildlife Management

Most wildlife managers dealing with people-wildlife conflicts have experienced that the human dimensions of such situations are the most difficult to manage. This has greatly increased interest in the human dimensions of wildlife management in recent years. Wildlife used to be mainly managed by biologists and ecologists. This management has gradually broadened in scope to include a scientifically-based understanding of people as an essential part of the management equation (Decker et al., 1992).

Manfredo et al., (1996) defined human dimensions of wildlife management as 'an area of investigation which attempts to describe, predict, understand and affect human thought and action...'. Specialists of human dimensions maintain that while traditional biological considerations are essential, managing people and managing the decision-making process itself are equally important for dealing with people-wildlife problems today.

A people-wildlife problem can potentially be any situation where: (1) the behaviour of people negatively impacts wildlife (this include human impacts on habitat); (2) the behaviour of wildlife creates a negative impact for some people or is perceived by some people to impact themselves or others adversely; or (3) the wildlife-focused behaviour of some people creates a negative interaction with other people, often as a clash of values. Thus, a people-wildlife problem can involve a people-wildlife interaction or a people-people interaction (i.e. a controversy), or both (Decker and Chase, 1997).

The dilemma for managers is that different stakeholders usually have different expectations of their interactions with wildlife—expectations that cannot be met simultaneously at the same place and time. The manager has to reconcile these competing interests, or stakes, in the wildlife resource.

Solving People-wildlife Problems

From a human dimensions, perspective, a people-wildlife interaction or people-people interaction problem can be considered to have been solved only when the stakeholders involved believe it to be so.

There is seldom a single answer to a people-wildlife problem. Rather, there is a range of more or less acceptable management objectives and actions. Solutions are primarily tied to the process used to make decisions, especially the extent of stakeholder input and involvement. To manage people-wildlife interactions successfully, stakeholders must be considered throughout all phases.

Decker and Chase (1997) outlined a typology of approaches that characterize most of the ways wildlife managers tend to address public input and involvement. They identified the following 5 types of approaches: authoritative, passive-receptive, inquisitive, transactional and co-managerial or delegatorial.

Some managers try to be good public servants. However, wildlife managers should be cautious not to become servantile to public opinion by relying too heavily on opinion polls to determine what they ought to do. Surveys and opinion polls should not become in effect surrogate referenda where what the majority want, the manager will automatically try to provide. To please the majority can sometime mean heading down a road toward devaluation of professional judgement and perhaps even to abrogation of professional responsibility (Decker and Chase, 1997).

These concerns notwithstanding, intelligent wildlife managers with the right attitudes and skills can develop acceptable approaches to solving people-wildlife problems and associated people-people conflicts.

Regardless of which general approach is adopted for development and application in a specific situation, it should include: (1) determining management objectives and selecting management actions that involve stakeholders in an inquiry or process that discovers and applies stakeholder weights in decisions; (2) involving stakeholders in evaluating management action implementation and resulting outcomes; and (3) using ongoing stakeholder input for adaptive management (Decker and Chase, 1997).

THREATENED MAMMALS

A 1996 study by the World Conservation Union (also known as the IUCN) revealed that the world is entering a period of major species extinction, rivaling five other periods in the past half a billion years. The IUCN found a much higher level of threat to several classes of animals than was generally thought. It found that an astonishing 25% of mammal species—and comparable proportions of reptilian, amphibian and fish species are threatened. Of five classes of animals, birds are the least at risk (see Fig. 8.10).

PERCENT OF SPECIES THREATENED

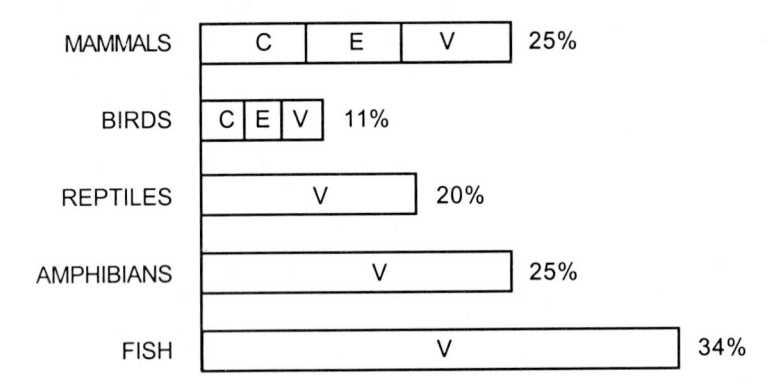

MAMMALS	C | E | V	25%
BIRDS	C|E|V 11%	
REPTILES	V 20%	
AMPHIBIANS	V 25%	
FISH	V 34%	

C = CRITICALLY ENDANGERED; E = ENDANGERED; V = VULNERABLE

Fig. 8.10 Threatened animals. Data for reptiles, amphibians and fish are insufficient to classify accurately the degrees of extinction risk (after Doyle, 1997).

Of the 4,327 known mammal species, 1,096 are at risk and 169 come in the highest category of critically endangered—extremely high risk of extinction in the wild in the immediate future. (The other two are endangered, i.e. very high risk in the near future and 'vulnerable,' a high risk in the medium-term future). Of the 26 orders of mammals, 24 are threatened, the most affected being elephants, primates, rhinoceroses and tapirs.

Habitat disturbance by humans increases the threat to mammals. There is a high proportion of endemic species, especially in the case of geographically isolated areas. Such regions have unique evolut-ionary histories and fixed boundaries to species ranges and, therefore, degradation of such habitats can take a heavy toll on animals. Striking examples are the Philippines and Madagascar, where 32% and 44%, respectively, of all mammal species are threatened. In both countries, over half the species are endemic and there is much habitat disturbance. In contrast, Canada and the US have respectively, 4% and 8% of mammal species threatened. Less than a quarter of the species in the US and only 4% in Canada are endemic. Habitat disturbance is moderately above average in the U.S. and very low in Canada.

The countries with the most threatened mammals are Indonesia, with 128 species and China and India, both with 75. These three account for 43 percent of the world's population and are among the most densely populated (Doyle, 1997).

NATURE CONSERVATION, HUNTING TOURISM AND SUSTAINABLE DEVELOPMENT

In recent times, nature conservation and ecological balance over large parts of the African continent have become incompatible with the utilization of nature by man and beast, since it is often becoming a matter of survival. Consequences of the dry tropical climate in general, together with the seemingly unstoppable growth of the population are the two chief obstacles in this context.

Nature conservation was introduced in Africa by the colonial powers. Selected areas were set aside for flora and fauna, administrators and supervisors were appointed. Unfortunately, nature outside the protected areas was largely ignored.

As a rule, the opening-up of conservation areas for tourism does not do justice to their declared objectives and could only be tolerated if the most severe restrictions were enforced. Some of the naturally attractive conservation areas have long been opened up for tourism, which has adversely affected many conservation areas, increasing the vulnerability of the flora to the damage caused by the increasing aridity. Suggestions have been made to partially close down conservation areas in rotation with access to open parts granted every 2-3 years (Reichelt, 1999).

Increasing growth of the human population has intensified the competition between humans and wildlife in Africa for scarce resources such as water and land for cropping, animal husbandry, and settlement (Vorlaufer, 1997). In some cases, this conflict is further aggravated by the proliferation of wild animals under stringent game protection regimes.

362 Environmental Technology and Biosphere Management

Securing game resources for future generations amid these various demands and conflicts has already become an important economic issue. Beyond this, species diversity and the abundance of wildlife are both the basis and an indicator of the biodiversity typical of Africa and an important component of any ecologically sustainable development (Vorlaufer, 1999). According to Vorlaufer, in large parts of Africa, hunting tourism can be an instrument for achieving sustainable development and securing widlife over the long term, provided that the indigenous population can participate in the planning and implementation of wildlife conservation and in the revenues from hunting tourism, in compensation for the opportunity costs incurred through their complete or partial abstention from the use of game, water, or land, and for the often considerable game damage. Figure 8.11 illustrates that the local population's deprivation since the nineteenth century of the use and profits to be had from game has frustrated the objectives behind colonial conservation laws of securing game (particularly elephant) populations. Nor can this be achieved in the future unless a steady interplay can occur between the human population, game protection and hunting tourism (Vorlaufer, 1999).

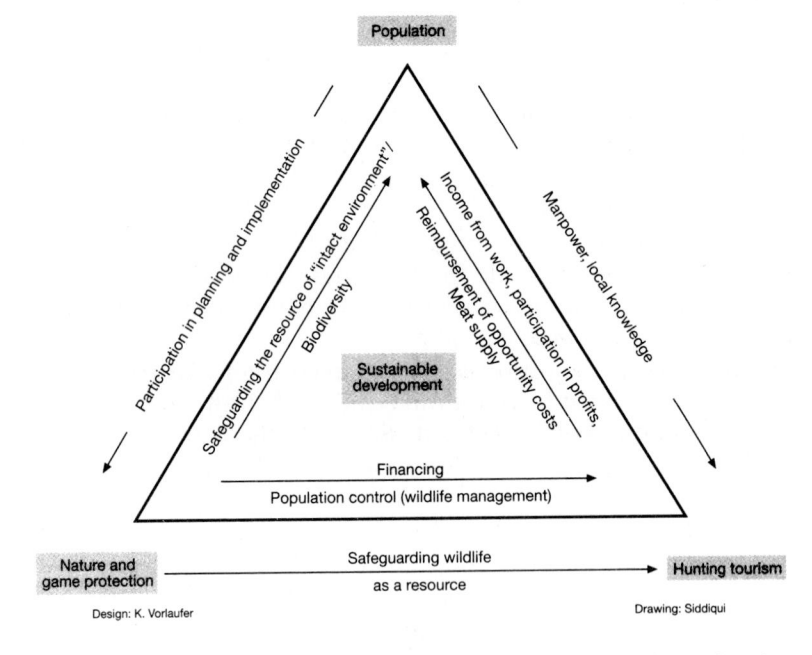

Fig. 8.11 Sustainable development as a result of interdependencies within the 'magic triangle' between population, nature conservation and hunting tourism (after Vorlaufer, 1999).

REFERENCES

Abel, H.J. Energieaufwand und CO_2 Ausstoß bei verschiedenen Formen der Lebensmitterlerzeugung. 16. Hülsenberger Gespräche, Travemünde, 5. bis 7. Juni, 153-161 (1996).

Chandra, J. Crop domage caused by blackbucks (*Antilope cervicapra*) at Karera Great Indian Bustard Sanctuary, and possible remedial solutions. *J. Bombay Nat. Hist. Soc.* 94: 322-331 (1997).

Decker, D.J., Brown, T.L., Connelly, N.A., Enck, J.W., Pomerantz, G.A., Purdy, K.G., Siemer, W.F. Toward a comprehensive paradigm of wildlife management: integrating the human and biological dimensions. In Mangun, W.R. (Ed.). *American Fish and Wildlife Policy : The Human Dimension*. pp. 33-54. Southern Illinois Univ. Press, Carbondale (1992).

Decker, D.J., Chase, L.C. Human dimensions of living with wildlife—a management challenge for the 21st century. *Wildlife Soc. Bull.* 25(4): 788-795 (1997).

Doyle, R. Threatened mammals. *Sci. Amer.* p. 32 (Jan. 1997).

Dutta, A.K. *Unicornis: The Great Indian One-horned Rhinoceros*. Konarak Publishers, Delhi (1991).

Erz, W. Tier- und Artenschutz aus fachpolitischer Sicht des Naturschutzes. In: Rahmann, H., Kohler, A. (eds.). Tier- und Artenschutz. *Hohenheimer Umwelttagung* 23: 21-33 (1991).

FAO. *Production Year Book*. Vol. 50. Food and Agriculture Organization. Rome (1996).

Flachowsky G. Animal nutrition in conflict with current and future social expectations and demands. *Animal Research and Development*. 49: 63-104 (1999).

Flachowsky, G., Schulz, E. Feed supplements and their significance for performance and ecology. *Animal Research and Development*. 46: 87-94 (1997).

Gadgil, M., Guha, R. *This Fissured Land, An Ecological History of India*. Oxford Univ. Press, Delhi (1992).

Gärtner, K. Sozialempirische Analysen der Mensch-Tier-Beziehung. In: Hardegg, W., Preiser, G. *Tierversuche und Medizinische Ethik*. Olms, Hildesheim (1986).

Grauvog, A. Tierschützerische Aspekte der derzeitigen Schweineproduktion. *Tiierarztliche Umschau* 51: 308-313 (1996).

Hartl, G.B., Kurt, F., Tiedemann, R., Gmeiner, C., Nadlinger, K., Khyne, U.M., Rübel, A. Population genetics and systematics of Asian elephant (*Elephas maximus*) : A study based on sequence variation at the Cyt b gene of PCR-amplified mitochondrial DNA from hair bulbs. *Z. Säugetierkunde* 61: 285-294 (1996).

Herre, W., Röhrs, M. *Haustiere - Zoologisch Gesehen*. Fischer, Jena (1973).

Horst, P. Animal breeding as a factor of influence on the conservation of natural resources and utilisation of genetic resources. *Animal Research and Development* 46: 63-74 (1997).

Hosken, L. Biodiversity: Green gold or sacred teacher? *Compas Newsletter* (2): 18-19 (1999).

Immelmann, K., Scherer, K.R., Vogel, L., Schmock, P. Psychobiologie: Grundlagen des Verhaltens. Fischer, Stuttgart (1988).
Kumar, H.D. *Biodiversity and Sustainable Conservation.* Oxford & IBH, New Delhi (1999).
Kumar, H.D. *Plant-Animal Interactions.* Affiliated East West Press, New Delhi (2000).
Kurt, F. *Naturschutz — Illusion und Wirklichkeit. Zur Ökologie bedrohter Arten und Lebensgemeinschaften.* Paul Parey, Hamburg (1982).
Kurt, F. Wild animals in a strife torn world - observations and reflections on 30 years of wildlife conservation in India. *Animal Research and Development* 50: 7-43 (1999).
Laurie, A. Das Indische Pancernashorn. In : Die Nashörner. pp. 95-114, Filander Verlag, Fürth (1997).
Mallinson, J.J.C. Zoo breeding programs - balancing conservation and animal - welfare. *Dodo.-J. Wildlife Preservat. Trusts* 31: 66-73 (1995).
Manfredo M.J., Vaske, J.J., Sikorowski, L. Human dimensions of wildlife management. In Ewert, A.W. (Ed.). *Natural Resource Management : The Human Dimension.* pp. 53-72. Westview Press, Boulder, CO. (1996).
Naegel, L.C.A. Development of small-scale sustainable farming systems in non-industrialized countries: New concepts are needed. *Animal Research and Development.* 43/44: 25-43 (1996).
Neidhardt, R., Grell, H., Schrecke, W., Jakob, H. Sustainable livestock farming in East Africa. *Animal Research and Development.* 43/44: 44-52 (1996).
Norton, B.A., Hutchins, M., Stevens, E.E., Maple, T.L. *Ethics on the Ark: Zoos, Animal Welfare, and Wildlife Conservation.* Smithsonian Institution Press Washington DC (1995).
Peters, K.J. Animal husbandry and securing the food supply — consequences for the environment? *Animal Res. Develop.* 49: 39-50 (1999).
Poetschke, J. The integration of animal breeding and husbandry into a sustainable, locally and environmentally compatible tropical farming system. *Animal Research and Development* 46: 7-13 (1997).
Reichelt, R. Nature conservation and ecology in Western Africa, climatic and anthropogenic obstacles. A general view including specific examples. *Applied Geography and Development.* 54: 93-105 (1999).
Sambraus. H.H. Befindlichkeiten und Analogieschluß. *Akt. Arb. Artgem. Tierhaltung* (KTBL) 370: 31-39 (1994).
Schulz, E. Möglichkeiten und Grenzen der N- und P-Absenkung in Schweinerationen. Vortragstagung "Aktuelle Themen der Tierernährung und Veredlungswirtschaft". 16/17.10.1991, Cuxhaven, 133-144 (1992).
Seidl, W. Intensification of animal husbandry effect on global warming and soil quality. *Animal Res. Develop.* 49: 7-13 (1999).
Singh, H.S. Population dynamics, group structure and natural dispersal of the Asiatic lion (*Panthera leo persica*). *J. Bombay Nat. Hist. Soc.* 94: 65-70 (1997).
Socolow, R.H. Nitrogen management and the future of food: Lessons from the management of energy and carbon. *Proc. Natl. Acad. Sci. USA* 96: 6001-6008 (1999).

Subramanya, S. Distribution, status and conservation of Indian heronries. *Bombay Nat. Hist. Soc.* 93: 459-486 (1996).

Sundstol, F., Owen, E. Straw and other firbous by-products as feed. *Dev. in Anim. and Vet. Sci.* Vll. 14, Elsevier, Amsterdam (1984).

Turner, M. The sustainability of rangeland to cropland nutrient transfer in semi-arid West Africa: Ecological and social dimensions neglected in the debate. *Proc. Internat. Conf.* Addis Ababa, Ethiopia, 22-26 November 1993, Internat. Livestock Center for Africa (1995).

Vijayan, V.S. Keoladeo National Park ecology study. *Bombay Nat. Hist. Soc.* Annual Report (1987).

Vorlaufer, K. Conservation, local communities and tourism in Africa. Conflicts, symbiosis, sustainable development. In : Hein, W. (Ed.) : *Trourism and Sustainable Development*. Schriften des Deutschen Übersee-Instituts 41, Hamburg, 53-123 (1997).

Vorlaufer, K. The Selous game reserve in Tanzania — nature conservation, hunting tourism and sustainable development in Africa's largest wildlife reserve. *Applied Geography and Development* 54: 106-132 (1999).

Chapter 9
Urbanization

INTRODUCTION

Today, roughly about one-half of humanity lives in urban areas. Eighteen cities in the developing world have over 10 million inhabitants. The population of Mexico City is greater than that of Yugoslavia.

Mega-cities impose a heavy strain both on their surroundings and their inhabitants. Every day, a city with a million inhabitants uses up some 625,000 tonnes of water, 2,000 tonnes of food and 9,500 tonnes of fuel and produces 500,000 tonnes of waste water, 2,000 tonnes of solid wastes and 950 tonnes of air pollutants (UNEP, 1990).

In 1985, half the world's urban population lived in areas where sulphur dioxide levels were potentially dangerous to health. A quarter of the world's city dwellers lacked access to clean water, two-fifths lacked sanitation. Some one-third of the urban populations of developing countries live in squatter camps, slums and shanty towns. The rural populations of developing countries are even worse off. Over half the world's villagers have no safe water, over four-fifths no sanitation and hundreds of millions live below the poverty line.

Urbanization is often a positive contributor to sustainable development. As a general rule, living in cities is more sustainable than rural living for the same level of material welfare. This is because the area of land required for urban living is less than for rural living. This, in turn, reduces interference with land-based ecosystems and the appropriation of agricultural land for settlement.

Urban patterns of land use offer several other opportunities to increase human welfare and lower demands for natural resources. Dense urban settlement patterns increase the affordability of piped treated water, sanitary waste collection and management, advanced telecommunications, health care, education and emergency services. Concentration of production and consumption in urban regions increases the possibilities for recycling of materials. Economies of scale and density of settlement allow for reduced fossil fuel use with the availability of public transport, heating and cooling, the use of multi-occupancy buildings with relatively low ratios of surface area to floor area, and the reduction in distances travelled.

Unsustainable land development practices adjacent to freshwater bodies, rich agricultural soils and sensitive coastal areas are exerting strongly adverse impacts on global fresh water resources, biodiversity and patterns of food supply. Some urban land use practices, as simple as landscaping or paving, can profoundly influence the nature of resource use and resource flows in their own cities and throughout the world.

There is no doubt that the sustainable management of land resources requires a much better coordination of land, transport, housing and economic development policies among national, subnational and local levels of government. As a first step, governments at all levels ought to undertake a coordinated country-level review of priority urban and management issues. Such reviews should ensure effective stakeholder participation. Guidelines for sustainable planning and management of urban land resources need to be framed and mechanisms to enhance sustainable urban land use through intergovernmental policy coordination need to be appraised.

A sick environment undermines the health of its inhabitants. In developing countries, contaminated food, dirty water and polluted air (Fig. 9.1) are the chief causes of death and disease, while fuel and food shortages caused by environmental degradation weaken people's resistance to sickness. Worldwide, chemical pollutants are associated with several diseases, including cancer.

Poor drying or storage of food in tropical and sub-tropical countries encourages fungi, which release potentially fatal mycotoxins.

Over 100 million people are infected with malaria every year, some 200 million suffer from schistosomiasis and sleeping sickness threatens over 50 million.

Reducing the use of agricultural pesticides not only protects the environment but also human health.

URBANIZATION AS A GLOBAL PROCESS

Urbanization is a complex, irreversible, global process. It cannot be curbed through policy-making. As it certainly has some impact on rural areas

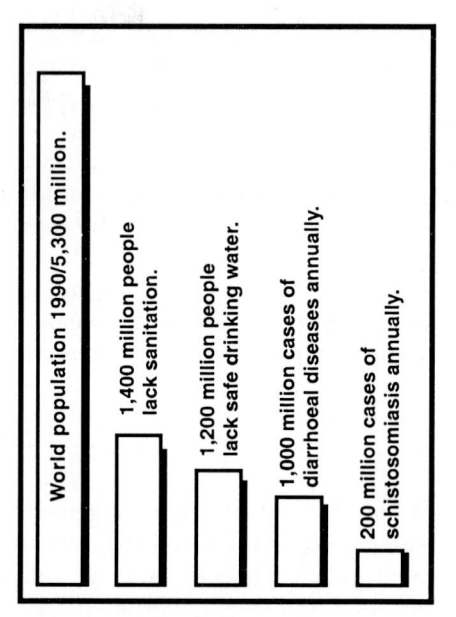

Fig. 9.1 Dirty water is the world's major cause of disease. Lack of safe drinking water leads to 1,000 million cases of diarrhoea every year—and the deaths of 4.6 million children under the age of five (UNEP, 1990).

also, rural development policy frameworks need to be broadened to include both the negative and the positive aspects of urbanization. But how to strengthen rural-urban linkages in a way that benefits both rural as well as urban populations, thereby supporting a sustainable and socially-just development process? Development and poverty alleviation strategies urgently need a spatial vision, so that solutions for urban problems do not cause rural problems, and vice-versa (Rabinovitch, 1999). Rabinovitch suggested the following dimensions of recent developments in rural-urban connections:

(1) the phenomenon of urbanization, including the increasing urbanization of poverty;
(2) agricultural transformations which have improved the potential productivity of rural areas;
(3) the demand for participatory approaches in governance and the growing recognition of social capital;
(4) decentralization trends in many developing countries; and
(5) globalization trends which highlight integrations between rural and urban areas in terms of worldwide systems of production, contracting, finance, merchandising and labour markets.

The following conclusions emerged from a recent international workshop on rural-ruban linkages, held on 10-13 March 1997 in Brazil:

(1) Rural-urban linkages add a crucial spatial dimension to understanding the key development issues of our time and to formulating effective policies and programmes to address them;
(2) Ensuring reciprocal benefits from rural-urban linkages requires a localization of planning and management capacities;
(3) The conventional concept of rural as being equivalent to agriculture does not reflect the reality of either rural regions or the rural component of rural-urban linkages, because rural areas are becoming more and complex;
(4) Rural-urban linkages add an important perspective to creating best practices (see Rabinovitch, 1999).

URBAN ECODEVELOPMENT

Urbanization Trends and Problems

Urbanization is perhaps the most important social transformation of our times. Whereas in 1800, only about 3% of the world population lived in the cities, urban dwellers probably outnumber the rural population.

By the year 2000, Latin America is likely to reach roughly the same proportion of urban population as Europe and Northern America (see Table 9.1). But the highest average annual rate of urban population growth is occurring in Africa, followed by South Asia (Sachs, 1986). In India, between 1971 and 1981, the urban population grew at the rate of 4.3% per year, almost twice as quickly as the overall population. During

Table 9.1 Per cent urban population by major regions, 1960-2025 (Source : Population Division, UN, 1983).

Region	1960	1980	2000	2025
World	33.6	39.9	48.2	62.4
Less developed	21.4	29.4	40.4	57.8
More developed	60.3	70.6	77.8	85.4
Africa	18.4	28.7	42.2	58.3
Latin America	49.3	65.4	76.9	84.4
Northern America	69.9	73.8	78.0	85.7
East Asia	23.1	28.0	34.2	51.2
South Asia	18.3	25.4	36.8	55.3
Europe	60.5	71.1	78.9	85.9
Oceania	66.3	71.6	73.0	78.3
USSR	48.8	63.2	74.3	83.4

those 10 years, 54 million people went to towns, equivalent to the total population of the United Kingdom (Mayur, 1983).

New technologies now allow us to design cost-effective integrated systems. Higher energy prices make it economical to capture the city's waste heat by internally consuming the energy and reduce our consumption of energy by increasing the efficiency of fuel use. Some buildings can be heated only by body heat and waste heat from appliances and light (Morris, 1986). Fuel and power sources can be moved closer to the final consumer.

The warmth of the soil and groundwater needs to be gainfully exploited. Heat pumps take advantage of the earth's furnace. Solar cells take advantage of the sun. The sunlight that falls on rooftop photovoltaic devices, pushes electrons across a junction, setting up an electrical current that can power devices within the house. The rooftops and sides of buildings can actually become power plants.

Underground or mined space can become a major development activity for some cities. Mined space has many useful applications. Medical diagnostic equipment works better there because of the lack of magnetic fields. Automated micro-electronic facilities work well there because of the lack of vibrations. They are excellent places for storage, especially for items that need good security.

Whether sunlight, water or soil, the self-reliant city uses human ingenuity to get the maximum amount of useful work out of abundant, renewable, local resources (Morris, 1986).

A striking feature of the present urbanization trends in the Third World is the increasing relative share of large and metropolitan cities (see Table 9.2 and 9.3).

Table 9.2 Average annual rates (%) of urban population growth by major regions 1960-2025 (Source : Population Division, UN).

Major	1960-1970	1970-1980	1980-2000	2000-2025
World total	3.0	2.7	2.5	2.2
Less developed	4.1	3.8	3.5	2.8
More developed	2.0	1.4	1.1	0.7
Africa	4.7	5.1	5.0	3.8
Latin America	4.2	3.7	2.9	1.8
Northern America	1.9	1.1	1.2	1.0
East Asia	3.3	2.5	2.1	2.2
South Asia	4.0	4.2	3.8	2.8
Europe	1.7	1.2	0.8	0.4
Oceania	2.7	1.9	1.5	1.3
USSR	2.7	2.0	1.6	1.0

Table 9.3 Percent of urban population by city-size class and major regions, 1960-2025 (Source: Population Division, UN, 1983).

Major area	Under 1 million				1 million - 4 million				4 million and above			
	1960	1980	2000	2025	1960	1980	2000	2025	1960	1980	2000	2025
World total	70.5	65.9	59.2	56.7	16.1	18.3	20.7	18.6	13.4	15.8	20.2	24.7
Less-developed regions	71.6	65.3	55.7	53.4	15.9	17.4	20.7	18.2	12.5	17.3	23.6	28.4
More-developed regions	69.6	66.6	66.1	67.6	16.2	19.3	20.6	19.9	14.2	14.1	13.4	12.5
Africa	85.0	72.6	55.5	50.5	15.0	22.1	24.6	15.6	0.0	5.3	19.8	33.9
Latin America	68.9	61.0	52.1	53.1	10.5	17.2	19.9	17.4	20.6	21.8	28.0	29.5
Northern America	59.0	53.0	53.3	54.0	21.8	27.6	30.2	26.5	19.3	19.4	16.5	19.6
East Asia	61.8	63.0	62.2	60.9	16.4	16.8	18.7	19.8	21.8	20.1	19.1	19.4
South Asia	78.2	67.2	54.7	52.4	15.8	13.9	19.1	18.5	6.0	18.9	26.2	29.1
Europe	69.4	69.9	70.5	72.7	18.5	18.7	17.5	17.3	12.1	11.4	12.1	10.0
Oceania	61.6	48.9	41.0	43.2	38.4	51.1	40.3	41.4	0.0	0.0	18.7	15.5
USSR	87.6	77.9	73.6	74.8	6.4	14.4	19.5	19.4	6.0	7.7	6.9	5.9

Future cities will have to be resource-conserving, even in the richest countries. There is considerable scope for energy conservation in the urban setting (Morris, 1982).

The Urban Bias. The U-City and the L-City

The urban revolution, started in the third millennium BC in Mesopotamia, has now virtually completed a full cycle from the agrarian community with practically the whole population busy in gathering and/or producing food to the modern urbanized society in which all but a few have non-agricultural occupations (Sachs, 1986).

Civilization was born out of the appropriation by the urban elites of the economic surplus produced by the rural masses. Throughout the ages, the 'urban' bias' (Lipton, 1977) has taken many forms according to the needs of different types of 'producer cities' and 'consumer cities' (Weber, 1982), involving asymmetric flows of goods and people. Yet secondary sources of accumulation are scanty and the disproportion between the urban and rural population makes it utterly unrealistic to rely for the functioning of the cities on the surplus extracted from the contryside (Fig. 9.2).

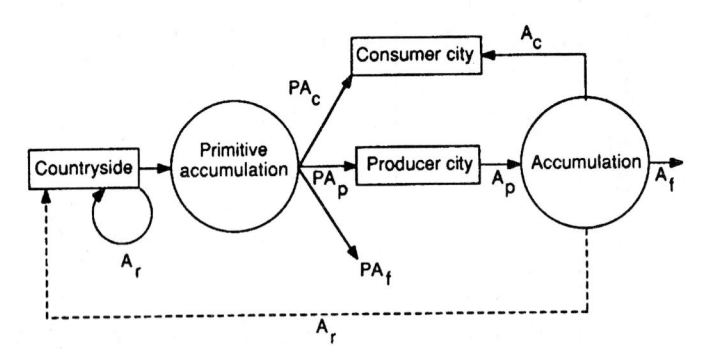

Fig. 9.2 Cities and countryside—primitive and secondary accumulation. The primitive accumulation achieved at the expense of the countryside has been misused: a substantive part of it has been flowing abroad (PAf), the share taken by the consumer city (PAc) is greater than that of the producer city (PAp); investment in agriculture (Ar) has been too low. The same structural weaknesses affect the secondary accumulation obtained in the producer city. Part of it is drained abroad (Af), the capital invested in the consumer city (Ac) reduces the outlays for the producer city (Ap). The much needed transfer of resources to the agriculture (Ar) does not occur. A further structural weakness is that Ac goes to the U-city at the expense of the L-city. There is over-investment in the U-city and severe under-investment in the L-city (after Sachs, 1986).

Although accumulation and investment have an important role in the process of economic growth, they do not represent the only possible source of growth — Kalecki (1972) emphasized the need to carefully screen the non-investment opportunities for growth.

The resources saved, thanks to the lowering of the capital-output ratio, the extension of the useful life of equipment and a more intensive use of the same, if properly channelled, could be used as an additional fund for the development of the U-city. The problem has been restated accordingly as one of identifying at the micro-economic level inside the urban ecosystem, considered as a production system, concrete steps in this direction, of implementing them and of using the resources thus released for socially legitimate purposes (Sachs, 1986).

Growth is a necessary—but by no means sufficient condition—for development. Left to uncontrolled market forces, it can equally sustain maldevelopment. Patterns and uses of growth need to be subordinated to the triple criterion of social usefulness, economic viability and econological sustainability or, in short hand, ecodevelopment (Sachs, 1980a and 1980b).

The Urban Ecosystem as a Resource Potential

Many economists look at cities as the site of many enterprises, whose concentration creates both positive and negative externalities but requires a costly infrastructure. Human ecologists advocate the concept of cities as ecological systems (Boyden, 1984).

The analogy between the urban and the natural ecosystem as resource potentials holds good up to the extent to which resources do not exist as such. They are merely portions of the natural and urban environment that people learn to use for some specific purpose. Knowledge about environment or, if one prefers culture, is thus an essential component of the very concept of resource.

The human and physical resources potentially available in the urban ecosystem are shown in Figs 9.3 and 9.4. Figure 9.5 shows a classification of the forms of wealth production and uses.

The Food-Energy Nexus

Among the various goals, greater food and energy self-reliance deserve emphasis. Daily provisioning of food and cooking fuel as well as expenditure on transportation take a disproportionately high share of family budgets (see Figs. 9.4 and 9.5). Figure 9.6 shows the food energy nexus in the city.

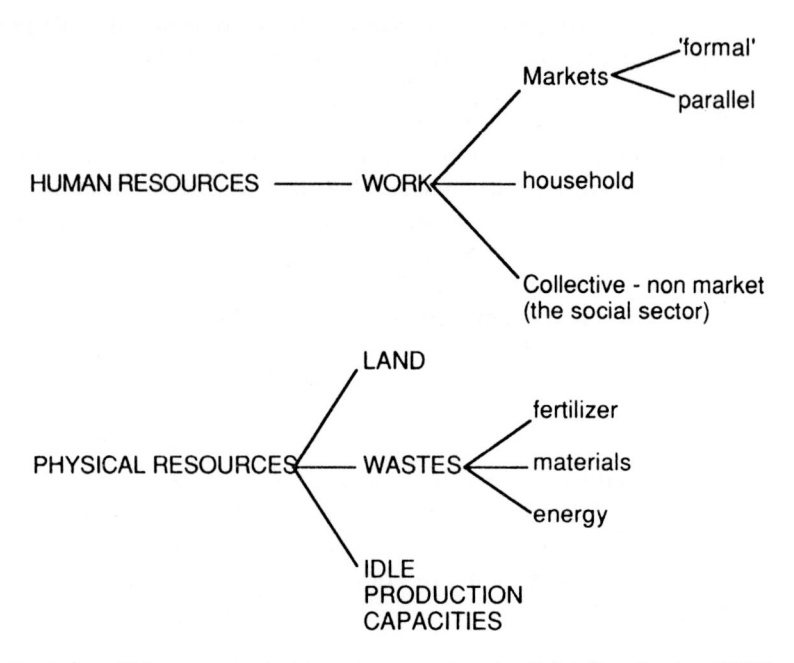

374 Environmental Technology and Biosphere Management

Fig. 9.3 Urban ecosystem as a resource potential (after Sachs, 1986).

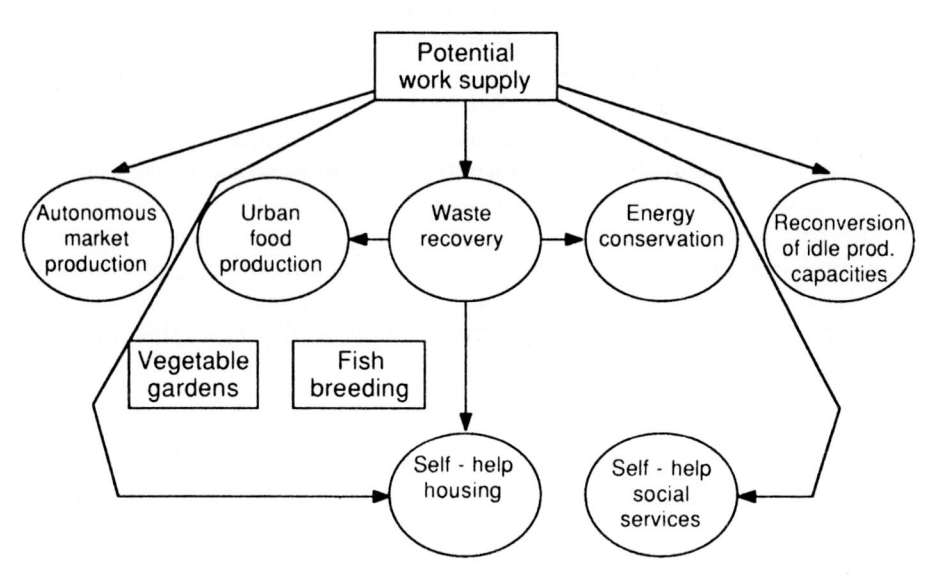

Fig. 9.4 Self-reliant strategies in the urban setting with a high potential (after Sachs, 1986).

Urbanization 375

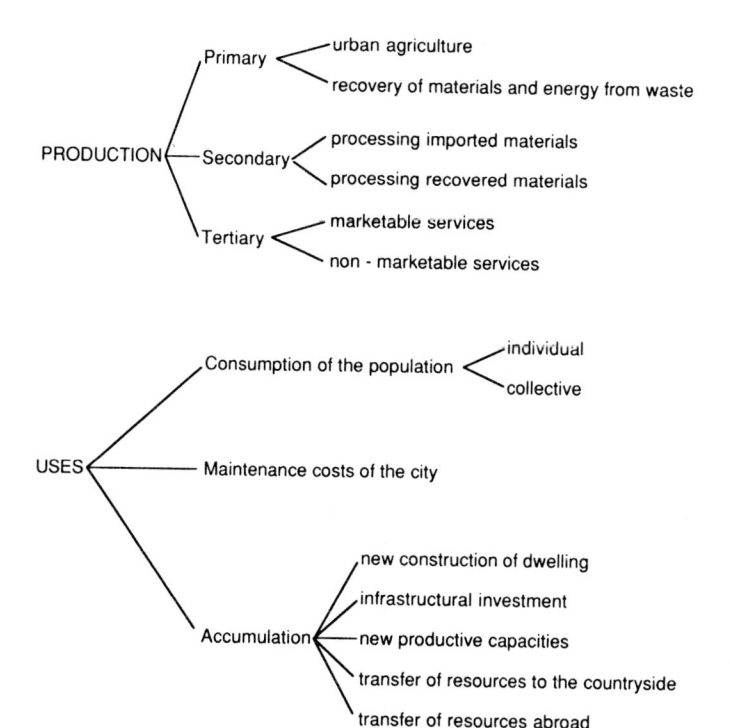

Fig. 9.5 Wealth production and uses in the city (after Sachs, 1986).

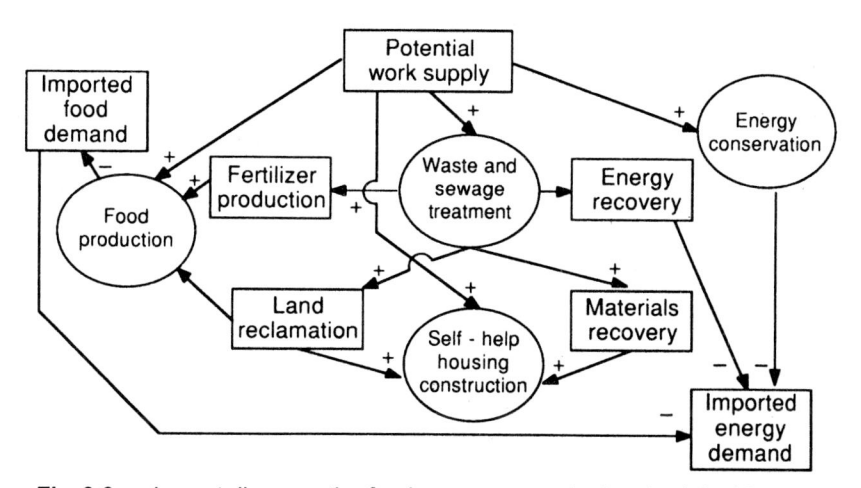

Fig. 9.6 Impact diagram: the food-energy nexus in the city (after Sachs, 1986).

Mega-cities

There are over 20 large mega-cities in the world. A mega-city is defined as a city with population exceeding eight million (Kraas, 1997a, 1999). Mega-cities share the serious problems of large conurbations, rapid population growth, high and uncontrolled urban expansion, grave social problems of pressure on land use and residential and capital markets, inadequate infrastructure, serious shortcomings in public utilities and waste disposal, growing socio-economic disparities, extreme environmental burdens, and social rootlessness (Kraas, 1997, 1997a). Forceful development dynamics and scarce public finances increasingly make mega-cities prone to defy good governance.

Three aspects of mega-cities are particularly noteworthy:

(1) They are always at considerable risk in view of their high population.
(2) In crisis situations, they are highly vulnerable to sudden supply shortage and sharp environmental burdens. Any natural or unnatural catastrophe or disaster can quickly lead to serious bottlenecks or emergencies for a large number of people and aggravate those of the socially weaker people.
(3) Conflicts can quickly become multi-dimensional if the administrative machinery and planning are not well co-ordinated, and if there are growing socio-economic disparities and strong environmental burdens (Kraas, 1999).

Four central requirements of mega-cities include: (1) adequate water supplies for the population; (2) proper disposal of sewage: (3) reliable power supply; and (4) satisfactory roads. Increasing population concentrations and intensifying industrialization have converted the first two of the above areas into grave problems in many mega-cities. In some mega-cities in the developing world (e.g. Delhi) the additional menace of corruption among many government employees has also worsened the power supply position as well as roads and other infrastructure.

ENVIRONMENTALLY SUSTAINABLE TRANSPORTATION SYSTEM

A project called Euro-EST, where EST stands for Environmentally Sustainable Transportation System, has recently been started in Sweden. Special emphasis in this project is being given to appropriate instruments and environmental goals for sustainable transportation in Europe, particularly Sweden. It is being realized that bringing about an environmentally-adapted transportation system for Europe will require decisions to be made not only nationally and locally but also at the EU level.

The American auto industry is accelerating efforts to get the sulphur content of petrol and diesel fuel further reduced because a lower sulphur content will make it easier to lower emissions.

Recent studies have shown that the cost of lowering the sulphur content will not be too high. The American Environmental Protection Agency has proposed reducing it in petrol from 300 to 30 ppm, estimating the extra cost to be 2 cents per gallon. The car makers feel that it would only cost 2-3 cents more to bring sulphur down to 5 ppm (Source : *The Auto Channel*, 27 October 1999).

The European Municipal Green Fleets project, 'Buy Efficient' aims to find out how European cities can work together to promote low emission technologies and operating methods in their fleet operations.

Measures to reduce carbon dioxide emissions from the muncipal fleets and decrease fuel consumption include purchasing fuel efficient vehicles and reducing the number of vehicle kilometres travelled by municipal employees. Ancillary measures include reducing the size of the fleets, driver education and training and vehicle maintenance.

Helsinki (Finland) has developed a combined sustainable transport and land use policy which prefers and favours sustainable transportation options to conventional infrastructure development, with the overall goal of keeping the city-centre alive. Rather than building an extensive system of highways, Helsinki has developed a new metro line. Parking in the inner city has also been restricted. Today, over 70% of traffic during rush hours is being handled by trains, trams, buses and bicycles.

Bremen (Germany) is one of the first cities to start looking for solutions for regional public transportation issues through the successful development of a city car club and to encourage sustainable mobility. This includes local, regional, national and international frameworks for sustainable development of communities and the economy in an increasingly mobile society.

EMISSIONS FROM SEA TRANSPORT

International shipping is an important source of emissions of sulphur and nitrogen oxides. Unless something is done about these emissions from shipping in the northeastern Atlantic and the North and Baltic seas, it may well amount to almost half of the total EU emissions of sulphur, and more than a third of those of nitrogen oxides (see Table 9.4).

The figures for shipping, high though they are, apply only to ships in international trade. They do not include emissions from ships plying in inland and territorial waters.

In Sweden, a system of differentiated harbour and fairway dues has been introduced. It is designed so as to make it advantageous for

Table 9.4 European emissions of SO_2 and NO_x in 1990 and 2010 (Million tons).

	1990		2010	
	SO_2	NO_x	SO_2	NO_x
EU countries	16.3	13.2	3.6[1]	5.9[1]
Non-EU countries	21.6	10.2	9.9[2]	7.3[2]
International shipping	1.6	2.3	1.6	2.3
Sum total: Europe	39.5	25.7	15.1	15.5

[1,2] Projection
(*Source:* Elvingson, 1999)

shipowners to run their vessels on low-sulphur oil and install equipment for controlling emissions of nitrogen oxides. This Swedish system has already had an appreciable effect, especially in reducing the emissions of sulphur. Estimates show them to have come down by 60% since 1990. This figure, however, includes the emissions from ships whose owners had voluntarily switched over to using low-sulphur oil before the national system started (Elvingson, 1999).

The success of the system of differentiated dues in cutting down emissions may be largely accounted for by the fact that most of the traffic goes along fixed routes. The effect would, however, be greatly improved if similar systems were to be adopted by other countries. The problem is that many countries exact no fairway dues at all.

It has been proposed to remedy this through an EU directive requiring the member and candidate countries to introduce fairway dues that accord with the distance travelled. Using fairway dues only would be better, than trying to differentiate existing harbour dues because fairway dues can be made non-negotiable and completely transparent (Kageson, 1999).

Fairway dues need to be related to the amounts of pollutant emitted. They should be set at such a level as to make it worthwhile for owners to switch to low-sulphur fuel and invest in equipment for thorough cleaning of the exhaust gases from nitrogen oxides—at least in the case of ships plying regularly within the dues-paying area.

It is desirable for other countries with a seaboard to follow Sweden's example by introducing either differentiated fairway or port dues, or both. The more countries that do this, the greater will be the incentive for shipowners to change to a more environmentally friendly mode of operation. One disadvantage in the Swedish system, however, is that it does not take account of the distance travelled and that its cost-effectiveness is partly lessened by various fixed or negotiable discounts (Kageson, 1999).

URBAN AGRICULTURE

Food production inside the city and in suburbs depends on the availability of private and public land that can be converted into small individual kitchen gardens, or larger collective vegetable gardens in schools and factories, producing essentially for the self-consumption of the gardeners and their families. Production of fresh food for the local market can also be envisaged on the condition of being labour-intensive and high-yielding. Urban land is too valuable to be wasted. The nutrients recovered from organic and human wastes produced in the city constitute a valuable potential asset.

In most cities, possibilities exist for greatly expanding urban agriculture (including poultry and intensive fish culture in small ponds).

Urban agriculture in developing countries may be generally described in relation to rural agriculture as having the following characteristics: higher productivity per unit of space, low capital per unit of production, low energy consumption, low marketing cost and special fish and fruit. Though an urban agriculture technology exchange among developing countries undergoing rapid urbanization can be beneficial, this is not practised. Instead, urban agriculture in Asia has been largely pursued in a piecemeal and unco-ordinated manner. Nevertheless, the following commonalities can be seen (Yeung, 1986).

Fish is very important in the food basket of Asian cities. Especially in East and Southeast Asia, fish represents more than half of the animal protein intake. In South Korea and Indonesia, nearly 70% of the animal protein intake comes from fish. Although most of the fish is derived from capture in marine and freshwater, increasingly aquaculture is gaining importance. Aquaculture provides almost 50% of the edible fish in China.

Secondly, a perceptible trend has occurred in many countries over the past three decades, in response to changing food consumption patterns, to divert from traditional food grain production to cash crops and livestock products. This trend has been especially marked in countries that have experienced rapid economic growth, such as South Korea, Japan and Hong Kong. In Shanghai, a 60% income gap between growing vegetables and grain has activated a shift to the higher value crop. Higher yields in Shanghai also permitted diversification.

Thirdly, excepting some Chinese and Indian cities, urban forestry for fruit production is a rare phenomenon in Asia. In Bangalore (India), a large number of roadside trees are grown by the Department of Horticulture. A quarter of these trees bear fruit, and many provide fodder for animals at the same time. The potential of many trees such as Eucalyptus, Subabul (*Leucaena*) and Neem (*Azadirachta indica*) for multiple use in Indian cities is equally feasible for several other Asian cities.

Fourthly, no matter where it is practised in Asia, urban agriculture can be very intensive and highly successful. In several large Chinese cities, well over 85% of the vegetables consumed by the urban population are produced within the bounds of the municipality. Vegetable production, highly structured spatially, has evolved as part of the traditional ecological complex tied to pig breeding, and recycling of night soil and rubbish produced by the urban population for application to vegetable fields (Yeung, 1986).

In Kolkata, India, fisherfolk and scavengers have supported themselves in ecological niches in the extensive wetlands to the east of the city. Here, ecological processes sustain occupations that provide a living for a million people and sustain significant production of vegetables and fish for the metropolitan area. There are about 4,500 hectares of ponds stocked with fish of three main types—common carp, exotic carp and *Tilapia* (Panjwani, 1986). The average annual catch over the last few decades has been about 6,000 tonnes.

Growing fish in ponds receiving biological wastes improves the water quality and increases the ponds' capacity for waste absorption.

There are some 800 hectares of gardens at garbage dump sites. Cauliflower, spinach, radish, pumpkin, cucumber, maize and eggplant are grown here. The rich soil makes it possible to grow several varieties of vegetables simultaneously. An estimated, 25,000 people work with garbage compost at the dump and adjacent areas. The fresh garbage is also prized by private farmers who may arrange to buy cartloads directly from the Corporation refuse trucks.

Fishing and vegetable farming are the lynch pins of production in this resource-conserving environment. Other aspects include duck raising, an activity that is quite compatible with fish farming. Washermen use the ponds for their trade, believing that the waters are particularly good for laundering.

Deprivation and Development : Peri-urban Agriculture

A few centuries ago, the world suffered from acute situations of deprivation as a result of the low productivity of societies and the unfair distribution of wealth. The growth of production and productivity in the last two centuries has changed this picture radically. Now, in a greater proportion, the conditions of participation in production and distribution are the determinants of poverty and not the lack of productivity or production (Gutman, 1986). Take, for example, the case of food. According to the World Bank (1980), at the global level, if incomes were distributed differently, present output of grain alone could supply every man, woman and child with more than 3,000 calories and 65 grams of protein per day —far more than required. Eliminating malnutrition would require redirecting only about 2% of the world's grain output to the mouths that need it (see also Sen, 1982).

For a long time, modern development has been associated with an increasing division of labour. Economists believed that producing only a few things in large quantities is the way to take advantage of the increases in productivity found in specialization and large scale. Modern cities are both the outcome and source of the division of labour. Thus, urban development was the opposite of the traditional rural milieu, in which semi-isolated units produced mostly for self-consumption (Gutman, 1986).

The industrial revolution generated much efficiency in material production. It led to the belief that productivity is only achievable through specialization and large-scale production characteristic of modern capitalist and socialist societies.

This model of growth and specialization has been increasingly challenged in the last two decades. Several 'counter models' have emerged as fairly useful frameworks for new urban initiatives that attempt to overcome the trend of growth and specialization of the modern city. Three of them are: (1) The treatment and management of the city as an ecosystem; (2) The self-reliant city; and (3) self-production, despecialization and partial delinking from market circuits (Gutman, 1986).

The treatment and management of the city as an ecosystem involves an ecological interpretation of the city, stressing the importance of mastering the energy flows, the life cycles and the interest in transforming the unidirectional flows (farm—city—sink) to circular flows through recycling and reduction of losses. This alone can reduce the inputs requirements. It also searches for increasing complementarity, linking resources and needs and looking for satisfiers with less material and energy inputs (Gutman, 1986).

Urban and peri-urban food production fits perfectly in this framework. Indeed, it is now seen as a central piece of a strategy of food self-reliance for many cities. Hong Kong is said to produce 40% of its vegetable consumption; Singapore 80% of its poultry demand; Shanghai its entire vegetable consumption (Gutman, 1986).

Not long ago, peri-urban commercial agriculture was the chief source of most fresh food consumed in the cities. This pattern has been changing over the last two centuries. Many rural areas were converted into cities. Transport systems and food conservation techniques improved. New and highly productive natural regions were colonized, remote from consumption centres. Urban diets and consumption patterns also became more varied, demanding exotic products or traditional products all year round (both making it necessary to depend on distant ecosystems) (Gutman, 1986).

Recently, the role of peri-urban agriculture has been re-evaluated for several reasons: less energy consumption in transport and conservation; urban employment opportunities; possibilities for a richer and more varied interaction between natural and urban spaces and, last but not least, as a measure to control the urban growth (Gutman, 1986). Urban gardens for self-consumption constitute a positive strategy, when combined with other actions.

Myths about Urban Agriculture

Some common myths about urban agriculture (UA) are listed below:

1. *UA is a marginal activity or means of survival.* The contribution of urban agriculture is greatest in the worlds' poorer cities where the share of income spent by the low-income population on food and fuel is the largest household expense. UA is also a major urban economic sector that supplies some of the food consumed by a city and generates some income and jobs for women. It provides opportunities for the large number of part-time and low-skilled workers.
2. *It preempts higher land uses and cannot pay full land rent.* This is untrue because urban farming utilizes land that is unused and unsuitable for other purposes, or it makes usufruct use of land allocated for other uses, thus returning extra land rents. Most cities have a large area of such land that can be farmed. Indeed, some urban farming activities (e.g. peri-urban poultry coops) actually pay competitive land rent.
3. *UA is less efficient than rural farming.* According to this myth, UA adversely affects the incomes of rural farmers. But, in fact, urban farming thrives on products that are less suited to rural production or that might otherwise be too costly for many urban poor. By contributing to disposable urban income, it can help increase rural agricultural exports, while simultaneously reducing some of the pressure on marginal rural lands (UNDP, 1996).
4. *UA causes pollution and damages the environment.* Urban farming can cause pollution of the urban soil, water and air and affects open spaces adversely. But by providing guidance, it can be made a safer industry for farmers, consumers and the environment. Correctly practised, UA has many more potential environmental gains than problems. UA is one of the best, most sensible ways to dispose of much of the city's solid and liquid organic wastes by transforming them into a resource. Few activities contribute as efficiently to improving the urban environment while closing the urban open-loop ecological system of resources in, wastes out (UNDP, 1996).

5. *The 'garden city' is an archaic, Utopian concept that cannot be created today.* Many cities of developing countries are becoming garden cities. Meanwhile, concepts of modernity are actually holding back agriculture by defining industry as the activity for urban areas and farming as the activity for rural areas. Planning concepts of 'city beautiful' relegate farming to the position of an outdated, backward activity that is not fit for the modern city. In reality agriculture has an important and beneficial place in the contemporary city (UNDP, 1996).

CONSEQUENCES OF URBANIZATION

The world today is faced both by the massive degradation of the natural environment and by the accelerating decline in the quality of life of many of those who live in the built environment of cities. The two crises are related. Urbanization greatly contributes to the global environmental changes that threaten the very existence of life in the future and changes in the biosphere increasingly affect health and social conditions in the cities.

The urban environment—where the living conditions of hundreds of millions of people, particularly in the developing countries—creates problems that adversely affect their health, cause misery and have potentially catastrophic social consequences. The extent and implications of these problems are not so well recognized as are the long-term changes in the natural environment—global climatic change (especially the greenhouse effect); the destruction of rain forests; the pollution of seas, lakes and rivers; acid rain; and the extinction of plant and animal species. Yet the urban crisis has been destroying lives, health and social values (WHO, 1991). The crisis particularly affects all large cities in the world. The effects are felt most acutely by the poor. The enormous and increasing gap between human needs in urban areas and society's capacity to meet them has left millions of people with inadequate incomes, housing and services. Changing patterns of illness, accidents, crime and other forms of social pathology reflect the inability to manage urban change humanely and without destructive effects on the environment and on the availability of resources for sustainable development.

Without a healthy population, sound development is simply not possible. In the developed countries with falling population growth rates, the important factors affecting health and the environment are those associated with technological changes, increased consumption of energy and other resources, changing residential and transportation patterns, and the resulting decline in the environment's capacity to absorb wastes, so that many cities in such countries need to extend and replace collapsing infrastructures and to control pollution more effectively.

The following are the environmental changes having a major impact on health in developing countries:

1. urban population growth,
2. changes in the spatial distribution of the population, associated with the increased use of land in previously unsettled ecosystems, and the occupation of urban land subject to landslides, floods and other natural hazards;
3. overcrowding, congestion and high traffic flow;
4. the ever-growing numbers of people living in extreme poverty, many of them—especially women and children—exposed to high health and social risks;
5. increasing pollution of air, water and land as a result of industrialization, transportation, energy production and the increasing generation and improper disposal of commercial and domestic wastes; and
6. the increasing inadequacy of the financial and administrative resources of cities to meet the need for proper water supplies and sanitation, make suitable employment and housing available, manage wastes, impose environmental controls, and provide health and social services (WHO, 1991).

Further, large-scale changes in the local, regional and global environment also affect the health and living conditions of urban populations, even while urbanization itself contributes to these changes.

Environmental degradation, together with the social degradation resulting from widespread poverty, profoundly affects health. The resulting loss of human potential obstructs development and increases misery and social unrest which is aggravated in some countries by stagnant economies and a heavy burden of external debt.

Socioeconomic Development, Urbanization and Health

The health of a city's people is largely determined by physical, social economic, political and cultural factors in the urban environment. These can include the processes of social aggregation, migration, modernization and industrialization, and the conditions of urban living, which may vary with climate, terrain, population density, housing, income distribution and transport systems. The various factors interact synergistically with one another. Disregarding these factors can severely limit the impact of direct interventions by the health sector. For instance, deaths prevented by immunization may be offset by mortality resulting from poor nutrition or sanitation or from pathogens for which there are no vaccines. As many respiratory diseases are caused by air pollution, health can be improved by controlling traffic and vehicular emissions, the siting of industrial plants and production practices, but this will not affect pollution caused by the facilities used for domestic heating and cooking.

In general, the city's compactness, the interdependence of its populations, and economies of scale tend to optimize the factors that protect and promote health. But environmental factors often do not operate in this way; some types of deterioration of the urban environment threaten the health of all socioeconomic classes, not just of the poor people.

Factors in the physical environment that affect health include water supply, domestic and community sanitation, standing water, vector populations, pollution from industrial and domestic wastes, working conditions, housing quality and availability, use of chemicals, food and food safety, radiation, noise and the extent of gardens, parks and open spaces. The impact of these factors varies with the natural ecosystem in which the city is located (climate, altitude, terrain, vegetation and water resources), the character of social and technological development, urban-rural interactions, and the absorptive capacity of the region's air, water and land; high population densities aggravate the impact of both natural and man-made disasters. The impact on health also varies with socioeconomic conditions. Proper operation and maintenance of water-supply and sanitation systems become critically important with increased population density, as does effective vector control to reduce the incidence of communicable disease in slums and periurban settlements and the health risks to the general population (Harpham, 1986).

The physical environment conditions—and is conditioned by—the social environment. Social and cultural factors that affect health are low income, limited education, inadeqaute diet, overcrowding, poor hygienic practices, and social instability and insecurity. Poverty is the most serious threat to health both in cities and villages.

A few examples of economic conditions that adversely affect health in cities in developing countries are the following:

1. Structural adjustments to national and local economies that, *inter alia* involve unplanned and uncontrolled industrialization which, in turn, produces harmful economic, environmental and social side-effects including rising prices, declining real incomes, job losses and reduced public expenditures.
2. Limited prospects for the rapid revitalization of stagnant economies, with the major burden being borne by the poor through high consumer costs.
3. Disruption and degradation of rural areas, sometimes as the result of economic development projects, which tend to induce people to

migrate to cities, where they may be at increased risks themselves and may increase the risks to others.
4. Failure of social and economic development to keep pace with the natural population increase and rural-urban migration, so that environmental controls aimed at protecting health become more and more weak leading to deterioration of the infrastructure and more uncontrolled pollution.
5. Inadequate access to stable employment and income for many urban dwellers, so that families are unable to obtain adequate diet, housing and services.
6. Gross inadequacy and maldistribution of resources, services, and the economic status, creating an underclass that must perforce survive through the informal economy and whose members may be 'illegals' with respect to their tenure of land and shelter, access to utilities and services, and eligibility for employment and education.
7. Inefficient economic and financial management of cities, with failure to allocate available resources correctly.
8. Poor planning and control of land and water resources, industrial development, transportation, and working conditions, to the detriment of health and the standard of living, resulting, for instance, in the need to satisfy the demand for water by obtaining supplies from distant sources and by treating polluted water (WHO, 1991).

Impact of Urban Development on Health

Generally, many aspects of life in cities are quite favourable to improved health. Yet the odds against good health are often greater for city-dwellers who are severely exposed to the hazards of malnutrition, inadequate shelter, poor sanitation, pollution, poor transportation to and from the workplace, and various other kinds of stress.

Communicable diseases flourish where the environment fails to provide barriers against pathogens and the risks are increased by overcrowding, the importation of pathogens to which people are not resistant, and increases in vector populations resulting from rapidly changing settlement patterns and the disruption of ecological relationships. Some environmental conditions conducive to the spread of communicable disease include lack of an adequate and safe water supply, insanitary disposal of excreta, inadequate disposal of solid wastes, absence of or inefficient drainage of surface waters, inadequate personal and domestic hygiene and structurally inadequate housing (WHO, 1991).

There are the following five notable patterns of communication that lead to an increased incidence of parasitic diseases in many cities in the developing world.

1. Infected people enter non-endemic areas;
2. Infected people enter non-endemic areas where vectors are present;
3. Non-infected people enter endemic areas,
4. Natural foci of zoonotic disease undergo urbanization or domestication; and
5. Vectors and parasites enter non-endemic areas, through agricultural products and the belongings of migrants.

Noncommunicable diseases and injuries are associated with exposures to toxins and hazards that are often intensified by the urban living conditions. Structural hazards in shelter, transport systems and workplaces increase the incidence of noncommunicable disease, accidental deaths and injuries from poisonings and burns.

People living in many cities of north India are at considerable risk from bad roads (many of which are virtual death traps) and from badly designed and maintained stairways, windows, heating and cooling devices, floors, and escape routes from fire and explosions. Use of unsafe building materials, such as lead-based paints, asbestos, creosote and synthetic materials that give off toxic fumes is also hazardous. Another serious hazard is traffic systems.

The incidence of many respiratory diseases can be significantly reduced by effective ventilation for the removal of such air pollutants as nitrogen oxides, carbon monoxide, radon, formaldehyde, tobacco smoke, ozone, mineral fibres and sulfur dioxide. Apart from the hazards of cigarette smoking, complex air pollutants are produced inside the home by the burning of biomass fuels in open fireplaces, which exposes to risk some 500 million people in developing countries; the risks are increased by inadequate ventilation.

High-rise buildings, built in order to make the best possible use of expensive land in densely populated urban areas, may present special hazards to physical and mental health.

The location of settlements in relation to areas used for industrial and domestic waste disposal is extremely important in protecting people against air, water and soil pollution. In this respect periurban squatters fare badly, as industrial plants and dumps are likely to be located at the city's periphery. Squatters are also exposed to flooding, mudslides and other natural hazards when they are forced to settle on unsuitable sites such as hillsides and flood plains (WHO, 1991).

Management of Urban Development and Environmental Health

With continuing urbanization, tremendous social, economic and environmental problems are being created with profound effects on people's health and well being. The need for effective responses to these problems is increasing commensurately. Urban change challenges the capacity of a community to make such responses. This challenge affects all levels of government and social organization, because economic and environmental systems form an integral whole. Health development is inseparable from urban development.

In many advanced countries, there is a continuous struggle to resolve conflicts between development and the protection of the environment, to organize community action and to solve concrete problems of finance and technology. The situation is more serious in the cities of developing countries, where the infrastructure is either very weak or even worsening, where change is more rapid, the problems are less well defined, the pressure of need is more intense and the capacity to respond to them is more limited (see Hanlon, 1969).

Since decisions about urban development affect the various economic and social interests differently, urban development is, politically, a problem in trade-offs between competing demands and objectives, which often tends to discourage the taking of decisions for the long term.

Preventing pollution, minimizing the environmental impact of agriculture and industry and otherwise protecting people and ecosystems from damage requires scientific knowledge, specialized engineering technology, and design skills as well as planning. The need for environmental management increases as population density increases and with the introduction of more advanced and dynamic modes of production and distribution.

The application and effectiveness of environmental management technologies have usually failed to keep pace with the growth of environmental problems. Neither science nor politics has been geared to anticipating the untoward side-effects of new technologies. Recycling, process modification and waste treatment have been essentially a reaction to such side-effects. Governments have been unwilling to restrict economic development by requiring that, before they are introduced or new facilities built, new technologies and products must be shown to be safe and not to have harmful effects on the environment. Environmental assessment technology has developed over the last three decades mainly through applications in industrialized countries. Waste management technologies should give due attention to ecological, economic, political, cultural and behavioural considerations, in addition to the technical problems. Lack of suitable sites may preclude the burial of solid wastes, while incineration increases air pollution. For preventing the pollution of land and water by toxic and radioactive wastes, such wastes must be collected, contained, transported and stored at remote sites that may not be physically available in small countries or politically available when the population concerned objects to this; some developed countries have tried to solve their problems by exporting such wastes to poor countries.

The goal of optimizing well-being among urban dwellers, including the reduction of the incidence of communicable, chronic, traumatic and psychosocial diseases and disabilities, can be achieved largely through certain changes in the physical and social environments of cities, which call for the following four types of action:

1. The prevention, whenever possible, of environmental health hazards.
2. Where hazards cannot be avoided, the reduction or elimination of their ill-effects.
3. The creation of healthy surroundings (good housing, recreational facilities, etc.), especially in densely populated areas.
4. The creation of urban and regional infrastructures that reduce risks and vulnerability and support sustainable health and urban development, while at the same time establishing a better balanced relationship between cities, their immediate environs, and more remote hinterlands.

According to WHO (1991), health is better protected and promoted in the physical environment by making provision for:

1. Nutritious and safe food, primary and higher levels of health care, public facilities and areas of unspoilt natural beauty.
2. Housing, surroundings and infrastructures that adequately protect people against: (a) the pathogens and vectors of communicable disease, including adequate, safe water supplies, excreta disposal and domestic sanitation; (b) avoidable injuries and poisonings; (c) toxins implicated in chronic diseases; and (d) threats to personal and psychological security so that the domestic environment fosters personal and family growth and development.
3. The reduction of exposures to various pollutants and their health effects in the work and community environment, with special reference to the control of ubiquitous air, water and land pollutants and the hazards of toxic wastes and ionizing radiation.
4. Environmental safeguards against the pathogens and vectors of parasitic diseases, control of vector breeding in standing water, waste dumps and domestic refuse.
5. Open spaces to meet the need for recreation, rest and aesthetic satisfaction, thereby contributing to the creation of a favourable urban climate.
6. Protection against the hazards of work and transportation; the latter should provide safe, pleasant and comfortable access to facilities.

Likewise, in the social environment, health can be better protected and promoted where:

1. The population is educated in the measures necessary to protect personal and community health and to minimize environmental degradation.
2. An effective system of incentives and sanctions is in place for responsible social and economic behaviour.
3. Public policies support strategies for meeting the basic minimum needs for a decent standard of life for the entire population.
4. Groups and individuals in the community are enabled and empowered to take action to improve their living conditions.
5. Access is provided to health and social services, including emergency services.
6. Appropriate technology is applied to meet community health needs.
7. The resources needed to increase protection against environmental hazards to health throughout the community are provided, and all sectors, including the health authorities, work together to that end.
8. Opportunities are available to participate in community development, to foster social cohesion, to improve the built and natural environment and to enjoy urban life (Oberai, 1989; WHO, 1987; WHO, 1991).

HUMANE URBAN DEVELOPMENT

Humane urban development requires that greater use is made of the productive capacity of cities, while at the same time, the health of their people is protected and promoted. Cities have a key role in the development process. It is in urban areas that people, resources and economic activities are most strongly interconnected. Yet cities in many developing countries are often unhealthy, unproductive, inefficient and inequitable.

For deciding a good policy framework for urban development, key criteria for assessing important environmental problems are: (1) health impacts; (2) economic impacts; (3) the reversibility (or otherwise) of environmental deterioration or damage to ecosystems; and (4) whether resource-use patterns can be sustained. The use of such criteria helps in determining priorities and in planning longer-term actions.

Urban policies are influenced by macro-economic policies and pricing structures. Therefore, administrators need to have a much better awareness of the economic role of cities and the basic needs of their citizens.

Protecting and promoting health is an implicit objective of society in both rural and urban settings. It cannot be attained without the preservation and improvement of the urban environment. All activities, whether of individuals or organizations, make positive or negative contributions to the state of the community environment, and many both directly and indirectly affect the health of others, as well as that of the individual and the family.

A society's health objectives, therefore, can be attained by the optimization of the environmental and behavioural determinants of health affected by such activities. This optimization should take account of the many trade-offs involved in meeting health needs, e.g. between pollution costs and the need for people to have an income adequate to provide them with food and shelter, between competing uses for land and public funds, and between capital accumulation and current consumption.

Besides protecting health, it is highly desirable to promote healthy lifestyles and improved performance of social and economic roles. The urban environment should provide equitable access to recreational and cultural facilities, 'friendly' structures and surroundings (including safe and healthy buildings and open spaces), and activities that counter tendencies to alienation, violence and crime, especially among young people. The social environment should help people to avoid isolation and enhance a sense of belonging to, and participating in, a community. Community support of child care, particularly in situations in which women must work and traditional family structures have broken down, is important in child development and protection, as well as in enhancing the well-being of families (WHO, 1991).

URBANIZATION AND SUSTAINABLE DEVELOPMENT IN DEVELOPING COUNTRIES

The twentieth century may be regarded to have been the century of the urban transition, the last phase of which is taking place in the developing countries. The magnitude of this transformation of the developing world— the sheer number of people involved—has no precedent in history. However, only very recently, after a few decades of rapid urban growth in the developing countries, has the international community begun to grapple with the social and economic development policy implications of this transition.

The urban transition represents a profound economic transformation produced by the processes of modernization and development, of which the recent dramatic demographic shifts are only the mere superficial manifestations. In 1950, less than 300 million people lived in towns and cities in the developing countries. By 1985, this number had increased to 1.1 billion. Increasing at a rate of 50 million a year, the urban population of the developing countries is expected to reach some 2 billion by about

2000, when two-thirds of the world's urban population will be living in the developing countries of Africa, Asia, Latin America, the Caribbean and Oceania. These trends are irreversible. Contrary to popular belief, the growth of urbanization and city population cannot be explained only in terms of migration from rural to urban areas. The image of cities swamped by rural migrants is undeniably a powerful one, but it constitutes an inadequate basis for explaining what is happening. Only in a few developing regions is migration the dominant element of urban growth. For most developing countries as early as the 1960s, natural increase accounted for about 60% of the growth of urban areas, and this percentage has been increasing ever since.

While cities have significantly contributed to the output of developing countries, their physical and living environment is rapidly deteriorating. This deterioration manifests itself in a number of ways—growth of slums and squatter settlements, inceasing traffic congestion, air and water pollution, deteriorating infrastructure and shortfalls in service delivery, etc. These problems not only reduce productivity but also undermine social stability and public health.

Concerted attempts are called for to improve existing urban-management practices and procedures, particularly in key areas such as financial and land management, infrastructure operations and maintenance and service delivery.

Urbanization means an escalating demand for shelter. The poor people should be enabled to construct and improve their own shelters. This move would support a strategy of urban economic growth, particularly in the informal sector. It can also generate employment and promote not only social but also economic development.

Environmental considerations will have to be taken into account if urban growth over the coming decades is to be truly sustainable. Urban management and planning practices must attempt to minimize resource depletion in the urban-expansion process and ensure that environmental resources are used wisely. Urban policies should be developed in the developing countries to avoid the public health risks that accompany pollution of all types.

ENVIRONMENTAL PROBLEMS

Since times immermorial, people have been unconsciously or intentionally acting as the controllers, regulators, operators and modifiers of their environment through their activities and, in turn, have been influenced by the environment. To some extent, the advances in science and technology and the intensive industrial development along with excessive utilization of resources have brought about the current global environmental crisis.

In the early days of human evolution, human activities sometimes fouled only the local habitats to a small extent. But these foulings now have extended to the whole world society. Man, through his own activities, has jeopardized the ecological system upon which his existence and survival basically depend and has been responsible for much environmental degradation. Natural hazards such as landslides, erosion, droughts and floods, desertification, atmospheric pollution and cultural hazards like deforestation, overpopulation, poverty, slums, noise and crimes are the manifestations of the growing ecological imbalances.

Reckless plunder of natural resources by humans has scarred the face of the earth's surface and increased the area under desert and wasteland.

Man's assaults upon the environment have contaminated the air, earth and water with toxic materials (see Fig. 9.7). Natural vegetation is being destroyed and cut down and land is being spoiled. Soil erosion is chiefly

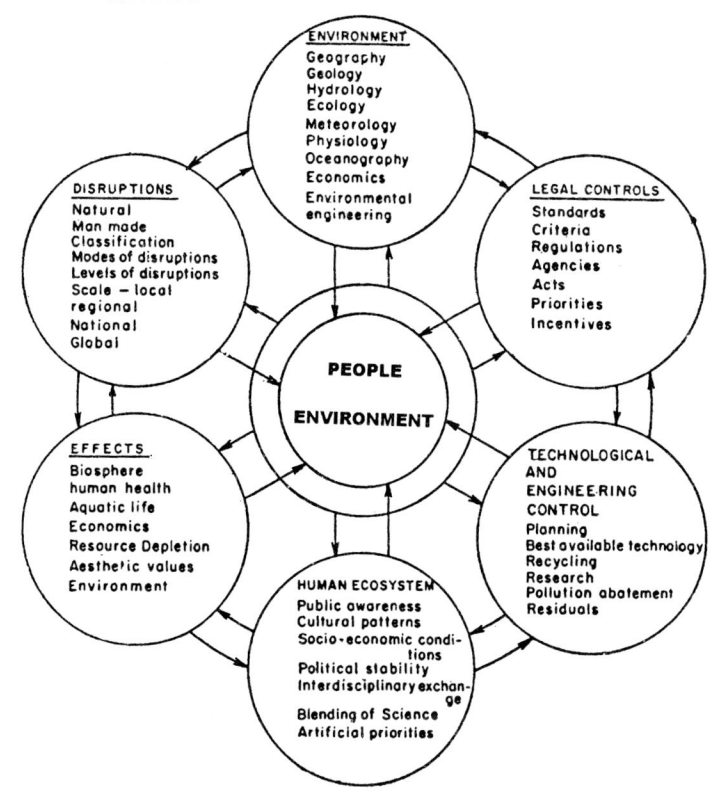

Fig. 9.7   Various facets of environmental pollution and control measures (after Lohani, 1984).

being caused by the activities of man, such as alteration in vegetation and soil conditions and faulty landuse practices. Soil erosion has assumed alarming proportions in India. Every year, the country is losing 6,000 million tonnes of top soil, taking with it more than 6.0 million tonnes of valuable nutrients (Lohani, 1984).

The world is now facing several environmental problems of different scales. Some of the environmental problems are very serious and likely to be increasing. The most serious problems in cities of the developing world, for instance, are: a lack of piped water systems both for homes and businesses; inadequate sanitation and the disposal of solid and liquid wastes; and governments which are either unable or unwilling to penalize polluters and provide basic services to poorer groups. These problems are responsible for millions of preventable deaths every year. Third World citizens find it difficult to share Western concerns regarding global warming—questions of survival 20 or more years into the future do not interest those whose survival is at stake today (see Schteingart, 1988).

There is evidence that the most pressing environmental problems can be greatly reduced at a relatively modest cost. It is not necessary to divert resources from rural to urban areas, but a much higher priority should be given to those environmental problems which impact most on the health and living conditions of the poorest citizens. Many city-based environmental problems are linked to rural problems and therefore the two should be considered together.

An 'environmental problem' may be taken to mean either an inadequate supply of a resource essential to human health or urban production (for instance, sufficient fresh water) or the presence of pathogens or toxic substances in the human environment which can damage human health or physical resources such as forests, fisheries or agricultural land.

Environmental problems and their consequences can be considered at five different geographic scales:

1. Within or around home and the ill-health, disablement and premature death arising from contaminated water and inadequate sanitation;
2. Within the wider neighbourhood or district environment, and the diseases and accidents which result from lack of drainage, bad roads and no garbage collection;
3. The city environment, with rising quantities of toxic wastes for which no special provision is made for their disposal and rising levels of air and water pollution;
4. The region surrounding the city and the rural-urban interlinks between the two, with a widening circle of damage caused by city-based consumption and city-generated wastes;

5. The global level, and the contribution of city-based production and consumption to climatic change, especially global warming.

INDOOR ENVIRONMENT

A large proportion of the third world's urban population lives and works in very poor conditions. Despite the many different forms of housing used by poorer groups, almost all have two environmental problems. The first is the presence in the human environment of pathogens because of a lack of basic infrastructure and services—no sewers, drains or services to collect and dispose of solid and liquid wastes. The second is crowded, cramped living conditions.

Many health problems are linked to the quantity and the quality of water available, the ease with which it can be obtained and the provisions made for its disposal after use. Millions of urban dwellers have no alternative except to use contaminated water—or at least water of uncertain quality.

Removing and safely disposing of excreta and waste water is also an important health requirement. The absence of drains or sewers to take away waste water and rainwater can lead to waterlogged soil and stagnant pools which transmit diseases like hookworm. Pools of standing water act as breeding grounds for mosquitoes which spread filaria, malaria and other diseases. Inadequate or no drainage often means damp walls and damp living environments. Around two-thirds of the Third World's population have no hygienic means of disposing of excreta and an even greater number lack adequate means to dispose of water wastes.

Removing and disposing of excreta in ways which prevent human contact is central to reducing the burden of disease. There is overwhelming evidence that the economic burden of disease and ill health, which is in large part the result of deficiencies in both water supply and sanitation, is very great in the Third World—particularly for the poor. Around one-tenth of each person's productive time is sacrificed to disease in most Third World countries.

The crowded houses used by poorer sections are conducive to high incidence of diseases such as tuberculosis, influenza and meningitis, which are transmitted from one person to another—their spread is aided by low resistance among the malnourished inhabitants.

Environmental problems that occur in the workplace are also a major problem in many third world cities. They can be seen in workplaces from large factories and commercial institutions down to small. Among the hazards are dangerous concentrations of toxic chemicals and dust, inadequate lighting, ventilation and space, and inadequate protection of workers from machinery and noise.

Some one third of those working in asbestos factories in Mumbai suffer from asbestosis while many of those working in cotton mills suffer from byssinosis (brown lung) (see CSE, 1983). A study of workers in the Mumbai Gas Company found that 24% were suffering from chronic bronchitis, tuberculosis and emphysema (see CSE, 1983).

NEIGHBOURHOOD ENVIRONMENT

In the absence of sewers to remove excreta and waste water, open drains carry these wastes from houses and pose health dangers for the neighbourhood. Three problems are important within the neighbourhood environment—dangerous sites, no collection of household garbage and inadequate site infrastructure.

CITY ENVIRONMENT

In the past, it used to be assumed that environmental problems such as air and water pollution are less pressing in the third world than in the West for two reasons: (1) a smaller proportion of the population lives in cities; and (2) the third world is less industrialized. Some major rural and agricultural environmental problems are deforestation, soil erosion, water pollution and deaths and disablements from biocide use (and over-use). These may seem more urgent even if some of these have important links with cities. Thus, the great appetite of industries and cities for water (and its subsequent pollution and disposal) and problems of air pollution and solid waste disposal may perhaps be less of a problem in developing countries than in the more urbanized and industrialized West.

However, the above general picture conceals the fact that there are hundreds of third world cities with high concentrations of industries. China, India, Mexico and the Republic of Korea figure prominently amongst the world's largest producers of many industrial goods. Third world cities or city-regions with high concentrations of heavy industries suffer comparable industrial pollution problems to those in Europe, Japan and North America. Indeed, in some cases, problems may be even more serious. Industrial production has increased very rapidly in many third world countries in the last four decades in the absence of an effective planning and regulation system. Very few governments have shown much interest in controlling industrial pollution.

Another reason why pollution can be particularly serious is the concentration of industry in relatively few locations or industrial areas. No doubt, some governments have supported a decentralization of industry away from the largest cities, yet many new industrial plants have been set up outside the main city but still within or close to its metropolitan area.

Citywise environmental problems also arise from some activities other than industrial pollution. Very congested traffic and inefficient and poorly maintained engines in most road vehicles contribute greatly to air pollution. Many households use inefficient heaters and cookers (especially those using solid fuel).

Certain Southeast Asian cities have long periods in the year with very little wind to help disperse air pollution; with increasing numbers of automobiles and an abundance of sunshine, photochemical smog has become an increasing problem.

Toxic/hazardous Wastes

All over the world now, one comes across the familiar list of pollution problems — heavy metals (e.g. lead, mercury, cadmium and chromium), oxides of nitrogen and sulphur, petroleum hydrocarbons, particulate matter, polychlorinated biphenyls (PCBs), cyanide and arsenic, as well as various organic solvents and asbestos. Some of these and certain other industrial and institutional wastes are termed 'hazardous' or 'toxic' because of the special care needed for storage and disposal, so they are isolated from contact with people and stored in ways which prevent them from contaminating the environment. Most come from chemical industries although others such as primary and fabricated metal and petroleum industries, pulp and paper industries, transport and electrical equipment industries and leather and tanning industries also produce significant quantities of hazardous wastes (UNCHS, 1989) (see Table 9.5).

Water Pollution

As regards impact on human health in Third World cities, the dangers from most toxic industrial wastes are probably more localized and more open to swift and effective government control than those from other industrial pollutants. There are usually four different sources of water pollution : sewage, industrial effluents, storm and urban run-off and agricultural runoff. This last type is often an 'urban' problem too since water sources from which an urban centre draws may be polluted with agricultural run-off and contain dangerous levels of toxic chemicals from fertilizers and biocides.

Along the River Ganga alone, 114 cities each with 50,000 or more inhabitants dump untreated sewage into the river each day. DDT factories, tanneries, paper and pulp mills, petrochemical and fertilizer complexes, rubber factories and a host of others use the river to get rid of their wastes (see Kumar, 1995).

Similarly, Hooghly estuary is choked with the untreated industrial wastes from more than 150 major factories around Calcutta. Raw sewage pours into the river continuously from 361 outfalls.

Table 9.5 Examples of toxic chemicals, their use and their potential health impacts (after UNCHS, 1989)

Chemical	Uses	Health Impacts
Arsenic	Pesticides/some medicines, glass	Toxic/dermatitis/muscular paralysis/damage to liver and kidney/possibly carcinogenic and teratogenic
Asbestos	Roofing insulation/air-conditioning conduits/plastics/fibre/paper	Carcinogenic to workers and even family members
Benzene	Manufacture of many chemicals/gasoline	Leukaemia/chromosomal damage in exposed workers
Beryllium	Aerospace industry/ceramic parts/household appliances	Fatal lung disease/lung and heart toxicity
Cadmium	Electroplating/plastics/pigments/some fertilizers	Kidney damage/emphysema/possibly carcinogenic, teratogenic and mutagenic
Chromates	Tanning/pigments/corrosion inhibitor/fungicides	Skin ulcers/kidney inflammation/possibly carcinogenic/toxic to fish
Lead	Pipes/some batteries/paints/printing/plastics/gasoline additive	Intoxicant/neurotoxin/affects blood system
Mercury	Chloralkali cells/fungicides/pharmaceuticals	Damage to nervous system/kidney damage
PCBs	Electric transformers/insulators in electric equipment	Possibly carcinogenic/nerve, skin and liver damage
Sulphur Dioxide	Sugar/bleeding agent/emissions from coal	Irritation to eyes and respiratory system/damage to plants and buildings
Vinyl Chloride	Plastics/organic compound synthesis	Systemically toxic/carcinogenic

Air Pollution

Four major sources of air pollution are: industry, fuels for heating and electricity generation, solid-waste disposal and motor-vehicles. The scale of the problem and the relative contribution of the different sources varies greatly from city to city.

High levels of respiratory diseases reflect high levels of air pollution within a city although in the case of airborne lead, it shows up in blood tests. If the incidences of, say cancer and respiratory diseases are drawn on a city map, this often pin-points certain areas where residents suffer most from air pollution — directly linked to one industry or one industrial complex.

Many cities have unacceptably high concentrations of air pollutants. For example, Kuala Lumpur has a concentration of suspended particulates in the air 29 times the desirable goal recommended by the Malaysian Environmental Quality Standards Committee. In Cairo, the prevailing winds blow north or south; one brings toxic fumes into the city from the lead and zinc smelters in Shubra al Khaymay; when the winds shift, they bring poisonous pollutants from steel and cement factories in the south of Helwan, an area noted for its many dead trees (Khalifa and Mohieddin, 1988).

In many Indian cities, open fires or inefficient solid fuel stoves are also major contributors to air pollution. Domestic fuel burning probably generates about one-third of Delhi's air pollution.

REGIONAL IMPACTS AND RURAL-URBAN INTERACTIONS

Cities are the major centres of production and consumption. They demand a very high input of resources such as water, fossil fuels and other goods and materials. The more populous and spread out the city and the richer its inhabitants, the larger its demand on resources is likely to be and the larger the land area from which these are drawn. Cities are also major centres for resource degradation. Water, after being used in various ways, is returned to aquatic bodies at a much lower quality than that originally supplied. The solid wastes collected from city households and businesses are usually dumped in land sites around the city. Much of the uncollected solid waste finds its way into water bodies and pollutes them. All these effects exemplify regional impacts.

The demand for rural resources from city-based enterprises and households pre-empts their use by rural households. Some common land that was previously used for gathering wild produce and grazing has been taken over by monoculture tree plantations. The high demand for fuelwood from cities can contribute to deforestation (and the soil erosion which usually accompanies it).

Deforestation is usually considered as a rural problem but it is often intimately linked to the demand for fuelwood or charcoal from city inhabitants and enterprises. Electricity demand in cities can also cause problems for rural areas. The environmental impacts of large hydroelectric dams are usually rural even if most of the electricity will be consumed in urban areas.

Complex population movements (migrations) in many countries link rural areas, towns and cities. Such movements have a major impact on the population growth and demographic structure of cities and so on their environment. Migration to cities is promoted by high population densities in rural areas, shortage of cultivable land, declining soil fertility, increasing commercialization of agriculture and agricultural markets, inequitable land ownership patterns and exploitative landlord-tenant relations as well as government support for cash crops. The livelihood of rural artisans has frequently been destroyed or damaged by the increasing availability of mass-produced goods made or distributed by city-based enterprises.

Three kinds of regional-level impact by cities are noteworthy: uncontrolled physical expansion; solid- and liquid-waste disposal; and air pollution.

Liquid wastes from urban activities produce environmental impacts that extend beyond the immediate hinterland. Fisheries can be damaged or destroyed by liquid effluents from city-based industries with hundreds or even thousands of people losing their livelihood as a result.

GLOBAL IMPACTS

Some important global issues relate to depleting the atmosphere's ozone layer, to increasing carbon dioxide concentrations in the atmosphere (mainly through fossil fuel combustion and through the contribution to deforestation as centres of fuelwood consumption), and protection of endangered species.

Issues relating to the changes in global climate have come to the top of the agenda of many industrialized nations. One reason for this is that most of the regional, urban, neighbourhood and housing-related problems in the West have been fairly successfully addressed, so therefore the western countries direct attention to the global-level concerns as they consider third world environmental problems. Actually, greater attention should be given to the environmental problems which daily take a high toll on human health in the Third World—inadequate and unsafe water supplies, inadequate sanitation and drainage, high levels of air pollution at home and at work, contaminated rivers, coasts and land sites. Effective action to provide a healthy living environment today should not rank second to encouraging such an environment for the future. It is equally important that governments and environmental groups in the third world should not dismiss the global concerns promoted by the West.

Much of the problem with the changing global climate is the result of overconsumption by Europe, North America and Japan. It is their consumers and businesses which have been responsible for most of the increase in carbon dioxide concentration and ozone depleting chemicals. It is, therefore, fair that the West should compensate the underdeveloped world by providing technical and financial incentives to switch over to cleaner fuels and better energy sources.

SUSTAINABLE DEVELOPMENT AS A TRANSFORMATION PROCESS

Sustainable development is a transformation process—the transformation of economies, institutions, infrastructure, environments and human hearts and minds. Its actual direction and progress depends on how this process of change is implemented at the local level. A variety of tools and methods are being employed by local governments worldwide in their movement toward a sustainable twenty-first century. To make this move effective and meaningful, it is desirable to review the social and environmental legacy of the twentieth century and critically examine local environmental priorities and challenges so as to establish objectives for the future.

The Brundtland Report was in a way too cautious and internally contradictory. It neglected the impact of transnational corporations and ignored the role of women. It proposed that economic growth continue everywhere, negating its own goal of a more just international development. In reality, growth has not benefited the poor. Brundtland implied the further deterioration of the environment and the increase of waste of natural resources. If the concept of sustainable development, which is the core notion of the report, is to be taken seriously, the whole philosophy of economics and the characteristics of cultivation economy need to be reappraised. The role of the unpaid labour, primarily that of women, as well as nonmonetized subsistence and household economies needs to be accepted as an integral part of human economy.

This WCED Report did indeed assess the situation accurately but the recommendations of what should be done, and how, fell far short of the drastic diagnosis, probably due either to political considerations or to a lack of analysis and insight.

The strongest criticism of the report related to its stand concerning economic growth. It proposed continuous economic growth everywhere, in rich as well as poor countries and even made it a prerequisite of sustainable development. This notion is not consistent with existing knowledge about and experience of international development—the growth of the rich has not benefited the poor. In a way, The Brundtland Commission made the same mistake as the report to the Club of Rome, *Limits to Growth*, made in the 1970s. This report failed to distinguish between where economic growth should be halted and where it was still necessary. The WCED report committed the same mistake but in reverse —it did not differentiate industrialized countries from Third World countries, but proposed that economic growth continue everywhere.

The growth of production and consumption of goods in rich countries today and in the future only implies increasing wastage of natural resources and destruction of the environment, not sustainable development. Continuing enrichment of the more advanced and affluent countries and sustainable development are mutually exclusive goals. The damage done by unchecked and indiscriminate economic growth cannot be restored by additional growth.

The Brundtland Report referred in passing to the dangerously rapid consumption of finite resources but did not see any need for radical measures, let alone de-development to frugal conserver lifestyles. Rather, according to Brundtland, only new policies in urban development, industry location and housing design etc. were needed. The Report's essential prescription for the solution of the Third World's problems was more of industrial development, e.g. a 5 to 10 fold increase in manufacturing capacity. The Report did not discuss whether such an increase is really achievable, possible, or necessary, or what it would mean for the environment, resource and energy problems. Above all, it did not state anything about where the increased Third World exports were supposed to be sold in a world of chronic gluts, protection and trade wars with even the richest countries struggling to solve their huge deficit and balance of trade problems by increasing exports (see Trainer, 1990).

According to Trainer, on no issue is the Report as seriously mistaken as on Third World development. It takes for granted the bankrupt 'indiscriminate growth and trickle down approach' (although it might seem to deny this) and its fundamental recommendation is to gear up that approach to faster growth, without realizing that this would amount to accelerating the very mechanism responsible for the disaster that is Third World 'development'.

The Report's solution to many serious problems facing the Third World was economic growth, of at least 3% p.a. What's more, to make this possible, at least 3-4% p.a. growth was also needed in the rich countries. It assumed that such growth is possible in view of resource and environmental constraints and ignored the vast literature condemning the growth strategy as not only having failed to improve the living standards of the Third World's poor majority but also identifying the mechanisms whereby that strategy is in fact the prime cause of the Third World's major problems. Following is a brief summary of the main categories of evidence and argument (Trainer, 1988).

The basic principle in the conventional growth approach is the trickle down mechanism. The aim is to maximize the rate of growth of business turnover, i.e. to increase investment, sales and exports as fast as possible, on the assumption that the more the wealth that is produced, the more wealth there will be to trickle down to enrich the poor. However, the reality is that hardly anything ever trickles down. This approach to development does result in rapid increases in national wealth—but the already rich virtually get most or all of it.

The one-fifth of the world's people who live in rich countries take four-fifth of the world's energy. Resources are flowing into the production of items for relatively high-income earners and especially for export to the rich countries. Similarly, the wrong industries tend to be set up in the Third World. The problem is not lack of development in the Third World, but rather *inappropriate* development. Simple tools, cheap housing, satisfactory roads and clean water are needed, but capital, land and labour are drawn primarily into developing export plantations and baseball and car factories. Inevitably, there is a head-on clash between the growth strategy and an *appropriate* development strategy. The latter would ensure that scarce development resources went into establishing the most needed industries *contrary to market forces*. Many items, such as cars, were not produced, whereas things that would add little or nothing to GNP were developed, e.g. village forests that would enable peasants to have little involvement in the cash economy. The essential principle of appropriate development is the maximization of village economic independence and self-sufficiency (Trainer, 1990).

In the conventional development model, the growth approach by nature *contradicts* equity, social justice and environmental sustainability. If growth that does respect these values is the objective, then the conventional model has to be abandoned in favour of an *appropriate* development model in which the distribution of resources, goods, land and capital is governed by stringent controls derived from considerations of everyday basic needs. The only way to enhance the resource base, to restore the environment and establish justice and equity in the world is by: (a) restricting growth to those few things that we need more of; and (b) cutting back or eliminating the luxurious and wasteful production and consumption taking place in the affluent countries.

Our chances of achieving a just, peaceful and environmentally-sustainable world order depend *not* on growth and affluence but on how soon we can shift from the growth and greed society to a conserver society (Trainer, 1990).

RURAL-URBAN INTERFACES

The industrial revolution in Europe was accompanied by growing urbanization, which led to two viewpoints regarding the rural-urban divide: an anti-urban view and a pro-urban view. The former view idealizes and regrets the disappearance of, rural life. It considers urbanization as a destructive process, leading to the breakdown of social cohesion, lack of good services and the creation of a variety of socio-economic problems, including poverty, crime, vice, disease and environmental pollution (Mutizwa-Mangiza, 1999).

The pro-urban view sees urbanization as a progressive process. It regards it as one of the key forces underlying technological innovation, economic development and socio-political change. In this view cities, being national repositories of scientific and artistic knowledge, are both the loci and agents of innovation. Cities are viewed as agents of innovation diffusion and socio-economic transformation. Proponents of this view believe that the history of scientific and technological innovation, and that of civilization in general, cannot be separated from that of towns and cities. These people believe that urbanization has positive impacts on fertility, mortality and other demographic trends, particularly in the recently urbanizing nations of the developing world. The current resurgence of interest in and attention to urban management, and the view that cities are the engines of national economic growth and of development in general, owes much to this pro-urban perspective.

On the issue of rural-to-urban migration, perceptions are split. To some, policies aimed at reducing rates of rural-to-urban migration should be supported. To others, 'urban-containment' policies designed to curtail rates of rural-to-urban migration have failed, and rapid urbanization should be accepted as inevitable, with no energy wasted on attempting to reduce the growth rates of cities and towns (Mutizwa-Mangiza, 1999).

A third view of rural-urban linkages (or rural-urban continuum) occupies an intermediate position in relation to the above two views. This view has a long history in the study of economics, geography and regional planning.

Rural-urban linkages are essential for economic growth and sustainable development of both rural and urban areas. Urbanization is a process that transforms patterns and styles of living, not only within urban areas, but also within rural areas. Villages and cities are connected physically (by infrastructural networks such as roads and telecom-munications), economically (by exchange of processed and unprocessed commodities, with each area acting as a market for the other), and environmentally (by their joint exploitation of some natural resources, by their competition for land, and by the environmentally polluting consequences of productive activities in both areas).

Urban Markets and Rural Producers

Rural-urban interactions may be divided into two broad categories: 'spatial', which include flows of people, goods, money, information and wastes; and 'sectoral', including 'rural' activities taking place in urban centres (such as urban agriculture) and activities often classified as 'urban' (such as manufacturing and services) taking place in rural areas.

Spatial policies of regional development have attempted to build on the interactions between rural and urban areas. The 'virtuous circle' model of rural-urban connection emphasizes efficient economic linkages and physical infrastructure connecting farmers and other villagers with both domestic and external markets. However, the assumption that proximity to markets and availability of infrastructure would benefit all farmers has turned out to be misguided. In many cases, smallholders within the area of influence of large centres do not benefit from proximity to urban markets, as low incomes make it difficult to invest in cash crops, let alone compensate for their lack of land through intensification of production (Tacoli, 1999).

Peri-urban areas constitute the transition zone between fully urbanized land in cities and areas in predominantly agricultural use. The interactions between cities and countryside are most intense here, with changes in land use and farming systems, changing patterns of labour force participation, social change, changing demands for infrastructure and pressure on natural resource systems to absorb urban-generated waste. This land conversion from agricultural to urban-industrial use is often the cause of conflict between rural and urban sectors, often at the expense of both urban and rural poor people (Tacoli, 1999).

Increasing competition for land in both urban and peri-urban areas also affects 'sectoral' interactions, e.g. urban agriculture.

Despite growing pressures on the living conditions of large parts of the rural population, there has been a general increase in the number of non-agricultural activities, especially in trade of agricultural produce, agricultural tools and inputs and consumer goods, while services and manufacturing activities have tended to decline. The growth in non-agricultural activities has generally reduced rural poverty but has increased rural income disparities, as large farmers also tend to operate the most profitable non-agricultural activities.

The growth of non-agricultural activities is rapidly changing the structure of rural-urban linkages in many developing countries. In some areas, it has catalyzed rapid growth in the number of small towns which serve as important links in rural-urban interaction.

Local Development

In the developed world, debates among planners are progressively moving away from urban-rural linkages and focussing on the relationship between local development and globalization. Economists are now considering local economies more positively and the crucial role of urban territories is being recognized (Barcelo, 1999).

We are now entering the era of the so-called 'local globalization' of the economy. Local governments are now trying to attract national and international investments.

In many-less developed countries, local development strategies still remain limited to the traditional and archaic urban-rural issue: urban and rural development decision-makers are unfortunately often opposed to each other.

Active urbanization is a crucial phase in the development process of developing countries and successful agricultural development depends on how well urbanization is managed. Urbanization also increases the productivity of the rural sector.

According to Barcelo (1999), the costs of urban growth need to be compared to the benefits of an economic and social transformation based on an urban productivity, which is three to four times greater than rural productivity.

REFERENCES

Barcelo, J. Modern approaches to local development. *Habitat Debate (UNCHS)* 5 (1) : 10-11 (1999).

Boyden, S. Integrated studies of cities considered as ecological systems. In: *Ecology in Practice*. Tycooly Ltd., Dublin/Unesco, Paris (1984).

CSE. (Centre for Science and Environment). *The State of India's Environment — a Citizen's Report*. Delhi (1983).

Elvingson, P. A proposal for wider control. *Acid News* No. 4: 1-4 (1999).

Gutman, P. Feeding the city : potential and limits of self-reliance. *Development* 4: 22-26 (1986).

Hanlon, J.J. An ecologic view of public health. *Amer. J. Public Health* 59: 4-11 (1969).

Kageson, P. Economic instruments for reducing emissions from sea transport. Swedish NGO Secretariat on Acid Rain. Göteborg (1999).

Kalecki, M. *Selected Essays on the Economic Growth and the Mixed Economy*. Cambridge University Press, Cambridge (1972).

Khalifa, A.M., Mohieddin, M.M. Cairo. In Dogon, M., Casada, J.D. (eds.). *The Metropolis Era*. Sage Publications, New Delhi Vol. 2. (1988).

Kraas, F. Managing resources in mega-cities: Water as a bottleneck factor in Bangkok. *Natural Resources and Development* 49/50: 117-127 (1999).

Kraas, F. Megastädte: Urbanisierung der Erde und Probleme der Regierbarkeit von Metropolen in Entwicklungsländern. In Holtz, U. (Hg.): *Probleme der Entwicklungspolitik*. Cicero-Schriftenreihe Bd. 2, 139-178, Bonn (1997).

Kraas, F. Ressourcenmanagement in der Megastadt: Wasser als Engpaßfaktor in Bangkok. In Hoffman, Th. (Hg.) *Wasser in Asien. Elementare Konflikte*. 174-182, Osnabrück (1997a).

Kumar, H.D. *Modern Concepts of Ecology*. 8 ed. Vikas, New Delhi (1995).

Lipton, N. *Why Poor People Stay Poor: Urban Bias in Developing Countries*. Temple Smith, London (1977).

Lohani, N. Environmental impact studies. *Environ. Quality Management* pp. 263-322 (1984).
Mayur, R. Urban crisis in India. *Populi* 10: (1) (1983).
Morris, D. *Self-Reliant Cities, Energy and the Transformation of Urban America.* Sierra Books, San Francisco (1982).
Morris, D. The city as nation. *Development* 4: 72-73 (1986).
Mutizwa-Mangiza, N. Strengthening rural-urban linkages. *Habitat Debate (UNCHS)* 5 (1): 1-6 (1999).
Oberai, A.S. *Problems of urbanisation and growth of large cities in developing countries: a conceptual framework for policy analysis.* International Labour Office, Geneva (1989).
Panjwani, N. Calcutta's backyard: food and jobs from garbage farms. *Development* 4: 29 (1986).
Population Division, UN. Urbanization and city growth. *Populi* 10(2): (1983).
Rabinovitch, J. Placing rural-urban linkages in the development debate. *Habitat Debate* 5(1): 12 (1999).
Sachs, I. Work, food and energy in urban ecodevelopment. *Development: Seeds of Change* 4: 2-11 (1986).
Sachs, I. Ecodéveloppement: concept, application, enjeux. Texte introductif pour l'Atelier de développement au 5é Congrés mondial de sociologie rurale, Mexico (1980b).
Sachs, I. *Les Stratégies de l'Ecodéveloppement.* Editions Economie et Humanisme et les Editions Ouvriéres, Paris (1980a).
Schteingart, M. Mexico City. In Dogon, M., Casada, J.D. (eds.). *The Metropolis Era.* Sage Publications, New Delhi Vol. 2. pp. 286-293 (1988).
Sen A.K. *Choice, Welfare and Measurement.* Blackwell, Oxford (1982).
Tacoli, C. Straddling livelihoods. *Habitat Debate (UNCHS)* 5 (1) : 8-10 (1999).
Trainer, F.E. *Developed to Death: Rethinking Third World Development.* Marshall Pickering, London (1988).
Trainer, F.E. A rejection of the Brundtland Report. *IFDA Dossier* 77: 72-84 (1990).
UNCHS (United Nations Centre for Human Settlements). *Urbanization and Sustainable Development in the Third World: An Unrecognized Global Issue.* UNCHS (*Habitat*), Nairobi (1989).
UNDP. Urban Agriculture : Food, Jobs and Sustainable Cities, United Nations Development Programme, Publication Series for Habitat II, Vol. 1, New York (1996).
UNEP. *UNEP Profile.* United Nations Environ. Prog., Nairobi (1990).
WEBER, M. *La Ville.* Anbier-Montaigne, Paris (1982).
WHO. *Improving environmental health conditions in low-income settlements.* Geneva, World Health Organization, 1987 (WHO Offset Publication, No. 100).
WHO. *Environmental Health in Urban Development.* World Health Organization, Geneva (1991).
World Bank. *World Development Report.* Washington, D.C. (1980).
Yeung, Y. Urban agriculture in Asia. *Development* 4: 27-28 (1986).

Chapter 10
Biodiversity, Biotechnology and Food Security

INTRODUCTION

By 2030, the world is expected to require about 55 grams of animal protein per person for four billion Asians, or they will destroy their own tropical forests to produce it themselves. It will not be possible to ward off disaster for biologically-rich areas unless we continue to raise farm yields (McCalla, 1994; Avery, 1997).

Low-yield farming means burning and plowing tropical forests and driving wild species away from their ecological niches.

The International Rice Research Institute in the Philippines is improving the rice plant to get 30% more yield. Attempts are underway to put another 10% of the plant's energy into the seed head (supported by fewer but larger stalk shoots). Biotechnology techniques are used to increase resistance to pests and diseases. The new rice has been genetically engineered to resist the tungro virus; this is humanity's first success against a major virus. The US Food and Drug Administration has approved a pork growth hormone, which can produce pigs with half as much body fat and 20% more lean meat, using 25% less feed grain per pig. Investments in agriculture have given handsome dividends in productivity gains. As it is universally acknowledged that in the coming decades we will need more food, there is need for further increasing investment in agricultural research.

AGRICULTURE, ETHICS AND EQUITY

Till quite recently, the role of agriculture in food production was considered to be intrinsically good and its overall norms and values were not questioned. However, in the quest for ever-higher productivity needed to feed an increasing world population, exploitation of natural resources has started to cross boundaries of sustainability (see Table 10.1). Progress in science has increased the ability to genetically manipulate plants and animals beyond perceived natural states. The consequences of these developments have started to raise ethical and moral questions about the integrity of nature, about the socioeconomic effects of development, and about access and ownership of knowledge and the natural resource base in an increasingly globalized and competitive world economy (Hardon, 1997).

Table 10.1 Objectives of traditional and modern agriculture (after Hardon, 1997).

Traditional agriculture	Modern agriculture
Sustainability, yield/food security	Profitability, efficiency, income
Maintain resources	Exploit resources
Long term	Short term
Community interest	Self interest

In the new millennium, work on agriculture should focus around a set of values that stem from the following four ethical premises: (1) The scientific ethic which underpins the quality and unique contribution of research in agricultural science. (2) Equity and poverty alleviation, motivated by a moral concern ultimately linked to the value of human life and the dignity of human beings. (3) Ethical use of genetic resources to create new biological products which can transform and improve the lives of people. (4) The ethics of conservation based on the value placed on the total biosphere as the best means to ensure continued progress and maintenance of the quality of human life.

Ethics suggests a set of standards by which a particular group or community decides to regulate its behaviour—to distinguish what is legitimate or acceptable in pursuit of their aims from what is not. Hence, we talk of business ethics, medical ethics or agricultural ethics, etc.

Ethics is broadly defined as the science of morals and human duty. Ethics needs to be rigorously applied to agricultural research, especially genetic improvement and genetic resources conservation and use. Ethical principles need to be identified and, in turn, should translate into guidelines for the benefit of those working with genetic resources.

The need of the hour is to reconcile the scientific ethic, the conservation ethic and the development ethic and adequately guide the choices required to safeguard the resources for the future benefit of humankind and target groups such as the rural poor and women who may be increasingly marginalized from the access to resources and the benefits of their exploitation.

Much of agriculture has changed from a sustainable system to maintain resources in the long term with community interest to a commercial system which exploits the resources in the short term for material gain and self interest. Modern methods also require high investment, but can give high returns with high risk.

Timely steps should be taken to define guidelines for the conservation and use of genetic resources because the loss of any species and even gene could limit our options for the future. Equity and ethics are inseparable. No system of ethics can be accepted that is not fully committed to the attainment of equity. And equity cannot be pursued except on a fully ethical basis.

The greatest responsibility falls on the humans to ensure that whatever they do is not harmful at any level—to themselves, other species or the environment. That ethical responsibility is a strong challenge indeed because it seeks to fulfill the objectives of equity: alleviating poverty and reducing hunger. In the short term, a productivist upsurge based on new agricultural technologies can release millions of people from the bondage of hunger and the fear of famine—as the Green Revolution did. But over the long term, how can people protect the natural resources on which productivity is based, not just for themselves but for future generations and for Mother Earth to whom they owe a primary loyalty? Under no circumstances an agriculture that is not sustainable should be promoted. We need to be as concerned with the ethics of tinkering with nature, the ethics of patenting and the ethics of preserving life's natural systems, as we are with eliminating the inequities of hunger and poverty (Serageldin, 1997). The defining elements of a good ethical vision can be: (1) liberation of the deprived and disadvantaged from hunger and poverty; (2) responsible and creative management of natural resources and (3) wide application of people-centred policies for sustainable development.

Ethics in the modern world cannot be separated from economics. Conservation involves trade-offs and determination of costs and benefits; sustainable use is based on ensuring that the benefits arising from use are greater than the costs; and equitable sharing of benefits requires an explicit economic assessment of winners and losers. Some important economic principles that need to be incorporated in ethical guidelines may include equitable sharing of benefits, sustainable use, incentives and subsidies, externalities, opportunity costs (see McNeely, 1997) and indigenous knowledge and traditions. The scale of enquiry is another important consideration. The agricultural researcher is often faced with ethical difficulties, even conflicts, because the research results can give unpredictable differential benefits to various groups at various scales. Understanding the competing interests and values of such groups requires an understanding of social processes as well as that of how ecosystems provide benefits to different interests. Research may need to become more decentralized and better adapted to specific local conditions; it should ensure operational transparency, build on a variety of basic concepts, and be acceptable to a broad political base (McNeely, 1997).

Indigenous communities may be using resources the world needs to conserve, but in a sustainable way. Protected areas need to allow access to those communities that depend on the resources for their livelihood (see Kumar, 1999). In terms of scale, biotechnology is beyond the level of understanding of most farmers. Different ethics may be needed to guide choices and actions at different levels of scale. The focus should be on a process by which ethics can be addressed on a variety of scales. Some guiding principles may be:

1. follow the CBD;
2. go with diversity—expand species;
3. work at system, not at crop level;
4. work to build adaptability into agroecosystems. The code of ethics should recognize the dynamism inherent in today's world, seeking ways to make feedback systems effective; and
5. give particular value to cross-disciplinary research as a way of ensuring broader representation of ethical viewpoints. For example, aspects of law, economics and rural sociology should be built into agricultural research.

Conservation of genetic resources is a critical component of food sustainability in the long term. For contributing to conservation, focus should be on a process by which the ethical issues that may arise at various scales can be addressed.

BEYOND PARADOX

Progress across the developed world in the past few decades has aroused the hope for a vision fulfilled. Much of the world is now a world of ennobling endeavour, of plenty, of outstanding scientific advances and technological breakthroughs. On the other hand, we are also witnessing conflict, the near-suicide of nations, economic uncertainties, degrading backwardness and poverty. In many parts of the developing world (e.g. Uttar Pradesh and Bihar in India), there is no electric supply available for 8-10 hours everyday and the condition of the roads is the worst in the world. Students are forced to study for their examinations in candle light when there is no power supply. For millions of people on the Indian subcontinent and Africa, the only vision is a persistent, real-life nightmare. It is neither equitable nor ethical for people to succumb to these contradictory tendencies. Yet nothing whatsoever is done to grasp opportunity and control crises—except, of course empty speech making and hypocrisy at which many politicians have become past-masters.

Ironically, majority of the poor and hungry people live in rural areas where food is produced. They are concentrated in poor, slow-growing areas, often in regions with poor agricultural potential. Lack of opportunity and poverty have pushed the rural poor into cities. Yet, a century of migration from the countryside to cities has not reduced the number of poor people living in rural areas.

The concentrated poverty and hunger of rural populations is certainly unethical and inequitable and needs to be eliminated at the earliest possible. This requires a strong focus on agricultural and rural development, especially in slow-growing and food-deficit areas. Transforming agriculture is the key to achieving that goal; it can become the starting point of the ascent from poverty in most of the world's developing regions.

PROTECTING THE ENVIRONMENT

To neglect the environmental aspects of development is to invite new problems. Another set of inequities, deriving from an imbalance between societal needs and nature's protection, can dominate the development agenda and could even overshadow and neutralize the progress achieved. The integration of the environmental dimension as a component of development is a necessity that cannot be overemphasized.

Conservation of biodiversity is a very important aspect of the environmental component of development. Genetic resources for food and agriculture, including those which underpin global forest and aquatic harvests, lie at the heart of the poverty, food security and environment nexus (Serageldin, 1997; Kumar, 1999). They are crucial to solving some of the most pressing issues confronting humanity today—the burgeoning human population, widespread destruction and degradation of the environment, and persistent poverty. The wise and sustainable use of genetic resources is fundamental to meeting these challenges, yet these resources themselves may be more threatened today than at any time in the past.

Global systems relating to conservation and sustainable use of genetic resources are now in a state of transition. Whereas global food production may keep pace with the growth in population and effective demand in the short term, long-term projections are alarming. Soil degradation, pollution and unsustainable use of water, loss of forests and biodiversity, and the possibilities of adverse changes in climate and sea level all emphasize the fact that the loss of every species, and even gene, can potentially limit our options for the future (Serageldin, 1997).

Agriculture exerts strong effects on the global environment in terms of land and water use and may also have some impact on climate change, biodiversity loss and pollution of international waters. There is need to find out what, if any, is the likely long-term effect of a small but cumulative effect over many years.

Not everything that is technically feasible is ethically desirable. Understanding technology's potential and risk involves science and economics. The ethical issues have multifarious complexities. The ethical dimension of depriving the poor of the advantages that biotechnology (with adequate safeguards) can bring must be weighed against ethical concerns about tinkering with nature through biotechnology (Serageldin, 1997).

The concept that informal innovation by tribal and rural women and men in the field of genetic resources conservation and improvement needs recognition and reward has found universal acceptance in the past few years. Women in particular have played a key role in seed and plant selection and preservation. The term Farmers' Rights denotes the rights arising from the past, present and future contributions of farmers in conserving, improving and making available plant genetic resources, particularly those in the centres of origin/diversity. Farm women and men have developed and continue to help develop and maintain genetic diversity. Some farmers consciously select and breed their crops. Ethnobotanists around the world have chronicled the invaluable contributions of tribal and rural people in the conservation and enhancement of genetic diversity in plants (Swaminathan, 1997).

SYSTEMS OF BIODIVERSITY MANAGEMENT

There are three chief types of biodiversity management systems, viz., in situ, in situ/on-farm and ex situ. The systems in place and the personnel involved in operating them in India are depicted in Fig. 10.1. The precise methods of benefit-sharing may have to be tailored to suit the practices and procedures adopted in their conservation and utilization. The area relating to the conservation and use of genetic material in plant and animal breeding and biotechnological enterprises has received the maximum attention.

In the new millennium, most developing countries have no option except to produce more food and agricultural commodities from diminishing per capita arable land and irrigation water resources. They will have to achieve higher production through a vertical growth in productivity than through a horizontal expansion of area. It has now become possible to enhance the yield potential of major food crops through novel genetic

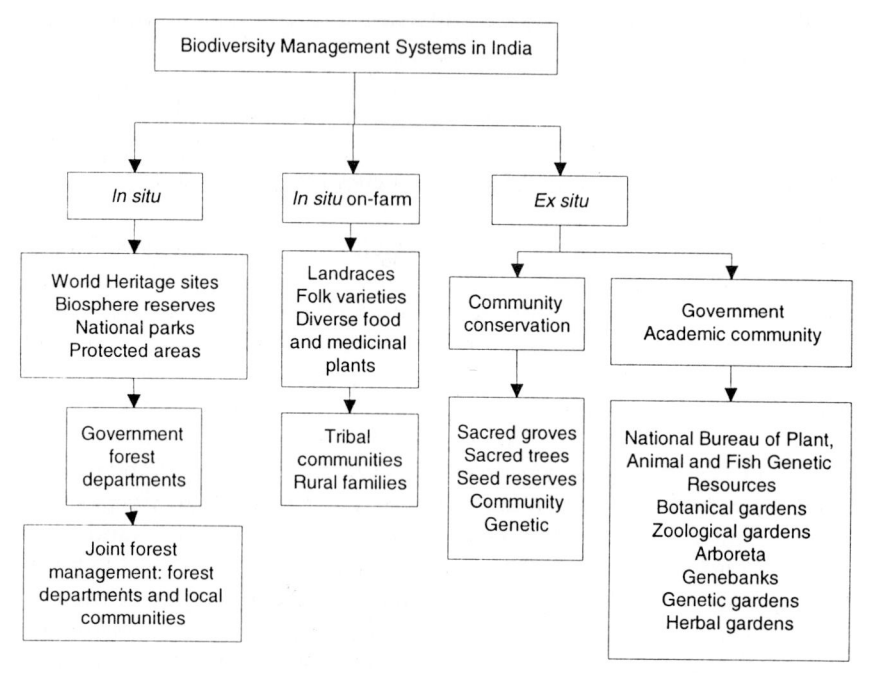

414 Environmental Technology and Biosphere Management

```
                    Biodiversity Management Systems in India
                                    |
          ┌─────────────────────────┼─────────────────────────┐
          ▼                         ▼                         ▼
       In situ               In situ on-farm               Ex situ
          │                         │                         │
          ▼                         ▼                 ┌───────┴───────┐
  World Heritage sites       Landraces                ▼               ▼
  Biosphere reserves         Folk varieties      Community        Government
  National parks             Diverse food       conservation   Academic community
  Protected areas            and medicinal           │               │
          │                   plants                 ▼               ▼
          ▼                      ▼            Sacred groves   National Bureau of Plant,
     Government               Tribal          Sacred trees    Animal and Fish Genetic
      forest               communities        Seed reserves       Resources
    departments           Rural families      Community       Botanical gardens
          │                                    Genetic        Zoological gardens
          ▼                                                     Arboreta
   Joint forest                                                 Genebanks
 management: forest                                          Genetic gardens
 departments and local                                        Herbal gardens
    communities
```

Fig. 10.1 Biodiversity management systems in place and the persons involved in operating them in India (after Swaminathan, 1997).

tools such as recombinant-DNA technologies. Rice illustrates this well. The gradual improvement in yield potential of rice has been achieved through crosses between *indica* and *japonica*, exploitation of hybrid vigour using a male sterile strain originally identified in the Hainan Island of China and, more recently, through distant hybridization among species of *Oryza*. To achieve a new plant type yielding over 10 tonnes of rice, there is need for germplasm from a wide range of countries and agro-ecological regions (Table 10.2).

In a highly unequal world with an environment that presents several socially-tempting opportunities to exploit the 'less equal' party, use and conservation of genetic resources become vulnerable to unethical practice and inequitable consequence—these resources acquire greater value in the 'eyes of the beholder', but not necessarily in the 'eyes of the holder'. Genetic resources, at first, used to be regarded as the common heritage of mankind but later became a common concern; and now they have become a serious matter of national sovereignty. In the meantime, Intellectual Property Rights, Breeders' Rights and Farmers' Rights have become the hotspots of controversy in biodiversity. Farmers' Rights, considered to be desirable are not yet implementable and have already become very politicized and divisive (Castillo, 1997).

Table 10.2 Donors for various traits being used for developing the new plant type in rice (tropical japonicas) (after WCMC, 1992)

Trait	Donors	Country of origin
Short stature	MD2, Sheng-Nung 89-366	Madagascar, China
Low tillering	Merim, Goak, Gendjah, Gempol, Gendjah Wangkal	Indonesia
Large panicles	Daringan, Djawa Serang, Ketan Gubat	Indonesia
Thick stems	Sengkeu, Sipapak, Sirah Bareh	Indonesia
Grain quality	Jhum Paddy, WRC4, Azucena, Turpan4	India, Philippines, Thailand
Donor for resistance		
Bacterial blight	Ketan Lumbu, Laos Gedjah, Tulak Bala	Indonesia
Blast	Moroberekan, Pring, Ketan Aram, Mauni	Indonesia
Tungro	Gundil Kuning, Djawa Serut, Jimburg, Lembang	Indonesia
Green leafhopper	Pulut Cenrana, Pulut Senteus, Tua Dikin	Indonesia

There has been some debate in recent years on whether conservation of biodiversity should focus on protected areas (where humans are excluded) or on sustainable development in areas where people live, a strategy termed community-based conservation. In fact, both protected areas and community areas are complementary strategies, and neither is self-sustaining. Although protected areas are essential for the conservation of some species, they only represent a small fraction of the world's biota. Also, most biodiversity (such as insects and microbes) lies in regions where local communities cannot use it for self-sustaining commercial enterprise. Integrated Conservation and Development Projects (ICDPs) offer a compromise, in which conservation agencies supply aid in return for local stewardship of the biota. However, local communities usually stop compliance once the aid projects are completed, and both governmental and commercial interests often override local authorities (Sinclair et al., 2000).

Past experience with ICDPs precludes the assumption that a legislated protected area will automatically save biodiversity. All such areas are suffering in terms of loss of area, habitats, and species. They become degraded through illegal resource use. Active replacement of lost area is essential if these reserves are to be maintained over the long term. Conservation also needs to focus on solving the underlying social problems. If local communities are to successfully manage protected areas, then outside economic assistance is needed to remove poverty and illiteracy, allocate resources equitably among all community members and stabilize populations in regions surrounding the protected areas. Unlike the philosophy of all current donor programmes for most conservation projects, sustained and flexible outside support will be required. It will be necessary to curb harmful and excessive resource extraction.

ETHICS OF HUMANITY'S COMMON FUTURE

This refers to the global value of biodiversity which accrues to the global community at large rather than being initially channeled through any particular group. Values flow to the society at large rather than to the individual farmer (Swanson, 1996).

The concept of intergenerational equity has emerged because of our preoccupation with sustainability. Because that which is not sustainable will not reach the future, the coming generations will not benefit from it. If genetic resources are renewable and can be maintained, we bring equity to our grandchildren.

The routes to equity are varied, complex and often indirect. However, indirect should not mean neglect. In some cases, the indirect routes have resulted in more equitable but less acknolwedged benefits. Interwoven with ethics and equity are also such issues as transparency, role of intermediaries, and erosion of reciprocity (Castillo, 1997).

LAND DEGRADATION

According to UNEP (1990), every year the world has about 80 million more mouths to feed—and over 20 billion tonnes less topsoil on which to grow food. This global environmental disaster threatens food security in many developing countries.

Soil is a rich but fragile ecosystem. A few grams of good earth contains millions of micro-organisms, which ensure and sustain fertility. A centimetre can take centuries to develop but can be lost forever in a year —blown away by wind, scoured off deforested slopes by rain, sterilized by salts, poisoned by chemicals, bled dry of nutrients, or buried under buildings.

Every year, soil erosion robs Ethiopia of 1.5 million tonnes of grain. In Australia, six tonnes of topsoil erode away for every tonne of produce grown; India loses 6 billion tonnes of soil annually.

As soil and land degrade in arid, semi-arid and dry sub-humid areas, desertification ensues. In the early 1980s, this process affected 4.5 billion hectares of land and over a tenth of the world's population. Every year, over 20 million hectares lose their productive capacity.

The African continent has largely abandoned the traditional patterns of grazing and shifting cultivation, which gave its soil time to recover between crops. As more and more people try to scrape a living from the land, the earth is grazed bare by their herds and trees and shrubs are cut down. Exposed to the rain, wind and sun, precious topsoil is carried away and the remaining soil bakes dry. Soon, it cannot absorb even what little rain falls. Meanwhile, as firewood runs scarce, people are forced to burn dung, robbing the land of the only fertilizer they can afford. In dry areas, productive land turns into desert; in more humid areas it degrades into unproductive wasteland (UNEP, 1990).

Habitat loss and fragmentation are the greatest threats to the world's biodiversity. For unknown reasons, the local extinction of plant species from habitat fragments is quite common. Fragmentation influences both birth- and death-related procsses. The disruption of plant reproduction, especially pollination and seed production, is particularly important (see Bruna, 1999). Seeds of many tropical rainforest plants rarely survive in seed banks for more than a year because they are highly susceptible to fungal pathogens, seed predation and burial under leaf litter. Bruna showed that seeds planted in forest fragments are less likely to germinate than those in continuous forest—a finding that can have negative demographic consequences because it reduces the emergence of seedlings.

If inbred seeds are less likely to germinate, genetic effects could further reduce the recruitment of seedlings into forest fragment populations, explaining why plant populations in habitat fragments often fail to persist in the long term (see Laurance and Bierregaard, 1997).

FOOD SECURITY

Food security exists when all people at all times have access to sufficient food to meet their dietary needs for a productive and healthy life. Food insecurity exists when the availability of nutritionally-adequate and safe foods, or the ability to acquire acceptable foods in socially acceptable ways, is limited or uncertain.

The problem of food insecurity is widespread in the developing world but the total number of undernourished people is unknown and estimates vary widely (see Tables 10.3 to 10.5). For example, estimates for 58 low-income, food-deficit countries range from 576 million people to 1.1 billion people (GAO, 1999).

In November 1996, a world summit was held in Rome, Italy, to address a global commitment to ensure that all people have access to sufficient food to meet their needs, referred to as 'food security'. Participants set a new interim goal of reducing undernourishment by 50% by 2015.

This summit brought together delegates from 185 countries and the European Community to discuss the problem of food insecurity and produced a plan for reducing undernutrition and to assist developing countries to become more self-reliant in meeting their food needs by

Table 10.3 FAO Estimates of Chronically Undernourished People in 93 Developing Countries, 1990-92 (after GAO, 1999)

Country	Number of under nourished (millions)	Number of undernourished as per cent of total number of under nourished for all countries	Cumulative per cent
China	188.9	22.5	22.5
India	184.5	22.0	44.5
Nigeria	42.9	5.1	49.6
Bangladesh	39.4	4.7	54.3
Ethiopia	33.2	4.0	58.3
Indonesia	22.1	2.6	60.9
Pakistan	20.5	2.4	63.4
Vietnam	17.2	2.1	65.4
Zaire	14.9	1.8	67.2
Thailand	14.4	1.7	68.9
Philippines	13.1	1.6	70.5
Afghanistan	12.9	1.5	72.0
Kenya	11.3	1.3	73.4
Peru	10.7	1.3	74.6
Tanzania	10.3	1.2	75.9
Sudan	9.7	1.2	78.2
Brazil	9.7	1.2	78.2
Mozambique	9.6	1.1	79.3
Mexico	7.2	0.9	80.2
Somalia	6.4	0.8	80.9
Subtotal	678.9	80.9	80.9
Second 20 countries[a]	91.0	10.8	91.8
Third 20 countries	49.8	5.9	97.7
Fourth 20 countries	16.6	2.0	99.7
Last 13 countries	2.4	0.3	100.0
Total	838.7	100.0	100.0

Note: Countries are ranked in descending order based on the number of undernourished.
[a] Aggregate number of undernourished people for the next 20 countries with the largest number of undernourished people.

promoting broad-based economic, political, and social reforms at local, national, regional and international levels. It was agreed that achieving food security is largely an economic development problem. A willingness on the part of food-insecure countries to undertake broad-based policy reforms is a key factor affecting whether such countries will actually

Table 10.4 FAO Estimates of Incidence of Chronic Undernourishment in Developing Countries by Regions of the World, 1969-71 to 1994-96 (after GAO, 1999)

Region	Year (3-year averages)	Total population (millions)	Undernourished Percentage of total population	Undernourished Persons (millions)
Sub-Saharan Africa	1969-71	268	40	108
	1979-81	352	41	145
	1990-92	484	40	196
	1994-96	543	39	211
Near East and North Africa	1969-71	182	28	51
	1979-81	239	12	29
	1990-92	325	11	34
	1994-96	360	12	42
East and Southeast Asia	1969-71	1,166	43	506
	1979-81	1,418	29	413
	1990-92	1,688	17	289
	1994-96	1,773	15	258
South Asia	1969-71	711	33	238
	1979-81	892	34	302
	1990-92	1,137	21	237
	1994-96	1,223	21	254
Latin America and the Caribbean	1969-71	279	20	55
	1979-81	354	14	48
	1990-92	440	15	64
	1994-96	470	13	63
Total	1969-71	2,609	37	959
	1979-81	3,259	29	938
	1990-92	4,078	20	822[a]
	1994-96	4,374	19	828

[a] In May 1998, FAO provided revised estimates of the number and percentage of undernourished people by regions of the world for 1969-71, 1979-81, and 1990-92 and for the first time provided estimates for 1994-96. According to FAO, the revised numbers reflect the reestimation of historical population figures by the U.N. Population Division. (As examples of the changes, FAO previously estimated the number of undernourished people for all developing regions at 917 million in 1969-71, 905 million in 1979-81, and 839 million in 1990-92.) However, FAO did not release data on a country-by-country basis for either its revised or new estimates. As a result, other tables in this report that are based on individual country data use FAO's previous estimates.

achieve the summit goal. Other important factors that could affect progress toward achieving the summit goal are: (1) the effects of trade reform; (2) the prevalence of conflict and its effect on food security; (3) the sufficiency of agricultural production; and (4) the availability of food aid and financial resources.

Table 10.5 Distribution of Chronically Undernourished, 1990-92 (after GAO, 1999)

Percent of country's population chronically undernourished	Number of countries	Total number of chronically under nourished (millions)
1-10	18	40
11-20	17	255
21-30	24	267
31-40	15	146
41-50	9	50
51-60	3	9
61-70	5	53
71-73	2	19
Total	93	839

Summit participants approved an action plan that included 7 broadly stated commitments, 27 objectives and 181 specific actions. The plan highlighted the need to reduce poverty and resolve conflicts peacefully. While recognizing that food aid may be a necessary interim approach, the plan encouraged developing countries to become more self-reliant by increasing sustainable agricultural production and their ability to engage in international trade, and by developing or improving social welfare and public works programmes to help address the needs of food-insecure people.

Achieving improved world food security by 2015 is largely an economic development problem. Developing countries will need huge quantum of financial and other resources to achieve the level of develop-ment necessary to cut in half their undernutrition by 2015.

Many developed countries feel that the private sector is a key to resolving the resources problem. Whether the private sector will choose to become more involved in low-income, food-deficit countries may depend on the extent to which developing countries can implement policy reform measures and radically improve their infrastructure. Private sector resources provided to the developing world have grown dramatically during the 1990s, and by 1997, the private sector accounted for about 75% of net resource flows to the developing world, compared to about 34% in 1990. However, according to the OECD, due to a number of factors, most of the poorest countries in the developing world have not benefited much from the trend and will need to rely principally on official development assistance for some time to come (GAO, 1999).

Among factors that may affect whether the goal of food security is realized are trade reforms, conflicts, agricultural production and safety net programs and food aid. It is generally felt that developing countries should increasingly rely on trade liberalization to promote greater food security. The summit plan called for full implementation of the 1994 Uruguay Round Trade Agreements (URA). As trade liberalization may result in some price volatility that could adversely affect the food security situation of poor countries, to help offset these possible adverse effects, a Uruguay Round decision on measures to mitigate possible negative effects needs to be implemented. The URAs have the potential to strengthen global food security by encouraging more efficient food production and a more market-oriented agricultural trading system. Reforms that enable farmers in developing countries to grow and sell more food can help promote increased rural development and improve food security. Trade reforms that increase the competitiveness of developing countries in nonagricultural sectors can also lead to increased income and, in turn, a greater ability to pay for commercial food imports. However, trade reforms may also adversely affect food security, especially during the short-term transitional period, if such reforms result in an increase in the cost of food or a reduced amount of food available to poor and undernourished people. Reforms may also have adverse impacts if they are accompanied by low levels of grain stocks and increased price volatility in world grain markets (GAO, 1999).

As world price and supply fluctuations are of special concern to vulnerable groups in developing countries, food-exporting countries should: (1) act as reliable sources of supplies to their trading partners and give due importance to the food security of low-income, food-deficit importing countries; (2) reduce subsidies on food exports in conformity with the URA; and (3) administer all export-related trade policies and programmes responsibly to avoid disruptions in world food agriculture and export markets.

There is a general feeling that developed countries have been quite slow in removing their trade barriers and that this makes it difficult for developing countries to further liberalize trade. Without an open trading environment and access to developed country markets, developing countries cannot derive full benefit from producing those goods for which they have a comparative advantage. Without improved demand for developing countries' agricultural products, for example, the agricultural growth needed to generate employment and reduce poverty in rural areas cannot be achieved. This is critical to food security. For developing nations to adopt an open-economy agriculture and food policy, they must be assured due access to international markets over the long term (Anonymous, 1997).

Agricultural markets are likely to be more volatile as the levels of world grain reserves are reduced, an outcome to be expected when trade reforms are implemented. Some observers feel that most countries, including food-insecure developing countries, are better off keeping only enough reserves to tide them over until they can obtain increased supplies from international markets (see Tweeten and McClelland, 1997), since it is costly to hold stocks for emergency purposes on a regular basis and other methods might be available for coping with volatile markets. Others believe that ensuring world food security requires maintaining some minimum level of global grain reserves (CWFS, 1998) and that developed countries should establish and hold reserves for this purpose.

The need to increase agricultural production and rural development in the developing world, especially in low-income, food-deficit countries, cannot be over-emphasized.

Perhaps, a very critical factor affecting the realization of this need is the willingness of food-insecure countries to undertake the kind of economic policies that encourage, rather than discourage, domestic production in the agricultural sector and their willingness to open their borders to international trade in agricultural products. An 'enabling environment' that favours domestic investment and production in the agricultural sector is a must in this context.

FUTURE FOOD SUPPLY PROSPECTS

According to certain projections of supplies of commodities (in terms of area and crop yields), equilibrium prices and international trade volumes to the year 2020 made by Evenson (1999), it appears that a 'global food crisis,' as would be manifested in high commodity prices, is unlikely to occur. The same projections show, however, that in many countries, 'local food crisis,' as manifested in low agricultural incomes and associated low food consumption in the face of low food prices, will occur. Delays in the diffusion of modern biotechnology research capabilities to developing countries may probably exacerbate local food crises. Similarly, global climate change will also exacerbate these crises again highlighting the importance of bringing strong research capabilities to developing countries (Evenson, 1999).

'Local' effects can drastically differ from 'global' effects. A 'global food crisis' is not likely to occur in the next 25 years or so. But that does not mean that 'local food crises' will not occur in some places. The projections showed general improvement in local indexes of malnutrition, but even under the most favourable simulations, malnutrition will continue to be a real problem for much of the developing world (Evenson et al., 1999).

Likely global-level effects are summarized in Tables 10.6 and 10.7 which includes the base case 2020/1990 ratios for production, area, trade, and prices by commodity. 2020/1990 price ratios are reported for three policy scenarios plus a 'worst-case' scenario.

Table 10.6 Global base case and policy scenarios by commodity (after Evenson, 1999)

	Base case 2020/1990 ratios					Some policy-price 2020/1990 ratios			
Commodity	Production	Area	Trade	Prices	Delayed industrial-ization	Delayed biotech-nology	Global warming	Worst case
Wheat	1.58	1.06	1.62	.85	.90	.94	.86	1.16
Maize	1.56	1.13	1.36	.77	.81	.90	.78	1.03
Rice	1.66	1.07	1.70	.80	.86	.96	.81	1.17
Other grains	1.48	1.09	1.60	.75	.81	.82	.75	1.04
Soybeans	1.77	1.14	2.20	.90	.91	.92	.91	.98
Roots/tubers	3.28	1.15	1.30	.82	.86	.95	.84	1.09
Beef	1.53	1.35	2.87	.94	1.01	.95	.95	1.31
Pork	1.83	1.53	1.64	.90	1.04	.91	.91	1.38
Mutton	1.98	1.36	1.84	.96	.99	.97	.98	1.13
Poultry	1.80	1.53	3.27	.90	.92	.91	.90	1.01
Eggs	1.92	1.06	5.81	.75	.75	.76	.75	.75
Milk	1.53	1.15	3.60	.93	.93	.93	.93	.93

Table 10.7 Base case for all cereals (for selected countries only) (after Evenson, 1999)

Countries/regions	Growth rates (%) 1993-2020					
	Area/no.	Yield	Production	Demand	Food	Feed
United States	0.12	0.96	1.08	0.81	0.64	0.84
Western Europe	0.04	0.42	0.46	0.39	0.10	0.53
Japan	-0.49	-0.03	-0.52	0.29	-0.05	0.62
Australia	0.12	1.75	1.88	1.01	0.84	1.07
Other developed	0.07	1.09	1.16	1.07	1.16	0.99
Eastern Europe	0.09	1.02	1.11	0.24	-0.19	0.43
Sub-Saharan Africa	1.17	1.67	2.86	2.96	3.00	2.25
India	0.07	1.42	1.49	1.53	1.4	3.04
Pakistan	0.19	1.54	1.73	2.92	2.92	2.96
Bangladesh	0.02	1.36	1.39	1.65	1.65	2.41
Other South Asia	0.12	1.72	1.84	2.73	2.74	2.61
Indonesia	0.09	0.98	1.07	1.44	1.17	3.34
Thailand	-0.07	1.00	0.93	1.39	0.45	2.77
Malaysia	-0.04	1.00	0.95	2.22	1.96	2.48
Philippines	0.10	2.06	2.17	2.26	1.88	3.00
Other Southeast Asia	0.14	2.22	2.37	2.14	2.14	2.07
China	0.02	0.98	1.00	1.32	0.58	3.22
Other East Asia	-0.47	0.84	0.36	1.57	0.67	2.49
South Asia (excluding India)	0.11	1.50	1.61	2.42	2.42	2.88
Southeast Asia	0.08	1.30	1.38	1.61	1.34	2.94
East Asia	0.00	0.98	0.98	1.34	0.59	3.13
Asia	0.05	1.16	1.22	1.51	1.14	3.09
Developed	0.06	0.94	1.00	0.57	0.19	0.71
Developing	0.29	1.20	1.49	1.71	1.43	2.63
World	0.20	1.06	1.27	1.27	1.21	1.40

The base case production scenarios reveal that global crop production will increase by about 60% by 2020. The area planted with crops will expand by roughly 10%. Most production gains will come from yield gains. These yield gains are roughly similar to the post-Green Revolution period gains (Evenson, 1999).

Animal production will increase more than crop production and most of this increase will be caused by increased animal units.

World trade will increase for all commodities, in the form of increased exports by developed countries and increased imports by developing countries. There will be continued declines in world prices for all commodities, being highest for rice and other grains.

Delayed industrialization will mean that prices will be higher than in the base case (by roughly 5%-6%). This is because of reduced private-sector R&D spillovers to agriculture. Delayed biotechnology for developing countries also has significant price effects.

In the worst case, prices of most crops will rise over the 1990 levels but not so high as to become a 'world food crisis'.

GLOBAL FOOD PROSPECTS TO 2025

Despite some fears to the contrary, in recent years there has been continued progress toward better methods of feeding humanity with the sole major exception of Sub-Saharan Africa. Today, roughly half of the world's cropland is devoted to growing cereals. If their direct intake (e.g. as cooked rice or bread) is combined with their indirect consumption, in the form of foods like meat and milk, then cereals account for approximately two-thirds of all human calorie intake. According to Dyson (1999), the continuation of recent cereal yield trends should be sufficient to cope with most of the demographically driven expansion of cereal demand that will occur until the year 2025 (Mitchell et al., 1997). However, because of increasing mismatch between the expansion of regional demand and the potential for supply, there will be a major expansion of world cereal (and noncereal food) trade. Other consequences for global agriculture arising from demographic growth include the need to use water much more efficiently and an even greater dependence on nitrogen fertilizers (e.g. in South Asia). In the coming years, farming everywhere is going to depend more and more on information-intensive agricultural management procedures. Moreover, in spite of continued general progress, there still will be a large number of undernourished people in 2025 (Dyson, 1999).

World population growth has been outpacing cereal production since 1985. But much of the recent decline in world cereal production has occurred in relatively better-off regions. And it does not account for the fact that the regional composition of humanity is changing. In particular, most demographic growth is taking place in parts of the world with low levels of per-capita cereal consumption and, other things being equal, this fact tends to weight downward the average level of world per-capita cereal consumption (and hence production) (Dyson, 1999) (see Fig. 10.2).

According to Dyson (1996), in 2025, the world's farmers will be producing roughly 3 billion tons of cereals to feed the human population of around 8 billion, which will require an average world cereal yield of about 4 metric tons/ha. It is likely that some regions will, for different proximate reasons, experience an increase in their harvested cereal area.

World agriculture will have to use its water supply much more efficiently in the coming decades as well as price it better.

No doubt, we will have new crops and improved seeds. But most of the required increase in the world's harvest will come from the application of

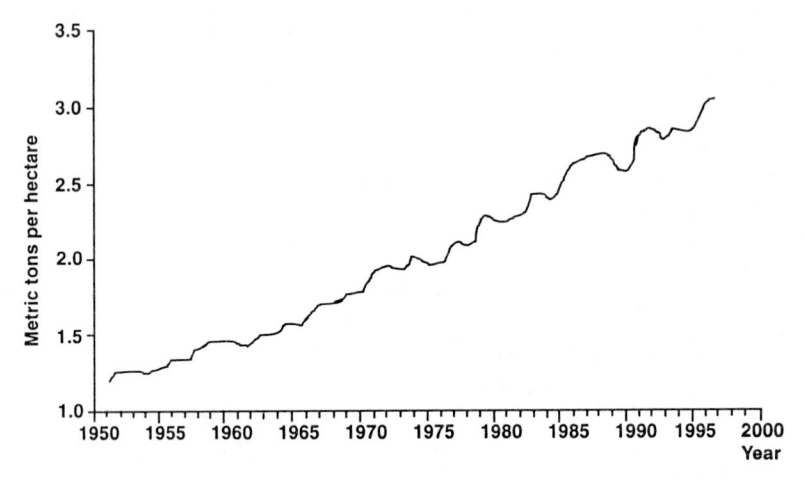

Fig. 10.2 World cereal yield, 1951-1997 (after FAO, 1998).

existing procedures and knowledge to the current world harvested area. Humanity will depend even more on synthetic nitrogen fertilizers for its food supply; there may have to be an approximate doubling of global use of synthetic nitrogen to produce 3 billion tons of grain (Dyson, 1996, 1999). Farming everywhere will involve much greater dependence on information-intensive farm management procedures, as well as greater attention to variation of conditions within individual fields.

Many biotechnology proponents believe that genetic engineering is the key to ending world hunger. They accuse biotechnology opponents of selfishly denying life-saving benefits to developing countries. But in international negotiations, developing countries have demanded protections against uncontrolled trade in these potentially dangerous products. Some countries have raised issues concerning the influx of genetically engineered (GE) foods and seeds and have desired cross-border non-governmental organization (NGO) collaboration.

Both corn and its progenitor teosinte originated in Mexico. For generations, individual farms and small farming collectives in Mexico have bred and preserved corn varieties that are adapted to specific local conditions within the country. Because of the skill with which Mexican farmers have tended these varieties, Mexico is also home to great genetic diversity in cultivated corn.

In the United States, concerns about genetically engineered corn include the possible contamination of conventional foods, changes in soil chemistry, and harm to nontarget organisms. But in Mexico, there are even greater risks since GE corn could contaminate corn's progenitor species or the local varieties developed by farmers over millennia. Such contamination could alter the range of genetic variability available to farmers, and threaten the future food security of millions of people around the world who depend on this crop. Genetic diversity in corn is crucial not only for traditional farmers but also for the industrialized agriculture that produces much of the food in the industrialized countries.

Mexico currently produces genetically engineered cotton, soybean and tomatoes, and has allowed field tests of a variety of genetically engineered crops, including corn and potatoes. Commercial (as opposed to experimental) cultivation of GE corn is not allowed in Mexico at this time, but large quantities of corn are imported from the US for use as food and animal feed; some of the imported varieties were genetically engineered. Many small farmers in Mexico purchase corn to use both as food or fodder and for planting. So they may have unknowingly planted GE corn in their fields—potentially contaminating surrounding fields and wild corn relatives.

BIODIVERSITY, FOOD SECURITY AND INDIGENOUS PEOPLE

Indigenous knowledge is certainly the largest single knowledge base that needs to be mobilized in the development enterprise. It is the practical relevance of indigenous knowledge, and the contribution of indigenous knowledge that can allow a better insight into the reality of sustainability.

The solutions to environmental, political or other problems that ail the world can only come through a patchwork quilt. Each piece must have its own integrity. Finally, the patches are sewn together. But there can be no quilt unless each patch is complete. William Blake, in his book *Jerusalem*, put it very eloquently: 'He who would do good to another must do so in Minute Particulars; General Good is the plea of the scoundrel, hypocrite and flatterer; For Art and Science cannot exist but in minutely organized Particulars.' For those striving to protect the rights of indigenous peoples or farmers, the message is that these rights must have integrity first within these communities before others attempt to codify them within international laws or conventions. The ultimate understanding and implementation of these rights is built from the 'minutely organised particulars' of each community. Those struggling to protect biological diversity and the ecosystem need to draw inspiration from Blake and the quilt-makers. The security of the environment will only be assured if each eco-niche is secured by those who know it best. For those fighting the cause of world food security, the real challenge is not to create new technologies (though this can be very useful)—it is to build food security from the family to the farm to the community to the nation and on to the world. If somebody's proposed solution does not make sense in each patch and particular, then the 'General Good' will only serve the world's scoundrels and charlatans.

If a tree falls in a forest, those who feel it first are the rural poor who depend on that tree in many ways and who nurture biological diversity for their own survival. If a tree falls in the forest, it is not likely to die alone. Genetic erosion is not confined to crops or even forests but also threatens livestock and soils. There has been going on a flow of micro-organisms found in both marine and soil environments from South to North. The value of this genetic material can be enormous and its commercial use around the world is already quite considerable. Beyond micro-organisms, the rate of erosion of livestock breeds—all based upon species originating in the South—makes crop genetic erosion look modest by comparison. According to some estimates, we are losing close to 1% of our rainforests every year, 2% of our crop genetic diversity and 5% of our rare livestock breeds. Ten per cent of our soils have vanished in the past 50 years, and 70% of our coral reefs could be gone in the next 50 years (Mooney, 1997).

The most alarming fact is that human genetic diversity is also at risk and is eroding rapidly. When the culture disappears, so does the hope of ecological agriculture and of genuine food security. Human diversity, too, has become a matter of corporate profit and patent speculation.

The starting point for all work and policy is the poor. If environmental or development proposals fail to stregthen the poor in their quest for community self-reliance, they will ultimately not work.

Biological diversity offers the poor the means of satisfying nine-tenths or more of all their basic survival needs. Only about a half or less of this essential diversity comes from formal cultivation for food or fibre. At least as much is usually protected and even nurtured, by the poor. Safeguarded along the banks of rivers or the borders of fields, in family gardens or in forests, this diversity provides medicines, cleaning agents, flavours, fuel and aesthetic value to rural communities. Even in countries with highly sophisticated cropping systems, as much as one-third of rural nutritional requirements are met by 'uncultivated' species (Mooney, 1997).

There used to be a time when the environmental movement enthusiastically championed the cause of tiger on the belief that, since it was on top of the food chain in the ecosystem, the campaign to save the tiger would mean saving the whole ecosystem, as per the old 'trickle-down' theory. But trickle down has not worked. Indeed, the poor have become tired of getting trickled down upon. We should recognize that rural communities are the species at the top and that strengthening these communities—rather than cutting their bioresources out from under them—may be the best way to protect not only their future but also of humanity at large (Mooney, 1997).

As the global climate changes, as the ozone depletes, as the soils erade into seas and as innumerable unrecorded species become extinct, the importance of the remaining genetic diversity increases exponentially.

Biodiversity is far from being just a raw material. It is, more often than not, the protected and improved genetic resource of rural innovators. And the 'tool box' not only includes the techniques of biotechnology but also the collective wisdom and skills of farm communities. Both biodiversity and rural knowledge have economic value.

The important need for bio- and genetic variability is only now beginning to be appreciated in the socioeconomic context of world food security and national self-sufficiency. Peasant farmers need variability to survive and the rich need it to become richer—and also, increasingly, to survive as well.

Many errors and misconceptions related to agriculture have been unwittingly spread by scientists and researchers during the last few decades. Many of the agricultural success stories were nullified years later, when unexpected long term adverse effects became known. A possibly exaggerated claim surfaced in 1998 in the news media with regard to 'miraculous' gene-engineered high yield seeds, which produce the insecticides and pesticides themselves and thus would alleviate hunger for ever (Hagen, 2000).

Environmental deterioration manifests itself in many ways—for instance, deterioration of nature in general, like soils, climate, and water and the linkage of biodiversity loss with agricultural productivity.

Hagen (2000) has reviewed many aspects of environmental deterioration and loss of biodiversity, limits of sustainable and fair human development, high value of indigenous knoweldge, questionable ecological projects, globalization of the green revolution, sustainable agro-output through high biodiversity, soil protection through root associated fungi, successful biofarming, mini dykes and terraces, soil protection, flood control and irrigation and genetically modified crops. According to Hagen, agriculture without independent and self-responsible indigenous farmers will be ecologically disastrous. The best environment-compatible agriculture and soil protection is achieved through a combination of agriculture with integrated soil conservation measures such as terracing, wind hedges, mixed cropping, mixed farming, local water control, composting and mulching. However, all these activities, commonly known as biofarming, are very labour-intensive. Western style agriculture is capital-intensive, but not labour-intensive. In the USA, less than 5% of the population is engaged in farming, and at very high labour cost. The labour force required for integrated soil protection measures is simply not available. Consequently, the American type of modern farming cannot serve as a model for sustainable agriculture in developing countries, in which still up to 75% of the population is engaged in farming.

The small hill farmers in the Swiss mountains are well aware of this fact and have already taken to biofarming based on ancient indigenous knowledge in order to survive with their small holdings in rough mountain areas. Any sustainable development must be centered on men and women—not on science and technology alone. Rural development is simply not possible without full participation and responsibility of the farmers concerned.

In November 1999, the results of a large and ten-year long project on biodiversity and agricultural output financed by the European Union (EU) in eight countries of the EU were published. The findings proved that the propagated monocultures with loss of biodiversity would have an adverse long term impact on yields. Only high biodiversity, both in natural vegetation and crops, ensures a sustainable agricultural yield. Other similar research projects have been recently completed in Basle (Switzerland) and Guelph (Canada) and show that subsurface root-associated fungi have a markedly favourable effect on biodiversity.

Haverkort and Hiemstra (1999) presented new insights about knowledge of rural people, hoping to stimulate development agencies to take indigenous knowledge seriously. Based on the experiences and insights of some fifteen organizations in drastically ten countries, they went beyond technical knowledge embodied in traditional farming, land use and health practices. They dealt with ancient world views and the role of traditional leaders and drew conclusions about the holistic nature, strengths and also limitations of this knowledge. In various countries in Asia, Africa, Latin America and Europe, some development agencies have been supporting rural people in carrying our practical experiments based on local concepts.

Wolverkamp (1999) appraised 17 case studies from around the world to describe responses to threats and opportunities created by the struggle of local communities and their supporters for: the preservation of the forest; recognition of local rights; and ecological and socio-economic circumstances that impact on local forest conditions and the welfare of communities.

Considering the current impasse in international forest negotiations and the inertia of most governments, it is crucially important to give due importance and support to such local citizens' initiatives. It is desirable to re-establish community control over forest lands and preserve them for the future. The virgin forests of the world owe much to their symbiotic relationship with the indigenous peoples who live in and on the margins of the forests. Such forests, occurrring in Asia, Africa and South America reflect the global nature of the phenomenon.

FROM SEEDS TO GENES (1970s TO 1990s)

It was a popular belief in the 1970s that some Northern governments and corporations were hoarding the South's germplasm, embargoing its exchange and patenting the best material to sell back to the poor farmers who had created the germplasm in the first place.

By the late 1980s, the following basic points had emerged:

- Both the process and structure of decision-making for the global conservation and circulation of plant genetic resources were unclear.
- What Northern facilitators call 'the stakeholders'—the farmers, the plant breeders, the corporations, governments and UN agencies involved—were not sure of their roles or how they related to one another. Consequently, there was some suspicion.
- Ownership was a chief concern. The debate over the science and practice of genetic resource conservation was heavily skewed by legislative initiatives surrounding intellectual property rights.

Most agreed that more money and more science were needed to conserve an extremely important resource (Mooney, 1997).

BIODIVERSITY AND AGROFORESTRY

For a long time, biodiversity used to be preserved by concentrating on species diversity and genetic variability. Later on, higher levels of organization (populations, biocoenoses and ecosystems) were included. Next, the landscape was taken as a graphic and suitable spatial parameter for describing the structural and functional relationship of temporal and spatial patterns, for example, of ecosystem diversity. Today, approaches to the preservation of biological diversity shift back and forth between being utilitarian on the one hand (the premise that diversity is preserved solely through management and use) and the concept of exclusively ecosystem-oriented conservation measures, on the other. This latter strategy aims at protective measures without integrating any anthropogenic influencing factors. The wide range of autochthonous strategies of resource use in many ecologically stable ecosystems, as well as the diversity of cultural landscapes and the associated biocoenoses are also indirectly affected by this (Backes, 1999). Here, cultural and biological diversity are intimately linked with each other. For example, broad sections of the savanna landscapes and grassland communities are viewed as anthropogenically-shaped cultural landscapes, being characterized by extensive grazing and regular fires.

In many villages of developing countries, the adverse effects of current land use on the natural environment are being increasingly studied from the view point of the degradation of natural habitats and the threat faced by them. Since large sections of the tropics are used for crop growing or for pastoral purposes, a protective strategy per se (excluding all human use) quite often conflicts with the subsistence-dominated, traditional-farming methods of the local population. In this context autochthonous land use strategies are quite important for the *in situ* preservation of biological diversity. An evaluation of various forms of farming could emphasize those forms of land use which, as an integral component of intact cultural landscapes, also exemplify cultural and socio-economic practices of how landscape diversity and biological diversity are being preserved (Backes, 1999).

In view of the fact that the population density is increasing in many places, the question is: How intensified forms of agricultural land use can be combined with the preservation of biotic diversity. Population growth, changes in economic practices and the intensification of agriculture are by no means modern phenomena. The issue is: How different population groups dealt with such transformational situations in the past, and how they have considered biodiversity as a resource. The increasingly intensified forms of land use have had a wide range of effects on biotic and landscape diversity, and it would be an oversimplification to assume that this has resulted in a uniformly linear decline in biodiversity (Swift et al., 1997). In order to extend our knowledge of the complexity of current resource management systems, Slikkerveer (1995) considered it prudent to combine retrospective, historically-oriented analysis with anthropological studies and participative research, including links between the current and historically effective factors of intensified land use, on the one hand and floristic diversity, on the other (Backes, 1999).

The contribution made by agroforestry to preserving species diversity can be attributed mostly to the indirect protection of virgin rainforest regions, due to the fact that sustainable methods of production have prevented massive emigration of the population (Sanchez, 1995). There are some convincing arguments in favour of undertaking critical studies into the in-situ diversity of traditional agroforestry systems. The types of use and the management of the agrobiological and natural diversity found in the traditional land use systems, which have suffered increasing pressure in the subtropical region, are considered to be elements of potential solutions; (a) for regenerating degraded regions; (b) for developing a buffer zone strategy in the catchment area of the national parks and biosphere reserves or as an alternative to shifting cultivation. Also, the recognition of systematic correlations between society and an environment shaped along culturally specific lines is being emphasized so also the protection of cultural diversity, the socio-cultural integration and the appropriateness of land-use methods or; (c) the means to prevent the loss of autochthonous agriculture methods, and the agrobiological diversity strongly linked with them. The protection of agrobiological and cultural landscape diversity is intimately connected with the preservation of autuchthonous forms of land and resource use (Backes, 1999).

In the course of mapping the semi-natural vegetation stands outside the farms, Backes (1999) divided the intensity of use into three categories (A, B and C) (Table 10.8). This makes it possible to evaluate anthropogenic impact on the distribution of the individual species and on the species spectrum of the respective formations.

Table 10.8 Assessment of the intensity of human use according to disturbance frequency and disturbance intensity (after Backes, 1999)

Use category	Type of use	Intensity of use
A	Occasional use of individual species; sometimes widespread, extensive use followed by a long-term regeneration phase	Low anthropogenic impact
B	Regular use of individual species or occasional widespread use over relatively long periods of time	Moderate anthropogenic impact
C	Frequent or permanent use of species and/or regular widespread use	Intensive anthropogenic impact

FORESTS AND FARMING SOCIETIES

Forests have always been at the root of most environmentalist causes. Scientific arguments that the conservation of food plants must also mean concern for the wild and weedy relatives of these species, most of which are in forests, have sometimes not been well received by those dedicated to conserving pristine rainforests.

The success of the Green Revolution has generally been attributed to the work of the FAO, but by 1990, an opposing view also emerged—that the FAO was to be blamed on the ground that the Green Revolution had, in fact, contributed to world hunger! Although many in the UN agency would have claimed credit for the spread of high-response seeds around the South, the truth was the FAO had stood on the sidelines watching the Consultative Group for International Agricultural Research (CGIAR) lead the way. Some 70% of the South's annual harvest of wheat and rice and well over 10% of its maize, potatoes and sorghum come from the plant varieties of CGIAR breeders. Overall, the largest and best-run crop gene banks in the world have been built and provisioned by CG scientists. The CGIAR is, by any consideration the most important—if invisible—agricultural agency in the world.

The stated mandate of the CGIAR is to strengthen food security in the South. Some have estimated that there are a billion people eating today, who would go unfed were it not for CGIAR. The Green Revolution has revolutionized agriculture in the South. But it came with tremendous socio-economic and environmental costs. It trampled over the ancient wisdom and knowledge that resided—and still resides to some extent—in the farming communities of the South.

Table 10.9 lists some primary concerns voiced by critics of the CGIAR and of the Green Revolution, which saw the widespread introduction of high-response varieties (HRVs) of seeds. It also sets out the responses and counter-responses to the particular issues raised.

The International Centres that comprise the CG system have been making a strong impact on agricultural development not only in the North but also in the South. For instance, according to IRRI (Manila), the CG's rice centre, its improved germplasm has been effectively used as genetic building blocks for US rice varieties, and that US scientists have received more than 2,800 rice genotypes from the Centre. Researchers have identified 16 US rice varieties based on IRRI material and report similar benefits to Australia as well as Japan. It is impossible to precisely quantify the germplasm and intellectual contribution of the South's farmers to the North's agriculture. But some estimates are available, e.g. for wheat from CIMMYT (the International Maize and Wheat Improvement Centre in Mexico). Research institutions in the USA, Australia, New Zealand and Italy came up with figures that total almost US$ 1.5 billion as the annual contribution of CIMMYT wheat to agriculture in those 4 countries alone. In 1984, barely a third of the US crop incorporated CIMMYT material. By 1994, two-thirds of the American wheat crop used the Centre's germplasm. Included in this is the country's entire source of rust-resistance genes—contributed from Africa. Almost 90 percent of the Australian wheat crop is based on CIMMYT research.

The estimated value to the North of CG-derived wheat, rice, beans and maize is summarized in Table 10.10.

There is no doubt that the North is benefiting substantially from Southern farmers. Together, these crops account for about one-third of all CGIAR research. Extrapolating from the figures available, it becomes evident that the North benefits from CG germplasm by more than US$ 4.8 billion per year for these four crops alone, which account for only 39% of the CG's core research budget and the estimate excludes highly transferable research on livestock (US goat producers have improved milk yield and livestock exports through CG-funded research from Kenya), potatoes, barley, triticale, soybean, groundnut, forage grasses, legumes and fisheries.

As regards the CG contribution to the South, forty million hectares of Southern lands are sown to CIMMYT wheat material, for example, which represents 56% of all developing country wheat lands (excluding China).

Table 10.9 The ongoing debate between Civil Society Organizations (CSOs) and CGIAR (after Mooney, 1997).

First Critique by CSOs	Belated response by CGIAR	Comeback by CSOs
Big farmer bias? Fixed costs mean that large-scale farmers adopt high-response varieties (HRVs) faster than small-holders. Big farmers then gain unassailable market advantage that drives small farmers off the land.	But studies show that smallholders are catching up and their participation is roughly equal to that of larger farmers. Smallholders also sow a larger proportion of their land to HRVs than their big neighbours in order to equalise fixed costs.	But smallholders tend to be on less suitable lands (soil and slope), meaning that HRVs cannot perform as well. If they are increasing the ratio sown to HRVs, they are taking risks in order to minimize fixed costs.
High-yielding for all? HRVs are bred to need fertilisers and irrigation to increase yield.	No, HRVs equal or out-perform folkseeds under virtually all conditions. However, fertiliser bias may have been overdone.	But HRVs mine the soil if poor farmers cannot provide fertiliser or irrigation so yields are not sustainable.
Gender bias? Failure by CGIAR to examine the issue.	What's gender?	To the extent that HRVs leas to the capitalization and mechanization of tasks traditionally done by women, women are marginalized.
Nutrition? Emphasis on a handful of base crops has taken research away from poor people's crops that are often more nutritious.	IARCs have shifted from focus on wheat, rice and maize to about 25 crops crops offering the world well over 90% of its food requirements.	But, the spread of HRVs has taken land from home gardens and minor but key nutrient crops important to poor families.
Disease-resistance? HRVs rely heavily on pesticides rather than breeding to combat disease and pests.	Yes, but yield maintenance is now a major focus in breeding programmes. Some IARCs have major successes to report which have been beneficial to small farmers.	But IARCs consistently over-estimate chemical requirements and have endangered farmers by recommending high use of class I and II chemicals which are most dangerous.
Vulnerable groups? HRVs push families into the market economy, forcing sales of food needed at home. Women and children tend to	Studies show that new surpluses on the farm reduce intra-family food competition, helping pregnant and lactating	There is growing evidence that folk varieties offered a wider range of nutrients than HRVs. Studies of benefits to vulnerable

Contd.

Table 10.9 Contd.

be the first to suffer.	women and infants more than anyone else.	groups are weak. More information is needed.
Multi-purpose crops? Folk varieties were bred to meet a multitude of needs not met by HRVs.	A larger, more stable and less expensive food supply is most important to the poor.	Semi-dwarfs mean straw can no longer be used as fuel or for households. Some varieties take longer to cook, so fuel requirement goes up and forests come down. Dung has to be used for fuel rather than fertiliser. Special-purpose seeds for medicine, ceremony or certain recipes or lost.
National research? Green Revolution is an external input that truncates national research and perverts national priorities.	IARCs have trained well over 60,000 scientists. ISNAR (international Service for National Agricultural Research) was created specially was created specifically to support national programmes. NARS (national agricultural	But host countries often let IARCs do the work and both the type and focus of research is biased by IARCs. NARS expansion was probably due to other factors.

Table 10.10 Estimates of value to the North for selected CGIAR crops (US$, million) (Source: Mooney, 1997)

Crop	Known data	The North's total extrapolation
Wheat	1,436.5 (4 countries)	4,063.5
Rice	126 (USA)	655
Beans	60 (USA)	111
Maize	20 (USA)	29
Total		4,858.5

Roughly, North and South benefit about equally from the CG in the case of CIMMYT's wheat.

IRRI and CIAT (Internat. Centre for Tropical Agriculture) rice varieties are harvested on close to 70% of developing country paddies. Here, the balance is definitely towards the South, as rice is far more important in the South.

The point at issue is that the commercial value siphoned Northward is not acknowledged and not compensated. The situation worsens massively when governments of industrialized countries allow the patenting of material wholly or partially derived from farmers' varieties, freely given and held in trust in the CGIAR Centres. As private companies move into the South's seed markets, farmers risk paying for the end product of their own genius. In that context, the monopolistic Northern companies take freely given germplasm from the South to win patent monopolies themselves. Cooperative innovators—including indigenous and other rural societies—deserve to improve their farming systems, rather than undermining and replacing them.

In recent years, it has become abundantly clear that the North believes food security to be a matter of trade and investment, not a matter of domestic self-sufficiency. This is an extremely dangerous attitude. Food security must be built and maintained on the foundations of rural communities propelled by the wisdom of indigenous peoples. The tragedy is that these farmers and their communities are gradually disappearing. It is important to encourage a seed conservation strategy supported by farmer-curators, as has been practised in China since times immemorial.

For half a century, the multilateral community has been approaching food security from the wrong angle—as though it were an item of mechanical manufacture; as if it only required external intellectual and material input. More than half of the world's people continue to live in rural areas closely concerned with food production. About 20% of the rural population is composed of indigenous peoples with vast innovative experience and an intimate knowledge of foods and ecosystems. For half a century, we have neglected to work with the millions of researchers with their millions of field laboratories and billions of annual experiments that should be the starting point of food security.

In the light of a half-century of failed adventurism in development theory and destroyed wisdom, the recent resurgence of interest in food security needs to be welcomed with caution. The old notion of Food Security has the merit of building other securities around a centrepoint in the daily needs and environmental realities of local communities. Local cultures, food, health and the environment are quite often virtually indistinguishable. But the focus is food—staple foods. Medicinal plants are mostly non-staple foods, used less frequently for the same purpose. The availability of both staple and non-staple foods depends on the health and diversity of the community ecosystem. There are complex inter-relationships in fields, forests and waters between cultivated species and 'partner species' or 'wild species'. This latter term 'wild species' is certainly erroneous and misleading as it connotes a license to expropriate resources with impunity. These partner species must be nurtured by the community. Most of humanity's knowledge is highly specialized for the survival of the community in a specific, ever-changing ecosystem. The traditional Northern approach to rural and community development has ignored this reality. It has divided (food, health, etc.) and destroyed (local knowledge through literacy compaigns).

MODERN FARMING AND FORESTS

Some recent researches have raised questions about sustainable economic growth. Sustainable growth has been a concept dear to both environmentalists and development specialists. Both camps have generally assumed that improving agricultural practices in the developing world should help relieve pressure to cut down nearby forests. But the factual equation has turned out to be more muddled: In Brazil, for example, a new strain of soybeans planted by farmers wound up accelerating the destruction of the tropical forest, while in the Philippines an irrigation project protected a tropical forest elsewhere on the same island (Helmuth, 1999).

It appears that the key factors determining how the new technologies affect the labour market and migration, are whether the crops are sold locally or globally, and how profitable farming is at the boundary between cultivated land and forest.

Before the mid-1980s, most conservationists tended to be antigrowth. More recent thinking, in early 1990s, suggested that economic development and environmental conservation could be complementary: as farmers earned more from their existing plots—thanks to better irrigation, new crops, investment in tools, and easier access to markets—they would become less and less motivated to clear marginal land.

But it was recently observed that growth and conservation are only sometimes compatible.

One important variable is: how much labour an agricultural system requires. Brazilian soybean cultivation is highly mechanized and large plantations of a new strain that thrives in the tropics have displaced small southern Brazilian farmers who had cultivated grains, vegetables and coffee. These farmers were forced to the agricultural frontier, where they cleared forests to eke out a living (Helmuth, 1999).

In contrast, projects that create employment can relieve deforestation pressure. An irrigation project in the Philippines has drawn wage labourers to newly-created rice fields in the lowlands and reduced pressure to cultivate forested areas. Creating opportunities elsewhere can pull people away from a forest—or even from farming itself.

Some people have also assumed that easier access to markets makes farmers' crops more profitable, thus allowing them to farm less land and spare the forest. But if one can produce twice as much, it makes more sense to produce more on more land.

The simple assumption that has guided many development projects in the past was that poverty is the cause of deforestation. But it seems that big plantation projects are more likely to contribute directly to deforestation than are small farmers. The challenges are to foresee how specific development strategies will impact a region's environment—displacing workers or making forest-clearing profitable, for example—and to identify projects that achieve both economic and ecological objectives (Helmuth, 1999).

Ultimately, it appears that high-tech farming in the tropics should reduce the over-all amount of land devoted to agriculture. But that may or may not be relevant for saving the forest today.

FOOD, FARMING AND BIODIVERSITY

According to Avery (1997), unless the growing world food demand is met with improved farm productivity, much undeveloped land with its biodiversity will be lost to agriculture. Contrary to common wisdom, saving the environment and reducing population growth are likely to happen only if governments strongly support high-yielding crops and advanced farming methods, including the use of fertilizers and pesticides.

The gravest risk facing the world's wildlife may neither be pesticides nor population growth but the potential loss of its habitat. Conversion of natural areas into farmland is the major impact of people on the natural environment. It seriously threatens biodiversity. The vast majority of the known species extinctions have occurred because of habitat loss.

In many industrialized countries, farms now occupy less and less of their land. Worldwide, however, the opposite is true. According to the World Bank, cities take only 1.5% of earth's land whereas farms occupy 36%. As world population rises to around 8.5 billion in 2040, it will become even more clear how much food needs govern the world's land use. Unless we strengthen our efforts to produce high-yielding crops, a plow-down of much of the world's remaining forests for low-yield crops and livestock might become inevitable.

By and large most, environmentalists seem unaware of how crucial the green revolution has been in preventing famine and preserving biodiversity.

The Green Revolution was very crucial in forestalling famine and simultaneously saving the environment. By maximizing land use, its high-yield crops and farming techniques proved vital in preserving wildlife. By effectively tripling world crop yields since 1960, they saved an additional 10 million square miles of wild lands (Avery, 1997).

The green revolution, however, received a setback after the publication of Rachel Carson's *Silent Spring* in 1962, which cautioned that modern farming can kill wildlife, endanger children's health and poison the topsoil. The organic gardening and farming movements gained popularity. However, organic and natural farming techniques and the crop varieties they favour require large amounts of relatively fertile land supporting small numbers of people, and they can be highly destructive to soil and forests, degrading biodiversity quickly and irrevocably. Slash-and-burn agriculture is perhaps the most harmful to the environment.

In many parts of the developing world, hunger confronts the population every day. Finding new ways to meet our global need for food, while maintaining ecological balance, might be the greatest challenge we face in the next century.

We all share the same planet and the same needs. Biotechnology advances can prove potentially valuable for agriculture—healthier, more abundant food; less expensive crops, reduced reliance on pesticides and fossil fuels; also a cleaner environment.

To feed the world in the present century, we will need more food that is more affordable than it is today. With more productivity needed from less tillable land, new ways will be needed to yield more from what is left—after development and erosion have taken their toll. To strengthen our economies, we need to grow our own food as independently as we can. Agricultural biotechnology will play a major role in realizing the hope we all share. Accepting this science can make a dramatic difference in millions of lives because securing food for our future begins a better life for us all.

Use of pesticides became popular to increase foodgrain production. Farmers started using pesticides because this improved crop yield. In India, pesticides have been used mostly on cotton, plantation crops and vegetables. But there is need to reduce the use of pesticides. This need can be met by developing GMOs (genetically modified organisms). It is not possible to rule out the use of pesticides in India if the growing food demand in the country is to be met.

It is confusing to link the issue of food security with GMOs. Very few GM crops mostly, cotton, soybean and tomato, are being grown at present in the world. Not many GMOs have reached the field level. Therefore, GMOs are not likely to change the situation of foodgrains in India for another decade or two.

Saving the Soil

Since times immemorial, soil erosion has been the biggest problem affecting farming sustainability. Modern high-yield farming changes that situation greatly. Tripling the yields on the best cropland automatically cuts soil erosion per ton of food produced by about two-thirds.

Conservation tillage and no-till farming are making a big difference. The former leaves crop residues into the top few inches of soil, creating millions of tiny dams against wind and water erosion. Besides saving topsoil, conservation tillage produces many more earthworms and subsoil bacteria than any plough-based system. No-till farming involves no plowing at all. The soil is never exposed to the elements. The seeds are planted through a cover crop that has been killed by herbicides. The use of these systems can cut soil erosion per acre by 65 to 95%. These powerful conservation farming systems are already being used on hundreds of millions of acres in the United States, Canada, Australia, Brazil and Argentina. They have proved successful in Asia and tested successfully in Africa (Avery, 1997).

According to Avery, the model farm of the future will use still more powerful seeds, conservation tillage and integrated pest management along with even better veterinary medications. It will use global positioning satellites, computers and intesive soil sampling ('precision farming') to exactly apply the seeds and chemicals for optimum yields, with no leaching of chemicals into streams. Such farming will offer only near-zero risk to either the environment or to humans, which will be more than offset by huge increases in food security and wild lands saved.

Feeding the world's people while simultaneously preserving fertile land will require two things: more agricultural research and freer world trade in farm products (Avery, 1997). In order to use the world's best farmland for maximum output, farm trade needs to be liberalized. Farm subsidies and farm trade barriers have drained billions of dollars in scarce capital away from economic growth and job creation and they now pose serious danger to preservation of biologically diverse lands. The key factor in the farm-trade arena is Asia's present and growing population density. Without an easy flow of farm products and services, densely populated Asian countries will be tempted to try to rely too much on domestic food production—an extremely diffucult task. By 2030, Asia will have about eight times as many people per acre of cropland as will the Western Hemisphere. Asia already has the world's most intensive land use. Countries reduce their food security with self-sufficiency because droughts and diseases that reduce crop yields are regional, not global.

A renewed emphasis on high-yield farming aimed at preserving biodiversity appears to be the best option for saving nature's legacy.

AGRICULTURAL BIOTECHNOLOGY

Hammond et al., (1999) have reviewed the latest research on using genetically engineered plants and plant viruses to produce new products for medicine and industry. The three main technologies for engineering plants are *Agrobacterium*-mediated transformation; particle bombardment transformation; and plant viral vectors. Strategies are now available for producing medically important products such as vaccines, human enzymes, monoclonal antibodies, and other therapeutic proteins in plants. Hammond et al. have discussed both the potential benefits and risks involved in producing pharmaceuticals in plants and the challenges of bringing such products to market.

There is no doubt that the problem of food security should be recognized as a challenge facing all of humanity, and that concerted global efforts in agricultural research be intensified. Several successful examples of effective use of genetic diversity for the development of insect-resistant crops are now known and show that close, effective and long-term collaboration is required among different players, including germplasm curators, plant breeders, molecular biologists, entomologists, ecologists and social scientists (see Clement and Quisenberry, 1999).

In recent years, regional and local farm seed variety has been reduced because of increased population, agricultural science and technology and the integration of the world's many diverse cultures. Because of this, diversity on individual farms across wide regions is threatened by modern crop varieties that have been bred for broad adaptation, resistance to disease and other risk factors such as their ability to better use water, fertilizer, and higher yields. The concern of the farmers to maintain production levels and income is often incompatible with those whose focus is on the maintenance of viable and sustainable ecosystems and maintaining genetic diversity. Exploring and understanding these different concerns is an essential starting point for answering some of the key questions about the implementation of 'on farm' conservation and the role of local cultivators in sustainable development (see IDRC/IPGRI, 1999).

Sadly, tropical crops such as cowpea, yam, plantain, and cassava have not attracted much research focus even though besides rice, maize, wheat and potato they are important as primary or secondary food staples in the developing world. The modern tools of molecular and cellular technology allow not only to advance knowledge of these crops, but also to overcome some of the obstacles which presently restrain both the genetic improvement and the productivity of these crops in tropical farming systems. Increased nutritional value of these crops, reduced post-harvest losses and lower costs of production are some of the advantages from a biotechnology perspective (Hohn and Leisinger, 1999). Engineered genetic resistance would also allow to drastically reduce the use of pesticides, which are expensive or unavailable for farmers in developing countries and can create environmental and health hazards.

According to Farrington and Greeley (1989), the likely effects on less-developed countries (ldcs) of agricultural biotechnology cannot be examined in isolation from broader trends in agricultural technology, production and trade. Among many least-developed countries (ldcs) in the 1970s, higher rates of agricultural production stimulated such rapid increases in demand for food that they could not be met by domestic production. This led to increase in food imports.

Biotechnology transcends the techniques hitherto specific to crop or animal improvement. In both areas, it exploits biochemical, molecular and cellular techniques for germplasm manipulation for improvement of the genetic stock and for better control of pests and disease. Suitable biological techniques are being developed across a variety of activities ranging from the improvement of farm inputs to the processing of crops and by-products. This means that biotechnology should be viewed in a food industry context, not merely in that of crop or animal production. The growing shift in food production from the farm to the factory reinforces this wider perspective.

Biotechnology has strong linkages across industrial sectors which are found only to a limited degree in earlier seed, agrochemical and mechanical technologies. They include linkages in materials, processes and techniques.

Materials

1. The use of agricultural, animal and forestry biomass in food processing, pharmaceutical, chemical and energy industries.
2. The reduction of agricultural biomass to intermediate products (e.g. glucose, fructose, dextrins, lactose) by chemical or biological processes and the design of other processes in the same or different sectors to use one or more of these to provide feedstocks.

Processes

1. Enzyme catalysis in starch, detergent and dairy industries, and in reagents for laboratory analysis and clinical diagnosis.
2. Fermentation in food, drink, pharmaceutical, sanitation and energy industries.

Techniques

1. Genetic engineering.
2. Fusion and culture of cells, for embryo rescue and for inter- and intra-species hybridization (Farrington and Greeley, 1989).

Several recent advances in plant biotechnology have direct relevance to the developing world particularly in crop improvement, forestry, environment stabilization and for the future economic development of plant resources with potential to provide many novel sources of industrial compounds. The latter are used to produce various pharmaceuticals, agrochemicals, flavourants, enzymes and polymers. These precursors of core processes in the pharmaceutical, food, energy and the diagnostic industries represent strategic renewable resources for the Idcs (Mantell, 1989).

Nutritionally-fortified Food

Virtually, all countries in Latin America, Asia and Africa are either clinically or subclinically deficient in vitamin A (Kishore and Shewmaker, 1999). The best sources of vitamin A are the carotenes, particularly beta carotene, found in many fruits and vegetables. These carotenes become converted into vitamin A and have a higher safety than vitamin A itself. High-carotene fruits and vegetables are not usually available at affordable prices to poor people. Biotechnology can contribute to world health by producing crops naturally fortified with this important nutrient that people can grow in varied global regions.

Fortification within the seed enhances nutritional quality for all types of farmers. With fortification, local crops grown by subsistence farmers and best suited to their growing conditions naturally would include these nutrients. Large, commercial farmers can reap the same benefit.

Kishore and Shewmaker (1999) introduced the gene *phytoene synthase* into canola and showed that the expression of this gene results in high levels of beta carotene accumulation within the rape (mustard) seed.

Besides being rich in beta carotene, rape seed oil expressing the *phytoene synthase* gene has a higher level of alpha carotene, lutein. It contains a varying range of carotenes, vitamin E, and a healthy profile of fatty acids. The high beta-carotene canola is expected to be commercialized before 2005.

Improved Cereals

Golden Rice. A new genetically-modified variety of rice ('golden rice') has the potential to alleviate vitamin A deficiency around the world. But the tricky issues of intellectual property claims surround the technology and there are legal hurdles to be crossed before the rice can become widely available to farmers in developing countries.

Ingo Potrykus, a plant molecular biologist at the Institute for Plant Sciences of the Swiss Federal Institute of Technology in Zurich developed the new variety in collaboration with Peter Beyer of the Centre for Applied Biosciences at the University of Freiburg in Germany (*Science*, 13 August 1999, p. 994; and 14 January, p. 303). The variety has been patented by Monsanto Company, which has recently agreed to give royalty-free licenses to speed up the availability and access of small farmers to the golden rice (Normile, 2000).

Potrykus' team transferred into rice the multiple genes needed to create a synthesis pathway for β-carotene, which the body converts into vitamin A. In the feat, a trait requiring multiple genes was transferred into a plant. The enriched rice, which appears golden in colour, may help tackle the widespread public health problem of vitamin A deficiency, which afflicts

400 million people and can lead to vision impairment and increased susceptibility to disease.

Before the promise can be realized, however, the β-carotene pathway will have to be transferred to the different varieties of rice grown in each region. This process can be speeded up if and when the required germplasm for the particular area and the related technology become accessible to plant breeders, and legal hurdles and patent/IPR issues are sorted out.

Some 25 to 70 proprietary techniques and materials are involved in the gene transfer. Monsanto's most important claim is on the 35S promoter, which boosts the expression of the genes introduced into the rice. It is this change that increase the b-carotene level.

But some pieces of the puzzle lie outside Monsanto's control. Incorporated into the golden rice, for example, are two 'bits of DNA' acquired under separate agreements that require the donors' consent to be passed on to third parties. The agreements forbid distributing the germplasm to any of the rice breeders interested in working with it (Normile, 2000).

Evangeline Villegas and Surinder Vasal of CIMMYT, Mexico City, have produced a new maize that produces up to twice as much lysine and tryptophan (two essential amino acids) as most modern varieties of tropical maize.

Villegas and Vasal used traditional trial-and-error breeding techniques to incorporate a newly discovered gene into plants that had the same yield and pest resistance as traditional strains. Their perseverance produced a 'quality protein maize' that looks like traditional varieties, the only difference being the superior (nutritional) quality.

Farmers have already planted at least 1 million hectares of the maize—which can boost harvests by 10%—in 10 countries in Asia, South America and Africa (see *Science*, 289: p. 1871, 2000).

Biotechnology Management

The environmentally sound management of biotechnology focuses upon the need for: (a) increasing the availability of food, feed and renewable raw materials; (b) improving human health; (c) enhancing protection of the environment; (d) enhancing safety and developing international mechanisms for cooperation; and (e) establishing enabling mechanisms for the development and the environmentally-sound application of biotechnology.

Biotechnology refers to any technique that uses living organisms or parts of organisms to make or modify products, to improve plants or animals, or to develop micro-organisms for specific use. Commer-cialization of biotechnology ranges from research to products and services. It is supported by complementary bioprocess engineering to help translate new discoveries of life sciences into practical products and services. Biotechnology needs to be considered as an integration of the new techniques of modern biotechnology with the well-established approaches of traditional biotechnology such as plant breeding, food fermentation and composting.

Emerging biotechnologies can help strike a balance between development needs and environmental conservation. A wider diffusion of the technology is the key to directing its positive impacts onto the global society. Biotechnology is continuously and rapidly developing in several sectors that improve the effectiveness of the way in which products and services are provided. For several developing countries, unfortunately, funds, means, infrastructure and other conditions required for the transfer and development of biotechnology in an environmentally-sound manner are not readily available.

Biotechnology, if properly managed, can play an essential role in fostering the economic and social development of both developed and developing countries. Biotechnology development and applications have continued to grow at a very rapid rate, leading to an expanding range of products and processes across several sectors, that began with pharmaceuticals and health care, was extended to agriculture and, more recently, to the environment. Many biotechnological products such as insulin, diagnostics and vaccines have already reached the market. In agriculture, products such as diagnostics, biopesticides, improved seeds and bovine growth hormone are being used commercially. From a global perspective, it has been forecast that major impacts can be expected on health, pharmaceuticals, agriculture, food and the environment within the next 3 decades.

The tendency of most developing countries is to acquire biotechnologies aimed at improving agriculture, food and pharmaceutical production and converting low-cost or marginalized raw materials into high value-added products and marginalized lands into more productive areas. Biofertilizers, tissue culture, vaccines and some new diagnostics can be utilized despite relatively low levels of resources and technological capacity. In addition to the appropriate use of traditional and intermediate biotechnologies, several developing countries are attempting to integrate more advanced biotechnologies.

As regards the level of biotechnology development and applications, there is a great variation among developing countries. The People's Republic of China, India, Korea, Singapore, Brazil and Cuba have given high priority to biotechnology for development. They have encouraged fermentation industry and pharmaceutical products. Modern biotechnology research programmes have also steadily increased, especially in agricultural sectors such as biofertilizers, biopesticides, virus-free seedlings and plant tissue cultures. Transgenic crops and improved seeds are being raised in China. Traditional and intermediate biotechnologies are being used in food fermentation and nitrogen fixation in less develop-ed countries.

Case studies on biotechnology and sustainable agriculture in Kenya, Zimbabwe, India, Thailand, Colombia and Mexico identified common constraints in the diffusion of environmentally-sound biotechnologies, especially to small farmers, weak collaboration between the private and the public sectors, and inadequate financial resources as well as mechanisms for the effective exploitation of emerging technologies. A number of major breakthroughs in crop, animal and forestry research and development have been noted.

In highly developed countries many biotechnological products and services have already reached the market and are widely used, especially those in the pharmaceutical sector. Over 1,700 clinical trials and 1,000 field tests are in progress. A recombinant rabies vaccine has been developed. In situ bioremediation is being applied to reclaim contaminated soil. In the USA, there is a shift towards increasing public acceptance of the development and use of biotechnology-based growth hormone for increasing milk yield and the genetically engineered tomato. Similarly, the pressure to decrease dependency on chemical pesticides is expected to drive the growth of biopesticide production and use, estimated to reach US$ 150 million in the USA alone as compared to the US$ 6.8 billion for conventional pesticides (UNIDO, 1995).

Both governmental and non-governmental organizations have assisted, encouraged and cooperated in stimulating and supporting biotechnology in the past several years. Some lessons learned from these assistance/cooperation programmes include:

(a) Long-term commitment is vital to achieving sustainable capacity building and to enable a country to reach a critical level in self-reliance for further biotechnological development. For instance, the Indo-Swiss project initiated in 1974 has led to pilot commercial production of biopesticides in India.
(b) A networking arrangement among institutions within the country and regions is a most cost-effective way to maximize limited resources.
(c) Access to or provision of modern scientific equipment and key biomaterials for research are important components of successful and equitable strategies for collaborative research.
(d) The financial commitment of developing/recipient country government is critical to successful collaboration. This commitment can include in-kind contributions.

Financial contributions from the private sector for commercial biotechnology development have so far been relatively low, mainly due to the high business risk involved with modern biotechnology enterprises, but also because of an unfavourable policy environment. In view of the relatively high risk associated with biotechnology product development and commercialization, more risk capital needs to be found. In developing countries, an alternative approach being increasingly adopted to promote biotechnology development and commercialization is for the private sector to form partnerships with the governmental enabling institutions such as science and technology parks. Venture capital funds, such as the Transtech Venture Fund in Singapore, can serve as successful models for fund mobilization from banking institutions and industrial subscribers.

To meet the growing demands of increasing global population, not only food production has to be increased but also its distribution systems have to be improved. These challenges may be met through the successful and environmentally safe application of biotechnology in agriculture. Most of the investment in biotechnology so far has been in the industrialized world. But now, international organizations are supporting several new efforts in biotechnology in the developing world. The FAO is assisting more than 30 developing countries in the use of advanced but relatively conventional biotechnology for increased yield and quality of food and feed crops, cash crops and livestock. Assessment and pilot-testing of appropriate biotechniques and products to enhance crop and livestock productivity through the development and promotion of in vitro culture techniques and embryo transfer are underway in several developing countries. Rice biotechnology is being supported by the Rockefeller Foundation. Biotechnology for improved production of buffaloes, flax and nuts is being pursued. Advanced agricultural biotechnologies may prove helpful to impoverished farming communities while being environmentally-friendly. Being more adaptable than mechanical innovations and Green Revolution technology, they are more accessible to small producers. These biotechnologies can reduce farmers' dependency on environmentally degrading agrochemicals whilst decreasing crop losses. Several international agricultural research institutes have important research programmes to increase the yield of major crops through the study of plant stress resistance, tolerance to herbicides and resistance to some specific pests and toxins, and through the study of lignin biodegradation aimed at the recycling of agro wastes as feed stock.

Tissue culture and artificial seed biotechnologies are contributing significantly to agricultural productivity gains in Asia and Africa, and to the afforestation of marginal lands in China.

FAO's biotechnology work in the field of animal production covers three chief areas: (i) better disease diagnosis; (ii) better and safer vaccines for disease prevention; and (iii) genetic manipulation of the germline of economically important livestock to improve specific disease resistance.

Specific FAO projects in Asia and the Middle East encourage the application of fermentation methods for the large-scale production of bacterial aerobic and anaerobic vaccines. Deadly pests and diseases have been eradicated through the development and use of the Sterile Insect Technique, in particular in the management of the African tsetse fly that causes trypanosomiasis.

Human Health

Continuing environmental degradation, compounded by poor and inadequate development, adversely affects human populations. Biotechnology may be used to combat major communicable diseases, in promoting good health, in improved programmes for treatment of and protection from major non-communicable diseases, and in developing appropriate safety procedures. Biotechnology products in health care are now widespread (UNIDO, 1995).

DNA technology offers novel approaches towards the design and production of drugs, vaccines and diagnostic tools. Even its limited application in these areas has proved very successful.

Environmental Protection

There is urgent need to prevent, stop and reverse the effects of environmental degradation through the safe uses of biotechnology. Production processes that make optimal use of biotechnologies for the rehabilitation of land and water, waste treatment, soil conservation, reforestation and afforestation, should be vigorously promoted. Advances in biotechnology offer powerful tools for the conservation, evaluation and use of genetic resources.

To promote the application of biotechnologies for the conservation and sustainable use of biodiversity and to prevent, halt and reverse environmental degradation, UNEP has funded several regional Microbial Resources Centres (MIRCENs) for the: (i) collection and maintenance of microbial genetic resources in view of the tremendous potential of microbial germplasm for economic development and environmental management and protection; and (ii) training in, research on and pilot application of environmentally sound biotechnologies. Examples include increasing food production and soil fertility through biological nitrogen fixation, biodegradation of persistent chemicals used in agriculture and industry, bioremediation, biocontrol of insect pests and disease vectors, bioleaching, and bioconversion of agricultural residues and surpluses into useful products.

Applications of modern biotechnology for bioremediation of contaminated land and water have aroused much global interest and increased demand by developing countries for technical advice and assistance from

UNIDO, which has ongoing activities on waste minimization and industrial effluent treatment.

The development of environmentally-sound alternatives and improvements for environmentally damaging production processes, including the environmentally sound management of hazardous wastes generated in various sectors of industry and commerce became a specific component of the recommendations to European countries on the five R policies (reduction, replacement, recovery, recycling and reutilization) of industrial products, residues or wastes, adopted in 1992. Guidance is available for the control of pollutants in industrial processes, the application of the best available technology for containment and treatment of hazardous substances and the substitution of potentially hazardous substances in industry, trade and services (see UNIDO, 1995).

Enabling Mechanisms

Enabling mechanisms for the development and the environmentally sound application of biotechnology have been established by the UNIDO/United Nations with the following objectives:

1. Stregthening endogenous capacities of developing countries, including employment opportunities;
2. Socio-economic impacts of new biotechnology on conventional production systems;
3. The contribution of indigenous peoples and their share in economic and commercial benefits accruing from biotechnology;
4. Intellectual property rights with respect to biotechnology and bioresources;
5. Increasing access both to existing information about biotechnology and to facilities based on global databases and;
6. Creating a favourable climate for investments, industrial capacity building and distribution marketing.

Programmes dealing with biosafety have been extended beyond the scientific sector to include the concerns of workers exposed to biological agents in the workplace, farmers and the general public. Programmes on marine, agricultural and industrial biotechnology are forging closer cooperation among various UN agencies and now involve the private business sector, financial institutions and NGOs.

Various UN and other international organizations are actively attempting at the global, regional and national levels to help developing countries to take advantage of opportunities offered by rapid advances in biotechnology.

UNDP places great emphasis on sustainable human development and encourages the participatory involvement of local organizations and people in the planning and implementation of scientific endeavour. It seeks to avoid heavy dependence on the transfer of technology, an approach shared by other UN agencies. It is hoped that in the future it will be possible for biotechnology to be regarded as one major constituent of a broader multi-sector planning approach (UNIDO, 1995).

There is tremendous scope in many countries for poductivity gains, fo impovement in the quality of food and agricultural poducts, and for consevation of the environment, using existing technologies which are available but are not being applied. A key issue for developing countries should be selectivity in determining whether biotechnology might povide the most effective solution, from both the cost as well as the social points of view. Most countries can take advantage of new technology either through endogenous development or through international technology transfer or, more often, a combination of the two, depending on national conditions and policies. Although external cooperation can facilitate technology development and diffusion, it can only complement—not substitute for—national efforts.

In developing countries, the successes achieved in transfer to and development of biotechnology were based on strategic alliances with institutions in developed countries either at the development stage or at both the research and development stages. These alliances can involve public institutions or the private industrial sector, or both. For successful development of biotechnology as also private support, and support from NGOs, is needed.

In most advanced (developed) countries that have increasingly privatized biotechnology research and development, rapid progress has occurred in several sectors so much so that they are now expanding from the pharmaceutical and health sector to the agricultural sector and the environmental sector. Environmental biotechnology in the future will focus increasingly on the conservation, protection and sustainable utilization of the world's scarce natural resources. Biotechnology can enable nations to achieve sustainable development while safeguarding the environment. However, biosafety is a crucial issue in all countries to establish standards for the development, handling and commercialization of biotechnology products, to protect human and animal health and to safeguard the environment (UNIDO, 1995). It extends well beyond the conservation and sustainable use of biological diversity.

Guided by several reviews, assessments and lessons learnt, UNIDO (1995) identified the following:

1. The key role of the private sector—business, industry and the banks—in promoting and applying biotechnology for sustainable development.

2. The need to integrate biotechnology concerns into national sustainable development policies for making and building national capacities.
3. The need to achieve and demonstrate safe and viable results for sustainable development in the application of biotechnology.
4. Biosafety and dependable institutional mechanisms for the further development and implementation of international policy on biosafety.
5. Patenting and intellectual property rights, with special reference to the effective participation of developing countries in the process towards adopting realistic standards for intellectual property rights on biotechnology, taking into consideration crucial role of biotechnology in sustainable development, its potential impact on human society, and opportunities for mutually-beneficial collaboration and cooperation.
6. The need to promote greater public awareness of biotechnology issues.

In the light of the above priority issues, necessary action on the following matters needs to be taken:

1. Enhancing the contribution of the private sector to sustainable development, in which governments seek to involve business, industrial and banking interests more actively in safely applying and promoting biotechnology to meet the objectives of sustainable development.
2. Integrating biotechnology concerns into national-level development policy making and building national capacities, in which governments, supported by the United Nations and other appropriate intergovernmental bodies, act to ensure the participation and contribution of all major groups in the integration process; enhance public awareness through promoting and disseminating an accurate understanding of biotechnology that includes the issues associated with progressive trends in its development; support cross-fertilization of ideas among major groups to enable decision makers to identify problems to be solved and to recognize the appropriateness, feasibility and sustainability of perceived biotechnological solutions; stregthen environmentally and economically sustainable capabilities in the sound management of biotechnology, including matters relating to intellectual property rights; establish national databases on information relating to biosafety; and assess the need for advice and assistance in promoting appropriate biotechnology and biosafety regulations where these do not already exist or need to be stregthened, with a view to designing effective programme building wherever possible on existing capabilities (UNIDO, 1995).

3. Promoting 'best environmentally-sound and viable practices', in which governments with maximum support from UN and other organizations as well as the private sector cooperate to identify and exchange information, especially at the regional level, about examples of 'best practice' viable and environmentally sound applications that have demonstrably resulted in meeting the sustainable development objectives.
4. Encouraging the environmentally sound application of biotechnology for sustainable development.

PRECISION FARMING

In the same way that agricultural technology available in the 1940s could not have helped to meet the demand of food for today's population, in spite of the Green Revolution, it may be assumed that food requirement for the population of 2020 AD will not be supplied by the technology of today. To meet the future demand, it will be necessary to adopt new technologies for sustaining and augmenting our agricultural productivity.

In the post-Green Revolution period, agricultural production is stagnating and horizontal expansion of cultivable lands has become limited due to burgeoning population and industrialization. In 1952, India had 0.33 ha. of available land per capita, which has probably already declined to 0.15 ha. Increasing application of fertilizers and pesticides has increased production but our agriculture has become chemicalized. This situation warrants development of eco-friendly technologies for maintaining crop productivity.

It is well-known that crops and soils are not uniform within a given field (see Pierce et al., 1996; Stafford, 1997). Technical methods have now been developed to make use of modern electronics to respond to field variability. Such methods are known as spatially variable crop production, geographic positioning system GPS-based agriculture, site-specific and precision farming (PF).

The management philosophy of PF identifies situations where yield is limited by controllable factors and determines intrinsic spatial variability. The variations detected in crop or soil properties within a field are recorded and appropriate management actions are taken to account for the spatial variability within that field. Development of geomatics technology in the later part of the 20th century has aided in the adoption of site-specific management systems using remote sensing (RS), GPS and geographical information system (GIS). This approach is called PF or site-specific management (Mandal and Ghosh, 2000). PF represents a paradigm shift from conventional management practices of soil and crop for small farms. In PF, management decisions are adjusted to suit variations in resource conditions.

PF requires special tools to discern the inherent spatial variability associated with soil characteristics and crop growth and to prescribe the most appropriate management strategy on a site-specific basis. PF critically considers the conditions for agricultural production as determined by soil, weather and prior management across space and over time. The inherent variability dictates that management decisions be specific to time and place, rather than rigidly scheduled and uniform.

Conventional agriculture is practised for uniform application of fertilizers and biocides and irrigation, without considering spatial variability. PF alleviates the ill-effects of over- and under-usage of inputs. Site-specific management of spatial variability of farm is developed to maximize crop production and to minimize environmental pollution and degradation, leading to sustainable development.

Technology

Spatially variable crop production is largely technology-driven and uses advances in electronics and computers such as RS, GPS and GIS. Technologies used in PF involve data collection, analysis or processing of recorded information, and recommendations based on available information.

The generation of maps for crop and soil properties is the most important first step in PF. These maps record spatial variability and provide the basis for controlling it. Data are collected precisely on location coordinates using the GPS, grid soil sampling, yield monitoring, RS and crop scouting. During crop production, the data are collected through soil probes, electrical conductivity and soil nutrient estimations. Optical scanners help detect soil organic matter and weeds. All these data are stored in a computer system for future action.

Remote Sensing

This involves acquisition of precise information about an object from a distance, without coming into contact with it. RS senses visible and invisible properties of a field or a group of fields and converts point measurements into spatial information, to monitor temporally dynamic plant and soil conditions. The visual observations are recorded through a digital notepad and geo-referenced to GIS database, the most commonly used RS device; aerial photography and videography can also be used. Images can be taken from satellites such as LANDSAT. These images allow mapping of crop, pest and soil properties for monitoring seasonally variable crop production, stress, weed infestation and extent within a field.

Use of RS for PF allows for providing input supplies and variability management through decision support system. The point data of soil test results can be translated into spatial coverage based on geostatistical interpolation, which gives chemical properties of the soil, nutrient status, organic matter, salinity or moisture content (Mandal and Ghosh, 2000). This information on spatial variability proves useful for identifying both seasonally stable and variable units, based on which management strategies can be developed. Space technology, together with satellite RS and informatics, provide valuable timely information like early warning, occurrence, progressive dangers, damage assessment, quick dissemina-tion of information regarding disaster and decision support to mitigate it.

Geographic Information System

GIS techniques deal with the management of spatial information of soil properties, cropping systems, pest infestations and weather conditions. They are used to integrate spatial data coming from various sources in a computer.

GIS-aided techniques are badly needed for sustainable food production and resource utilization in the face of continuing deterioration of the environment. GIS technology helps both farmers and scientists in decision making as it generates precise field information. GIS techniques make weed control, pest control and fertilizer application site-specific, precise and effective; they may also prove useful for drought monitoring, yield estimation, pest infestation, monitoring and forecasting. GIS, along with GPS, microcomputers, RS and sensors is used for soil mapping, crop stress, yield mapping, estimation of soil organic matter and available nutrients.

Control Strategies

The documented spatial variability in maps is used to control the variability of soils, crops or pests through field operations. One common response to soil variability within fields involves scheduling the fertilizer applications in a spatially variable manner. Soil moisture map is used to control irrigation. Crop yield and pest infestation maps are used to control the application of irrigation, fertilizer and patch spray of pesticides.

TRANSGENIC PLANTS AND PEST MANAGEMENT

Genetically-engineered crop plants showing resistance to insect pests offer an environmental friendly method of crop protection. Positive results have been obtained with the expression of *Bacillus thuringiensis* (*Bt.*) and other toxin genes in several crops. However, both exotic and plant-derived genes have some performance limitations and some failures have occurred in insect control through transgenic crops. The production and deployment of transgenic crops for pest control should carefully consider the issues related to impact of the transgenic crops on the insect pests, ecological cost of resistance development, effects on the nontarget organisms, availability and distribution of the alternate host plants, and the potential for introgression of genes into the wild relatives of crops.

In the near future, most of the increase in food production will have to come from increased yields of major crops grown on existing arable lands. One way of increasing crop production is to minimize the pest-associated losses, estimated at 14% of the total agricultural production (Oerke et al., 1994). Besides the direct losses, insects also cause indirect losses through their role as vectors of various plant pathogens. Additional costs accrue from pesticides applied for pest control, running into billions of dollars annually. Massive application of pesticides not only leaves harmful residues in the food, but also causes adverse effects on non-target organisms and the environment.

Modern genetic transformationt techniques make it possible to insert exotic genes into the plant genome that confer resistance to insects. Amongst these, the bacteria such as and *B. sphaericus* have been used successfully for pest control through transgenic crops on a commercial scale. Insecticidal genes such as *Bt.*, trypsin inhibitors, lectins, ribosome inactivating proteins, secondary plant metabolites, vegetative insecticidal proteins and small RNA viruses have been used either alone or in combination with *Bt.* genes in transgenic plants for pest control (Sharma and Ortiz, 2000). Several transgenic crops with resistance to the target pests have been developed and these have proved promising in reducing insect damage, both under laboratory and field conditions, and thus reducing the need to use pesticides for pest management. Genes conferring resistance to insects have been inserted into maize, rice, cotton, potato, tobacco and soybean.

Modern techniques of biotechnology provide access to novel molecules, ability to change the level and pattern of gene expression. Development and use of transgenic plants with insecticidal genes can lead to: reduced exposure of farmers, farm labour and non-target organisms to the pesticides; increased activity of natural enemies because of reduction in pesticide sprays; reduced amounts of pesticide residues in the food and food products; and a safer and cleaner environment.

Host plant resistance (HPR) reduces the need to apply pesticides as it is compatible with biological control and other methods of pest management in an integrated pest management (IPM) programme. Pesticides are highly toxic to the natural enemies, pollinators and other non-target organisms. Conventional HPR slows down the rate of increase of pest populations and exposes the pests for prolonged periods to the natural enemies. The introduction of transgenic plants reinforces HPR which can potentially influence the tritrophic interactions (Sharma and Ortiz, 2000).

Synergism has been detected between *Bt.* toxins and the HPR for *Trichoplusia ni* and it needs to be exploited for important crops and their pests in the semi-arid tropics (SAT) to achieve satisfactory control of the target pests, thereby avoiding the need to use pesticides. There appear to be no adverse effects of transgenic plants on the performance of the natural enemies.

Whereas sprays based on *Bt.* formulations are unlikely to displace chemical insecticides because of their limited spectrum of activity and lower efficacy compared to synthetic chemicals, transgenic plants are usually sufficiently effective to either displace chemicals or to be used along with chemical insecticides or other methods of pest control. Simulation models using data from the diamondback moth (*Plutella xylostella*) and the Indian meal moth (*Plodia interpunctella*) have shown that transgenic plants bearing only one *Bt.* gene may be more effective than the sprays for delaying the development of resistance to *Bt.* A Colorado potato beetle strain that survives *Bt.* sprays cannot develop successfully on transgenic plants, not even on plants showing very low levels of *Bt.* gene expression (Sharma and Ortiz, 2000). Simulation models have also suggested that transgenic plants may be much more durable than sprays of similar efficacy when more than two genes are deployed.

The above findings should, however, not mislead us to believe that transgenics are a panacea for solving all the pest problems. With the deployment of transgenic crops: the secondary pests will no longer be controlled in the absence of sprays for the major pests; the need to control the secondary pests through chemical sprays will kill the natural enemies and, thus, offset one of the major advantages of transgenics; the cost of producing and using transgenics can be quite high; proximity of transgenic crops to sprayed fields and insect migration can reduce the effectiveness of transgenics; and development of resistance in insect populations may limit the usefulness of transgenic crops for pest management and can often bring down the number of pesticide applications significantly even by as much as two-thirds to one half.

Environmental Impact of Transgenic Crops

The number of pesticide applications on a crop such as cotton varies from 10 to 40, most of the sprays being directed against the key pests such as *Heliothis* and *Helicoverpa*. Reduction in pesticide application would stimulate the natural enemies, while some of the minor pests may tend to attain higher pest densities in the absence of sprays applied for the control of major pests. The introduction of transgenic crops can have a major impact on the abundance of some insects—the effects being negative for some and positive for others.

Efficacy of transgenic crops for controlling non-target pests should be determined in each region. Some pests maintain high densities on alternate hosts, e.g. cereal stem borers on the wild relatives of sorghum. Potential impact of transgenic crops on the beneficial insects can occur through reduction in the number of eggs and larvae of the natural hosts, which may also affect the activity of natural enemies.

A *Bt.* toxin, which is highly effective against one insect species, may be weakly so or ever ineffective against some other insects. We do not know how effective the engineered crops would be in providing protection against the target pests, and there is no information on the key pests of the econonically important crops. The technologies used in the developed countries may not prove suitable for the developing countries as the latter have complex cropping systems and a more diversified farming system. Some issues that need to be considered while introducing transgenic crops for pest control include: (i) effects on population dynamics of target and non-target insects; (ii) evolution of new insect biotypes; (iii) insect sensitivity; (iv) performance limitations; (v) gene escape into the environment; (vi) secondary pest problems; (vii) environmental influence on gene expression and failure of insect control; (viii) effects on non-target organisms; and (ix) impact on natural enemies (Sharma and Ortiz, 2000).

In general, no negative effects of *Bt.* proteins have been recorded in mammals. There are no specific receptors for CryIA(b) protein present in the gastrointestinal tract of mammals, including man. Oral exposure to transgenic *Bt.* tomatoes poses no additional risk to human and animal health. No significant differences in survival or body weight have been seen in broilers reared on meshed or pelleted diets prepared with *Bt.* transgenic maize and similar diets prepared using control maize (see Hedley et al., 1996). Broilers raised on diets prepared from transgenic maize showed better-feed conversion ratios and improved yield of the breast muscle.

The quality of produce from the transgenic plants is broadly similar to that from the nontransgenic plants of the same cultivar. The levels of the antinutrients gossypol, cyclopropenoid fatty acids and aflatoxin in the seed from the transgenic cotton are similar to or lower than the levels present in the parental variety and other commercial varieties. The seeds from the *Bt.* transformed cotton lines are as nutritious as those from the parental and other commercial cotton varieties.

Some proteins involved in the defence mechanisms of food plants are allergens which have potential use in molecular approaches to increase resistance to insect pests. These include α-amylase and trypsin inhibitors, lectins and pathogenesis-related proteins. Several self-defence substances made by plants are highly toxic to mammals, including humans. In these cases, the source of the transgene has no relevance in assessing the toxicological aspects of food from transgenic plants. This may result in a trade-off situation between nature's pesticides produced by transgenic plants, synthetic pesticides, mycotoxins or other poisonous products of pests.

Trypsin inhibitors and plant lectins contribute to a plant's defence mechanism in nature and can be potentially used in developing insect resistant transgenic crops. But these compounds can have some adverse effects. No crop plants expressing these genes have been deployed for commercial cultivation. Rats fed on purified cowpea trypsin inhibitor in a semi-synthetic diet based on lactoalbumin showed a moderate reduction in weight gain in comparison with controls, despite an identical food intake.

Resistance Management

Insect pest populations usually develop resistance to chemical pesticides. Most of the transgenic *Bt.* crops express only one toxin gene and lack the complexity of the commercial *Bt.* formulations. Also, the plants continuously produce the toxins, and the insects are exposed to the *Bt.* toxins throughout their feeding cycle; this imposes on the insect population a heavy selection pressure. With the development of resistance to *Bt.* toxins, the value of microbial insecticides based on *Bt.* proteins declines due to weakening sensitivity of the target pest to the *Bt.* formulations. One consequence of this is that the farmers have to return to broad spectrum insecticides, which cause environmental hazards. Most of the transgenic plants produced so far have *Bt.* genes under the control of cauliflower mosaic virus (CaMV35S) constitutive promoter—a system that develops resistance in the target insects as the toxins are expressed in the plant. Toxin production may also decrease over the crop growing season. Decreasing levels of toxin production can lead to resistance development not only to the toxin but also to other related *Bt.* toxins to which the insect populations may initially be quite sensitive. Low doses of the toxins eliminate the most sensitive individuals of a population, leaving a population in which resistance can develop rapidly.

The ability of insects to overcome host plant resistance poses a strong risk. There are several reports on the development of resistance to *Bt.* in many insect species; e.g. diamondback moth *Spodoptera exigua, Trichoplusia ni, Ephestia kuehniella, Plodia interpunctella, Christoneura fumiferana, Chrysomella scripta, Leptinotarsa decemlineata, Aedes aegypti, Culex quinquefasciatus, Drosophila melanogaster* and *Musca domestica*. This raises the real challenge of developing a strategy for deployment of transgenic plants for sustained protection of crops from insect pests. Different insect species react to *Bt.* toxins differently. The resistance can develop quickly in *Diatraea saccharalis*, as some larvae can survive up to a week on transgenic maize (see Green et al., 1990).

Reduced binding of *Bt.* toxins to midgut epithelium is one mechanism of resistance in *Plutella xylostella*. There is also a broad-based mechanism of resistance to *Bt*. Complete degradation of *Bt.* toxins by the proteolytic enzymes is the principal mechanism of resistance in some insects. Development of resistance may be due to changes in insecticidal crystal protein (ICP) receptors, and alterations in ICP receptors are a general mechanism by which insects can adapt to *Bt*. The absence of cross-resistance to ICPs other than those present in the selecting agent, and the finding that these ICPs bind to distinct receptors indicate that the use of ICP mixtures or multiple ICPs expressed in transgenic plants may be a valuable resistance management strategy.

Induction of proteinase activity probably is one way by which insects that feed on plants overcome plant proteinase inhibitors (PIs). Herbivorous insects can overcome the activity of PIs by secreting inhibitor-resistant enzymes. The insect's midgut contains a number of different proteins with trypsin-like activity, some of which are susceptible to inhibition by PI, while other trypsin(s) are not susceptible to inhibition. When inhibitor-resistant insects ingest PI, the level of activity of inhibitor-resistant trypsin(s) is enhanced in the midgut, thus allowing the insect to digest dietary protein in the presence of PI. Conceivably, a group of PIs are required to inhibit the majority of proteolytic activity in the midgut of the target organism, and thus reduce insect growth and development. Once the PIs have been identified, their genes can be transgenically inserted into plants to enhance phytochemical resistance against herbivorous insects. An adaptive mechanism in *Helicoverpa* elevates the levels of other classes of proteinases to compensate for the trypsin activity inhibited by dietary PI.

Use of transgenic plants should be based on the overall philosophy of IPM, and consider not only gene construct, but alternate mortality factors, and reduction of selection pressure. Populations should be monitored for resistance development to design more effective management strategies. One good strategy to minimize the rate of development of resistance in insect populations to the target genes is through: (i) use of resistance management strategies from the beginning; (ii) gene pyramiding; (iii) gene deployment; (iv) regulation of gene expression; (v) development of synthetics; (vi) refugia; (vii) destruction of carryover population; (viii) control of alternate hosts; (ix) use of planting window; and (x) use of economic thresholds and IPM (see Sharma and Ortiz, 2000).

Gene pyramiding: Many of the candidate genes so far used in genetic transformation of crops are either too specific or only weakly effective against the target insect pests. Therefore, to make transgenics effective in pest control (e.g. by delaying the evolution of insect populations resistant to the target genes) it is desirable to deploy genes with different modes of action in the same plant. Severel genes for trypsin inhibitors, secondary plant metabolites, vegetative insecticidal proteins, plant lectins, and enzymes that are selectively toxic to insects can be deployed along with the *Bt.* genes to prolong the durability of resistance. Several advances have been made in introducing and expressing multiple transgenes in crops. For instance, Cry1A(c) and Cry1F can be expressed together in transgenic plants for effective control of some pests to increase the durabiltiy of resistance. Activity of *Bt.* in transgenic plants can be enhanced by serine protease inhibitors, or in combination with tannic acid. Transgenic poplars expressing proteinase inhibitor and *CryIIIA* genes show reduced larval growth, altered development and increased mortality compared to the control.

Gene deployment: Appropriate strategies need to be developed for gene deployment in different crops or regions depending on the pest spectrum, their sensitivity to the insecticidal genes, and interaction with the environment. The deployment of different genes and their level of expression should be based on insect sensitivity and level of resistance development. High levels of Cry1C production can protect transgenic broccoli not only from susceptible or Cry1A(R) diamondback moth larvae, but also from those selected for moderate levels of resistance of CrylC. The CrylC-transgenic broccoli is also resistant to two other lepidopteran pests of crucifers (cabbage looper and imported cabbage worm).

Development of synthetics: One targetted deployment strategy is the development of synthetics. By incorporating various constitutively expressed Cry toxins into lines adapted for specific environments, synthetics can be formed quickly and prove effective against the pest complex and are compatible with the natural enemies. Once released to the farmers, the synthetics can be maintained as narrow-based populations at the farm level by removing the plants showing insect damage.

Refugia: One of the main strategies to manage the deployment of resistance to *Bt.* toxins is high dose and production of refugia, in which a certain percentage of the crop consists of non-*Bt.* plants (4-20% in maize, and 20-40% in cotton). The non-*Bt.* plants produce the susceptible insects, which have a probability of mating with those emerging from the *Bt.* crops nearby, and thus dilute the frequency of the resistant individuals. The growers have a contractual obligation to grow the non-*Bt.* crops. The refuges can be sprayed or unsprayed. In the latter case, the area under non-*Bt.* crop has to be much larger than that under unsprayed conditions. Refugia improve the durability of transgenic plants. The optimal spatial and temporal scale of refugia is likely to be unique for each insect-plant interaction. Refugia should be located closer to the transgenic plants so that the moths producd in the transgenic plants can mate with the insects produced on the transgenic plants. The refugia also enhance the capacity of biological control agents. For some polyphagous pests which feed on several field crops and alternate hosts in the wild, there may not be any need to maintain the refugia in the SAT.

Separate refuges are usually superior to seed mixtures for delaying resistance.

Use of planting window: Recourse to a planting window that allows the crop to escape pest damage or avoid peak periods of insect abundance can be useful in maximizing the benefits from transgenic crops or to prolong the life of these crops. Observing a close season and planting the crop with first monsoon rains have been effective in controlling the damage by sorghum shoot fly (*Atherigona soccata*) and sorghum midge (*Stenodiplosis sorghicola*). Similar strategies may be feasible for prolonging the effectiveness of transgenic crops.

Use of economic thresholds and IPM: Crop growth and pest incidence should be monitored carefully so that appropriate control measures can be initiated in time. Care should be taken to use control options such as natural enemies, nuclear polyhedrosis virus (NPV), neem or entomopathogenic nematodes and fungi, which do not disturb the natural control agents. Use of pesticide formulations such as soil application of granular systemic insecticides and spraying soft insecticides such as endosulfan may be considered to suppress populations in the beginning of the season. Broad-spectrum and most toxic insecticides should be used only during the peak periods of the pest. Pesticides with different modes of action should be rotated and repetition of insecticides belonging to the same group or the insecticides that fail to give effective control of the pests, should be avoided.

INTEGRATED PEST MANAGEMENT

Despite dramatic advances in pest control technology over the last several decades, pest control continues to be a serious constraint in agricultural production. Some major pests of crops and animals are insects, pathogens and weeds. Strategies include cultural control, biological control, pest resistant crop varieties, and chemical control.

In the 1800s, farmers used to depend almost exclusively on cultural methods such as crop rotation in their efforts to control pests. Chemical controls became available in the 1880s with the development of arsenical and copper based insecticides. Use of biological control dates from the late 1880s with the introduction of the vedelia beetle (from Australia) to control a California citrus pest, the cottony cushion scale. Work then began to identify, develop and introduce pest-resistant crop varieties and animal breeds.

The development of dichlorodiphenyl-trichlorethane (DDT) in the late 1930s was a milestone in the history of pest control. DDT was used during World War II to protect American troops against typhus. Early tests showed DDT to be effective against almost all insect species and relatively harmless to humans, animals and plants. It was effective at low application levels and was also relatively inexpensive. Chemical companies rapidly expanded their research on synthetic organic insecticides as well as chemical approaches to the control of pathogens and weeds.

Unfortunately, some serious problems appeared shortly after the introduction of DDT. When DDT was introduced in California to control the cottony cushion scale, its introduced predator, the vedelia beetle, turned out to be more susceptible to DDT than the scale. In 1947, just one year after its introduction, citrus growers, confronted with a resurgence of the scale population, were forced to restrict the use of DDT. In some places, cotton boll worm quickly developed resistance to DDT and other chlorinated pesticides. Producers then turned to the more recently developed and much more toxic organophosphate insecticides, which were again selected for resistant strains of the boll worm. In the meanwhile, natural predators were almost completely exterminated. Cotton production collapsed. It could be revived only after a programme to regulate insecticide use was implemented (see Ruttan, 1999). The adoption of the high-yielding 'green revolution' cereal varieties in developing countries was associated with a marked increased in pesticide use. When yields were low, there was no benefit from pest control. As yields rose, the economic incentive to adopt chemical pest control technologies also increased. It rapidly emerged that the benefits of the pesticides introduced in the 1940s and early 1950s were obtained at high costs, which included not only the increase in resistance to pest control chemicals in target populations and the destruction of beneficial insects, but also the direct and indirect effects on wildlife populations and on human health.

The solution to the above pesticide crisis offered by entomologists has been termed Integrated Pest Management (IPM). It involves the integrated use of some or all the pest control strategies referred to above. It is more complex for the producer to implement than spraying by the calendar. It requires skill in pest monitoring and an understanding of insect ecology. It involves cooperation among producers for effective implementation. When IPM began to be promoted as a pest control strategy in the 1960s, there was no IPM technology available to be transferred to farmers, but by the 1970s enough research had been conducted to provide the knowledge to successfully implement a number of important IPM programs (Conway and Pretty, 1991; Pimentel, 1991).

Integrated approaches to weed management evolved later than for insect pests, partly because emergence of resistance to chemical herbicides occurred much more slowly than resistance by insect pests to insecticides. By the mid 1990s, however, the development of genetically engineered herbicide-resistant crop varieties aroused some new concerns. In some cases, herbicide-resistant crops may have beneficial effects on the environment—when, for instance, a single broad spectrum herbicide that breaks down rapidly in the environment is substituted for several applications of pre- and post-emergence herbicides, or for a herbicide that is more persistent in the environment. When a single herbicide is used repeatedly it endangers selecting for herbicide resistant weeds.

Indigenous management of agriculture and use of natural resources in pest management is practised by farmers in remote rural areas, where chemical pesticides are not available. For instance, over fifty species of plants and indigenous techniques are used in Nepal to help protect field crops and stored grains.

The influence of one crop on another (non-host crop) can reduce the pest buildup significantly in a given agroecosystem. Farmers generally go in for polyculture to maximize crop yields. In a mixed or intercropping pattern, farmers have been directly or indirectly utilizing the mechanism of resistance through non-preference.

In several developing countries, e.g. Nepal, more than 75% of the pesticides used on food crops are either wasted or misused. They fail to target specific pests. They are not applied on time and in proper doses. Farmers in low-income groups face serious consequences with regard to began crop protection when applying these pest chemicals. Health hazards threaten those farmers who handle chemical pesticides in a careless manner.

INTEGRATED DISEASE MANAGEMENT

Integrated disease management (IDM) may be defined as a suitable approach to the selection, integration and use of methods for prevention and control of diseases on the basis of their anticipated economic, ecological and sociological consequences. The various groups of scientific, legislative and environmental aspects involved in IDM are shown in Fig. 10.3. In the integrated control approach, all the available control techniques are blended to minimise the economic damage caused by diseases, with a minimum disturbance to the common environment. At the farm level, this means that resistant varieties, sanitation, changed agronomy and disease monitoring for operational disease management needs to be done, besides making timely sprays. All these vigilant actions lead to integrated disease management.

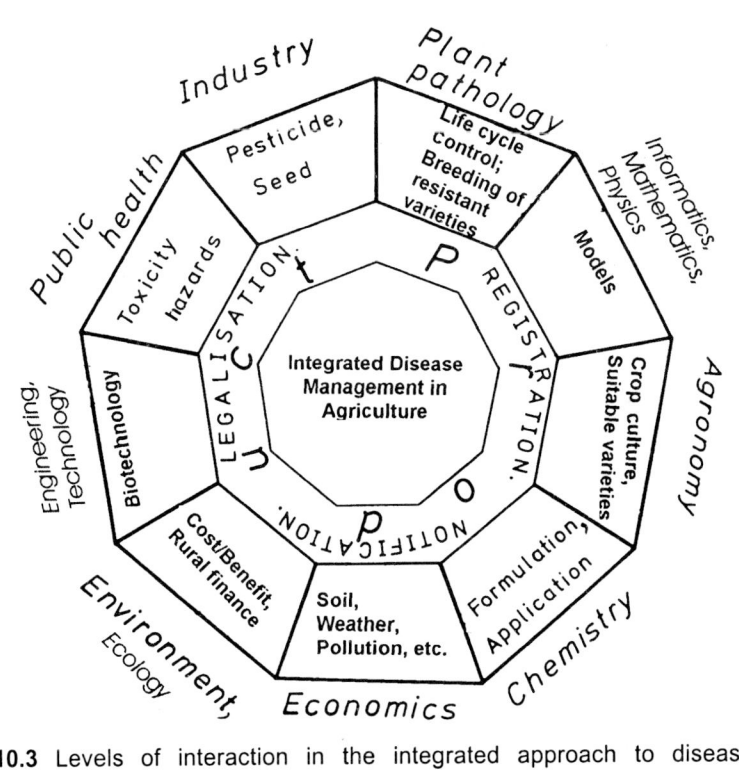

Fig. 10.3 Levels of interaction in the integrated approach to disease management in agricultural systems.

REFERENCES

Anonymous. *Rural Development : From Vision to Action.* World Bank, Washington, D.C. (Oct. 1997).

Avery, D.T. Saving nature's legacy through better farming. *Issues in Science and Technology* 59-64 (Fall, 1997).

Backes, M.M. Floristic and biocultural landscape diversity in autochthonous agroforestry systems — A case study from Western Kenya. *Applied Geography and Development* 54 : 7-27 (1999).

Bruna, E.M. Seed germination in rainforest fragments. *Nature* 402: 139 (1999).

Castillo, G.T. Whose ethics and which equity? : Issues in the conservation and use of genetic resources for sustainable food security. pp. 19-31 In IPGRI, Rome (1997).

Clement, S.L., Quisenberry, S.S. *Global Plant Genetic Resources for Insect Resistant Crops.* CRC Press Boca Raton (1999).

Conway. G.R., Pretty, J. *Unwelcome Harvest: Agriculture and Pollution.* Earthscan Publications, London (1991).

CWFS. *Assessment of the World Food Security Situation.* Committee on World Food Security, Rome (Apr. 1998).
Dyson, T. *Population and Food: Global Trends and Future Prospects.* Routledge, London (1996).
Dyson, T. World food trends and prospects to 2025. *Proc. Natl. Acad. Sci. USA* 96: 5929-5936 (1999).
Evenson, R.E. Global and local implications of biotechnology and climate change for future food supplies. *Proc. Natl. Acad. Sci. USA* 96: 5921-5928 (1999).
Evenson, R.E., Pray, C.E., Rosegrant, M.W. Agricultural Research and Productivity Growth in India. Research Report 109. International Food Policy Research Institute, Washington, D.C. (1999).
FAO (1951-1997). *Production Yearbook.* Food and Agricultural Organization, Rome (1998).
Farrington, J., Greeley, M. The issues. pp. 7–26. In Farrington, J. (Ed.). *Agricultural Biotechnology: Prospects for the Third World.* Overseas Develop. Inst., London (1989).
GAO. *Food Security: Factors That Could Affect Progress Toward Meeting World Food Summit Goals.* Report No. GAO/NSIAD-99-15. U.S. General Accounting Office (GAO), Washington, D.C. (March, 1999).
Green, M.B., LeBaron, H.M., Moberg, W.K. (eds.) *Managing Resistance to Agrochemicals.* pp. 3-16 ACS, Symposium Series, Washington (1990).
Hagen, T. Agriculture without farmers? A challenge for the next century. Pp. 1-36. In Jha, P.K., Karmacharya, S.B., Baral, S.R., Lacoul, P. (eds.) *Environmental and Agriculture : At the Crossroad of the New Millennium.* Ecological Society (ECOS), Kathamandu, Nepal (2000).
Hammond, J., McGarvey, P., Yusibov, V. *Plant Biotechnology: New Products and Applications.* Springer, Berlin (1999).
Hardon, J.J. Ethical issues in plant breeding, biotechnology and conservation: a review pp. 43-50. In IPGRI, Rome (1997).
Haverkort, B., Hiemstra, W. (eds.) *Food for Thought: Ancient Visions and New Experiments of Rural People.* Zed Books, London (1999).
Hedley, C., Richards, R.L., Khokhar, S. (eds.) *Agri Food Quality : An Interdisciplinary Approach.* pp. 23-26. Special Publication No. 179, Royal Society of Chemistry, Cambridge (1996).
Helmuth, L. A shifting equation links modern farming and forests. *Science* 286: 1283 (1999).
Hohn, T., Leisinger, K.M. *Biotechnology of Food Crops in Developing Countries.* Springer, Berlin (1999).
IDRC/IPGRI. *Genes in the Field: Conserving Plant Diversity on Farms.* CRC Press, Boca Raton (1999).
Kishore, G.M., Shewmaker, C. Biotechnology : Enhancing human nutrition in developing and developed worlds. *PNAS* (USA) 96: 5968-5972 (1999).
Kumar, H.D. *Biodiversity and Sustainable Conservation.* Oxford IBH, New Delhi (1999).
Laurance, W.F., Bierregaard, R.O. Jr. (eds.). *Tropical Forest Remnants : Ecology, Management, and Conservation of Fragmented Communities.* Univ. Chicago Press, Chicago (1997).

Mandal, D., Ghosh, S.K. Precision farming — The emerging concept of agriculture for today and tomorrow. *Current Science* 79: 1644-1647 (2000).

Mantell, S. Recent advances in plant biotechnology for Third World countries. pp. 27–40. In Farrington, J. (Ed.). *Agricultural Biotechnology: Prospects for the Third World.* Overseas Develop. Inst., London (1989).

McCalla, A. *Agriculture and Food Needs to 2025: Why We Should Be Concerned.* The World Bank, Washington, D.C. (1994).

McNeely, J.A. How the Convention on Biological Diversity can promote ethics and equity in the conservation of genetic resources. pp. 51-56. In *Ethics and Equity in Conservation and Use of Genetic Resources for Sustainable Food Security.* IPGRI (Internat. Plant Genetic Resource Inst.) Rome (1997).

Mitchell, D.O., Ingco, M.D., Duncan, R.C. *The World Food Outlook.* Cambridge Univ. Press, Cambridge (1997).

Mooney, P.R. *The Parts of Life : Agricultural Biodiversity, Indigenous Knowledge, and the Role of the Third System. Develop. Dialogue* Special Issue Nos. 1-2, 1996. (Published in 1997). Published by Dag Hammarskjold Foundation, Uppsala, Sweden, and Rural Advancement Internat. (RAFI), Winnipeg (Canada) (1997).

Normile, D. Monsanto donates its share of golden rice. *Science* 289: 83-84 (2000).

Oerke, E.C., Dehne, H.W., Schonbeck, F., Weber, A. *Crop Production and Crop Protection: Estimated Losses in Major Food and Cash Crops.* Elsevier, Amesterdam (1994).

Pimentel, D. (Ed.) *Handbook of Pest Management in Agriculture.* CRC Press, Boca Raton (1991).

Pierce, F.J., Robert, P.R., Salder, J., Searcy, S. (eds.). *The States of Site-Specific Agriculture.* pp. 213-218, Soil Science Soc. of America (1996).

Ranjan, R., Upadhyay, V.P. Ecological problems due to shifting cultivation. *Current Science* 70: 1246-1250 (1999).

Ruttan, V.W. The transition to agricultural sustainability. *PNAS* (USA) 96: 5960-5967 (1999).

Sanchez, P. Science in agroforestry. *Agroforestry Systems* 30: 5-55 (1995).

Serageldin, I. Equity and ethics : Twin challenges, twin opportunities. pp. 1-6 : In IPGRI, Rome (1997).

Sharma, H.C., Ortiz, R. Transgenics, pest management, and the environment. *Current Science* 79: 421-437 (2000).

Sinclair, A.R.E., Ludwig, D., Clark, C.W. Conservation in the real world. *Science* 289: 1875 (2000).

Slikkerveer, L.J. Indigenous agricultural knowledge systems in East Africa : Retrieving past and present diversity for future strategies. In : Bennun, L.A., Aman, R.A., Crafter, S.A. (eds.) : *Conservation of Biodiversity in Africa.* Proc. of Conference held at the NMK, 30 August-3 September 1992, Nairobi, 133-142 (1995).

Stafford, J.V. (Ed.). *Precision Agriculture.* Vol. 1, pp. 45-58, BIOS Scientific Publishers (1997).

Swaminathan, M.S. Ethics and equity in the collection and use of plant genetic resources : some issues and approaches. pp. 7-18. In IPGRI, Rome (1997).

Swanson, T. Global values of biological diversity: the public interest in the conservation of plant genetic resources for agriculture. *Plant Genet. Resour. Newsl.* 105: 1-7 (1996).

Swift, M.J., Vandermeer, J., Ramakrishnan, P.S., Anderson, J.M., Ong, C.K., Hawkins, B.A. Biodiversity and agroecosystem function. In : Mooney, H.A., Cushman,J.H., Medina, E., Sala, O.E., Schulze, E.-D. (eds.) : *Functional Roles of Biodiversity. A Global Perspective.* New York, 261-289 (1997).

Tweeten, L.G., McClelland, D.G. (eds.). *Promoting Third-World Development and Food Security.* Praeger Publishers, Westport, CT (1997).

UNEP. UNEP Profile. United Nations Environ. Programme, Nairobi (1990).

UNIDO. *Environmentally Sound Management of Biotechnology.* United Nations Industrial Development Organization, Vienna. Report No. v. 95-52609 (1995).

WCED. *Our Common Future.* Oxford University Press, Oxford (1987).

WCMC (World Conervation and Monitoring Centre). *Global Biodivasity: State of the Earth's Resources:* Chapman and Hall, London (1992).

Wolverkamp, P. (Ed.) *Forests for the Future — Local Strategies for Forest Protection, Economic Welfare and Social Justice.* Zed, London (1999).

WRI. *World Resources.* Oxford University Press, Oxford (1996).

Chapter 11
Sustainable Development

INTRODUCTION

It has become a generally accepted fact that a finite system, even when provided with renewable resources, can only temporarily absorb growing consumption. Yet, the principle of sustainable utilization in harvesting natural resources has usually been ignored in favour of ruthless overexploitation for short-term gain. Two prime requirements for the long-term subsistence of humanity in acceptable living conditions are that the world population must be controlled and prevented from exhausting the Earth's carrying capacity and that our present economic systems are geared and remodelled as a recycling economy that runs on renewable resources.

The idea of 'sustainability' originated from the field of forestry. With a view to ensuring that the existing potential of the forest resource is retained, the amount of timber felled and the natural losses sustained should, on average, not exceed the growth rate of the forest. The same or similar rule can be applied to any type of renewable resource. If the sustainable yield is exceeded, a long-term decline in the average yields will accrue because the resource will be so heavily overexploited that its renewal would be doubtful.

Well-regulated forestry systems undoubtedly have an important place in a sustainable future. Unfortunately, the rate of global deforestation is only decreasing in those areas where exploitable resources have already been exhausted. Except for these areas, deforestation is occurring unchecked and, is some cases, even accelerating as, for instance, in the ecologically highly sensitive regions like the forest vegetation of the eastern slope of the Tibetan Plateau (Winkler, 1998). Another example is the ecologically valuable remainders of near-primaeval forests, whose conversion to plantation land is still profitable because of the governmental subsidies. The possible adverse consequences of this practice (e.g. the loss of genetic resources) are utterly ignored. It is now generally agreed that attempts to establish sustainable management methods should concentrate on secondary (i.e. anthropogenically altered) forestland and on fallow land following its rehabilitation. This can relieve the exploitation pressure on primary forests, whose remnants must at all costs be protected against further destruction on conversion. The potential of secondary forests in the tropics and subtropics has been widely misjudged in terms of their economic value, ecological and sociocultural functions and development prospects.

Maintaining sustainability may mean having to restrict the immediate use of the resource. In the short term, this could inhibit or slow down the potential economic development of a region. In such a case, deliberately refraining from using natural resources should be viewed as an investment in the future.

In the light of this strict definition, the sustainable use of non-renewable resources is virtually impossible. But since the value of these resources is constantly changing, sustainable use is often defined as meaning that the exploitation of the resource runs approximately in keeping with the technical progress achieved in gradually replacing this resource. Some optimists believe that oil, for example, will gradually be replaced by alternative fuels so that consuming this resource does not mean that the energy supply of future generations will be restricted.

Another approach is based on the assumption that the natural resources can be replaced by man-made capital. In this case, development can be termed 'sustainable' if it is *purchased* by using natural resources, because these resources are gradually substituted by man-made capital stock.

In any case, 'sustainable' development requires that natural resources be included in the overall economic system. For example, in arid areas, the the regionally available water supply can be replaced by using more capital to build desalination plants, bring in water at great expense from elsewhere, or introduce some more efficient irrigation technology. Except for the latter case, all that would be achieved is to replace the local water resources with other natural resources, whether in the from of fossil fuels or water from other regions.

'Sustainable development' is a blend of the four sectors, viz., the ecological, social, political and economic sectors. Overall social development is only sustainable if no deterioration occurs in the medium and long term in any of these sectors. All these sectors are strongly interconnected. Sustainable development is, above all, a socioeconomic concept whose content varies according to culture, time and perspective; it can be understood only by adopting an integrated approach.

Some critics feel that sustainable development is meaningless in the strict sense as it is devoid of operational meaning. According to the generally accepted definition, sustainable development must meet the needs of present generations without compromising the ability of future generations to meet their own needs. According to the critics, how can one know with any certainty what future needs will be? In the absence of more information, one might suppose that future needs and present needs will be identical, but the logical consequence is that it would be improper of the present generation to make any use of a nonrenewable resource, e.g. lead. To do so would mean that future generations would be less free than we are to mine lead, but on the assumption that the goal of sustainable development is itself sustainable across the generations, future generations would also be forbidden from using lead. Nonrenewable resources would be untouchable unless one generation could persuade itself that its successors will have less need of such resources than we have.

Activities in relation to global climate change are central to sustainable development. Progress in this area has been made both within countries and internationally. Atmospheric concentrations and emissions of the major greenhouse gases continue to grow. Projections suggest that, in the absence of a concerted international effort, they will continue to increase, dooming us to virtually irreversible global environmental change—with no clear understanding of the consequences.

Scientific uncertainty and the link between improved living standards and greenhouse gas emissions are the primary impediments to action.

The three major areas that require much more attention by the international community are:

(1) Developing countries need greater assistance and encouragement for help in protection of global climate.
(2) A stronger focus on the linkages between individual issues is warranted. For example, many of the activities that involve phasing out of ozone-depleting substances also use significant amounts of energy. Some of the CFC substitutes (notably HCFCs) are greenhouse gases. Care must be taken to manage greenhouse gas emissions during the phase-out of the ozone-depleting substances.
(3) Environmental problems need to be tackled as resource management issues rather than in the traditional way of merely avoiding or correcting specific environmental impacts. Both ozone-depleting substances and greenhouse gases, for example, might be managed more effectively as part of the broader issues of industry and of energy production and consumption.

Only when the needs of the least advantaged members of society are met can development and environmental sustainability become truly sustainable. Poverty alleviation and environmental sustainability are by no means mutually exclusive goals—both have social roots and one is a prerequisite of the other (Oyen et al., 1996).

More than 20% of the world's 6 billion people currently live in absolute poverty. At the same time, our environment is deteriorating more rapidly. But poverty is not solely a phenomenon of the developing world. Hunger, the lack of adequate shelter and deteriorating ecosystems can readily be found in several industrialized societies.

The poor can be viewed as both victims and agents of environmental damage.

DEFINITIONS AND CONCEPTS

Sustainable development refers to the means by which 'development' is made to meet the needs of the present without compromising the ability of future generations to meet their own needs (WCED, 1987). Since the needs of future generations cannot be accurately defined and the potential for wealth generation of species and ecosystems also cannot be precisely estimated or judged, the term 'apparently' implies that total biological assets are not reduced, in the long-term, through use.

In a rural context, sustainable use includes such things as conservation of biological diversity, fauna and flora and maintenance of such ecological functions as soil quality, hydrological cycle, climate and weather, river flow and water quality. It also implies maintaining supplies of natural game, fish, fodder, fruits, nuts, resins, dyes, fibres, constructional materials and fuelwood; all these are essential to the livelihoods of local people.

The WCED definition of sustainability, with its emphasis on human needs and sustaining livelihoods, needs to be distinguished from other definitions, subsequently, adopted by many development institutions, whose more technical definitions are in terms of ecosystems' continued production of goods of services or the maintenance of biodiversity (see, for example, Pearce et al., 1989; World Bank, 1991). Many definitions downplay or ignore the social and political issues implicit in the notion of sustainability.

According to WCED (1987), to achieve sustainability, it is essential to radically transform the present-day economies. It requires a basic change in the way natural resources are owned, controlled and mobilized. To be sustainable, 'development' must meet the needs of local people. If it does not, people will be forced to take from the environment much more than what is desirable. Sustainability is strongly linked to concepts of social justice and equity (UNEP, 1989). Its realization implies major political changes.

According to the International Union for the Conservation of Nature and Natural Resources (IUCN), 'the people who live in and around tropical forests should control their management' (IUCN, 1989).

Sustainable development is a multi-dimensional concept that is open to a variety of definitions. It can have working sense only when its four pillars—economic, environmental, social, and cultural—are equally solid and linked with one another. So far, no country has been able to achieve such a balance, and growing disequilibria between the different pillars in certain parts of the world are threatening the prospects of advancing towards more sustainable ways of using natural resources, improving relations between people and the environment and between different types of people. In fact, a cultural change is necessary to move towards sustainability.

The concept of sustainable development (*susdev*) originated in 1980 (IUCN/WWF, 1980; IUCN/WWF, 1991), as the means by which biodiversity and natural ecosystems might be saved while allowing humankind to continue to prosper. Sustainable development was viewed as a means of balancing the demands of nature and people. But does it constitute a legitimate alternative to continued outright destruction of nature? The earlier definitions of *susdev* were largely anthropocentric because they focused on human aspirations and well-being, with the natural environment providing the means by which this was to be accomplished (Robinson, 1993). For example, the concept of development stressed that we should satisfy human needs and improve the quality of human life. Conservation of biodiversity was a way of ensuring that we give the greatest *susdev* to present generations while maintaining its potential to meet the needs of future generations. IUCN/WWF (1991) defined *susdev* as improving the quality of human life while living within the carrying capacity of supporting ecosystems.

As, in the above definitions, sustainable development is focused around one species—*Homo sapiens*, and promotes its continued and even expanded economic prosperity, these definitions fail to recognize the extraordinary complexities of nature and the diversity of life-forms, also to the services they provide to humanity. They also fail to appreciate the basic underlying problems that create a need for sustainability in the first place. As rightly pointed out by Korten (1991-92), the Brundtland (see WCED, 1987) report's key recommendation called for the world's economic growth to rise to a level five to ten times the then output and for accelerated growth in the industrial countries to stimulate demand for the /products of poor countries. This recommendation actually contradicted its own analysis that growth and overconsumption are the root causes of the problem. This kind of contradiction may have stemmed from the greater importance attached to political expediency than environmental needs, because it has been observed that wherever ecological reality conflicted with political feasibility, the latter usually prevailed. Interestingly, a subsequent addendum to the report (World Commission on Environment and Development 1992) stepped back from this recommendation for economic expansion and placed higher priority on population control (Goodland, 1995).

Any realistic definition of *susdev* must give due emphasis to both human population control and equitable environmental demands. In keeping with this idea, Viederman (1991, 1996) offered the following definition: A sustainable society is one that ensures the health and vitality of human life and culture and of nature's capital, for present and future generations. Such a society acts to stop the activities that serve to destroy human life and culture and nature's capital, and to encourage those activities that serve to conserve what exists, restore what has been damaged and prevent future harm (see Viederman, 1996).

Meffe et al., (1997) further improved the above view and defined truly sustainable development as *'human activities guided by acceptance of the intrinsic value of the natural world, the role of the natural world in human well-being, and the need for humans to live on the income from nature's capital rather than on the capital itself'*. They pointed out the distinction between sustainable *growth* and sustainable *development*: Growth is a *quantitative* increase in the size of a system while development is a *qualitative* change in its complexity. An economic, social, or biophysical sytem can develop without growing, and so can be sustainable. It can also enlarge without developing or maturing; this is *not* sustainable development. 'Sustainable growth' is a self-contradictory term. Continued, indefinite growth is a physical impossibility because eventually limits of some type (space, food, waste disposal, energy) must be reached; the point at which that will happen is the only debatable issue here. 'Sustainable *development*' is the issue of concern. Can we make *qualitative* changes in complexity and configuration within existing human systems that do not place increasing *quantitative* demands on natural systems, and are, in fact, compatible with their continued existence? (Meffe et al., 1997). To reach such a goal, it is necessary to understand the patterns of human behaviour and desires that brought us to this crisis in the first place. Viederman (1996) summarized these as follows:

1. The economic system is an open system in a finite biosphere. It is not a closed system, separate from the biosphere, as most traditional economists would have us believe. It requires inputs from and exports to living ecosystems, which impose real limits at both ends. Further, much of our attention has focused on resource constraints

and substitutability, rather than sink constraints, i.e. the disposal of wastes.
2. We have not realized that the environment is the basis for all life, including our own as well as production.
3. We have arrogantly held on to the erroneous belief that we can master and control nature.
4. Uncritical acceptance of technology cannot be the answer to all problems, despite a multitude of examples of today's problems being yesterday's solutions.
5. We have failed to differentiate clearly between growth and development, probably due to our belief in technology as saviour. Likewise, we do not realize that growth will not automatically lead to equity and justice within and among nations: an 'economic trickle down' effect is an unlikely and unfair assumption and simply an excuse for a few to amass personal wealth at the expense of many.
6. We are greatly mistaken in believing that market systems are the main mechanism for realizing social goods such as economic sustainability and justice.
7. We have tended to ignore any consideration of the needs of, and our obligations to, future generations—a concern that must surely lie at the core of any sustainability concept.

Viederman (1996) listed the following seven *principles of sustainability:*

1. Nature is an irreplaceable source of knowledge, from which we may extract potential solutions to many of our problems.
2. Environmental deterioration, human oppression and violence are inter-linked.
3. Humility and restraint must guide our actions.
4. The importance or 'proper scale' needs to be realized. Place and locality are the foundation for all durable economies, and must be the starting point of action to address our problems. Solutions are local and scale-dependent.
5. Sufficiency must replace economic efficiency because the earth is finite. Living within our needs on a planetary scale does not mean a life of sacrifice, but of greater fulfillment. We must distinguish among 'needs' 'wants' and 'greed'.
6. Community is essential for survival. The 'global community' should reflect and encourage diversity while being interdependent.
7. Biological and cultural diversity needs to be actively preserved, defended, and encouraged.

A few other similar hypotheses concerning sustainability as suggested by Bartlett (1994). Goodland (1995), and Meffe et al., (1997), are outlined below:

1. For today's average global standard of living, the current population of the earth exceeds its carrying capacity.
2. Increasing population size is the single gravest threat to representative democracy.
3. The costs of programmes to halt population growth are quite small as compared with the costs of population growth.
4. The time required for a society to make a planned transition to sustainability increases with increasing size of its population and the average per capita consumption of resources.
5. Social stability is a necessary, but not sufficient, condition for sustainability. Social stability tends to be inversely related to population density.
6. The burden of the lowered standard of living resulting from population growth and from the decline of resources is heaviest upon the poor.
7. Environmental problems cannot be solved or mitigated by increases in the rates of consumption of resources by society at large.
8. The environment cannot be enhanced or preserved through repeated compromises.
9. By the time overpopulation and shortage of resources become obvious to most people, the carrying capacity has already been exceeded. It is then too late to pursue sustainability effectively.

The concept of sustainability warrants that our consumption levels do not exceed the annual increment of renewable resources. Non-renewables need to be used very cautiously, and their reuse and recycling ensured. Resource-use should cause the most minimal and manageable damage to environment. This can be achieved by effective application of science and technology.

THE TRILOGY OF SUSTAINABILITY

All definitions of sustainable development imply the need for kinds of development that are sustainable in three senses; (1) Development should not damage or degrade natural resources and our basic life support systems such as air, water, soil and vegetation. (2) It should be economical and should ensure a continuous flow of goods and services from natural resources. More careful and efficient use of our water resources, soils and forests is required now than ever before. (3) Social systems should be sustainable at international, national and local levels to ensure equitable distribution of the benefits of the goods and services produced and of sustained life-support systems. The last two aspects of sustainability, viz., the need for economic and social systems, are human-centred whereas the first aspect applies to most forms of life on earth. However, all the three aspects are strongly interconnected.

The apparent dichotomy between environment and economic development has hindered developmental efforts in the past. Wise management of environmental resources should, in fact, be considered to be an integral part of economic development, not as an appendage.

GOALS OF SUSTAINING NATURAL SYSTEMS

Four important goals of sustaining natural systems for use of future generations are: (1) protection of life-support systems of the air, water and soils; (2) protection and enhancement of biological diversity; (3) maintenance or restoration of stocks of renewable natural resources and ecosystems; and (4) prevention of, or adaptation to, global change (see Bruce, 1992).

Several decades ago, there used to be a serious problem of local contamination of the atmosphere and water bodies in many countries. This has gradually led to regional and global problems, e.g. acid rain; acute scarcities of water in some regions; devastating droughts and desertification; increased incidence of natural disasters such as floods and tropical cyclones; airborne transport of toxic chemicals to distant seas and lakes; increasing greenhouse gases and the related climate change; depletion of the stratospheric ozone layer; oil pollution of seas and lakes; environmental emergencies from spills, fires and nuclear and chemical emissions (Bruce, 1992). Nevertheless, it is still possible to tackle some of these problems and strive towards nationally and globally sustainable development by resorting to the following measures:

- In the industrially-developed world, economic systems should be geared to use resources more sparingly and more efficiently and minimize waste discharges,
- In developing countries it should be ensured that economic growth to meet the expectations of growing populations occurs in a manner that will minimize resource depletion and environmental stress.

A vital consideration in any move towards the above goals should be to ensure that indicators which reflect resource depletion and environmental changes reflect the true costs to society of an economic activity. We should not simply determine the real cost of degradation of the quality of air or water, or of the use of a certain amount of water for a given activity, thus making it unavailable for other activities or for the maintenance of healthy ecosystems. Admittedly, it is virtually *impossible* either to calculate such environmental and resource-depletion costs, or to judiciously allocate water without the basic measurements and analyses of air and water quality and of the quantity and the flow of water.

Recent surveys of FAO suggest that political and economic liberalization in developing countries has not yielded its anticipated benefits for the rural poor and is not likely to succeed unless it provides them with the means of participation in development. No doubt, there is a strong consensus for liberalization, but for many nations, liberalization has turned out to be a fairly problematic process that has led to cuts in rural wages and employment, rising production factor costs and reduced public spending. These trends have been noted throughout Africa, Asia, parts of Europe, the former USSR and Latin America. According to a 1995 report of FAO, 'The relationship between economic reform, food production and rural poverty remains shrouded in seemingly contradictory evidence. Rural poverty has increased in many regions and continues to represent both a denial of a basic human right and a constraint on economic growth.' Economic and political reform may not succeed without strengthening rural institutions and social organizations to facilitate people's participation in development. Reforms should include support to small farmer production systems, human resource development and strengthening of local economic capacities. For many countries, this means giving high priority to agricultural development.

In many countries, while subsidies for rural production have been decreased or eliminated, those for urban consumers have not, leading to unequal terms of trade, and can lead to an increase in poverty, lower production and unsustainable cultivation practices. Removal of price controls is one way to boost output and reduce rural poverty.

SUSTAINABLE DEVELOPMENT AND PUBLIC POLICY

Sustainable development, appropriately defined, provides a broad goal and a frame of reference for public policy. It is a unifying concept embracing social, economic, and ecological aspects. The governing perspectives on sustainable development should specify and exemplify the nature and relationship of its common denominators namely economic growth, ecological integrity and intra- and inter-generational equity (Sadler and Jacobs, 1991). Key examples include studies of economic methods of resource valuation and the preparation of environmental accounts. Such research should be extended to social and community perspective and lead to the formulation of certain integrated indicators to assist decision makers, for instance, by clarifying the trade-offs among key variables and highlighting appropriate actions that promote sustainable development or redevelopment.

Environmental assessment is a necessary but not sufficient process for achieving sustainable development. But the process is usually too narrow in focus and limited in scope of application to capture cumulative environmental and social effects. These occur as a result of a multitude of technological changes, economic activities and resource use and management practices. Neither the market place nor government can satisfactorily account for the real costs of development or the true value of maintaining natural resources, such as clean air and water and wildlife habitat (Sadler and Jacobs, 1991). According to Sadler and Jacobs, research on improvements to EA processes in support of sustainable development should be linked to these broader issues of evaluation and decision-making. Sustainable development assessments (SDAs) should explicity address the economic, social and ecological interdependencies of policies, programs, and projects. These must be coordinated with other policy and management instruments as part of an overall approach to environment-economy integration.

No doubt there is less population growth in industrialised, countries, but still, these countries add a disproportionately high amount of carbon to the atmosphere each year as do the people in most developing countries. Thus, the developing world has to realize the importance of achieving a balance between population and the sustainable development of their resources, and similarly the developed countries must also change the way they consume resources. It also needs to be borne in mind that as incomes rise and lifestyles and technologies change in developing countries, their emissions of greenhouse gases will also rise beyond the current levels.

According to the 1990 *State of the World Population Report* by UNFPA, if carbon emissions in the world could be kept as low as one metric ton per person by the year 2025, and the population could remain below 8 billion, the impact upon carbon dioxide emissions would be equivalent to reducing deforestation by 97%.

The phrase 'sustainable development', has been used to express profound concern for both humanity and our environment (Westing, 1996). Unfortunately, the phrase has come to mean different things to different people, creating confusion (Frazier, 1997). Ecologists focus on the long–term persistence of biotic resources and the continuation of ecological processes. Resource managers are concerned with maintaining long-term yield. Social workers emphasize socio-economic development and economists perform cost-benefit analyses. Business leaders talk about sustainable companies and the 'greening' of industry (Gatto, 1995). In reality, the concept of sustainable development includes much more than either the indefinite availability of environmental assets or long-term business success (Lele and Norgaard, 1996).

A perusal of recent literature on the condition and improvement of environmental and/or human situations, as well as on the interactions between humans and their environment, shows that the term 'sustainable' has been used in conjunction with a wide variety of concepts and expressions generating such phrases as 'sustainable biosphere', 'sustainable ecology', 'sustainable landscape', 'sustainable pest management', 'sustainable mountain development', 'sustainable tourism', 'sustainable transportation', 'sustainable urban transport system', 'sustainable society', 'sustainable way of life', 'sustainable improvement' in the quality of 'sustainable future', 'sustainable planet' and 'sustainable world'.

The most obvious problem with the definition is that the use of the phrase sustainable development (*susdev*) is never accompanied by any precise definition of either 'sustainable' or 'development'. What is being sustained or developed? For whom? By what means? Why does it need to be developed? Why does it need to be sustained? (Shiva, 1992; Isbister, 1993; Orr, 1994; Korten, 1995). *Susdev*, defined in environmental terms, is very different from *susdev* focused on contemporary economic or industrial values. *Susdev* designed for the affluent people is not the same as that for the poor. In the phrase, 'development' is the noun, and 'sustainable' is the adjectival modifier. The term 'developed' is tightly linked with the concept of 'industrialized' or 'modernized'. *The World Conservation Strategy* (WCS, 1980) defined *susdev* to mean 'satisfy human needs and improve the quality of human life'. Quality of human life is commonly measured as access to manufactured goods. Human development is usually defined as some enlargement of people's choices. It is felt that short-term advances in human development are possible—but they will not be sustainable without further growth (Haq and Jolly, 1996)—the word 'growth' referring to the production and distribution of goods and services, especially through industry and manufacturing process (Frazier, 1997).

In a sense, 'development', as commonly used, presupposes that social progress associated especially with the industrial and post-industrial revolutions, is a necessary and natural outcome of human liberation from the constraints of 'primitive' traditional cultures 'dependent on nature' (Shiva, 1992). The basic idea is that 'modern' human societies have been 'freed' from the limitations of the geochemical and biological processes which restrain other forms of life on the planet (Bodley, 1988).

The term 'sustainable' refers to holding up, supporting, supplying or providing for, i.e. maintaining processes, as well as both physical and conceptual entities. Basic to the term is the concept of continuity, i.e., maintaining something indefinitely. A conjunction of the two words gives rise to the phrase sustainable development, of so much interest to all those concerned with the fate of natural resources, biodiversity and the environment, as well as those preoccupied with socio-economic progress, eradication of poverty, human rights and other matters of human welfare.

Largely, the basic idea of *susdev* is to assist the poor in acquiring better access to resources so that they will not remain so poor. Ideally, this should translate to *enabling* the poor to develop, not to develop the poor (Bodley, 1988, 1990).

For their progress, the poor will need access to far more resources than they now have. Any increase in their share of resources means a big change in accessibility, distribution and allocation. However, material reserves being resources and sinks, finally, human value systems have to change to reflect ecological realities. Humans need to accept limits, replace short–term thinking with long-term perspectives that include intergenerational equity, replace narrow anthropocentrism with a broader view, and incorporate ecological design principles as we move from an industrial age to an ecological age. Updated and appropriate value systems offer the best hope not only for conservation of biological diversity but also for sustainable development.

THE DEMOGRAPHIC IMPERATIVE

It seems impossible that our contemporary socio-economic systems can keep everyone satisfied (Brown, 1993; Durning, 1993; Orr, 1994; Korten, 1995; Lélé and Norgaard, 1996). The contemporary tempo of development, in terms of resource consumption and environmental disturbance, largely benefits a minority of the world's population and is untenable as a sustainable strategy. Just to *retain* present levels of material wealth and comfort *for* only 20% of the world population, major changes in human behaviours, attitudes and especially consumption patterns, will be required. To propose extending the contemporary level of development, which even today is unsustainable, to an even greater proportion of the world's population seems a herculean task indeed.

The human 'demographic imperative' is so overarching that all other success or failure may ultimately flow from it. Human population growth and its increasing impact on the planet must be slowed, stopped and reversed. The total impact is a function of population size, affluence and technology. Several countries have already reversed disastrous population growth trends in only a decade, demonstrating that it can be done.

One of the major forms of patently unsustainable and destructive practice that has caused so much environmental degradation is the so-called 'tragedy of the commons'. Tragedies of the commons occur when some common resource such as air, water, or wilderness is exploited for private gain at little or no expense to the user. The resources are typically overexploited to the benefit of a few, but the costs are spread among many. A solution is to have the costs of using these resources internalized to the users, rather than spread among the general public. At least five major actions on the part of humanity are needed to conserve biodiversity in the long run. The human population growth problem must be dealt with. Tropical forests and other hot spots of biological diversity must be protected from further destruction. We should develop a more global perspective on the earth's resources, but solve problems locally wherever possible. Growth oriented economics should be replaced with ecological, or steady-state economics; any significant change in allocation can mean increasing resources in some places while *limiting* or *decreasing* them in others (Shiva, 1992). The question is: can and will the rich and powerful wilfully relinquish access to resources?

During the next two decades, without any further reproduction, more than a third of the present-day human population, if it survives, will reach adulthood. These thousands of millions of people will naturally be searching for resources and a means of livelihood (Pinstrup-Andersen and Pandya-Lorch, 1996). Even the provision of basics, such as water, food, shelter, health, elementary education, security and peace, for the new additions to the adult population, is going to exert severe strain on existing patterns and rates of consumption and processes of resource distribution, to say nothing of waste production and disposal. If the population remained constant, the problem would be difficult enough, but there is absolutely no warrant that the numbers of people will really remain stable. There are demographic indications that our numbers will continue to expand fairly rapidly in the coming decades, with most of this increase occurring in Third World countries. Estimates are that by the year 2050, the world populations of today will have nearly doubled (Tolba et al., 1993). All these people will need basic resources, to say nothing of a 'developed' life style.

If underdeveloped societies were to progress economically and become developed, their requirements for resources and energy would be very high, even outstripping the demands of the present-day developed nations. Of particular concern are vast, irreversible modifications to the living environment, notably massive extinctions of plant, animal and microbial life and disruption to ecosystem services, the consumption of non-renewable resources, such as fossil fuels; the production of wastes, and changes in biogeochemical processes such as thermal characteristics of the atmosphere and the carbon cycle. All this means that the affluent and the rich people are not going to sacrifice their affluence for the sake of the poor, regardless of how loudly they talk of sustainable development!

PERSPECTIVES AND PRINCIPLES

Sustainable development is a strategic approach to integrating conservation and development. Although the concept of sustainable development has come into popular limelight only recently, its roots are fairly long and deep seated. In some countries, ideas of resource management and conservation, such as best use, wise use, and sustained yield, date back to the turn of the previous century.

At the first World Conservation Strategy Conference held in 1986, international experts rightly emphasized the need for continued development in harmony with environmental conservation, decrying the tendency of developed countries to focus on global conservation measures in isolation from the problems of under-development. Gardner (1991) mentions the following principles for sustainable development. These principles can also be interpreted as objectives, criteria, pre-conditions, desirable characteristics, components, parameters, or guidelines for sustainable development.

1. The satisfaction of human needs (Table 11.1). This principle contains the notion that biological sustainability depends upon the sustenance of the human culture that determines the way resources are used. Growth is necessary for the satisfaction of needs related to energy, water, food, jobs, and sanitation and for increased human prosperity and well-being. However, because needs can only be met by conserving and enhancing the resource base, economic growth will have to be less material-and energy-intensive and more equitable in its impact.
2. The maintenance of ecological integrity (Table 11.1-b). This principle relates to living resource conservation: maintaining essential ecological processes and life support systems; preserving genetic diversity; and ensuring the sustainable utilization of species and ecosystems.
3. The achievement of equity and social justice (Table 11.1-c).
4. The provision for social self-determination and cultural diversity.
5. Goal-seeking is a process-oriented principle.
6. Another process oriented principle is that analytical approaches to sustainable development must be *relational or systems-oriented.*
7. The third process principle enuciates that strategies for sustainable development must be *adaptive.* Adaptive approaches manage risk through anticipation and prevention while seeking balance in human and natural systems through monitoring and self-regulation.
8. The fourth such principle stipulates that organizational approaches to sustainable development should be *interactive.*

TRADITIONS

For centuries, tribal people in several Asian countries have preserved their forests to protect them and their animals against hunger, disease, injury and bad spirits. Having a forest belt around the village is supposed to

Table 11.1 Summary principles for framework assessment (after Gardner, 1991)

1. Ideology : Goal-seeking
Process-oriented
- 1a proactive, innovative, generates alternatives
- 1b considers range of alternatives and impacts
- 1c based on convergence of interests
- 1d normative, policy-oriented, priority-setting

Substantive
- A1 quality of life and security of livelihood
- B1 ecological processes and genetic diversity
- C1 equitable access to resources, costs and benefits
- D1 individual development and fulfillment, self-reliance

2. Analysis : Relational
Process-oriented
- 2a focused on key points of entry into a system
- 2b recognizing linkages *between* systems and dynamics
- 2c recognizing linkages *within* systems and dynamics
- 2d importance of spatial and temporal scales

Substantive
- A2 development as qualitative change
- B2 awareness of ecosystem requirements
- C2 equity and justice within and between generations
- D2 endogenous technology and ideas

3. Strategy : Adaptive
Process-oriented
- 3a experimental, learning, evolutionary, responsive
- 3b anticipatory, preventative, dealing with uncertainty
- 3c moderating, self-regulating, monitoring
- 3d maintaining diversity and options for resilience

Substantive
- A3 (growth for) meeting a range of human needs
- B3 maintenance, enhancement of ecosystems
- C3 avoid ecological limits and associated inequity
- D3 culturally-appropriate development

4. Organization : Interactive
Process-oriented
- 4a collaboration for the synthesis of solutions
- 4b integration of management processes
- 4c integration of societal, technical, and institutional interests
- 4d participatory and consultative

Substantive
- A4 organizations must respond to societal change
- B4 ecological principles guide decision-making
- C4 democratic, political decision-making
- D4 decision-making locally initiated, participatory

bring happiness to the community. Big trees are like mothers and little trees are like children, needing to be encouraged and nurtured.

The economic systems of Asia's forest peoples are very often backward and irrational. There is a mistrust of peoples who are neither subject to government control and taxation systems nor contribute greatly to the market economy. The Dutch have considered shifting cultivation in Indonesia as the 'robber economy'. In India, the British classified shifting cultivation areas as 'wastelands'—not because the practice laid waste the forests but because it provided no revenue to the Empire (Tucker, 1988). With the increasing pressure on natural resources such systems are also now criticized as being environmentally destructive (Agarwal and Narain, 1989). However, recent critical studies of these economies have suggested quite different conclusion. In many cases, the tribes consciously manage their resources to ensure sustained yield.

Practices among tribal communities to conserve resources, restore soil fertility, mimic biodiversity and protect watersheds have been fairly well documented throughout the region. Equally, studies reveal the immense reserve of practical lore in forest-based societies concerning their environment: to those who know, the forest is a rich treasure of drugs, medicinal herbs, spices, fruits, oils, resins, gums, dyes and many other products.

In summary, it appears that traditional systems of resource use are far more diverse, complex and subtle than believed hitherto. The social, cultural and institutional strengths inherent in traditional systems of resources use need to be built on to achieve sustainability and not dismissed as 'backward' and 'wasteful' (Colchester, 1992).

RESISTANCE TO DESTRUCTION

The intimate association between forest peoples and their land, and their determination to maintain their way of life is most obviously expressed in their opposition to imposed and destructive change. Across the region, hundreds of different indigenous movements have their roots in resistance to cultural, economic and political oppression. Most obvious of these have been the mass movements of forest groups that have mobilized to confront specific threats to their future.

In India, the tree-hugging 'Chipko' movement developed in the context of a long history of popular campaign against government control of forests in the Tehri-Garhwal area and has been gaining momentum during the last few decades. Eventually, due to the unrelenting determination of villagers and activists to protect the forests in the face of arrests and police harassment, the forest management policies in the area had to be changed. While logging has been finally curtailed, the Chipko grew into a tree planting movement to restore forests to denuded hill slopes. The movement has already extended to cover many other parts of India (Hegde, 1988).

Besides Chipko, similar mobilizations against hydropower programmes that are displacing thousands of tribal communities and flooding their forests and farmlands have sprung up all over the country. Mass marches of protesters have led, in some places, to the cancellation of proposed dams and, in others, have resulted in police firings and deaths. However, not all such movements have been environmentally benign. In Bihar, where the battle between commercial plantations and communal use of forests has been symbolized as a struggle between teak and sal trees, members of the Ho tribe, who have lost rights to forest lands, have mobilized against official forestry programmes and developed a 'forest cutting movement'. Despite having an ancient tradition of respect for forests, including the preservation of sacred groves for religious ceremonies, the Ho have turned to forest clearance as a means of asserting their rights to use the lands which forestry laws deny them (Colchester, 1992). There has also arisen some popular mobilization against forestry plantations in Karnataka state. Here, attempt to take over common lands for commercial plantations of fast-growing eucalyptus for paper, pulp and rayon led to a 'pluck and plant' movement, in which the eucalyptus seedlings were uprooted and replaced by indigenous species that provide products useful to the local peasants (Kanvalli, 1990). The movement led to the suspension of international financial aid from World Bank to 'social forestry' programmes centred on eucalyptus planting.

Instances of strong resistance to imposed development have occurred in several other countries such as the Philippines, Bangladesh, Papua, New Guinea, etc.

Figure 11.1 brings out some relationships involving development and rehabilitation strategies in lowland 'sal' forests in South Asia.

TRANSITION TOWARD SUSTAINABILITY

Many people have pioneered visions for achieving sustainable development based on a partnership between business and environmental scientists. Some people wrongly believe that ecologically destructive policies are a necessary, though unfortunate, outcome of achieving improved standards of living. An analogous view was held by several nineteeth century businessmen that slavery was an essential element in the economic system! Conversely, those who believe that business is not to be trusted and should be uniformly opposed, are equally wrong. The transition to sustainability requires active cooperation—not confrontation between business and environmental science (see Schmidheiny, 1992; NRC, 1999).

Sustainable Development   487

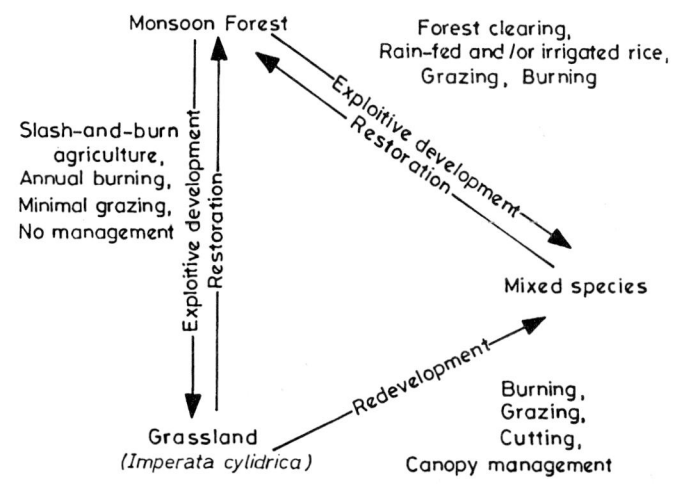

Fig. 11.1 Some development and rehabilitation strategies and tactics in lowland *Shorea robusta* (sal) forests in South Asia (after Dyer and Ishwaran, 1990).

With the increase in globalization and multinational corporations, the influence of governments in the management of natural resources has diminished. Most countries are unlikely to achieve the sustainable use of their resources without the active involvement of private enterprise. This is particularly evident in managing the Earth's natural capital so as to yield a flow of life-supporting services. The true social costs of conservation are quite small: indeed, the benefits outweigh the costs when the value of natural capital is properly calculated.

But governments alone will not be able to do much for environmental conservation. This is borne out by the observation that even well-managed nature reserves are vulnerable to rapid loss of species. Most governments lack the capacity to control the impact of human activity either within reserves (for example, poaching) or outside their borders (affecting water, fire and climate regimes, the influx of farm chemicals or invasive species). The impacts involve diverse contributors across all scales, from local to global, and the failure to control one impact often hampers the successful mitigation of others. Such circumstances typify the environmental problems of today and warrant a broad, integrated approach. There is a growing trend towards private-sector involvement in the funding and management of nature reserves (McNeely, 1999; Daily and Walker, 2000).

A strong private sector has an enormous capacity to achieve sustainable use of resources. It is industry, not law-makers, that manages and uses environmental resources, whether directly or indirectly.

Changing practices in industry exert cascading effects through all the determinants of environmental impacts.

The private sector is also innovative and adaptable. The challenge for society is to alter selection pressures so as to align profit-seeking with sustainability.

Thirdly, private enterprise is much more efficient as compared to government, its efficiency being directed towards minimizing private costs and thereby often imposing costs on society. This efficiency can be exploited to improve environmental conditions.

Lastly, private enterprise is pragmatic. The active participation of a critical mass of industries, which create and absorb particular environmental impacts, can catalyze the implementation of effective countermeasures.

Many businesses have found profitable ways of reducing environmental impacts within the existing economic system. Bigger steps—changing the system to align profit-seeking with sustainability—require at least four ingredients: a scientifically based vision of a 'sustainable' business enterprise; an understanding of the competitive advantage in achieving the vision; effective communication of the vision; and a means of incremental implementation. The first of these requirements is to where science can contribute most (see Socolow et al., 1994; Daily and Walker, 2000).

Businesses can profit by staying ahead of the policy maker and reducing environmental impacts voluntarily. Some new approaches facilitating this smart manoeuvring are outlined below.

The Natural Step programme, an overarching framework developed in 1989 in Sweden, is pre-eminent in Europe and now attracts major US corporations. It incorporates analytical tools from ecological footprint analysis, the ecosystem services framework and industrial ecology. There is also a growing number of environmental (and related) standards, basically compliance procedures for guaranteeing to the clients of a business that a particular set of pratices is really being implemented. Certification is voluntary and driven by the marketplace.

SUSTAINABLE MANAGEMENT OF ECOSYSTEMS AND NATURAL RESOURCES

Holling and Meffe (1996) defined 'pathology of natural-resource management' as a loss of system resilience when the range of natural variation of the system is reduced. This pathology results in less resilient ecosystems, following the manipulation and control by humans. It was identified as pre-consequence of a command-and-control approach to managing our natural resources. In attempting to control variation, commodity extraction and behaviour of natural ecosystems, these systems inevitably become less resilient when faced with extreme events such as periodic natural or human-induced disturbances.

Since the natural world has been traditionally considered to be ordered, segmented and mechanistic, with linear, cause-and-effect relationships, not surprisingly, the various resource management agencies compartmentalized themselves into specialties that employed a command-and-control approach to manage resources (Nelson, 1995). This strategy proved to be effective during an era of utilitarian management of natural resources. Viewing trees as lumber, wildlife as game and grass as forage, allowed the agencies to employ managers with new, advanced technology to produce more outputs and organize around a machine-model bureaucratic operation (Knight and Meffe, 1997). These management efforts faithfully maintained control of the resource. There was no cooperation with the resource users.

With the advent and adoption of ecosystem management agency attitudes and behaviours have been changing and evolving (Grumbine, 1994). Ecosystem management encourages partnerships, cooperation and risk-taking. It contrasts sharply with the linear command-and-control approach of traditional resource management that encouraged hierarchical decision-making and risk aversion.

Some concepts that illustrate differences between the traditional approach and an evolving perspective encouraged by ecosystem management are shown in Table 11.2. A stewardship approach to managing natural resources reflects general changes found in today's societal, political and economic systems, as well as a move in science away from certainty and positivism, and toward uncertainty and

Table 11.2 Comparison of management perspectives in natural-resource agencies driven by traditional command-and-control management versus an ecosystem approach to stewardship (after Knight and Meffe, 1997).

Tradtional management	Ecosystem-based approach
Top-down decision making	Input from all levels
Centralized, linear	Decentralized, with feedbacks
Risk-aversive	Risk-taking
Finality of decisions	Willingness to revisit, revise and admit error
Imposed vision	Shared vision
Within-administrative boundary	Across-administrative boundary
Control	Partnerships

pragmatism (Capra, 1991). By adopting ecosystem management, agency personnel have started to redefine their roles as land stewards rather than land controllers. The traditional management approach emphasizes command from above (top-down), rather than individual initiative and input from all levels. Because of the complexities of managing landscapes for a variety of worthwhile goals (commodities, amenities, biodiversity), this one-size approach issued from the top usually clashes with specific needs of individual projects. Ecosystem management acknowledges strengths associated with a decentralized approach to management, rather than the traditional, linear organizational structure. Ecosystem management encourages risk-taking rather than risk aversion.

There has been considerable progress in applying satellite-based remote sensing technology to the management of natural resources. Satellites can cover large areas repetitively and collect voluminous data which makes it possible to monitor regularly. Several satellites are already in orbit and providing data for study of the Earth's resources. Applications to water resources management have been especially very promising as satellite imaging can show ground water potential zones through association with geological and geomorphological features such as lineaments, flood plains, valley fills and rock outcrops. Indeed, this has already been accomplished for the whole of India using data from Landsat and IRS satellites. Borewell sites have been located for over 300,000 wells in remote and rural areas with a success rate of over 80%.

As sustainable development has to be approached in a holistic manner; merely a 'green' environment strategy is not enough. In some places, the green focus can be augmented with a 'blue' sector to extend environmental efforts to cover mangrove forests, coral reefs and fisherfolk in community-based coastal zone management activities. A 'brown' focus can be achieved by recognizing the importance of rapid urbanization and industrialization with concomitant environmental management capacity. In this sector, the business sector can be a strong partner.

The concept of the green, blue and brown foci may be illustrated (Fig. 11.2) with reference to municipal solid waste management (MSWM).

The environment impact assessment (EIA) system should be strengthened as the primary tool to strike a balance between development and environment by enabling the public to participate in decision-making through the requirement of social acceptability. All environmentally-critical projects should conduct EIA and include special sections on impacts on women and indigenous peoples, if relevant.

RESOURCE MANAGEMENT SUSTAINABILITY

There are varying interpretations of how social and economic issues and policies may be included in resource management decision making. Some

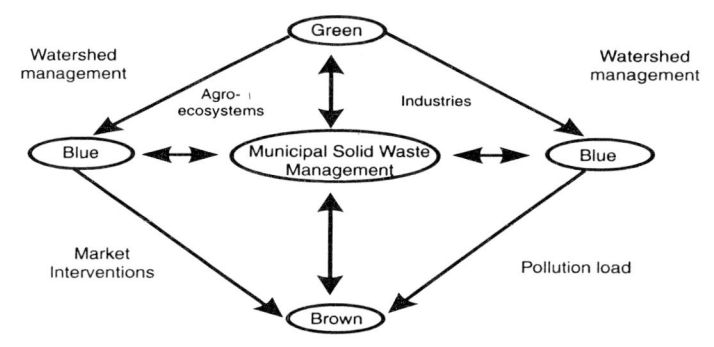

Fig. 11.2 Municipal solid waste management with multiple environmental foci.

give greater emphasis and precedence to the ecological aspects (the ecocentric approach) whereas others give precedence to the anthropocentric (social and economic) issues.

According to Fisher (1991), the ecological function is afforded a degree of priority over the management function, which is ancillary. This suggests that we should consider the 'ecological function' before making decisions related to the 'management function'. Nonetheless, according to Fookes (1992), the apparent preference in legal interpretation for the management function having primacy does not exclude the attainment of social and economic objectives. Rather, it simply means that such objectives must be achieved within biophysical/ecological objectives. Humans bring values into their considerations and these values can range from judgement as to the quality of the environment to the need for employment. The need to sustain natural and physical resources requires sufficient understanding of the physical limits or capacities of that environment to ensure that social and economic aspirations do not damage the environment. Conversely, socio-economic aspirations may require a quality of environment above that sustainable level. Any policy and plan, therefore, needs to identify these values in a transparent manner so that the various alternatives and consequences can be aptly judged.

For many issues, there is a range of options. Some options, stated either as objectives or policies, may address a social or economic agenda but affect some aspect of the environment quite adversely. Another option might still create social or economic benefits while not having dire consequences for the natural environment (Fookes, 1992).

The definition of environment is quite specific about how social and economic matters are to be considered. Environment has three parts to it: ecosystems (which may include people and communities when they are part of the ecosystem) natural and physical resources (which includes structures) and amenity values (which includes people's appreciation of the character of a locality). Social and economic factors are included in this context only to the extent that they affect, or are affected by, these three parts of the environment. In other words, social and economic factors are considered relevant 'effects' only to the extent they influence these specified aspects of the environment, or are influenced by these aspects.

REFORM AND REDEVELOPMENT OF DEGRADED REGIONS

Recent years have seen an emergence of the concept of sustainable redevelopment which may soon become a major policy for regions now suffering bad consequences of 'conventional exploitative development' (Fig. 11.3).

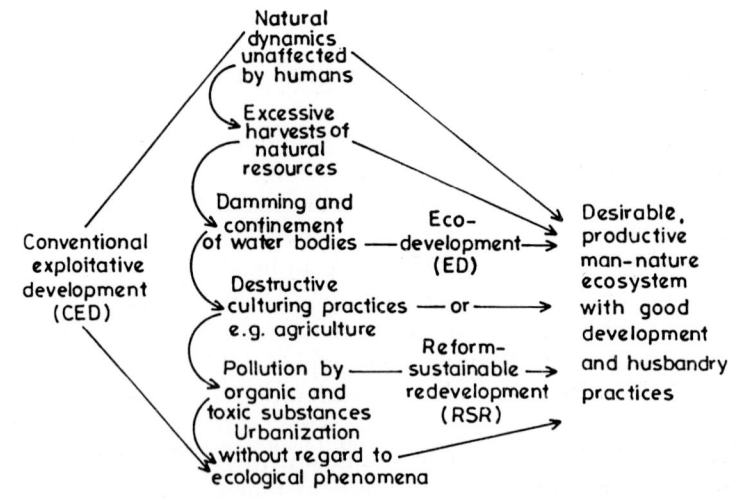

Fig. 11.3 The meaning of major policy options for sustainable development, eco-development and/or reform sustainable redevelopment (after Regier, 1991).

There exists a quite serious problem of the massive regional degradation arising as a consequence of conventional exploitative development.

For degraded developed regions of the world, sustainable development must mean 'cultural reform and sustainable redevelopment'. According to Regier (1991), sustainable development can have little in common with conventional exploitative development. With reform and redevelopment at various levels, we should:

1. seek to reverse the major abuses of the past and present;
2. foster ecosystem recovery even in the face of growing abuses or climate change;

3. substitute clean technology and ecosystem husbandry for degrading techniques (Regier, 1991).

GEOCHEMISTRY AND SUSTAINABLE DEVELOPMENT

By the year 2050, the human population is expected to be over 11 billion. Most of the additional people will live in the 'developing' nations and in urban environments (see WRI, 1997). The question is: whether we can support this large human population without destroying our most basic life-support system? Also, can we develop truly sustainable support technologies? Sixty per cent of Asian children suffer from malnutrition and 55,000 people die daily from polluted water and waterborne diseases. At least 40 nations suffer from a water supply crisis.

The most fundamental components of our life-support systems all necessitate the need for exact geochemical knowledge of the materials, the processes and the interactions amongst all parts of the Earth system from atmosphere to core. It is necessary to understand the processes that regulate the chemistry of the atmosphere, the climate, the interactions of the biosphere with the hydrosphere and soils, the factors which control bioproductivity, the materials and systems which provide our energy resources and all the material resources used by humankind, as also a proper management of our waste products (Fyfe, 1998). For many or most earth materials used on massive scales, such as energy systems, mining, agriculture, water treatment and waste systems, such data do not exist.

Some scholars (e.g. Abramovitz 1998) even doubt whether any river in the world will flow naturally to the ocean by 2050. By 1985, there were over 36,000 large dams in the world. Similar problems arise with soil erosion and degradation. The geochemical dynamics of a host of elements are changing dramatically (Nriagu, 1996). We use about 50,000 xenobiotic chemicals. Deforestation, soil erosion and genetic manipulation of the biomass (Edwards, 1996) can lead to basic alterations in many fundamental properties (such as the albedo) of our planet.

By 2050, the chemistry and biology of Earth's surface domains will certainly be very different from that of 1950. The changes will have to be monitored and their influences need to be understood (Fyfe, 1998). For really sustainable development, many of our major technologies need to be improved.

Today, oil, gas and coal provide about 90% of our energy systems, nuclear energy about 8% and hydropower 3% of total world energy. For decades to come, fossil carbon will continue to dominate world energy production, particularly in countries like China and India, which have large coal reserves and great need for more energy. Today's world is divided into countries that waste energy (N. America), those having energy reduction programmes (Europe, Japan) and those who need more energy (many developing nations) (Fyfe, 1998). Strangely, our knowledge of the geochemistry of coal is meagre—the engineer usually asks only what is the ash, moisture and sulphur contents. But coal is a very complicated and variable material, considering its bio-origins. It creates unique reducing environment in a sedimentary pile undergoing diagenesis. As regards the chemical composition of coals, almost every coal field, or even coal stratum, is unique (Fyfe, 1998). Coals can be rich in halogens, uranium, arsenic, nickel and even gold, as also some bio-essential elements. Coal can be mineralized and demineralized. In India, the coals tend to be ash rich, not very anomalous in toxic heavy metals, but have significant quantities of halogens (Sahu, 1991; Fyfe et al., 1993, Fyfe, 1998).

Great care needs to be exercised in the selection of fuels and planning the optimum coal-burning technology. Indian coal-ash wastes seem to have useful fertilizer potential (Young, 1994).

At the roots of the looming world food crisis are factors such as climate, biodiversity and water and soil quantity. Soil productivity depends on soil thickness, mineralogy and detailed chemical composition. Almost half of the elements have bioessential functions and their presence or absence can affect bioproductivity and the health of all species, plant and animal. Striking examples include iodine and the human thyroid, cobalt and the immune systems of animals (cf. vitamin B12), molybdenum and nitrogen fixation.

Across the world, careless use of mineral-based fertilizers has caused vast pollution problems; the same is true for pesticides, herbicides and many other agro-chemicals (Fyfe, 1998).

In tropical areas with nutrient-poor lateritic soils, simple mineral-based additives often produce spectacular sustainable results (see Fyfe, 1998). Very simple element additives can effectively enhance bioproductivity. Phosphate rocks can be rich in elements like uranium and cadmium and long-term or excessive use can lead to serious toxicity problems.

Today, more than a billion people lack access to clean drinking water. All our technologies for energy, mining, agriculture and urban systems inject wastes into the hydrosphere. The main villain is overpopulation.

To manage any wastes, their total chemistry needs to be accurately known. This can enable us to make judicious choices. Some wastes can be useful (e.g. metals, coal ash), and others must be isolated or destroyed (e.g. nuclear wastes). But, in many examples, the basic technology to eliminate wastes must be changed.

Geochemists should be actively involved in the problems of sustainable development and in the development of new clean technologies for the twenty-first century (Fyfe, 1998).

SUSTAINABLE DEVELOPMENT AND NUCLEAR ENERGY

One of the major challenges in the energy field is undoubtedly ensuring sustainability—a goal that requires improved management of natural resources and a reduction of the noxious emissions dangerous to health and the environment. The threat of global climate change due to such emissions is causing grave concern to many nations and reaching an international consensus will take some time. Carbon dioxide emissions have slowed only marginally in industrialized countries and have increased significantly in most developing countries owing to increase in energy demand and the increasing use of fossil fuels, which are the most readily available energy sources.

A potential application of nuclear power that has recently received particular attention relates to the growing problem of potable water in many countries. Considerable progress has been made on the evaluation of the technical and economic feasibility of seawater desalination using nuclear energy. The use of nuclear power plants for desalination is technically feasible and the costs compare favourably with those of fossil-fuelled plants.

Nuclear power plays an important role in reducing CO_2 emissions and other pollutants from the electricity sector. The health impacts from nuclear power plants are far lower than those from coal-fired plants.

There exist significant uncertainties in the risks associated with CO_2 emissions and the effects of these emissions on the average temperature, and the risks of small releases of chemical substances. However, if greenhouse effects are included in the overall assessment, hydro power and nuclear power are the only currently available large scale energy sources that have relatively low external costs. Unfortunately, strong messages about nuclear power are not reaching the decision makers and the public. Efforts should be made to present the results from comparative assessment studies in a more transparent manner so that they may be used more readily in decision-making processes and communicated to the media and the public.

Radiation-induced mutations followed by appropriate selection proced-ures have been successful in improving the performance of germplasm throughout the world, especially in China. In 1995, use of the sterile insect technique (SIT) eradicated the medfly pest from Chile.

Isotope techniques are being developed and used to monitor and control the impact of nutritional intervention programmes for overcoming 'hidden hunger', a term coined by WHO and UNICEF to describe the deficiencies of vitamin A, iron and other essential micronutrients that affect millions of people in developing countries.

SUSTAINABLE DEVELOPMENT AT THE LOCAL LEVEL

It is essential to keep the interest and welfare of local communities in mind while designing large-scale sustainable development and reform programmes Rural areas must adapt local practices that reflect the global sustainable resource directives suggested by international policy makers.

Sustainable development implies the challenging task of reconciling economic growth with ecological integrity from local to global scales. Global sustainability programmes also need to include considerations of social vitality and grassroots discretion over development. Successful harmonizing of economic, social and ecological values is not possible without examining the influence and effects that each has upon the others. To a great extent, such harmonizing has already been achieved in Europe (see Fig. 11.4), especially in relation to changes in landscapes and living conditions. In Phase I (early post-war years), emphasis was on the economies of larger scale, i.e. increased size and volume of enterprise, in the belief that the 'trickle-down' effects of increased economic prosperity would be conducive to improved living standards throughout society. The harmonization of economic and social goals posed a challenge for regional development and mechanisms had to be developed for the redistribution of income.

During the 1970s, concerns about ecology and the depletion of natural resources led to the emergence of another set of scale criteria. Oil crises warranted changes in lifestyles that had previously been taken for granted. Conflicts between economic and ecological reasoning dominated environmental issues during the 1980s, when it emerged that whereas good gains in productivity levels and living standards had been achieved, disparities still remained between core and peripheral regions. There was increasing concern over the environmental consequences of enlarged scales of production, consumption and circulation of products (Buttimer, 1998).

By the late 1980s, grassroots pressures from advanced regions and from nongovernmental organizations forced revision of policy directives on energy use. In the decade of 1990s, environmental dimensions were acknowledged in most if not all sectors, and sustainable development demanded harmonization of economic, ecological and social values at global, national and local scales. Indeed, the quintessence of sustainable development is the daunting challenge of reconciling the three previously mutually exclusive—sometimes even antagonistic—elements.

Sustainable Development 497

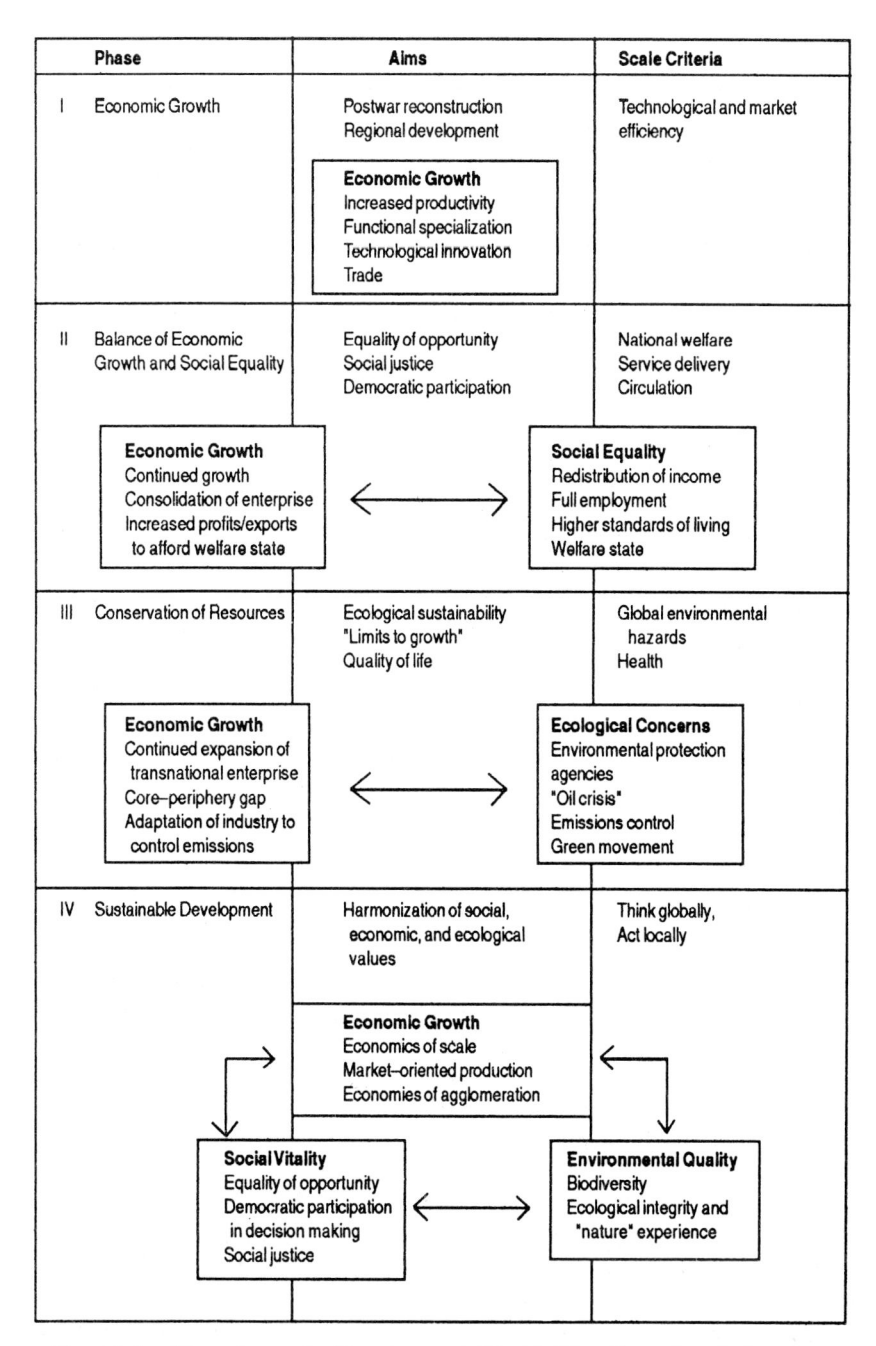

Phase	Aims	Scale Criteria
I Economic Growth	Postwar reconstruction Regional development **Economic Growth** Increased productivity Functional specialization Technological innovation Trade	Technological and market efficiency
II Balance of Economic Growth and Social Equality	Equality of opportunity Social justice Democratic participation **Economic Growth** Continued growth Consolidation of enterprise Increased profits/exports to afford welfare state ⟷ **Social Equality** Redistribution of income Full employment Higher standards of living Welfare state	National welfare Service delivery Circulation
III Conservation of Resources	Ecological sustainability "Limits to growth" Quality of life **Economic Growth** Continued expansion of transnational enterprise Core–periphery gap Adaptation of industry to control emissions ⟷ **Ecological Concerns** Environmental protection agencies "Oil crisis" Emissions control Green movement	Global environmental hazards Health
IV Sustainable Development	Harmonization of social, economic, and ecological values **Economic Growth** Economics of scale Market–oriented production Economies of agglomeration **Social Vitality** Equality of opportunity Democratic participation in decision making Social justice ⟷ **Environmental Quality** Biodiversity Ecological integrity and "nature" experience	Think globally, Act locally

Fig. 11.4 Changing priorities in post–World War II regional development in Europe (Source: Buttimer, 1995).

There are certain minimum essential thresholds of size, income and interaction for social vitality at local and national scales. Further, most ecologists are becoming increasingly convinced that 'human' elements must be included in definitions of ecological integrity. Sustainable landscapes and livelihood require identification of a suitable scale for action and interaction—a scale at which bottom-up and top-down interests could be balanced. However, the fact that social vitality greatly depends upon levels of economic prosperity and institutional reach raises a new challenge to determining the appropriate scale at which these priorities may be reconciled.

Whereas small-scale diversified economic activities are also justifiable socially, large scale, intensive, export-oriented production inevitably leads to population losses in rural areas. Without a dynamic social base, remote-controlled specialized activities seldom lead to integrated community development.

Land tenure patterns undoubtedly make up one of the more sensitive indices of the economic, social and ecological aspects of sustainable development in many areas. Land ownership is very important because farm size and tenure involve more than just access to the resources necessary for a viable economic base—they have implications for the sustainability of cultural traditions as well as for the rediscovery or maintenance of environmentally appropriate practices (Buttimer, 1998). The increasing size thresholds necessary for farm viability, accordingly, have strong social implications.

The ecological consequences of specialized and export-oriented production are now quite well known. For the agrarian sites, diversity of land use appears to be more conducive to ecological integrity than specialization. Diversity undoubtedly is the key to sustainability and appropriate scale. Critical measures of appropriateness must, therefore, include the spatial blend and intensity of land use as well as the temporal succession of land use such as the timing and cycle of planting and harvesting, crop rotation and land-use regimes.

Unfortunately, today, traditional researchers and policy experts usually operate in distinct, specialized channels, each one producing particular and often irreconcilable scenarios of rationality. Values of economic growth, social vitality and ecological integrity are usually voiced in separate meetings, each suggesting a different scale of success. Whereas the scale imperatives dictated by conventional economic theory during the 1950-90 period laid down good criteria for minimal size of enterprise, they did not do so for maximal scales in the production and circulation of products, short of a potentially global market (Buttimer, 1998). There has been a separation of the two dimensions of economic productivity on the one hand and marketing on the other. Several measures have been implemented to curb agricultural productivity (in some advanced countries), but the expansion of commercial enterprises continues to be virtually unchecked. This also points to other contradictions, such as the simultaneous rise in economic growth along with the decrease in employment and the 'flexible specialization' of transnational companies in peripheral regions that leads to a net export of people and wealth (Grimes, 1995; Buttimer, 1998). Commonly, many national governments have regarded the social and ecological consequences of such developments as ancillary matters. Several governments have failed to recognize that the 'ecological shadow' cast by the expanding scale of energy-consuming life styles in core regions causes spatial inequalities and socially undesirable effects even in places remote from the originating site. Transboundary environmental impact assessment must, therefore, involve the complete process of production, distribution and consumption. Citizens should be enabled to identify and assess complete paths of production and consumption, recycling or dumping of products across national or administrative borders. Effectively communicating issues of ecological integrity to the general public is the first step toward harmonizing the social, economic and environmental criteria needed for sustainable development, which implies a harmony between the advancement of objectives and the retention of established values. Creative solutions from grassroots sources may hold the key for globally sustainable development (see Giddens, 1990; Wilbanks, 1994).

TRANSITION TO SUSTAINABLE SOCIETY

The concepts of industrial ecology and industrial metabolism offer important insights for practical applications. Current approaches do not focus enough on underlying problems at the root of our environmental troubles. Unless some innovative ways are devised that can meet the challenges of the 1990s, we will lose the battle for the planet (Speth, 1992).

Since 1950, world population has doubled to about 6 billion. The output of the world economy has quadrupled. Economic activity on the planet today is greater by a factor of four than only a few decades ago. It took all of human history to build a world economy that produced about $600 billion in output in 1900. Today, the world economy grows by this amount every 2 years (Speth, 1992). For over a billion of people in the rich countries, this growth has brought material wealth unimaginable by earlier generations. But along side, it has also brought pollution, waste and consumption of the planet's resources on an unprecedented scale.

Meanwhile, in the poorer countries, the numbers of the poor have swelled dramatically. A billion people in the developing world live in hunger and poverty, destroying the fragile base of soils, water, forests and fisheries on which their future depends because no alternative is open to them. The world's deserts are advancing while its forests, with their immense wealth of life forms, are in retreat. On an average, an acre and a half of tropical forests disappears every second: four species are committed to extinction every hour (Speth, 1992).

For the first time, human numbers and impacts have grown so large that they are eroding on a global scale the natural systems that support life.

These challenges will multiply in the future. World economic activity is projected to be five times that of today in only six decades. If climate-altering gases, industrial wastes and other pollutants were to increase proportionately with the fivefold expansion in world economic activity projected for the middle of the next century, it will convert difficult problems into impossible ones.

Speth (1992) has advocated certain transitions in dealing with the root causes of environmental problems with a view to achieving environmental sustainability. These transformations include: (1) a demographic transition; (2) a technology transition that includes the 'green' automobile; (3) an economic transition to one in which prices reflect full environmental costs; (4) a transition in social equity; and (5) an institutional transition to different arrangements among governments, businesses and peoples. Businessmen and environmentalists are urged to work together in the next decade to make the environment a personal issue, to call for government action, to recognize the environmental challenges and to commit to accountability in order to leave a legacy of hope to the twenty-first century (Speth, 1992).

We must rapidly abandon the twentieth-century technologies that have contributed so abundantly to today's problems, and replace them with twenty-first century technologies designed with environmental sustainability in mind.

Today, everyone speaks positively of 'environmentally-sustainable development'. What this means in the context of pollution is technology transformation. It is a transformation that must begin today.

The needed fusion of economic and environmental objectives requires technologies that meet two criteria, both linked to the concept of industrial metabolism. First, the needed technologies must be able to transform industry and transportation from materials-intensive, 'high-throughput' processes to systems that use fuel and raw materials highly efficiently, rely on inputs with low environmental costs, generate little or no waste, recycle residual products and release only benign effluents. The need, in short, is for technological systems that are more and more environmentally 'closed'—that is detached as much as possible from natural systems (Speth, 1992).

Second, technological innovations must help societies move toward living off nature's income rather than consuming nature's capital.

For any economic system, environmental damage, over time, is a function of the consumption of inputs from environmentally unsustainable processes, the generation of pollution and post-consumption waste and other factors. This damage won't stabilize and decline until pollution per unit of output and materials consumption per unit of output decline rapidly enough to outweigh growth in economic output. Technological transformation for environmental sustainability can reduce environmental damage per unit of output fast enough to greatly outpace production increases.

Some of the following technologies are possible now: manufacturing processes and motors that cut energy needs in half; gas turbines that cogenerate electricity and heat 50% more efficiently than today's power plants; solar thermal and wind systems that are producing electricity today at prices competitive with nuclear power and photovoltaic power that promises to be competitive within a decade; manufacturing processes that make detoxification possible and waste elimination profitable; new microbial and other bioengineered products that can substitute for chemical pesticides and fertilizers, help treat effluents and other waste, promote vegetation growth on impoverished soils and increase the potential of biological sources of energy; miniaturization, microprocessors and computer-aided design and management that greatly improve efficient use of raw materials and reduce both waste and environmental pressures; and other computer and telecommunications application that can strengthen satellite remote sensing, monitoring instrumentation and environmental management through artificial intelligence (Speth, 1992). A few advanced countries are producing and marketing the 'green' automobile. There is probably no product that causes so much environmental damage as today's car. Both hydrogen and electric-powered vehicles are possible, and both hydrogen and electricity can be made from renewable energy sources, such as photovoltaic cells and wind power.

Many environmental laws favour old technologies over new ones and prescribe cumbersome administrative procedures that impede innovation. Relying on 'best available technology' standards has tended to entrench existing technologies at the expense of new ones, although that result was certainly not the original intent. Regulations often focus on only one medium (air, water, or land) rather than across the whole spectrum and tilt toward 'end-of-pipe' pollution controls instead of prevention options.

THE ENVIRONMENT-DEVELOPMENT INTERFACE

The earth, air, soils, oceans and forests on which our lives depend, are being increasingly degraded. Grossly unequal distribution of wealth and power threatens not only our immediate future but also the very existence of future generations.

Increasing poverty for two-thirds of humanity and environmental degradation are not separate problems. Neither of these can be solved in isolation from the other.

Whereas every minute of every day, some US$2 million are spent on an arms race intended to make the world 'safe', the real threats to the planet's survival are economic and environmental.

Humans are at risk because:

1. the environment on which we all depend is being destroyed;
2. the world's rich feast on global resources while two-thirds of humanity—the world's poor—struggle to survive on increasingly inhospitable lands; and
3. the populations of poor countries are trapped in a cycle of poverty as their incomes fall and debts rise.

The world's poorest people live on land which has been most seriously environmentally degraded. Quite often, the poorest of the rural poor are excluded from their traditional lands so that multinational companies can grow cash crops such as coffee, tea, cotton, tobacco, soybeans, peanuts, cocoa and fruit for export. It is the rich who are making the world poorer and poorer.

The consumer-oriented populations of the developed world use disproportionate amounts of the world's energy and resources. Many of those resources, being finite, cannot last forever.

The first priority in developed countries should be to reduce consumption and to change the industrial system, technology and culture, which maintains this constant drive for growth and more goods.

What is needed is a low-energy future. The world cannot afford more people consuming energy at the rates we take entirely for granted.

In the rich countries, people are drastically over-using the earth's precious, limited resources. Some important dangers of a continued high energy future are the following:

1. strengthening of the greenhouse effect (and further environmental destruction);
2. growing urban-industrial air pollution (aggravation of certain diseases);
3. high incidence of acid rain (damage to trees, lakes, fish as well as to buildings and monuments);

4. greater difficulties of waste disposal from nuclear energy use; and
5. higher risks of nuclear accidents with proliferation of nuclear energy and nuclear weapons.

People must choose a low energy future and must ensure that energy is used more equally within countries, and around the world. There should be a fair distribution of energy resources.

How to Save Energy?

1. By designing and creating small scale, renewable energy alternatives such as wind and solar power.
2. Increasing availability and use of public transport and of fuel efficient vehicles.
3. In rich countries, some 40-60% of all energy is consumed by industry. Skilful improvements in energy efficiency can potentially save as much as 20-30% of industrial energy.
4. The price of energy must reflect its real cost—including the cost of cleaning up the pollution it causes. Incentives to save energy must be built into energy pricing policy.

Each and every human being everywhere should strive to reduce his/her personal consumption of energy and resources.

Forests to Deserts

Every minute of everyday, over 20 hectares of tropical forest are being destroyed. In Africa, India and other dry areas of the Third World, people chop down the forests for firewood. Two-fifths of the world's people cook with wood they collect themselves because poor people have no other alternative. In South East Asia, the Pacific and Brazil, tropical forests are being cut down to meet the hunger of rich countries for paper and timber.

Forests are also being cut down to make more land available for growing cash crops. Cash crops are needed because the export revenue they produce helps to meet the interest on debt repayments from poor countries to rich countries and to pay for essential imports.

In 1988, deforestation in the catchment areas of the Ganga and the Nile rivers turned heavy rains into catastrophic floods. In Bangladesh, the Ganga floods left 25 million people homeless (Hunt et al., 1989).

Destroying forests reduces the number of trees releasing oxygen into the air. It also accelerates soil erosion, which leads to a lowering of the water table; leads to a huge loss of soil nutrients in humid tropical regions and destroys many species which depend on the forest ecosystem for their survival.

Besides deforestation, some other major causes of land degradation or 'desertification' are:

1. increased salinity (saltiness) of irrigated lands;
2. overgrazing on rangelands; and
3. planting of too many crops without sufficient crop rotation, often with the over-use of chemical fertilizers.

It is not fair to blame poor people for the damage to their environment. They have a right to live and survive.

RURAL POVERTY

There is widespread rural poverty in many regions of Africa and the Indian subcontinent. Strategies proposed to eliminate rural poverty include the development of alternatives to agriculture, for instance, in industry, mining and tourism, but also the commercialization of smallholder agriculture. An important challenge for the concerned governments is to foster smallholder agricultural production in ecologically marginal environments, in ways that will sustain or improve the quality of the resource base over the longer term.

NGOs and some development agencies play important roles in adaptive research and extension for the poorer smallholders, but their work often does not address such strategic research questions as integrated pest management or technologies for ecologically fragile environments. Consequently, smallholder producers in developing countries depend largely on public sector institutions for technological innovations and extension. The majority of smallholders are poor and have neither the institutional nor the economic power to ensure that their technology needs are met by the public sector research.

While developing novel technologies, researchers should recognize the conscious effort by smallholder producers to maintain subsistence food production, including indigenous and marginalized crops, varieties and livestock species. Given the risky economic environment and lack of secure markets, maintaining own food supplies can be an economically-optimal strategy, even if commercial production of, for instance, cash crops might lead to higher returns on land and labour. Therefore, a viable commercialization strategy needs to be developed that takes care of smallholder needs to improve productivity in both staple and market-oriented production (Anandajayasekeram, 1999). Assigning high priority to research that has a high pay-off in the short term can exert adverse effect on society in the long run. Public sector research would seem to have a comparative advantage in addressing those aspects of a smallholder sustainable agricultural research agenda that specifically targets persistent poverty, which other actors (such as private sector) are unlikely to address efficiently.

ENVIRONMENTALLY SOUND DEVELOPMENT

The environmental problems today are essentially the product of yesterday's technology choices. The sustainability of future development efforts depends on the capacity of rich and poor nations alike to make better choices based on technology assessments which partly determine the real costs of production—costs such as pollution, global warming and climate change—which future generations should not have to bear. The focus of efforts in environmentally-sound development needs to be to give developing countries the capacity to weigh the benefits of technologies against their potential environmental and social consequences. These concerns should be addressed under four main headings: the atmosphere; the biosphere; the geosphere; and general financing. Listed below are some specific recommendations that emerged from the deliberations at various workshops/symposia:

1. International Environment Funds (e.g. the $1.1 billion global environment facility of the World Bank, UNDP and UNEP) have already been created and have a strong technological important.
2. Tradeable permits and consumption rights, under which a global ceiling for a particular environmental concern, e.g. CO_2 emissions, is agreed upon by all nations and national quotas are allocated among them. Those who exceed their quotas could still remain legal if they would purchase their deficit from those who emitted less than their quotas.
3. Offset arrangements by which competing suppliers of goods and services to the developing world offer Third World purchasers special incentives such as investment, trade, research, training and technology transfers.
4. Debt conversions, such as debt-for-nature swaps.
5. Taxation mechanisms which give users and producers an incentive to respect environmental objectives. Examples include fuel taxes, fines, rebates and subsidies.
6. Grants and donations for environment-related activities.
7. Development assistance which integrates environmental concerns in aid projects.
8. Market mechanisms, such as greenhouse investment funds and royalties, which encourage environmentally friendly behaviour.
9. Regional development funds which address the protection of shared environmental interests such as river basins and regional deserts.
10. Insurance and reinsurance plans. To minimize risks, insurance companies would promote knowledge and training related to environmentally sound technologies. Reduced premiums would be offered to clients who were interested in absorbing such a package.

As each of the above individual items has its limitations, the package should be configured to the situation at hand.

TECHNOLOGICAL CHANGE AND GLOBAL SOCIETY

Many people view rapid technological change with hope and apprehension. On the one hand, they desire progress, want the best for themselves, their families and they societies; and the tools of modern technology have been instrumental in improving the material conditions of life for many. Yet, change can also be very unsettling, especially when the instruments of advanced technology cannot be understood or appreciated by most people.

Scientific progress has greatly helped humans in their collective struggle against the long-standing afflictions of mankind such as disease, poverty, hunger and ignorance. But what of the more subtle challenges? Can we foster true human values in the face of increasing social alienation? Can we achieve self-fulfilment, cultural identity and social stability in an age of consumerism, mass communications and globalism?

A few decades ago, in many parts of the world, there was great optimism regarding the capacity of advanced technology to solve human problems. Today, there is again a similar sense of optimism but, on the whole, we have become wiser and perhaps more humble. The tragedies of Chernobyl and Bhopal remind us that progress is painful. It should not be taken for granted. Modern technology is a complicating factor in social development—something to strive for, but certainly not a panacea.

In many parts of the world, local communities are worrying about the ecological and economic consequences of rapid technological change. People want greater control in shaping their destinies. The major problem revolves around the issue of personal responsibility. All of us attach fundamental value to scientific progress. Throughout history, our best minds have applied themselves to learning the rational laws of nature, endeavouring to push forward the frontiers of knowledge. Yet it is equally true that the search for knowledge cannot be sharply separated from concerns regarding how the fruits of discovery are applied. The problem is how to act upon this concern without being overawed by it (Akashi, 1991). We need not suppress the scientific spirit. Rather, there is need to think through the far-reaching implications of important technological developments and respond to them intelligently.

One happy development is that a global perspective is slowly emerging—one that sees 'national security' as a goal to be pursued, not at the expense of other nations, but rather in concert with them (Akashi, 1991).

The greatest danger in our present situation may be complacency. It would be unfortunate if the past naivete of societies regarding science were reborn in the form of excessive optimism about the progressive tendencies of political development. We are now entering a new, more complex and no less dangerous phase of history. The gap between rich and poor countries is widening at an alarming rate. This, surely, is an ingredient for instability. We must be on guard against it.

SYSTEMS ANALYSIS FOR SUSTAINABLE DEVELOPMENT

Systems analysis may be defined as the examination of the elements and linkages in a system, i.e. a regularly interacting or independent group of items forming a unified whole. Systems analysis can first be applied to the linkages among various elements of human development and environmental change, and then used to formulate and implement policies towards development pathways more environmentally sustainable than those that have been followed till date (Shaw and Öberg, 1994).

Systems analysis may be applied to natural and social sciences with a view to addressing the problems of environment and development. Firstly, a large, complex and dynamic model may be applied to various scenarios of population growth, resource consumption and waste generation. Secondly, a conceptual model may be used to connect the societal, ecological and economic components of the socio-ecological systems in which we live (Shaw and Öberg, 1994). This conceptual model brings out the interlinkages that could lead to unsustainable development.

The four important human needs are food, water, energy and the disposal of materials used in industry. None of the problems of food, water, energy and industrial waste management can be tackled individually. They can only be addressed in an integrated manner. There are intimate linkages among: population change; energy production; water; and the availability of food. Another example comes from the linkages among population change, energy production, emissions of acidifying gases and the capacity of the soil to absorb toxic heavy metals. Systems analysis can: (1) define ecological limits within which society should operate; (2) indicate how limited resources might be shared more effectively among different regions and among different generations; and (3) define management options for our socio-ecological system as it evolves and alters, partly as a result of human activities.

Conceptual Models

According to Miser and Quade (1988), some systems analysis tools for studying the environment/development nexus are: (a) conceptual models, needed for the initial analysis of a problem; (b) formalized models based upon conceptual models, but with linkages being more quantified; (c) information bases; and (d) geographical information systems (GIS).

A set of conceptual models of the environment/development nexus consist of three main subsystems:

1. *Societal.* The societal subsystem includes demographic characteristics, socio-political organization and related legal mechanisms and cultural influences. These characteristics determine the total demand of goods and services from the economy (per capita demand).
2. *Ecological.* The ecological subsystem is the basic provider of natural resources for development, waste assimilation capacity and of life-support functions affecting humans and other organisms.
3. *Economic.* The economic subsystem is made up of production and consumption processes, capital and labour demand, allocation systems and investment policies. Both production and consumption patterns affect the ecological subsystem. Consumption means not only the consumption of commercial products, but also that of natural goods and services such as air, water and fuelwood for cooking.

The ultimate aim of development is to improve the quality of life, equitably distributed within a given society. This quality is determined by both material factors (such as per capita consumption of food) and non-material factors (such as access to natural beauty in the environment).

Environment/Development Interactions

Being holistic should not mean that the analysis is rendered intractable. It is also important to look at those linkages that are most important for the problem at hand while at the same time including important cross-linkages (see Fig. 11.6). For instance, in any simplified operational model for food, linkages occur among: human factors (demography, dietary patterns); technological factors (e.g. food processing methodologies); and environmental factors (e.g. the area of cultivated land and yield per hectare). Demand is generated by a combination of population and dietary patterns. In a given region, the supply of food is met by net imports and agricultural production. Agricultural production in a region, in turn, depends upon land area and yield. Farming area is influenced by environmental factors such as climatic change, erosion and reforestation; likewise, yield is influenced by intensification, salinization and by sedimentation.

Food, energy and climatic change are intimately linked together. Climatic change is intimately linked both to the release of greenhouse gases and our need for energy. In turn, some energy is needed for the production, processing and transport of food, including the use of fertilizers, some of which come from the same fossil hydrocarbons that we use in various ways (Heilig, 1993).

Systems analysis can prove useful in defining ecological limits within which society should operate. These limits cannot be defined simply in terms of population carrying capacity with respect to food. For example, more stringent limits might be those associated with other human needs such as renewable water supply in arid regions, or the capacity of the ecological subsystem in industrialized regions to absorb wastes from energy generation and industrial processes.

Energy

Our use of energy is influenced by population, lifestyle and the available financial resources to follow and change some particular lifestyle. The emissions of acidifying and greenhouse gases, and the production of radioactive waste material, are currently mostly applicable to industrialized countries, whereas deforestation can be serious in developing countries where it is linked to poverty and international debt. Hydrodams which can disrupt habitat are relevant in both developed and developing countries.

Ayres (1989) compared the cycling of materials in nature and that in our industrialized world; the term 'industrial metabolism' may be used in this context. In nature's recycling, waste materials from one component serve as the input to another. In industrial metabolism, on the other hand, waste materials are often not used by another component but escape and are deposited in the natural system where it is difficult to monitor and control them. Reducing the flows within the metabolism by using less material and energy for a given product, a concept known as 'dematerialization', may partly relieve the problem.

As there is a strong link with the energy system, strategies to control toxic metals should include all parts of the pathways through the production, consumption and disposal parts of the life cycles. The capacity of the soil in landfill sites to absorb and store toxic metals may be decreased if it is acidified through the deposition of SO_2, NO_x, sulphates and nitrates resulting from burning of fossil fuels. This can suddenly release the stored toxic materials to the environment, acting like a 'chemical time bomb' (Stigliani, 1988).

Besides estimating the ecological carrying capacity of our environment, the use of limited resources should be made equitably and efficiently for sustainable development. Systems analysis can guide us in the sharing of our resources by taking into account linkages among various sectors of the socio-ecological system, among various geographical regions and among generations.

As our socio-economic system is dynamic, the ecological carrying capacity and the demands that humanity places on it can change with time owing to changes in population and environment (Fig. 11.5). Systems analysis can help us manage the socio-ecological system in a dynamic, anticipatory manner.

Figure 11.5 shows the relationship between the global requirement for calories, using indicators such as population and daily calorific consumption (a measure of lifestyle) and the global supply, as indicated by area of cereal-growing land and yield per hectare. The left-hand and lower axes of Fig. 11.5 show food requirements in terms of population and per capita calories; the diagonal line labelled '1.86 E9 tonnes grain' is the locus of combinations of population and daily calorific consumption that would have required 1.89 billion tonnes of grain being grown yearly, the estimated requirement in 1985. The circle '1985' is the actual demand situation in 1985: a population of 4.8 billion with an average daily calorific consumption of about 2700 calories/person (Shaw and Öberg, 1994). The upper right-hand axes show the supply side of the picture; the same diagonal line '1.89 E9 tonnes' represents the combinations of area of cereal-growing land and yield per hectare that would have produced 1.89 billion tonnes of grain annually. The circle '1985', here represents the actual 1985 supply situation of 700 million hectares of grain growing land and a global average yield of 2.5 tonnes/hectare (Shaw and Öberg, 1994).

The line labelled '3.21 E9 tonnes' shows the same type of information for the year 2025 when, for a projected world population of 8.5 billion and an assumed world average calorific intake of 2700 calories/day (the same as in 1985), about 3.2 billion tonnes of grain would be needed to be grown annually (Shaw and Öberg, 1994). If the growing area in 2025 remains the same as at present, the average global yield with have to be about 3.8 tonnes/hectare to meet global grain requirements (Shaw and Öberg, 1994).

The portions of the axes between the upper right-hand corner of Fig. 11.5 and their intersection with the two diagonal lines show the management options open to us to produce 1.89 and 3.21 billion tonnes of grain, respectively.

ENERGY USE AND SUSTAINABLE DEVELOPMENT

Energy use has increased more than 50 times since the nineteenth century, with profound effects on carbon, sulphur and nitrogen flows. The wealthy industrial economies of the world have developed on the basis of abundant cheap energy, mainly through the burning of coal, oil and natural gas. Each year's current consumption of fossil fuels represents a million years of accumulated fossil deposits. Known oil reserves will be exhausted over the next century with, at current consumption rates, the most readily accessible and cleapest reserves being depleted in 3 to 4

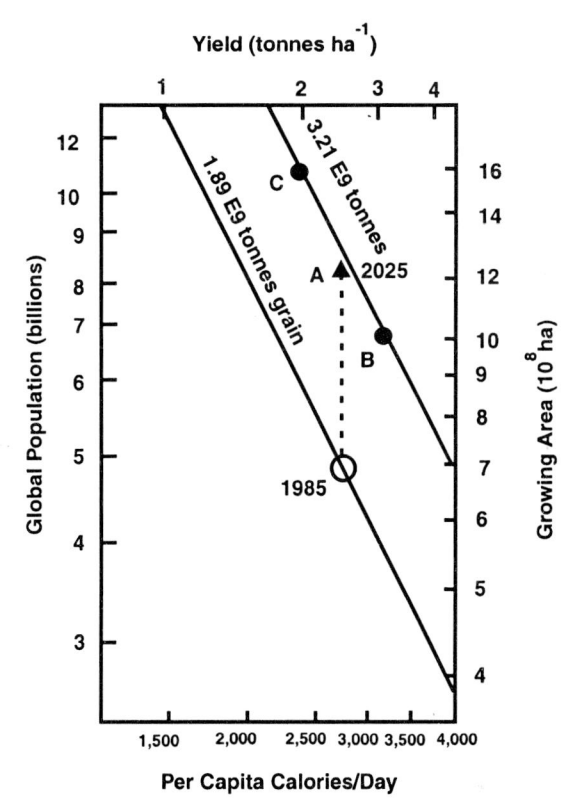

Fig. 11.5 Global population (left axis); global average/capita calorie consumption per day (lower axis); global cereal-growing area (right axis); and global average cereal yield (upper axis). Diagonal lines represent combinations of population and per capita calorie consumption that would require 1.89 and 3.21 billion tonnes of cereal and also the combination of growing area and yield that would be required to produce those amounts of grain. The large circle represents the actual situation in 1985. Point A represents a possible situation in 2025. Points B and C show the effects of possible uncertainties in yield due to climatic change (from Shaw and Öberg, 1994).

decades. Natural gas reserves and coal may last longer, but are likely to be consumed in the coming centuries at present rates of use. In short, the burning of fossil fuels is inherently a non-sustainable activity.

In contrast, renewable energy sources (solar, wind, wave, tidal power and biomass) make up a small proportion of the current global energy mix. Hydro-electricity is the only renewable source with a significant share (about 7%) of global commercial power production. Nuclear power is important only in Canada, France, Japan, Sweden and the USA, but represents only 5% of total commercial energy production. The daily demand for energy, the load forecast for an electrical utility, a gas pipeline operator, or a fuel oil supplier is highly dependent on the weather (Bruce, 1992).

In the present century, energy production and use are expected to account for some two-thirds of the radiative forcing due to anthropogenic greenhouse gases. The energy sector is also responsible for substantial proportions of other atmospheric pollutants. In several countries, the acid-causing emissions are mainly the result of energy production, but in Canada, metal smelters are a major source.

Not only is fossil-fuel burning inherently non-sustainable, but present energy practices over the world place a pollution burden of acid rain and large quantities of toxic metals on the atmosphere as well as greenhouse gases which can cause global warming. Reduction of the need for fossil fuels in industrial countries can contribute towards sustainable development.

The most widely used renewable energy sources are hydropower and biomass. However, solar and wind energy are gaining acceptance in special markets and improving technologies continue to lower production costs to more competitive levels (Fig. 11.6). Large-scale hydro-electric potential has been developed in various continents, but they are capital intensive and cause some environmental problems. Efficient design and operation of hydro-power plants require reliable data on river flows, as well as on precipitation and evaporation.

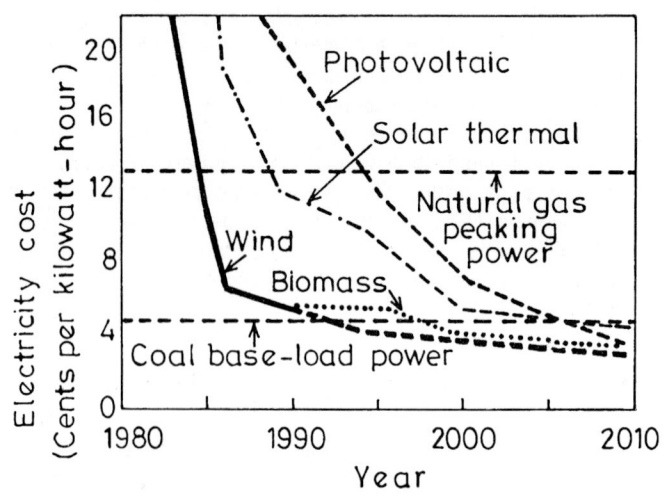

Fig. 11.6 Cost trends for renewable energy forms (after Weinberg, C.J., Williams, R.H., *Scientific American*, Sept. 1990).

The greatest long-term potential for renewable energy is undoubtedly through direct use of energy from the Sun. The two main approaches to converting the Sun's energy into useful forms are through photovoltaics, the production of electricity from the Sun's energy through solar cells and the solar-thermal approach in which the Sun's heat is converted to useable energy in various ways (Bruce, 1992).

The direct conversion to electricity of solar energy (i.e. photovoltaics) is very expansive as compared to other energy forms, but costs are rapidly falling with the use of new technologies (Fig. 11.6). In addition, photovoltaic techniques are well adapted to producing hydrogen from the breakdown of water, and hydrogen fuels are likely to become the main transport energy source towards the late twenty-first century.

Continuation of current world population growth rates is sure to exert severe pressures on available cropland. If population trends continue, the 0.28 hectare/person now devoted to crops will fall to 0.17 hectare/person by 2025. For 93 developing countries, nearly two-thirds of the increased production projected to the year 2000 has to come from higher production on existing land under cultivation with some 22% expected to come from increasing the area of cultivated land. But how to achieve such an increase in production without serious environmental degradation? This will require the use of improved techniques, including the maximum use of agrometeorological and hydrological knowledge and procedures. This applies to both rainfed and irrigated agricultural economies.

The most important hydrometeorological applications for irrigation include the following (see Bruce, 1992):

1. Use of hydrologic data to determine the quantity and quality of available water during the driest periods.
2. Use of climatic data in soil moisture budgeting to determine mean and extreme water demands of crops for planning of projects.
3. Hydrometric and meteorological data analyses for optimum design of dam and reservoirs for irrigation-water storage.
4. Operational soil moisture budgeting by daily precipitation and evaporation measurements to determine when plants need water, for optimum timing and amount of irrigation applications.

The world's forested area is currently estimated to be about 4 billion hectares. At present, tropical forests are under the greatest pressure. Deforestation has contributed greatly to the changing composition of the global atmosphere. However, the reduction of absorption of CO_2 by forest vegetation now appears to be a smaller factor than the burning of fossil fuels in increasing concentrations of atmospheric CO_2, although it was a major factor in earlier periods. Forest loss also affects the hydrologic regime of a region. While under some conditions, total runoff from a recently-deforested area may increase due to reduced evapotranspiration losses, the seasonal distribution is often drastically altered. The loss of trees usually results in floods and to more prolonged and intense low flow periods in the dry season. In addition, soil losses create large sediment loads in rivers and lakes especially when hilly slopes are denuded, and these sediments often contain pesticides, fertilizers and other water pollutants (Bruce, 1992). Due to the high mobility of the atmosphere, forest cover has generally little effect on local precipitation amounts, but over a very large river basin in tropical regions, the loss of forest and its replacement by grassland has the potential to reduce precipitation significantly.

THE ETHICS OF SUSTAINABLE DEVELOPMENT

The growth and development of human society has already passed the limits of sustainability. To achieve sustainable development, *every* cultural group—from rain forest foragers to urban elite development professionals in New York or Tokyo will have to make choices that seemed impossible to consider only a few years ago (see Tables 11.3 and 11.4).

We must move beyond arcane academic arguments about ethics that pit materialist against mentalist approaches, or arguments that reject ethics in the name of expediency or relativism. There is a growing need to balance scientific research to understand the parameters of sustainability with advocacy for human rights in an attempt to define the goals of development.

Both indigenous and modern sciences have clearly shown the intricate interrelationships in nature and human societies that support quite stable ecosystems. These sciences also document the loss of both ecological and cultural diversity, as local groups, their environment and indigenous knowledge are absorbed into industrial modernization. There are ample indications that the human toll on the planet has already pushed it beyond its 'carrying capacity', such that consumption levels and population numbers will have to be reduced to achieve sustainability.

Environmental sustainability is usually defined as resource management that does not degrade the environment for future generations. The social side of sustainability is more difficult to define, but must include a social system that does not destroy the natural world or our own species. Social values such as a human right to be free of hunger or political repression can be accepted by global consensus as desirable values (Cleveland, 1994).

Whereas today most chemical industries depend on natural gas, oil and coal for much of their feedstock, plants and animal-derived products

Sustainable Development 515

Table 11.3 A taxonomy of interventions in the energy sector

Stage option	Production/generation	Transmission and distribution	End-use
Reduce energy consumption of existing process by increasing efficiency	Refurbish/repower old power plants	Reduce transmission and distribution losses in electrical grids	Reduce energy intensity of basic materials production. Efficient motors and drives. Irrigation pumpsets. Vehicular fuel efficiency. Process heating. Space heating and cooling. Energy conservation.
Reduce emissions from existing processes	Reduce associated gas flaring. Use coalbed methane. Collect from fossil-fuel systems and store in depleted gas/oil fields or in deep ocean	Reduce leaks in natural gas pipelines	Install end-of-pipe emissions controls in wood-stoves and cars (e.g., catalytic convertors)
Switch to more energy efficient processes	Biomass gasifiers (see Fig. 12.9) and gas turbines. Advanced efficient gas turbine cycles. Clean coal technologies.	High-voltage direct current transmission. Promote inter-regional flows of natural gas and hydroelectricity	Lighting compact fluorescent lamps. Transport shifts (road to rail, personal to mass). Innovative technologies for appliances and vehicles. Improved cookstoves. Land-use planning. Infrastructure efficiency.
Switch to lower emission processes	Photovoltaics. Biomass. Wind farms. Solar thermal. Small hydro. Geothermal. Fuel cells. H_2 from non-fossil electricity. Methanol from flared gas. Nuclear (?). Magneto-hydro-dynamics.	Hydrogen as an energy carrier?	Solar water heating. Compressed natural gas transport. Electric vehicles. Natural gas fired engine-driven cooling systems.

Table 11.4 A taxonomy of interventions in non-energy sectors

I. FORESTRY SECTOR

Combatting deforestation
- Biomass combustion
- Incentives for maintenance of forests
- Alternatives to shifting cultivation

Greenhouse gas sequestration
- Carbon sequestration in growing forests and on currently degraded land
- Management of tropical forests

II. AGRICULTURE SECTOR

Reduction of emissions from
- Rice paddies
- Livestock management
- Nitrogenous fertilizers

III. WASTE MANAGEMENT SECTOR

Urban and rural waste treatment
- Collect, use, or flare landfill gas
- Biogas systems

IV. INDUSTRIAL SECTOR

Reduction of emissions from cement production
Halocarbons: CFCs, HFCs, HCFCs (reduce lifetimes and energy penalties of substitutes)

V. INSTITUTIONAL AND POLICY REFORMS

Improving performance through innovation
- Price and tax reform
- Least-cost planning
- Conversion of utilities to energy service companies
- Creation of new energy service companies
- Independent power companies
- Management of dispersed energy systems.

Technology transfer
Manufactured energy-efficient products in developing countries
Assessment of technology import versus domestic manufacture
Training and institution building
Market aggregation

once provided most of the basic ingredients. For instance, until the nineteenth century, plants and animal products formed the basis of most medicines.

In the twentieth century, industries started replacing their natural product raw materials with cheap and plentiful alternatives. A variety of synthetic raw materials were developed, e.g. synthetic rubber, which was cheaper and better than its natural equivalent and alcohols, which were replaced by petroleum-derived methanol and ethanol.

The consumption of petroleum-derived products and non-renewable petroleum has been rising steeply. There is now a need to move away from our dependence on fossil fuels and to look for new feed-stocks. Industry is now finding renewable alternatives. The principal use of fossil fuels globally is in providing energy for heating and air-conditioning in buildings, for operating production plants or as fuel for cars and other vehicles.

The speciality chemical industry, in particular, which depended on cheap byproducts of the petrochemical industry is looking for alternative feedstocks, e.g. cereals and starches (including potato, rice and wheat starch) as potential replacements for biocompostable polymers, for example; oilseed plants (such as rapeseed and soybean) for lubricants and raw materials for the oleochemical industry; fibres (including straw and flax) for structural materials such as composites; and plant extracts for flavours, fragrances and medicines (Gaskell, 1998).

Imperial Chemical Industry's (ICI) rigid polyurethane foam business exemplifies how a non-renewable feedstock has been successfully and cost-effectively replaced by alternative renewable raw materials. In the 1960s, the predominant feed-stock for rigid polyols was glycerol, originally derived from animal fats, later from petroleum. In the 1970s, glycerol was replaced by sorbitol, which performs better and is a renewable resource, being made by the hydrogenation of glucose. In the 1980s, sorbitol was replaced with another new, even cheaper feedstock, sucrose. More recently, ICI has developed a novel, high value-added end-use for waste straw. Low value straw that is left over when grain (rice, wheat) is harvested is used to form particle-board and fibreboard. The straw is ground up into fine particles, dried to a certain moisture content and then transported to a blender where glue (polymeric diphenyl–methane diisocyanate)—in place of urea formaldehyde—is sprayed on to it. The straw is then made into mats which are pressed into boards at 170–200°C. Applications of the new 'wheatboards' include furniture making and construction. The boards are almost as strong as wood.

While the mainstream chemical industry has been mainly based on fossil fuels, the oleochemical industry has always procured its raw materials from a virtually unlimited renewable supply of natural oils from crops such as rapeseed, coconut, palm kernel and soybean, as well as animal fats. The industry accounts for 14% of the 94 mt of oils and fats produced worldwide each year.

The oleochemical industry is the main supplier of raw materials for the soap and detergent industry. It is likely to benefit from recent researches on new raw materials. Animal fats have been a traditional feedstock source for the soap and detergent industry but the industry will have to find alternative raw materials when a European Union ban on the use of animal tallow products comes into force.

Some companies are examining the potential of using oat grain in non-food applications. Fractionated oat products have been produced with potential applications in the cosmetic and body care sector. For example, oat starch could replace talcum powder and hydrolysed oat protein could replace animal protein in shampoos and conditioners. In addition, oat oil and oat β–glucan have promising applications in sun creams and body lotions, while oat flour is a natural emulsifier (Gaskell, 1998). Figure 11.7 illustrates such a petroleum/grain analogy.

Amongst the alternative feedstocks being considered, cereals have the greatest potential because they are energy intense and environmentally benign, but to make the economical alternative feasible, processing routes such as fermentation are required. Unlike chemical processes, fermentations benefit from the use of complex, natural, raw materials. In the future, some industrial products may be extracted directly from the grain (e.g. oil and starch) and others produced by fermentation of grain flour. The production of a generic fermentation medium based on whole wheat flour is being attempted. The process includes a gluten extraction stage followed by continuous fermentation of the starch to glucose. Liquid fractions from the effluent of this fermentation are then used to further convert flour suspension into a glucose-rich stream. The solids pass on to a cell autolysis stage, which releases free amino nitrogen and other cell-based nutrients. Conceivably, the feedstock produced by such a process may have all the necessary nutrients for many different fermentations. Further, because it consumes no extra materials beyond the flour and produces no waste products, it is also environmentally-friendly.

TOMORROW'S SUSTAINABLE SOCIETY

According to The Swedish Environmental Protection Agency, tomorrow's sustainable society should be characterized in terms of the following ten parameters (see *Enviro* 14: page 36, Dec. 1992).

1. The stratospheric ozone layer will be preserved intact and the change in climate caused by man will be small enough to allow for natural adjustment.

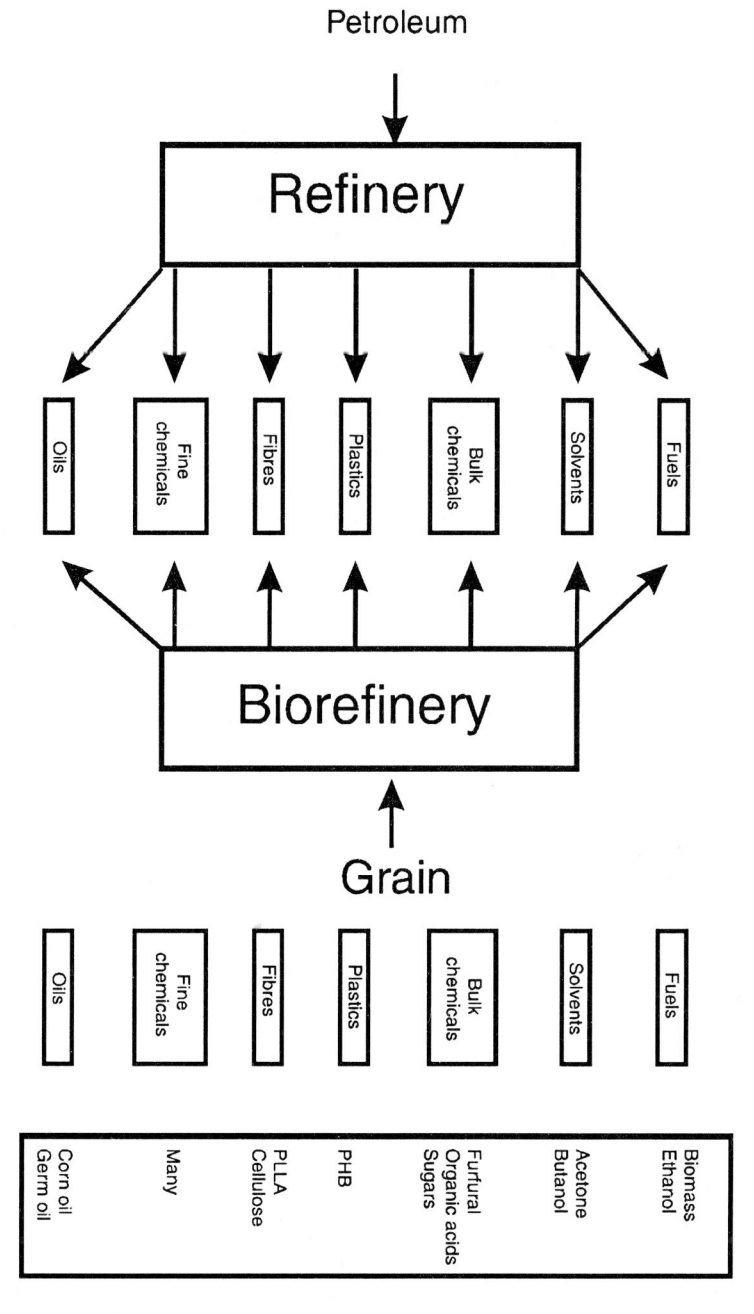

Fig. 11.7 The petroleum/grain analogy. As a raw material, cereal grains are analogous to petroleum for producing a comparable range of products to those from petroleum feedstocks.

2. Transboundary water and air pollution will be on such a small scale that every country will independently be able to determine the state of its own environment.
3. Levels of air pollution and noise will not impair people's health or well-being, and the threats to our cultural heritage will have been eliminated.
4. Lakes and seas will support viable, balanced populations of naturally occurring species, and their value for fishing, recreation or water supplies will not be impaired by pollution.
5. The productivity of agricultural and forest soils will be sustainable on a long-term basis. Pollutants will not be permitted to disturb natural biological soil procsses or restrict the use of groundwater.
6. Land and water will be used in ways that husband natural resources. Renewable resources will be used within the limits of ecosystem productivity, non-renewable resources sparingly and responsibly.
7. Natural species and populations will be able to survive in viable numbers. Particular care will be taken where native populations represent an important share of the world population of a species.
8. The country's most representative and valuable natural habitats and cultural landscapes will enjoy protection and be managed in accordance with that protection.
9. The potential of biotechnology will be harnessed for the benefit of environmental protection and its many applications in other areas will be scrutinized and controlled so that harm to the environment is avoided.
10. The flow of goods will be characterized by producer liability 'from cradle to grave' and economic growth will be used for consumption that spares natural resources and the environment.

REFERENCES

Abramovitz, J.N. Imperiled waters, impoverished future. Worldwatch Paper 128 (1998).

Agarwal, A., Narain, S. *Towards Green Villages*. Center for Sci. and Environ., Delhi (1989).

Akashi, Y. Technological change and global society. *Update Newsletter* No. 44 pp. 2-3. UN Centre for Sci. and Technol. for Development, United Nations, New York (Winter, 1991).

Anandajayasekeram, P. Poverty alleviation is the key issue for public agricultural research. *Biotech. Develop. Monitor.* 40: 24 (1999).

Ayres, R.U. Industrial metabolism. In : Ausubel, J.H., Sladovich, H.E. (eds). *Technology and Environment*. National Academy Press, Washington, D.C. (1989).

Bartlett, A.A. Reflections on sustainability, population growth, and the environment. *Pop. Environ.* 16: 5-35 (1994).
Bir, S.S. Plant exploration: My experiences. In: *Plant Wealth of India* (Mohan Ram, H.Y., ed.). *Proc. Ind. Natl. Sci. Acad.* B 63: 209-228 (1997).
Bodley, J.H. (Ed.) *Tribal Peoples and Development Issues: A Global Overview.* Mayfield Publishing Co., Mountain View, California (1988).
Bodley, J.H. *Victims of Progress.* 3 ed. Mayfield Publishing Co., Mountain View (1990).
Brown, L.R. A new era unfolds. In : *State of the World, 1993* (Ed. L. Starke) pp. 3-21, 201-204, W.W. Norton & Co., London (1993).
Bruce, J.P. *Meteorology and Hydrology for Sustainable Development.* WMO, Report No. 769. Geneva (1992).
Buttimer, A. Making sustainability work at the local level. *Environment* 13-40 (April, 1998).
Capra, F. *The Tao of Physics.* Shambhala Publ., Boston (1991).
Cleveland, D.A. Ethical dilemmas. Can science and advocacy coexist? *Anthropology Newsletter* p. 9 (March 1994).
Colchester M. Sustaining the forests : the community based approach in South and S.E. Asia. United Nations Res. Inst. for Social Develop. (UNRISD), Geneva, DP No. 35, Geneva (1992).
Daily, G.C., Walker, B.H. Seeking the great transition. *Nature* 403: 243-245 (2000).
Durning. A.T. *Guardians of the Land: Indigenous Peoples and the Health of the Earth.* Worldwatch Paper No. 112, Washington, D.C. (1993).
Dyer, M.I., Ishwaran, N. Redevelopment as a tactic in grazing land rehabilitation. In Wali, M.K. (Ed.). Environmental Rehabilitation : *Preamble to Sustainable Development.* SPB Acad. Publishing, The Hague (1990).
Edwards, R. Tomorrow's bitter harvest. *New Sci.* August 17, pp. 14-15 (1996).
Fisher, D.E. The Resource Management Legislation of 1991 : A Juridical Analysis of Objectives. Extract from *Resource Management.* pp. 1-30. Brooker and Friend, Wellington (1991).
Fookes, T. Social and economic matters. New Zealand Ministry of Enviroment, Wellington (Sept. 1992).
Frazier, J.G. Sustainable development : modern elixir or sack dress? *Environ.* 24: 182-193 (1997).
Fyfe, W.S. Toward 2050: the past is not the key to the future - challenges for the science of geochemistry. *Environmental Geology* 33: 92-95 (1998).
Fyfe, W.S., Powell, M.A., Hart, B.R., Ratanasthien, B. A global crisis : energy in the future. *Nonrenewable Resour.* 2: 187-196 (1993).
Gardner, J.E. The elephant and the nine blind men : an initial review of environmental assessment and related procsses in support of sustainable development. pp. 35-66. In Jacobs, P., Sadler, B. (eds.). *Sustainable Development and Environmental Assessment : Perspectives on Planning for a Common Future.* Environ. Assess. Res. Council, Hull, Quebec (1991).
Gaskell, D. Sowing the seeds of sustainability. *Chemistry in Britain*, pp. 49-50 (Feb. 1998).

Gatto, M. Sustainability: Is it a well defined concept? *Ecol. Applications* 5 : 1181–1183 (1995).
Giddens, A. *The Consequences of Modernity.* Polity Press, London (1990).
Goodland, R. The concept of environmental sustainability. *Annu. Rev. Ecol. System.* 26: 1-24 (1995).
Grimes, S. *Challenges of Development in a European Peripheral Region.* LLASS Working Paper No. 15, Department of Geography. University College Dublin (1995).
Grumbine, R.E. What is ecosystem management? *Conserv. Biol.* 8: 27-38 (1994).
Haq, M., Jolly, R., (coordinators). *Human Development Report 1996.* UNDP & Oxford University Press, Oxford (1996).
Hedge, P. *Chipko and Appiko : How the People Save the Trees.* Nonviolence in Asia Series, Quaker Peace and Service, London (1988).
Heilig, G. *Lifestyles and Energy Use in Human Food Chains.* Working Paper WP-93-14. International Institute for Applied Systems Analysis (IIASA), A-2361 Luxemburg, Austria (1993).
Holling, C.S., Meffe, G.K. Command and control and the pathology of natural-resource management. *Conserv. Biol.* 10: 328-337 (1996).
Hunt, J., Rollason, R., Ross, E., Salter, R. *One World or None : Making the Difference.* Austral. Council for Overseas Aid, Canberra (1989).
Isbister, J. *Promises Not Kept: The Betrayal of Social Change in the Third World.* 2nd Edition. Kumarian Press, West Hartford, Connecticut. (1993).
IUCN. (Internat. Union for Conserv. of Nature). *Guidelines for the Management of Tropical Forests.* Gland, Switzerland (1989).
IUCN/WWF. *World Conservation Strategy : Living Resource Conservation for Sustainable Development.* IUCN, Gland, Switzerland (1994).
IUCN/WWF. *World Conservation Strategy: Living Resource Conservation for Sustainable Development.* Iucn, Gland, Switzerland (1980).
IUCN/WWF. *Caring for the Earth: A Strategy for Sustainable Living.* Gland, Switzerland (1991).
Kanvalli, S. *Quest for Justice.* Samaj Parivartana Samudaya, Dharwar (1990).
Khoshoo, T.N. Plant Diversity in the Himalaya: Conservation and Utilization. Pandit G.B. Pant Memorial Lecture II. G.B. Pant Inst. Himalayan Environment and Development, Kosi Katarmal, India, pp. 1-29 (1992).
Knight, R.L., Meffe, G.K. Ecosystem management: agency liberation from command and control. *Wildlife Society Bulletin* 25(3): 676-678 (1997).
Korten, D.C. Sustainable development. *World Policy J.* 9: 157-190 (1991-1992).
Korten, D.C. *When Corporations Rule the World.* Kumarian Press, West Hartford, Connecticut (1995).
Lélé, S., Norgaard, R.B. Sustainability and the scientist's burden. *Conserv. Biol.* 19: 365 (1996).
McNeely, J.A. *Mobilizing Broader Support for Asia's Biodiversity: How Civil Society Can Contribute to Protected Area Management.* Asian Development Bank, Manila (1999).
Meffe, G.K., Carroll, C.R. et al. *Principles of Conservation Biology.* 2ed. Sinauer Assoc., Sunderland, MA (1997).

Miser, H.J., Quade, E.S. *Handbook of Systems Analysis: Craft Issues and Procedural Choices.* Wiley, Chicester (1988).
Nelson, R.H. The federal land management agencies. In Knight, R.L., Bates, S.F. (eds.). *A new Century for Natural Resources Management.* pp. 37-59. Island Press, Covelo, Calif (1995).
NRC. (National Research Council). *Our Common Journey: A Transition Toward Sustainability.* National Academy Press, Washington DC (1999).
Nriagu, J.O. A history of global metal pollution. *Science* 272: 223-224 (1996).
Orr, D.W. Twine in the baler. *Conserv. Biol.* 8: 931–933 (1994).
Oyen, E., Miller, S.M., Samad, S.A. *Poverty — A Global Review : Handbook on International Poverty Research.* Scandinavian Univ. Press, Oslo (1996).
Pearce, D., Markandya, A., Barbier, E. *Blueprint for a Green Economy.* Earthscan, London (1989).
Pinstrup-Andersen, P., Pandya-Lorch, R. Food for all in 2020: can the world be fed without damaging the environment? *Environ. Conserv.* 23: 226–234 (1996).
Regier, H.A. A focus on reform and on redevelopment of degraded regions. pp. 67-79. In Jacobs and Sadler (1991).
Robinson, J.G. The limits to caring : Sustainable living and the loss of biodiversity. *Conserv. Bull.* 7: 20-28 (1993).
Sadler, B., Jacobs, P. Conclusion and recommendations on further directions for research and development. pp. 169-173. In Jacobs, P., Sadler, B. (eds.). *vide supra* (1991).
Sahu, K.E. (Ed). Environmental impact of coal utilization. Indian Institute of Technology, Bombay (1991).
Schmidheiny, S. *Changing Course: A Global Business Perspective on Development and the Environment.* MIT Press, Cambridge, Massachusetts (1992).
Shaw, R.W., Öberg, S. Sustainable development: applications of systems analysis. *The Science of the Total Environment* 149: 193-214 (1994).
Shiva, V. Decolonizing the North. *The INTACH Environmental Series,* 18. pp. 1–16. Indian Trust for Art and Cultural Heritage, New Delhi (1992).
Socolow, R., Andrews, C., Berhut, F. & Thomas, V. (eds.). *Industrial Ecology and Global Change.* Cambridge Univ. Press, Cambridge (1994).
Speth, J.G. The transition to a sustainable society. *Proc. Natl. Acad. Sci.* USA 89: 870-872 (1992).
Stigliani, W.M. Changes in valued 'capacities' of soils and sediments as indicators of non-linear and time-delayed environmental effects. *Environ. Monitor. Asses.,* 10: 245-307 (1988).
Tolba, M.K., El–Kholy, O.A., El–Hinnawi, E., Holdgate, M.W., McMichael, D.F., Munn, R.E. (eds.) *The World Environment 1972–1992: Two Decades of Challenge.* UNEP & Chapman & Hall, London (1993).
Tucker, R.P. The British Empire and India's forest resources : The timberlands of Assam and Kumaon, 1914-1950. In Richards, J.H., Tucker, R.P. (eds.). *World Deforestation in the Twentieth Century.* Duke Univ. Press. Durham, pp. 91-111 (1988).
Viederman, S. Regulating Technolongy for a Sustainable future. *ISEE* (Intsnrt. Soc. for Ecol. Economics) *Newsletter* (Novr. 1991).

Viederman, S. Sustainable corporations : Isn't that an oxymoron? *Tomorrow : Global Environ. Business* 6 (4): July-Aug. 1996.

WCED. *Our Common Future.* OUP, Oxford (1987).

WCS (World Conservation Strategy) World Conservation Strategy: Living resource conservation for sustainable development. International Union for the Conservation of Nature (IUCN): Gland, Switzerland (1980).

Weinberg, C.J., Williams, R.H., *Scientific American,* Sept. 1990.

Westing, A.H. Core values for sustainable development. *Environ. Conserv.* 23: 218–225 (1996).

Wilbanks, T.J. Sustainable development in geographic perspective. *Annals of the Association of American Geographers* 84: 541–56 (1994).

Winkler, D. The forests of the eastern part of the Tibetan Plateau. *Plant Res. Develop.* 47/48: 184-210 (1998).

World Bank. *Forest Policy Paper.* Agriculture and Rural Develop. Deptt., Washington, D.C. (May 10, 1991).

WRI. *World Resources (1996-1997).* World Resources Institute, Oxford University Press, Oxford (1997).

Young, E. Waste does wonder for India's forests. *New Sci.* p. 19 (July, 1994).

Chapter 12
Biosphere Management

INTRODUCTION

In their Policy Forum 'International ecosystem assessment' (*Science's* Compass, 22 Oct., p. 685), Edward Ayensu et al., pointed out that future human welfare requires an integrated, predictive and adaptive approach to ecosystem management. They identified the types of information needed to support such an approach. Ayensu et al., called for a worldwide ecosystem assessment that might take 3 to 4 years, and be repeated at every 5- or 10-year interval. This assessment would build on other international activities and would ultimately be complemented by detailed local monitoring and assessment.

Ayensu et al., identifed two requirements; viz., (i) a new approach to ecosystem management, and (ii) development of an information base to support that approach. The major obstacle to sustainable management of the biosphere may be a lack of broad-based public understanding and political will. To build the necessary public understanding, biosphere management tools such as ecosystem assessment should invite the participation of numerous stakeholders.

An international process repeated at 5- to 10-year intervals, which would include catalytic local, national and regional ecosystem assessments, may be the best way to stimulate ongoing assessments at multiple scales. An international assessment can demonstrate the utility of the integrated approach, develop and test methodologies that could be used at multiple scales and build the capacity to undertake such ongoing assessments at local and national scales.

Due respect and consideration for the various resources essential to buffering vital life-supporting conditions for our species is absolutely essential for a sustainable management of the biosphere. According to a resource buffer theory (see Black, 2000), for every resource essential to a life form where a very small proportion is directly used for life processes of individuals, the vast remainder of the resource is indispensable to maintain conditions under which the population as a whole can survive. For instance, although we need only about 2 litres of water per person per day to remain alive, the total human population needs the balance of the water resources in the biosphere. These water resources help absorb and redistribute energy and waste products from life forms; by shielding us against the atmosphere's fluctuations in gaseous content; and by transportation of and provision of conversion sites for nutrients. If such resources become polluted or spoiled, conditions for human life will inevitably deteriorate—my 2 litres of water won't save me.

Some other resources that are disproportionately distributed in this way are the atmosphere, oxygen, carbon and biodiversity. The important role of the 'unused' portion of the resources critical to our continued existence needs to be included in the percentages of resource use, as Ayensu et al., (1999) suggested Intenational ecosystem assessment must include buffer reserves so that natural resource policy makers appreciate their role.

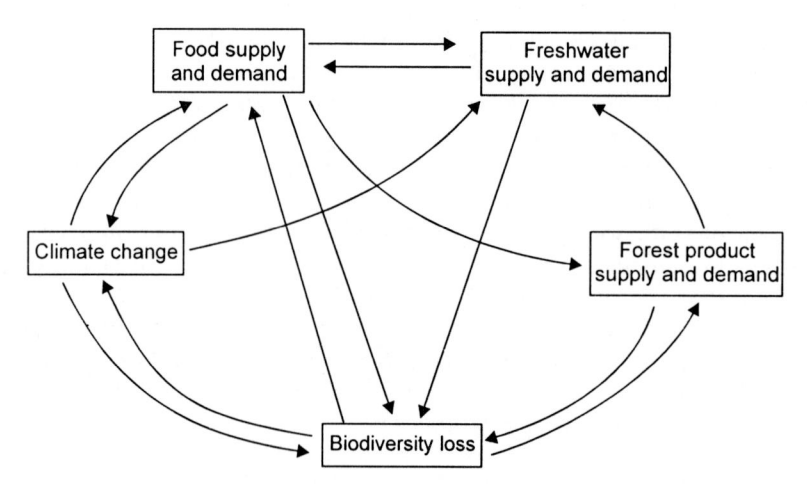

Fig. 12.1 The linkages between various ecosystem goods and services must be taken into account in ecosystem assessment and management (after Ayensu et al., 1999).

The sound management of lakes and reservoirs depends strongly on the cooperation of society to ensure preservation and wise use. A large number of urban and rural people use lakes and reservoirs as their main source of water for domestic, agricultural and industrial purposes.

Communities also enjoy these waters for such recreational activities as fishing, swimming and boati—ng—uses that sometimes adversely impact the water quality.

Large amounts of trash are thrown into waters, while piers and access roads continue to be built along shorelines without concern for their impacts. Boats spill fuel and waste and the sprawl of hotels and housing developments goes on without proper environmental planning. In some areas, wetlands are viewed as an obstacle to development, or simply a nuisance. Such thinking leads to the destruction of these invaluable ecosystems, with serious impacts on wildlife and freshwater water quality.

Inappropriate behaviour by the general public is, in part, due to carelessenss, but it is also the result of limited information and awareness. Such useful information as the characteristics of freshwater bodies, including how they function, their limited capacity to absorb waste and endure alterations, as well as the energy and costs involved in their preservation are often not well known to the layman.

There is an urgent need to educate the public on matters related to the management and preservation of lakes and reservoirs. This includes the need to promote awareness, more responsible behaviour and active participation to ensure preservation.

Global overspending on weapons research and arms is doubly tragic. Lavish military spending absorbs resources that are badly needed in the fight for a more sustainable world order. Vital research frontiers remain unexplored, while military research budgets soar. The diversion of capital, talent and energy to military matters undermines the ability of societies to undertake the investments and organizational reforms essential to their future ecological security. Here, even some developing countries also increasingly buy arms at levels far beyond their means, usually with the encouragement of developed-country arms manufacturers. It is ironical that the world is incapable of raising $80 million a day to provide clean water to all people, but gladly spends $1.5 billion a day on weapons!

VALUATION OF ECOSYSTEM SERVICES

For too long, ecosystem services have neither been captured in commercial markets nor sufficiently quantified in terms of economic and manufactured capital. This neglect can have an adverse impact on the sustainability and quality of human life the world over. Without doubt, the economies of the Earth would crumble without the services of ecological life-support systems, so their total value to the economy is infinite. Many workers have estimated the value of a variety of ecosystem services (Costanza et al., 1997). Costanza et al., have estimated values for ecosystem services per unit area by biome, and then multiplied by the total area of each biome and summed over all services and biomes.

Ecosystem functions refer to the habitat, biological or system properties or processes of ecosystems. Ecosystem goods (e.g. food) and services (e.g. waste assimilation) benefit human populations. Ecosystem goods and services may be clubbed together to mean ecosystem services. A large number of functions and services can be identified (Daily, 1997). Costanza et al. grouped ecosystem services into 17 major categories and included only renewable ecosystem services, excluding non-renewable fuels and minerals and the atmosphere. Ecosystem services and functions do not necessarily show a one-to-one correspondence. Occasionally, a single ecosystem service can be the product of a few ecosystem functions whereas, in other cases, a single ecosystem function may contribute to several ecosystem services. Many ecosystem functions are interdependent.

Capital means a stock of materials or information that exists at a certain time. Each form of capital stock generates certain services that may be used to transform materials so as to increase human well-being. Man's use of these services may or not leave the original capital stock intact. Capital stock can take diverse identifiable physical forms including natural capital, such as trees, minerals, the atmosphere, etc., and manufactured capital such as machines and buildings. Besides, capital stocks can exist as information stored in computers, in human brains and in species and ecosystems.

Ecosystem services include flows of materials, energy and information from natural capital stocks which combine with manufactured and human capital services to spell human welfare. The general class of natural capital is essential to human well-being. Zero natural capital means zero human well-being because it is not feasible to substitute, totally, purely non-natural capital for natural capital. As manufactured and human capital require natural capital for their construction, there is no sense in asking the total value of natural capital to human welfare, nor in asking the value of massive, particular forms of natural capital (Costanza et al., 1997). However, it does make sense to ask how changes in the quantity or quality of various types of natural capital and ecosystem services may have an impact on human welfare. Such changes can include small changes at large scale, as well as large changes at small scales.

Although it is extremely difficult to assign a value to ecosystem services, human life, environmental aesthetics, or long-term ecological benefits, yet we do so every day. For example, when we set construction standards for highway roads, bridges, etc., we value human life because spending more money on construction would save lives. Admittedly, however, such valuation is fraught with large uncertainties.

The valuation of services of some natural capital involves determination of the differences that relatively small changes in these services make to human well-being. Changes in quality or quantity of ecosystem services can have some value in so far as they may either change the benefits associated with human activities or alter the costs of those activities. These changes may either affect human happiness through established markets or through non-market activities. For example, forests provide timber through well-established markets, but the associated habitat values of forests are also felt through unmarketed recreational activities for urban dwellers. Forests not only provide timber materials but also hold soils and moisture, and create micorclimates, all of which contribute to human well-being in complex, non-marketed ways (Costanza et al., 1997).

According to some rough and uncertain estimates made by Costanza et al., (1997), ecosystems provide at least U.S. $33 trillions worth of services annually. The majority of the value of services they could identify is currently outside the market system, in services such as gas regulation ($1.3 trillion yr^{-1}), disturbance regulation ($1.8 trillion yr^{-1}), waste treatment ($2.3 trillion yr^{-1}) and nutrient cycling ($17 trillion yr^{-1}). About 63% of the estimated value is contributed by marine systems ($20.9 trillion yr^{-1}). Most of this comes from coastal systems. About 38% of the estimated value comes from terrestrial systems, mainly forests ($4.7 trillion yr^{-1}) and wetlands ($4.9 trillion yr^{-1}) (Costanza et al., 1997). Such uncertain estimates are, however, merely static pictures of the biosphere that is actually a highly complex and dynamic system. Although these estimates do have some utility, the need is to build regional and global models of the linked ecological economic system aimed at a better understanding of both the complex dynamics of physical/biological processes and the value of these processes to human welfare (Costanza et al., 1997).

Ecosystem services provide an important portion of the total contribution to human well-being on Earth. Unless we assign the natural capital stock that produces these services at adequate weightage in the decision-making process, human welfare may greatly suffer in the future. The annual value of these services may by $16-54 trillion, with an estimated average of $33 trillion. The real value is much larger. The practical use of the above estimates may help modify systems of national accounting to better reflect the value of ecosystem services and natural capital. Usually, because ecosystem services are largely outside the market and uncertain, they tend to be ignored or undervalued, leading to projects whose social costs greatly outweigh their benefits. As natural capital and ecosystem services are likely to become more stressed and scarce in the coming years, their value may be expected to increase.

MINERALOGY AND HUMAN WELFARE

Unlike physics, chemistry and biology which are focused sciences, ecology and environment are fairly diffused disciplines. Agricultural mineralogy is now attracting much attention in the context of food security (see Smith, 1999).

Modern techniques allow physicochemical characterization of mineral surfaces and adsorbed molecules and ions in soils. Plant growth depends on subtle interactions between mineral surfaces, fluids and microbes. Incorporation of non-toxic trace elements into food depends on the interaction of organic and inorganic components, as does that of toxic ones. Roughly, one quarter of the wheat and rice crops are lost to Mn-oxidizing bacteria. Soils become contaminated with such mobile toxic elements as Cd, Se and As, which can affect plant growth and food safety (Smith, 1999).

Mineral dust particles blown from drying geological basins pervade the atmosphere. Upon falling to earth, they can have both good results and bad ones. The good effects are exemplified by the loess soils that benefited early agriculture in Europe, China and North America whereas the bad ones appear, e.g. when air pollution affects lungs. Certain modern techniques allow analysis of tiny particles for distinguishing the natural and industrial components of aerosols.

Minerals in the oceans range from dozens of types in living organisms to various zeolites grown from volcanic ash and precipitates in ferro-manganese nodules.

Various physical and biochemical techniques are stimulating exciting studies on microbes which concentrate useful elements (including uranium) into ore bodies. Other such studies relate to how microbes interact with atmospheric gases to modify the climate. Some microbes form organic acids that accelerate mineral weathering to make soil minerals (beneficial action), and eat away outdoor statues (harmful action).

Some one billion people, mostly in developing countries, currently ingest harmful amounts of toxic elements.

Geologists now have access to modern chemical microscopes and other tools to study various toxic materials and can study at the atomic level those interactions between the inorganic and organic worlds that may help promote human welfare.

TRENCHLESS TECHNOLOGIES

Water and sewage infrastructure is a significant investment for many municipalities. If these systems are not properly maintained or if they become affected by land subsidence, problems with damaged or leaking water distribution and sewage collection pipes can be quite extensive. Conventional open trench methods to repair and rehabilitate these systems are costly and disruptive. Trenchless technology systems offer an innovative, cost-effective alternative. This technology is now being used in some cities in Poland. There is a need to create a greater awareness of trenchless technology methods, solutions, benefits and limitations; a better understanding of procedures and technology selection criteria; and identification of various mechanisms for technology cooperation, information exchange and targeted research.

ENVIRONMENTAL TECHNOLOGY PERFORMANCE VERIFICATION

Possible mechanisms and approaches for improving the quality of information about technology performance have recently attracted interest in the USA and Canada. While there are many different tools that can be used to assess technology performance, the Canandian and United States ETV programs both provide useful platforms for helping technology users understand the performance characteristics of environmental technologies.

Programs of environmental technology verification (ETV) have been developed only recently (see Fig. 12.2), with its genesis in North America. Just as eco-labeling provides guidance to consumers seeking to purchase 'environmentally friendly' products, verification programs are designed as a means of accelerating market acceptance of innovative technologies by providing technology users with information about performance, thereby minimizing the uncertainty about their purchasing decisions. Figure 12.2 gives an overview of some environmental assessment approaches.

	Voluntary	Mandatory
Self-Assessment	• Corporate Environmental Reports • Self-Labelling • Self-Declarations	• Hazard Warning Labels • Information Disclosure on Required Substances
Third Pardy Assessment	• Environmental Technology Verification (ETV) • Environmental Labelling Generated by Third Parties • ISO Certification	• Testing and Inspection for Regulatory Compliance

Fig. 12.2　Various mechanisms to qualify information about the environmental characteristics and performance of organisations, processes, technologies and products. The ETV approach focuses on verifying the performance of environmental technologies (after Neate, 1999).

Verification is the process of determining, through the application of guidelines or pre-determined criteria (and substantiated by investigation, statistical analysis and other means), that a programme, project or technology is technically sound and will produce the results described in a performance claim. Verification is not an isolated process but forms part of a larger system that includes monitoring, reporting, certification and accreditation. Verification guidelines outline the procedures and information requirements needed to verify a performance claim. These guidelines should reflect the demands of what is driving verification: a voluntary process or a regulatory process.

The salient characteristics of a verification system include:

Credibility—The process should involve credible organizations. This can be done in conjunction with internationally-recognized bodies that accredit competent organizations to verify and certify.

Transparency—The process should be open and transparent with information shared among interested parties.

Compatibility—Verification guidelines should be designed to be relevant for national and international applications.

Continuous Improvement—The system should be designed to accommodate continuous improvement, taking into account new, emerging information and knowledge (Neate, 1999).

'Verification' is often confused with 'certification,' but there is a subtle difference between the two. Verification involves the assessment and validation of performance claims by an independent third party. Certification goes one step further by guaranteeing that the technology, process or project meets specific standards or performance criteria.

BOTTOM SEDIMENTS AND THE SECONDARY POLLUTION OF AQUATIC ENVIRONMENTS BY HEAVY METALS

The accumulation of pollutants in the bottom sediments of water bodies and their remobilization are two very important mechanisms that regulate pollutant concentrations in an aquatic environment. Heavy metals (HM) are important chemical pollutants in natural waters (see Kumar and Häder, 1999). The HM pollution of aquatic ecosystems is often more obviously reflected in high metal levels in sediments, macrophytes and benthic animals than in elevated concentrations in water (Linnik and Zubenko, 2000). The ecological effects of HM in aquatic ecosystems and their bioavailability and toxicity are closely related to species distributions in the solid and liquid phases of water bodies. Whereas organic contaminants are degraded to various degrees, HMs do not undergo such transformations and are always present in aquatic ecosystems, redistributing among different components. Water bodies with slow water-exchange rates (e.g. lakes and reservoirs) accumulate HM in their bottom sediments in considerable quantities. This phenomenon has both positive and negative aspects. On one hand, the bottom sediments promote self-purification in the aquatic environment because of HM accumulation. On the other hand, HM accumulation in the bottom sediments is reversible. Often, bottom sediments act as a strong source of secondary water pollution. The release of HMs from bottom sediments is stimulated, for instance, by a deficit in dissolved oxygen, a decrease in pH and redox-potential (Eh), an increase in mineralization and in dissolved organic matter (DOM) concentration (Linnik and Zubenko, 2000).

The mobility of an HM depends on its form of occurrence in the solid substrates and pore solutions of the bottom sediments as well as on the physico-chemical conditions prevailing on the boundary of solid and liquid phases (Sager, 1992). HM flow from pore solutions is an important mode of exchange between bottom sediments and water.

SOME PASSIVE SYSTEMS FOR THE TREATMENT OF ACID MINE DRAINAGE

Acid mine drainage (AMD) generation occurs when pyrite and other sulphide minerals on exposure to oxygen and water and in the presence of oxidizing bacteria, such as *Thiobacillus ferroxidans*, oxidize to produce dissolved metals, sulphate, and acidity. The resulting solution interacts with other mineralogical constituents in secondary reactions such as acid-induced metal dissolution, ion-exchange and neutralization.

Acidic drainage is a serious environmental problem at many active and abandoned sulphide and coal mine sites. Untreated AMD pollutes receiving streams and aquifers and its effect on streams and waterways is often dramatic. In some cases, all aquatic life can disappear, river bottoms become coated with a layer of rust-like particles, and the pH decreases. Table 12.1 gives the typical composition of water effluents resulting from sulphide and coal mine operations along with the respective permissible levels for the mining effluents in USA (Gazea et al., 1995).

Besides pH and chemical composition, acidity and alkalinity, both of them expressed in mg $CaCO_3$/L, are important parameters for the characterization of mine waters. Acidity expresses the quantitative capacity of a water sample to react with a strong base to pH 8.3. The major components of mine waters acidity are protein acidity, associated with pH and mineral acidity related to dissolved metals content. Alkalinity is a measure of the water's acid-neutralising capacity. It is defined as the sum of all the titratable contained bases. The alkalinity of waters is primarily a

Table 12.1. Typical mine drainage composition and permissible levels for industrial effluents in the USA (after Gazea et al., 1995)

Composition (metal in mg/L)	Coal mines sulphide mines	Cu-Pb-Zn mixed effluents disposal in	Limits for industrial USA
pH	2.6-6.3	2.0-7.9	6-9
Fe	1-473	8.5-3,200	3.5
Zn		0.04-1,600	0.2-0.5
Al	1-58		
Mn	1-130	0.4	2
Cu		0.005 - 76	0.05
Pb		0.02 - 90	0.2

function of their carbonate (CO_3^{2-}), bicarbonate (HCO_3^-) and hydroxide content (OH^-), but contribution from borates, phosphates, silicates or other bases present is also included.

There are several methods for the treatment of mine waters, depending upon the volume of the effluent, the type and concentration of contaminants present. An effective treatment is one that generates water of neutral pH and low acidity and reduces the levels of the sulphates, iron and other metals present down to permissible limits.

The conventional mine drainage treatment systems involve neutralisation by addition of alkaline chemicals such as limestone, lime, sodium hydroxide, sodium carbonate or magnesia to water; as a result, pH is raised followed by the precipitation of metals. The equipments used are expensive.

Passive treatment schemes take advantage of naturally-occurring geochemical and biological processes in order to improve the quality of the influent waters with minimal operation and maintenance requirements. The pH of mine drainage is raised when the water mixes with alkaline water or through direct contact with carbonate rocks.

Some passive systems have mainly been used to treat acidic waters with low metals content. In some areas, wetlands are used for the treatment of coal mine drainage. Passive treatment schemes remediate the acid drainage quality with minimal operation and maintenance. The costs of passive treatment schemes are generally measured in their land requirements rather than labour and consumables, since these systems involve slow processes for contaminant removal and thus require longer retention times than the conventional systems and large areas to achieve similar results.

Many *Sphagnum* (a moss)-dominated bogs are natural examples of wetland treatment of AMD, and most of the early systems were attempts to construct such systems. *Sphagnum* proved quite sensitive to transplanting, abrupt changes in water chemistry and increased accumulation of iron. At most sites the, moss died within the first growing season. After repeated trials and much research, a wetland design evolved that proved tolerant to contaminated mine drainage and was effective at reducing the levels of dissolved metals. Most of these treatment systems consist of a series of small wetlands that are vegetated with cattails (*Typha latifolia*), occasionally with the addition of organic substrate in which the cattails root.

Anaerobic processes have also proven to be effective in metals removal. Ponds, ditches and rock-filled basins have been constructed that are not planted with emergent plants and, in some cases, contain no soil or organic substrate (Hedin and Naim, 1993). Pretreatment systems have been developed where acidic waters contact limestone in an anoxic environment before flowing into a settling pond or wetland system.

MECHANISMS OF CONTAMINANTS REMOVAL

Several physical, chemical and biological processes take place within wetlands to reduce the metal concentrations and neutralise the acidity of the influent water. The simplest mechanism is dilution. In some systems, major inflows of uncontaminated water can change the water chemistry. Even besides dilution, many constructed wetlands have considerable effect on the acidity of AMD and the concentrations of dissolved metals. Filtering of suspended material, metal uptake by live roots and leaves, adsorption and exchange by plants, soil and other biological materials, abiotic or microbially-catalyzed metal oxidation and hydrolysis reactions in aerobic zones, and microbially-mediated reduction processes in anaerobic zones all result in improvement of water quality. The above mechanisms as related to iron removal in a wetland are schematically presented in Fig. 12.3.

Some main processes encountered in passive treatment systems for the removal of contaminants are oxidation and hydrolysis; metal removal by plants, algae and organic substrates; reduction; and limestone addition (Gazea et al., 1995).

TYPES OF PASSIVE TREATMENT SYSTEMS

There are three chief types of passive technologies developed for acid mine drainage treatment: (1) aerobic wetland systems; (2) anaerobic organic substrate systems; and (3) anoxic limestone drains. Aerobic wetland systems involve mixed oxidation and hydrolysis reactions, and are mainly applicable to the treatment of net *alkaline* waters. Organic substrate systems enhance anaerobic bacterial activity that results in the sulphate

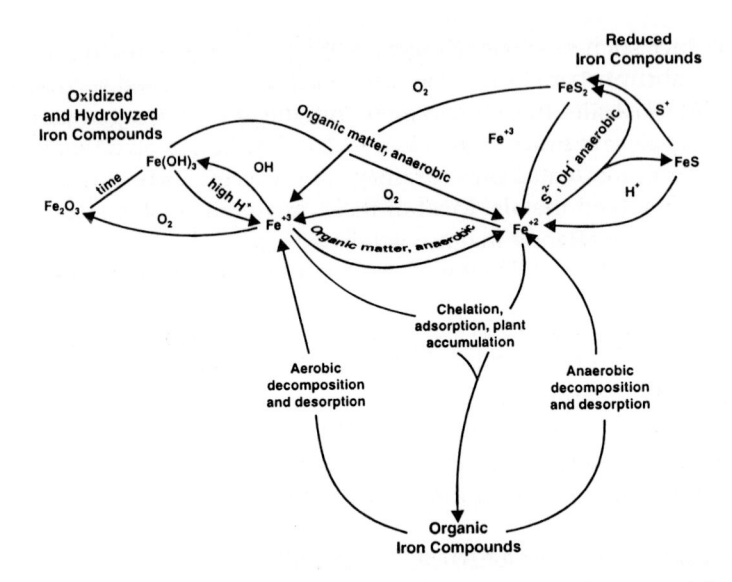

Fig. 12.3 Forms of iron in a wetland receiving mine drainage and the respective environmental conditions (after Hedin and Hyman, 1989).

reduction followed by the precipitation of metal sulphides and the generation of alkalinity. Anoxic limestone drains add alkalinity to the waters and may be used as a pretreatment stage before the wetlands.

Each of the three passive technologies is applicable to a specific type of mine water treatment scheme. The three are often most effectively employed in combination with each other. A model for the design of a passive treatment system based on the mine drainage chemistry is shown in Fig. 12.4; this model is suitable for the treatment of acidic mine waters where iron is the major contaminant. The design of the system becomes more complicated if other heavy metals such as manganese, aluminium, copper, lead, zinc, arsenic and mercury are present.

WATER HYACINTH

Many lakes, ponds and rivers all over the world are infested with water hyacinth (*Eichhornia crassipes*). This is a floating broad-leaved weed that produces beautiful purple flowers. Lake Victoria in Africa is heavily infested by water hyacinth, which currently covers some 100 square kilometres of the lake's surface.

Research has shown that under ideal conditions, the water hyacinth doubles itself once every two weeks. On Lake Victoria, wind and wave action tend to limit its growth, yet even the most conservative growth rate

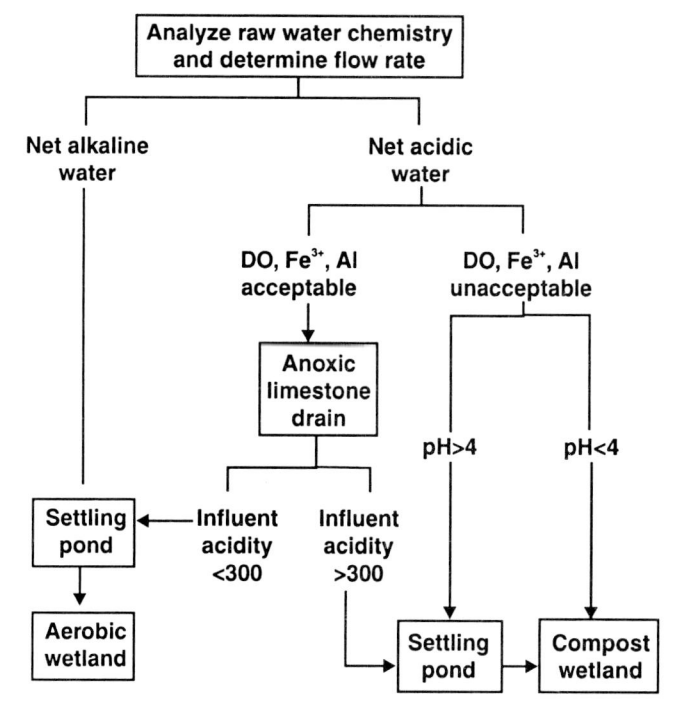

Fig. 12.4 Flow chart showing chemical determinations necessary for the design of passive treatment systems (after Gusek and Wildeman, 1995).

of 1% per day points to more inaccessible shoreline, and more unreachable fishing areas. When plant coverage depletes lake oxygenation, the entire fishing industry chokes along with the fish themselves. When wind-blown weed collects around the shores for several metres, harbours clog, blocking lake transportation. Mechanical harvesting, chemical and biological attacks are being considered and experimented with to cope with the weed.

Water hyacinth has also caused much havoc in India, Bangladesh and other developing countries. But the weed can be gainfully exploited for making ropes and even furniture.

Making furniture and ropes from water hyacinth requires little in the way of formal education and relies more on conscientiously following the following simple directions.

1. Cut off its roots and leaves and discard them.
2. Split the stem lengthwise and allow to sun dry for a day.
3. Scrape out the inner pith with a knife.

4. Allow stem to dry for another three days.
5. Soak in 10% solution of sodium metabisulphite for one hour.
6. Sun dry for 5-6 hours.
7. Cut stems lengthwise into strips (width depending on rope diameter required).
8. Prepare rope by braiding three strips of dried stem.
9. Optional—may be boiled in dye at this stage.
10. Cut off protruding strands from the rope.
11. Optional—may be varnished when used for craft making or furniture.

Simple wood frames for chairs and tables can be easily made from water hyacinth with papyrus used to wrap around the frames (Eliah, 1998).

Water hyacinth is also suitable for making paper. In Bangladesh, paper is being made by using 50% hyacinth stem and 50% waste paper.

Water hyacinth can also be viewed as a resource. It makes a good compost for raising mushrooms. Although water hyacinth is nutritionally comparable to other foraging food, cattle and water buffalo only occasionally browse the weed, only when no other food is available. However, a palatable silage mixture of four parts water hyacinth and one part maize bran can be made as a good cattle feed.

Water hyacinth has been found to be a good feed for pigs and rabbits but not for chickens.

It is also feasible to use the weed for lakeside power generation plants. Research in Bangladesh by Dr R. Eden (University of Warwick, England) has proved that 7.5 Megawatts of electricity may be produced at a single collection.

MOUNTAIN RESOURCES

High mountain areas constitute an important reservoir of resources and are important for the survival of mankind. The water resources of the Central Asian high mountain belt constitute the direct lifeline for a sizeable fraction of the world population. Vast areas in the arid forelands are irrigated by allogenous rivers from this extensive mountain belt.

Indeed, high mountain ranges can be viewed as water towers of modern civilization, whose function as storehouses of raw material must be preserved. The extensive irrigation oases on the edge of the Himalayan range (e.g. in Punjab) attest to the large-scale added value of water as a resource which has contributed decisively to the formation of trade and cultural centres of supra-regional importance (Kreutzmann, 1999). Arid zones can be colonized by areas producing gram or cotton for the world economy. The irrigation zones can be utilized through the construction of barrages, reservoirs and canal systems. Staggered construction of barrage dams can help regulate water releases for specialized agriculture and also generate power to supply the ever-growing demand for electricity in the mountain foreland.

TRADITIONAL RESOURCE MANAGEMENT

In Europe, human beings used to be regarded as being entitled to dominate and utilize nature at will. The European view recognized no limits to the exploitation and modification of ecosystems. This view has gradually changed since the mid-nineteenth century. Nevertheless the scientific methods of resource management that have since been developed have been applied mostly to single species populations in highly simplified ecosystems. On the other hand, a diversity of traditional cultures have elaborated management systems more consistent with traditional ecology (Gadgil and Berkes, 1991).

In recent years, ecologists have started appreciating how traditional peoples can use their resources without destroying them. Community-based resource management systems have worked mostly because of the presence of appropriate common property institutions. Traditional peoples have been interacting with their environment, modifying nature but actively maintaining it in a diverse and productive state.

The diverse traditional resource use practices make up a pool of human experience spanning many millennia. There is no doubt that the conservation of this rapidly diminishing pool of cultural diversity is as urgent as the conservation of biological diversity.

Traditional societies commonly regard physical and biological components of the environment and the human population as being interconnected in a web of complex relationships analogous to the modern ecosystem view of the natural world. The ecosystem concept means a holistic view of all ecosystem components and their interactions, including those involving human societies (see Table 12.2).

Restraints on Resource Use

Traditional ecosystem approaches are based on certain beliefs, including, e.g. restrained resource use. Some practices illustrating restrained resource use are outlined below :

1. A quantitative restriction by harvesters on the amount of harvest of a given resource stock of a species or from a given locality.
2. Abandoning the harvesting of a certain resource when the resource densities decline.
3. Stopping the harvesting from a certain habitat patch if yields from that patch are reduced.

Table 12.2 Some traditional ecosystem views (after Gadgil and Berkes, 1991)

S. No.	System	Country/Region
1.	Watershed management of salmon rivers and associated hunting and-gathering areas by tribal groups and kin groups	Amerindians of coastal British Columbia, Canada
2.	Delta and lagoon system management for fish culture and the integrated cultivation of rice and fish	South and Southeast Asia, especially Indonesia
3.	Land-water area and its water, soil, plants, animals and human occupants seen as an interrelated whole	Fiji, Solomon Islands
4.	The caste system in which different endogamous groups have exclusive rights and responsibilities for the use and management of specific resources	India
5.	The integrated resource management system in which farming, grazing and fishing territories are shared by different social groups through reciprocal access	Mali, Africa

4. Not harvesting a certain species in a certain season.
5. Not harvesting from a certain habitat patch in certain seasons or years.
6. Prohibiting the harvest of certain species or of their life history stages, by age, sex, size or reproductive status.
7. Certain habitat patches may either never be harvested or subjected to very low levels of harvests through strict regulation.
8. Prohibiting or regulating the use of certain methods of resource harvest.
9. Certain age-sex classes or social groups may be banned from employing certain harvesting methods, or utilizing certain species or habitat patches.

Sustainable resource-use develops when a particular group has both control and responsibility for a particular resource (Berkes, 1989). In the modern urban-industrial mega society, such limits to exploitation are achieved through government regulation. But in traditional societies and some modern rural groups, resource use is regulated through social controls. Restraint is more likely to evolve in groups of smaller numbers of individuals in repeated social interactions (see Gadgil and Berkes, 1991).

It is paradoxical that modern science seems largely unable to halt and reverse the depletion of resources and the degradation of the environment, partly because scientific resource management and Western reductionistic science in general, developed in the service of the utilitarian, exploitive 'dominion over nature' world view of colonialists and developers. It is best suited to the efficient utilization of resources as if they were boundless.

This is why modern resource management science is well suited for conventional exploitive development, but not for sustainable use (see Berkes, 1989, pp. 110-126). The need of the hour is to develop a new resource management science that is better adapted to serve the needs of ecological sustainability and the people who use these resources. This may possibly be achieved by conserving both biological and cultural diversity, which are closely tied together. There also is a further need to conserve the diversity of traditional resource management practices and systems (Gadgil and Berkes, 1991).

CONSERVATION AND MANAGEMENT OF RESOURCES

Resources are generally defined in terms of their usefulness in fulfilling human wants and social objectives. A resource is something on which one relies for aid, support or supply. It is something that is capable of satisfying human need and aspiration. The concept of resources is highly dynamic and includes the ever-changing patterns of want and technology in relation to nature. Any changes in cultural and technological state also change the notion and value of resources.

Natural resources are the basic earth materials: water, soil, plants, animals and minerals, all of which serve our needs. They are essential to our existence, as our continued prosperity and well-being depends upon the care and wisdom with which we use them. Resources are assets that sustain our economy and development.

Various resources differ depending on how they are used and what happens to them in the process of development and use or misuse.

Classification of Resources

Two major kinds of resources are natural resources and human resources. Natural resources have been historically treated as if they are free gifts of nature. But if we overexploit them or disturb their equilibrium in the environment, we have to pay some price. In the second category, man himself is the prime resource. He may make proper use or misuse of resources. His ability and wisdom are essential for proper development, management and use of resources.

Conservationists usually divide resources into four groups. Air and water exemplify perennial or perceptual resources. Plants, animals and soils can be called as 'renewable resources' because they respond to man's manipulation. Minerals are non-renewable resources because man is not able to induce their natural accumulation. The fourth group is less tangible but perhaps even more important than others—the aesthetic 'amenity resource', i.e. the desirable attributes of our cultural environment.

The main objective of resource conservation is the maximum use and maximum benefit for the longest term. The production of various renewable resources depends on the environmental conditions; if these conditions are favourable and suitable the resources will continue to flourish. If the conditions become unfavourable and unsuitable, production of resources would normally cease or decrease. The best management for any particular resource depends upon its quantity, mutability, and reusability. Omen (1971) proposed the resource classification shown in Table 12.3.

Table 12.3 Classification and management of resources (after Omen, 1971)

Resource	Comments
A. Inexhaustible	
(a) Immutable	Can suffer much adverse change through human activities.
(b) Misusable	Little danger of complete depletion but when improperly used, their quality may be impaired.
B. Exhaustible	
(a) Maintainable	Resources in which permanency depends upon methods of use
(i) Renewable	Living or dynamic resources whose perpetual harvesting is dependent upon proper planning. Their improper use can exhaust them with adverse socioeconomic consequences.
(ii) Non-renewable	Once gone, they cannot be replaced.
(b) Non-maintainable	Mineral resources (quantity static). Wasting assets. When destroyed they cannot be replaced.
(i) Reusable	Minerals whose consumptive usage is small and reuse potentialities are high
(ii) Non-reusable	Minerals with high or total consumptive use and whose exhaustion is a certainty.

Conservation, Development and Management of Resources

The conservation of resources does not mean their total non-use. Rather, it means their scientific and rational utilization for the greatest good of the maximum number for the longest time.

Conservation implies both the development and protection of resources, the one as much as the other. It is the wise use of resources according to which society divides its wants now and in relation to that which it feels it should leave for future needs.

Conservation of natural resources means their fullest possible use without abusing, without destroying needlessly, and without neglecting anything that can be used. It does not mean hoarding or storing anything for possible use in the future if we can make reasonable constructive use of it today. Resource management is concerned with the best possible use of a resource, but not when we let it remain idle. We conserve renewable resources when we use and reuse them without destroying or exceeding their regenerative capacity.

People's active participation is crucial for addressing the various local-level issues involved in defending or improving the natural resource base, which form a central part of the current debate on sustainable development. People's participation should preferably be an organized effort on the part of hitherto excluded groups and movements to increase control over resources and regulatory institutions in given social situations, so as to enhance the ways in which people take sustainable development into their own hands, either by working to maintain their traditionally sustainable resource management systems or by acting to resist projects or policies which will adversely affect their livelihood by degrading the environment. Special attention needs to be focused on addressing specific questions of policy relevance, including the structural factors which influence the outcome of resource-related programmes initiated from outside the community, the viability of traditional management systems and common property regimes and their potential for informing sustainable development policy, and the dynamics of movements which arise in resistance to development projects or private initiatives which threaten the environment (see UNRISD, 1991).

It has been seen that threats to traditionally-sustainable resource management practices often result in the development of local initiatives to protect livelihoods as, for instance, the popular movements in the Himalayan region, opposition to deforestation and dam construction in Brazil, movements to combat urban pollution in Mexico and traditional fishworkers' opposition to commercial overfishing in India.

It has also become apparent that because of their extensive ecological knowledge, societies based on sustainable-environmental management practices can accurately assess the true costs and benefits of ecosystem disturbance much better than any evaluator coming from outside the locality. Such societies are quick to realize that 'development', which results in environmental degradation, will rarely yield net benefits in the long run, and the emergence of popular opposition to a project is a strong indication that it is going to have adverse environmental consequences.

It is also emerging that the need for activism on local environmental issues has prompted some organizations not previously concerned with those issues to put sustainable resource management on their agenda. In urban-based non-governmental organizations, for example, earlier concerns with housing rights and other basic needs have developed into environmental activism when industrial pollution threatens the water supply of the community or when atmospheric pollution affects the health of the population. Likewise, in some support groups working with rural people, responses to environmental degradation have included adoption of new programmes which encourage a return to drought resistant crops, promote organic farming methods or develop afforestation programmes (UNRISD, 1991). Communities not only have high incentives for managing local resources sustainably, but also have often succeeded in developing effective means of doing so. As local level degradation is a major component of global environmental problems, the inability of communities to participate in resource management decisions has an important bearing on the potential for sustainable development. Struggles for greater participation are essential elements of a foundation for sustainable development (see Kumar, 1999).

ASSESSMENT OF FOREST CONDITION

This requires accurate data from field inventories as well as a good professional ludgment. In general, the following parameters are used to describe forest stands in figures:

Basal area, indicating stand density in m^2 ha^{-1}; diameter distribution and stem numbers, indicating stand structure in number of trees per hectare in diameter classes; and growing stock above a certain threshold diameter (normally 10-20 cm), indicating timber volume in m^3 ha^{-1}.

For timber production forests, two additional parameters for the assessment of timber related aspects of sustainability are: Increment (i), normally the net growth in m^3 ha^{-1}. a^{-1} of (commercial) trees above a threshold diameter indicating the production capacity of the forest; and annual allowable cut (AAC) indicating the planned harvesting target in m^3 a^{-1} for the whole management unit.

The AAC is normally stipulated for a certain period of time (usually 10 years) by consideration of the actual forest condition, its present growth potential and a desired future forest condition to be defined. This desired condition should be described using the quantitative key-parameters mentioned above and is not necessarily identical with the condition of the unlogged forest.

FOREST MANAGEMENT

Traditionally, foresters have looked upon forests in terms of maximum yields of lumber, without any consideration of ecological principles. Utility has been a primary concern in forestry, especially relating to the use wood and paper. However, some forest ecologists are now recognizing noncommodity values in forests so as to look for ways to obtain products from forests without jeopardizing these values. Natural forests certainly have values beyond those of just lumber or wood (see Kumar, 1999). Plantation forests have to be viewed as being ecologically quite different from natural forests.

Plantations are stands of trees planted by humans. They are usually dense, monospecific stands of conifers or eucalyptus in straight rows. They are not aged beyond a few decades before harvest. The species planted may be either native or exotic to the region. Even when native species are used, genotypes may be foreign, or the species may be planted on sites where it would not normally occur.

Impact of Forest Conversion on Biodiversity

Conversion of forests to anthropogenic habitats constitutes a serious threat to forest biodiversity worldwide. Replacement of natural forests by cropland, pasture, or urban developments involves removal of trees. The conversion of natural forests to plantations is very harmful for biodiversity. Somewhat less extreme, but still quite significant, is the conversion of virgin forests to second growth (Table 12.4).

Loss of structural diversity in plantations results in lower biodiversity and reduced abundance of many species. Cavity-nesting birds and mammals, bats and other animals that depend on holes, loose bark, or other aspects of standing dead trees become victims of any management strategy that fails to retain and recruit large snags. Large down logs contribute to diversity of invertebrates, small vertebrates, nonvascular plants, algae, fungi and bacteria. Natural forests of all ages are more diverse than plantations.

Long-term stability of old growth forests is conducive to enhanced arthropod diversity. Clearcutting followed by slash burning can drastically reduce total arthropods in the soil. Spiders also decline with clearcutting and site preparation; recovery to a typical forest spider species composition can take many years on wet sites and even longer on dry sites.

Differences in diversity appear to account for greater stability of natural forests relative to plantations, both in ability to maintain themselves against stresses and ability to bounce back after a disturbance. The loss of diversity caused by the conversion of natural ecosystems to artificial systems is usually accompanied by a loss of stability. Most tree plantations require considerable energy and labour to maintain their productive simplicity against the natural processes that would diversify them and decrease their timber value (Noss and Cooperrider, 1994). It appears that the higher diversity of natural forests, especially in

Table 12.3 Classification and management of resources (after Omen, 1971)

Characteristic	Virgin forest	Secondary natural	Plantation
Origin	After natural disturbance	After timber harvest of other human disturbance that cleared pre-existing vegetation	Usually, same as secondary natural but also may be established on existing natural vegetation
Regeneration	Natural reseeding, recolonization, germination from seed bank, or vegetative propagation from remnant structures	Same as virgin forest	Planting of seeds or seedlings
Within-stand structural diversity	High with abundant standing and down dead wood and high vertical complexity (multicyered)	Initially low, but gradually increasing with stand age and increased rates of gap formation	Low unless rotations are abnormally log, In which case, diversity increases as stand ages and disturbance, mortality injury and self-pruning occur
Within-stand horizontal patchiness	High due to heterogeneity in physical site conditions; canopy gap dynamics and horizontal variation in disturbance intensity	Variable but Initially lower and gradually increasing with stand age and gap formation	Usually low, except in long rotations, where patchiness increases over time with disturbance and differential mortality
Landscape diversity (between-stand patchiness)	Variable but generally high	Intermediate between virgin forests and plantations, but increasing with time since abandonment of intensive human activities	High if variety of age classes present, but because stands are structurally similar and monotonous in pattern, overall landscape diversity is quite low

Functional diversity (processes)	High diverse	Low initially but becomes high after recovery from human domination	Low
Species diversity	Varies but generally high in early and late seral stages (often lower in mid-succession)	High in early seral stages with influx of weedy colonists; low in mid-succession; high in old growth	High in early stages, lower thereafter
Animal population density	Varies but higher for most species than in plantations	Probably intermediate	Episodiclty high for some specless (e.g herblvorous insects); otherwise generaly lower than in natural forests
Stability	Variable, but probably maximum for region	Intermediate	Low
Aesthetics	Beautiful, sublime, inspirational	Generally similar to virgin forest	Dull, monotonous and ugly, but liked by slivicuturists and merchants
Conservation value	Extremely high	Moderate to high	Low

predaceous arthropods and other insectivores, enables these forests to resist pest outbreaks. Disease problems are also more prevalent in plantations than in natural forests.

NEW FORESTRY

In recent years, the adoption of certain novel approaches to multiple use forest management has led to the concept of 'New Forestry', which has largely been developed by J. Franklin of the University of Washington. New Forestry is based largely on emulation of nature in forest management. Traditional management practices have resulted in fragmentation, edge effects and structurally simplified plantations that differ greatly from stands appearing after natural disturbance. New Forestry focuses on aggregation of cutting units to reduce fragmentation, minimal road building, and retention of coarse woody debris and scattered live trees or their clumps in harvest areas. A site harvested in this way broadly resembles an area after a natural fire more than it resembles a clearcut.

New Forestry changes natural resource management philosophy from regulation of undesirable uses to sustained yield management (of e.g. wood and other commodities) and sustained ecosystem management.

However, although New Forestry holds great silvicultural potential, it may not be an effective substitution for protection of natural forests.

In some places, declining forest health is attempted to be 'cured' through large scale salvage logging which may really be even worse than the malady. For proper forest restoration, the following ideas deserve serious consideration.

1. Some areas may need no management at all (wildfires may be allowed to burn).
2. In some cases, there may be no management except fire suppression, or no management except prescribed burning (various treatments).
3. Light non-commercial thinning of small live and dead trees that may have invaded after fire suppression, followed by periodic prescribed burning that emulates the natural fire regime.
4. Dead trees may be salvaged in various amounts, proportions and size classes but live trees are not touched. Salvage to be followed by prescribed burning regularly.
5. Non-commercial thinning of small live trees, plus salvage of dead trees as in # 4.

To be on the safe side, greater emphasis may be on the less intrusive treatments (# 1-3). Those few natural and near-natural stands that remain in fairly good health should be completely protected, as they are best for restoration of degraded areas. Intrusive treatments should be confined to accessible stands, avoid roadless areas, and harm no mature or old-growth trees (Noss and Cooperrider, 1994).

The only way to protect biodiversity while practising forestry may be to reduce the intensity of forest management for commodities. The best way to manage multiple-use forests today may be ecological restoration. Most forests have suffered from human activities and require healing. Restoration should not imply returning to some pristine natural condition, but go in for restoration forestry which means, *inter alia*, minimizing human impacts and restoring biodiversity to the extent possible, by recovering viable populations of rare species and reintroducing extirpated ones. Restoration forestry simply means reversing the landscape changes that have been associated with loss of biodiversity.

The following are some of the conservation and restoration priorities for multiple-use forests: saving of virgin forests, protection of many regenerated forests, giving due consideration to landscape ecology, reducing road networks, stopping of clearcutting and intensive site preparation, protection of sensitive sites and minimizing mechanical operation, allowing natural regeneration, planning for long term rotations, due encouragement to alternative uses, emphasis on adaptive management and, above all, curtailing of consumption of wood products as far as possible.

TROPICAL FOREST SYNERGIES

The *El Niño*-Southern Oscillation (ENSO) is the cyclical expansion of warm waters in the equatorial Pacific Ocean that is responsible for the weather patterns known as *El Niño*. Quite often, the havoc caused by ENSO as drought or heavy rainfall adds up synergistically with human activities to devastate crops and forests in the tropics. Curran et al., (1999) reported that the unfortunate synergy between ENSO events and logging in some tropical forests, greatly reduces the regenerative capability of the ecologically and economically important tree family Dipterocarpaceae. In some of these forests, it has been documented that fire damage is much more severe in logged than in unlogged areas because logging debris provides a ready source of fuel.

Members of the family Dipterocarpaceae dominate Southeast Asian tropical forests. In some parts of Borneo, dipterocarps constitute as much as 70% of the canopy tree biomass and 80% of the tallest canopy trees. Many dipterocarp species reproduce irregularly in a manner called 'mast-fruiting' (Hartshorn and Bynum, 1999). This process involves the synchronous production of large numbers of single-seeded fruits once every 3 to 4 years, with very little seed production in the intervening periods.

Curran et al., (1999) demonstrated a strong correlation between the onset of dipterocarp mast-fruiting and ENSO events during the same time period. More than 50 dipterocarp tree species dispersed seed within a 1- to 2-month period every 3 to 4 years during ENSO events. ENSO exerts powerful effects not only on the flowering and fruiting of trees in tropical forests, but also on the population fluctuations of seed-eating birds and mammals.

The density of dipterocarps in Southeast Asian forests is commercially as well as ecologically important. Commercial logging companies cut large volumes of timber per unit area. Intensive logging leaves behind large areas of abundant debris that, in a negative synergy, with drought conditions related to ENSO, increase the probability of fires that further decimate the forests. Commercial logging has severely damaged the ability of residual stands of dipterocarps in some forests to reproduce.

Copious fruit production through masting is thought to satiate vertebrate predators such as the bearded pig and orangutan.

It seems that successful recruitment of dipterocarps may be dependent upon large, intact forested landscapes over which nomadic seed predators can move freely. In the longer term, conservation of dipterocarps and, indeed, all tropical forests, may only succeed in the context of large, connected, protected areas and the enforcement of existing laws governing forestry and other extractive activities (Hartshorn and Bynum, 1999).

Our knowledge of the dynamics of tropical forests is changing rapidly with the realization that their environment is variable on the decadal to century scale. Fluctuating climatic conditions partly determine tropical forest structure, species composition and dynamics. Tropical communities are highly contingent in space and time with respect to site, climatic and historical factors. Tropical forests have most probably experienced this disturbance regime to some degree in the past, and at the local scale species have probably become well adapted to it. However, climatologists now predict increasingly frequent extreme events in the future, and the combination of continuing deforestation and land-use conversion of anthropogenic origin plus an increasingly variable environment means an exacerbated situation that could be fairly difficult to manage.

The role of stochasticity is being recognized in the study of complex systems in fields such as climatology, ecology and economics. Simple deterministic equilibrium models rarely reflect nature and are of limited use in understanding the dynamics of species-rich communities such as tropical forests. Predictability would enhance if physical disturbance could be measured reliably and community change was correspondingly quantified. With what probability, for instance, does a volcanic eruption, a strong drought, a fire, or a cyclone occur with a given magnitude in x

years at a particular site or within a given region? This is becoming the science of extreme events. Long-term data sets are needed for tackling this problem.

Much tropical research has focussed on issues concerning the maintenance of high biodiversity in tropical communities. The multifarious factors involved vary in importance from site to site but a general driving factor in all vegetation systems is the extent, frequency and mode of disturbance. For the tropical forest perhaps the most recent, far reaching and unique source of disturbance is human activity which needs to be placed in the context of other 'natural' disturbances, with recognition that deforestation may also feed back on regional climate (Newbery et al., 1999). Biological interactions determining biodiversity have already been fairly well studied, yet they too are subject to the effects of disturbance and climatic events.

During the past two decades, there has been a notable shift in focus from disturbance processes internal to the forest towards the disturbance processes which are externally driving the ecosystem.

Recent work has shown that reactions and responses to disturbance can be understood, provided there is a long-term commitment. Major events of different magnitude may occur perhaps once every decade or century and trees live typically for 100-350 years. Animal populations are highly dependent on the existence of this forest structure. Furthermore, the dynamics at the community-ecosystem level is complex due to spatial variation in species and age-class composition, intrinsic time-lags and feedback processes. Scaling up from a plot to a region is not yet possible but by testing predictions at other sites and having longer runs of data from established ones, a better picture could emerge in the next few years. In the coming years more attention should be given to canopy-level processes and forest water balance (Newbery et al., 1999).

FOREST DISTURBANCE AND LAND COVER IMPACTS IN SOUTHEAST ASIA

For the last three decades, Southeast Asia has been a major source of timber for the international tropical hardwood trade. The majority of the accessible lowland forests have already been harvested by selective logging at least once, or have been cleared for other land uses. In some ways, commercial exploitation offers the best opportunities for improved forest management. Change is easier to introduce into a large concession managed by a single agency than into a forest occupied by a variety of social groups using forest resources in conflicting ways and under pressure to supply fuel wood and timber to expanding local towns and cities (Douglas, 1999).

Southeast Asia differs from the main tropical rainforest areas of the Amazon and Congo in that the latter occur on sedimentary basins on ancient Gondwana shield rocks whereas Southeast Asia has much greater geological diversity, with the island arcs of active tectonism and volcanicity. In Southeast Asia, population densities are quite high. These diverse conditions provide cautionary lessons on the degree to which solutions that have worked in one tropical rainforest may be extrapolated to another.

Water resources management, provision of urban water supplies, irrigation development and exploitation of hydroelectric potential have for long driven hydrological studies in Southeast Asia, with special reference to the risks of soil erosion and land degradation affecting crop and timber production.

For most countries in Southeast Asia, considerable information is now available on changes in the rainfall-runoff relationship due to forest disturbance or conversion, some lesser information on the impacts on sediment delivery and erosion of hillslopes, but quite little about the dynamics and magnitude of nutrient losses. Improvements have been made in the ability to model the consequences of forest conversion and of selective logging and there are exciting prospects for the development of better predictions of transfer of water from the hillslopes to the stream channels using techniques such as multilevel modelling. Understanding of the processes involved has advanced through the detailed monitoring made possible at some permanent field stations such as that at Danum Valley in Sabah (Douglas, 1999).

Shifting Cultivation

In areas of traditional cultivation, where the pressure on land resources is rather light and fallow periods are long, both field inspection and aerial photography give little evidence that large quantities of sediment are eroded from newly-cleared, burnt fields prepared for hill rice planting. Work done in Malaysian Borneo indicates that under shifting cultivation with long fallows, soil erosion differs little from that under natural forest (see Teck, 1992).

Elsewhere, in Southeast Asia, where conditions are different, there is often too little land for long fallows to occur. Wherever the length of fallows decreases, continued cultivation accelerates land degradation. For instance, for Thailand, Samrit et al., (1997) reported that some 68,2000 km^2 of the country subject to shifting cultivation, rubber, orchard, field crop and agroforestry production experience soil erosion rates of 10-10,000 t km^{-2} yr^{-1}, with a further 62,600 km^2 of steep hill country under shifting cultivation and field crops experiencing extremely severe losses in excess of 10,000 t km^{-2} yr^{-1}. Clearly, there is great diversity in soil loss from shifting cultivation, often due to contrasts in lithology, soil and climatic conditions, also reflecting differences in farming practices, and in the methodology of plot studies (Douglas, 1999).

Although soil erosion is a natural process, it is greatly accelerated when forest disturbance exposes the soil to raindrop impact. We know a great deal about the way hydrological and earth surface processes work in individual forest research areas, but it is difficult to reliably make predictions at the scales at which forest managers work. Many techniques of regulating water flows and reducing soil loss are known but much less is known of why people use land in ways that cause erosion and land degradation. Research is needed on how the detailed plot and small catchment studies can be upscaled to larger units where results can be used at the management scales of the annual logging coupes and for integrated catchment management. Nothing has been done on the effectiveness of management responses, e.g. the building of water bars across logging tracks, or to determine the appropriate spacing of cross-drains on logging roads of a specified gradient on a particular type of parent material. Such practical, applied studies should attract priority to ensure that the suggested remedies achieve the erosion mitigation for which they are designed (Douglas, 1999).

MANAGING EXCESS CO_2

Environmental engineers have started devising schemes for funneling excess CO_2 into the deep ocean, unminable coal seams, abondoned oil fields, and isolated aquifers. One idea is to pipe CO_2 waste from power plants to the bottom of the sea. The Norwegian oil company Statoil, since August 1996, has shunted 1 million tonnes per year for the last 2 years of CO_2, a byproduct of natural gas production, into a salt aquifer in the North-CO_2 that otherwise would have added 3% to Norway's carbon emissions. The CO_2 should eventually form a bubble under the formations' shale roof.

However, the paucity of enough aquifers in convenient places to hold all the world's extra CO_2 has prompted scientists to test another approach on a pilot scale in the Pacific Ocean. In December, 1997, the US, Japanese and Norwegian governments agreed to launch a giant experiment off Hawaii's Kona Coast in the summer of 2000. Engineers extended a 2- to 3-kilometer pipe down a steep slope, so that it reaches about 1000 meters below the ocean surface. Liquid CO_2 sprayed out of a nozzle mixed with seawater and was swept away by deep currents. Scientists followed the fate of the CO_2—300 tons over 30 days—with sea-floor instrumens and from remote vehicles.

About 90% of the CO_2 produced on Earth already gets sucked up by oceans eventually, so, in a way, this experiment emulated nature.

But major uncertainties loom over this approach. For example, solid hydrates formed from CO_2 and water could clog the pipe, and CO_2-laced water—more acidic than normal seawater—could harm marine life. CO_2 could also resurface if bottom currents which ought to circulate the CO_2 from power plants—scrubbing off impurities and pressurizing the gas—could require up to 30% of the energy that produced the original CO_2.

TOTAL ASSESSMENT AUDITS

Traditionally, energy, waste reduction and productivity audits for a manufacturing facility have been conducted independent of one another. Auditors make recommendations for improvement based on their specialized expertise (energy, waste reduction, productivity, etc.) without regard to how those recommendations may impact other, sometimes less obvious, subsystems or processes within the facility. The audits are typically done in isolation from the plant upper management and often without adequate knowledge of how inherent, inter-related operational constraints may influence the success of audit recommendations.

The Total Assessment Audit (TAA) concept originated from the belief that a manufacturing facility is better served using a consilient, holistic approach to problem solving rather than the more conventional isolated approach. The TAA methodology partners the upper management team of a company with a multi-disciplined team of industry-specific experts to collectively ascertain the core opportunities for improvement in the company and then to formulate a company-oriented continuous improvement plan. Productivity, waste reduction and energy efficiency objectives are intimately integrated into a single service delivery with the TAA approach. Nontraditional audit objectives that influence profitability and competitiveness such as business management practices, employee training and human resource issues are also subject to evaluation in a TAA. The underlying promise of this approach is that the objectives are interrelated and that simultaneous evaluation will produce synergistic results. Ultimately, the TAA approach may motivate a manufacturer to implement improvements it might not otherwise pursue if it were focused only on singular objectives (Haman, 2000).

Industrial manufacturers can potentially prosper during the current expanding economic climate. The internal resources of manufacturers are often stretched just trying to maintain their current level of productivity. Management teams in small-to-medium sized companies are often taxed with the day-to-day tasks and emergencies and lack sufficient opportunity to assess their operations or complete strategic planning.

Four critical ingredients crucial to the success of the TAA concept are the following. First, the presence of environment for change in the company and the offer of an audit should be seen as a mechanism for facilitating improvement actions. Second, the ability to benchmark and analyze the company's operations and practices quickly. Third, the ability to work with the company in a partnership relationship throughout the process and implementation. Fourth and perhaps most important, the ability to establish credibility of the entity offering the assistance (Haman, 2000).

The TAA model can impact industrial energy efficiency through a nontraditional method that focuses on improving the productivity of the company. The TAA concept offers a mechanism whereby specialty assistance can be tailored and channeled for the specific needs of industry. Specific needs may be fulfilled by either simply offering pre-existing assessment tools, training and information or by integrating more technical and research resources for the development of a comprehensive strategic plan (Haman, 2000).

There has been considerable debate over environmental risk assessment—whether and how, in the absence of unequivocal evidence, human and ecological harms from toxic chemicals and other concomitants of industrialization may be estimated. How we assess them determines whether some concerted action is taken to reduce these potential harms, intervene in more measured ways or simply do nothing.

The thesis underlying current methods of risk assessment essentially is that inference can substitute for observation, especially when preventable and often irreversible harm to human health, safety or the natural environment is at stake. Thus, regulator authorities assume that the probability of harm to humans exposed to low doses of chemicals, radiation and other stimuli can be estimated from experimental results on laboratory animals (or from inadvertent 'experiments' on human subpopulations) exposed to somewhat larger concentrations.

It is well known that people differ widely in their susceptibility to cancer and other possible effects of exposure to toxic substances. Numerous but relatively rare conditions make some individuals very sensitive to 'threshold' toxicants or put them at high risk from exposures to small concentrations of carcinogens. For example, asthmatics can suffer severe respiratory distress from ambient levels of sulfur oxides too low to affect non-asthmatics; those with a genetic polymorphism causing slow acetylation are at substantial risk of bladder cancer when exposed to arylamine dyes at levels that pose minimal risk to many others. Recent evidence about more common predisposing conditions (nutritional status, immune function and expression of cellular oncogenes, for example) brings out that such less readily apparent differences can cause two 'normal' individuals to differ significantly in their susceptibility. Current methods of extrapolating risks from animal bioassay data or occupational epidemiologic studies only estimate risks to a hypothetical individual of average susceptibility.

Physiological changes caused by pollutant exposure occur against a substantial background of similar effects. If we were not already in a constant state of adaptation to physiologic stresses from the natural and man-made environment, 'minuscule' increments of exposure might be harmless. The same background that makes additional increments of risk small in relative terms, however, also makes them much more important in absolute terms.

Finally, it is not logical or sensible to discuss risks without considering the magnitude of the benefits they confer or the identity and conduct of any beneficiaries. Consumers were not irrational, for example, to tolerate small risks from the fungal aflatoxin naturally present in peanut butter and yet to reject risks of similar magnitude from the growth regulator Alar used on apples. The latter hazard was veiwed as an intentional transfer of cost from producers to consumers.

HUMAN HEALTH EFFECTS

The term **effect** may be defined as: (a) any positive or adverse effect; (b) any temporary or permanent effect; (c) any past, present or future effect; (d) any cumulative effect which arises over time or in combination with other effects—regardless of the scale, intensity, duration, or frequency of the effect, and also includes: (e) any potential effect of high probability; and (f) any potential effect of low probability which has a high potential impact. Effects constitute the basis for environmental management.

Guidelines have been suggested in various countries to meet the objective of protecting human health from the adverse effects of air pollution.

The guideline values, which differ from country to country, serve as a baseline for regions to develop air quality objectives.

The guidelines help to meet the aim of safeguarding the life-supporting capacity of air, water, soil, and ecosystems. Sustainable management means managing the use, development and protection of natural and physical resources in a way, or at a rate, which enables people and communities to provide for their social, economic, and cultural well-being and for their health and safety while: (a) sustaining the potential of natural and physical resources to meet the reasonably foreseeable needs of future generations; (b) safeguarding the life-supporting capacity of air, water, soil and ecosystems; and (c) avoiding, remedying, or mitigating any adverse effects of activities on the environment. From clause: (a) minerals are sometime excluded. The additive or synergistic effects of simultaneous exposure to the air we breathe are not yet understood. This means that air quality that meets the guideline levels for several pollutants separately. It may still, when these are combined, lead to adverse effects in sensitive groups in the community.

Studies of the effects of air pollution on vegetation also have been made in many countries.

Fixing objectives to avoid adverse effects on the environment requires an in-depth knowledge of how various levels of pollutants affect animals and plants. To sustainably manage the environment, regional objectives should be based upon an integration of pollutant levels to protect both ecosystems and humans.

No detailed critical research on the sensitivity of native Indian species to air pollutants has been made, although several researches of a preliminary, exploratory nature have been done. Air pollution emitted from sources, even in those areas which have met the guidelines, can sometime result in other adverse effects, e.g. on amenities. Amenity values include elements such as the ability to see across vistas. Air quality at guideline levels may not protect such values.

The carrying capacity of the air is limited. This becomes particularly important in areas where conditions allow any existing pollution to accumulate.

In developing regional standards for air quality, the guidelines need to be used as a starting point. With the setting up of monitoring networks, data will become available on actual air quality.

Marine Pharmaceuticals

Many of the active components of today's medicines are natural products or their semi-synthetic derivatives. The biosynthetic capabilities of marine organisms are opening up new prospects for research into the drugs of the future. A huge diversity of unique chemical structures exists in marine plants and animals, and this fantastic resource is being explored for the identification of novel drugs.

The development of new pharmaceuticals usually starts with the recognition of a so-called lead structure, a chemical that contains structural elements essential for pharmacological effects. The properties of such compounds are then optimized in order to obtain new drugs. Several natural products not only directly serve as drugs but also as models for the development of novel products for therapeutic use. In the past, natural product researchers focused primarily on metabolites from terrestrial plants and micro-organisms. Through research into marine natural products, completely different types of organisms such as sponges, corals and tunicates have attracted interest as producers of novel natural compounds (König, 1999).

Arabinose nucleosides, isolated in the 1960s from sponges, became leads for the development of the drugs Ara-A and Ara-C, now being used against viral infections and cancer. Since then, efforts have yielded many highly active compounds, some of which are currently being evaluated in clinical trials. Marine metabolites frequently contain halogens (chlorine or bromine). One such compound, obtained from the red alga *Portieria hornemanni*, is currently in pre-clinical development on account of its marked and selective toxicity towards tumour cells. Anti-inflammatory and immune suppressors are also being explored from marine organisms. The preferred locations for collecting marine biota are tropical reefs, where there is a high level of biodiversity and multifarious ecological interactions. The reef habitat is particularly promising because the organisms residing there have learnt to defend themselves with highly effective natural products. Pharmaceutical sciences can profit from these natural substances since what nature employs for its own purposes may well prove to be of value for medical therapy.

Application of the nuclear magnetic resonance (NMR) technique is now permitting the investigation of very small extracts of natural substances. There is scope for research into the potential of marine natural products as anti-infection (antibiotic) agents. Malaria therapy, as well as that of other infectious diseases, is becoming more and more problematic with the development of resistance to conventional drugs. Novel drugs are, thus, needed and natural products have been shown to be promising model compounds in this context. At very low concentration, a terpenoid from the sponge *Acanthella klethra* has been found to inhibit the growth of the malaria parasite *Plasmodium falciparum*. Minute amounts of extracts from the tropical sponge *Cymbastela hooperi* inhibit the in vitro growth of plasmodia, and also are effective against drug-resistant plasmodia (König, 1999).

ENVIRONMENTAL HEALTH AND ITS MANAGEMENT

'Health', according to the World Health Organization, 'is a state of complete physical, mental and social well-being and not merely the absence of disease or infirmity'. In reality, this concept is illusory. The general state of health in societies has largely to be extrapolated from our knowledge of diseases and deaths. Luckily, data on life expectancy and major causes of death show much about prevailing social and environmental conditions.

Nearly all diseases are linked to the environment. Although heredity determines susceptibility to afflictions, the individual's interactions with his social and physical milieus influence whether or not that potential is actually realized.

Focusing on the underlying sources of disease and death patterns points the way toward healthier societies. Much of the suffering and early death occurring in the world today results from circumstances over which humans do wield considerable influence, both in rich and poor countries. However, the social and physical conditions of the latter exact an infinitely higher human price.

Poverty and wealth have different environmental impacts. This is revealed when we compare life expectancy figures from different countries. The global average life expectancy at birth in the late 1980s was around sixty-two years. In all the developed countries and a few more advanced developing countries, life expectancy exceeded seventy. In nearly every country of Sub-Saharan Africa, as well as some of the poorest countries of Asia (e.g. Afghanistan, Bangladesh and Nepal) life expectancy was in the forties. In India, Pakistan and Indonesia, the figure was near fifty—comparable to life spans in Great Britain and the United States around 1900.

Deaths among infants in their first year of life, more than anything else, account for international variations in life expectancy. For the developed world as a whole, the loss is no more than one in fifty. In the poorest African and Asian countries, one in five infants dies.

Half of all Third Word deaths are among infants and children under the age of five, compared to only 2-4% of deaths in Western countries (see Berg, 1973).

In general, incomes correlate with health status. People of richer countries enjoy higher life expectancies and within countries, income levels usually affect health.

The collective impacts of malnutrition, infectious spread by human excreta, airborne infections and parasites account for most or all of the childhood deaths in developing countries. Rooted in the social ecology of poverty, these threats can seldom be countered with medical aid alone.

Undernutrition's effect on health is subtle (see Puffer and Serrano, 1973). A lack of adequate amounts and quality of food increases both the frequency and severity of diseases in children.

The pernicious effects of undernutrition occur even before a baby's birth. Women who themselves were undernourished and unhealthy during childhood, are more apt to give birth to underweight babies—which, in turn, are more apt to die in infancy than stronger infants. An inadequate diet during pregnancy increases the probability that a woman will give birth to an underweight baby.

Once born, infants are exposed to diverse infectious agents. The less sanitary the habitat, the more numerous and varied the exposures; the less well-nourished the baby, the greater the likelihood of serious illness (Eckholm, 1991).

Only improvements in the environment, in sanitation and in nutrition can bring down the alarming frequency of illnesses in the Third World. Immunization against measles, polio and other specific diseases have saved many lives. But many of the afflictions killing Third World children cannot be prevented or cured with medical technologies alone. Infectious agents thrive in dirty habitats.

Reasonably clean and plentiful water, good sewage disposal facilities and the adoption of sanitary principles are the three most important basic, interconnected essential pre-requisites for better health and deserve the same high priority as the prevention of undernutrition.

On top of the omnipresent threats of infection and undernutrition, people in developing countries are exposed to several parasitic and other types of tropical diseases. Research on the prevention and cure of six tropical diseases (malaria, schistosomiasis, filariasis, trypanosomiasis, leprosy, leishmaniasis) and others needs to be intensified.

Malaria's resurgence has been especially marked in South and Southeast Asia and in Central America. In India alone, the number of reported cases rose from 40,000 in 1966 to 6.5 million in 1976. According to WHO, malaria kills at least one million Africans under age fourteen each year.

For a growing number of the world's poor, modern insult is being heaped on traditional injury—they not only lack access to pure water but what they do drink is often laced with heavy metals. Plantation workers and their families often live and work amid overly-applied, overly-dangerous biocides. Those lucky enough to land a factory job often face exposures to chemicals or deadly fibres at levels that were banned in the West long ago. City dwellers breathe foul emissions from cars and from factories whose products and profits may never touch their lives. These unlucky people have to struggle simultaneously against ancient wants and the hazards of a modern age that has passed them by (Eckholm, 1991).

In developed countries, vascular afflictions and cancer account for over two-thirds of all deaths. The timing and overall incidence of cardiovascular diseases are unquestionably influenced by personal lifestyles.

Cancer accounts for about one-fifth of the deaths in developed countries. Its toll in poor countries is significant too and about one-tenth of the world's fifty million annual deaths are attributable to cancer. More than any other disease, cancer appears to be associated with environmental causes. The environment of concern here is an all-embracing concept, including all the natural and the man-made materials that people eat, breathe, or touch; sunlight (whose radiation causes skin cancer); even childbearing patterns (which affect women's hormonal balances and in turn the odds of certain types of cancer). The chief proven cause of cancer is cigarette smoke, which alone may cause about 20% of the cancers in many developed countries. Eight or nine out of ten lung cancers are attributable to smoking, which also promotes cancers of the mouth, throat, oesophagus, bladder, kidney and pancreas. Tobacco smoke acts synergistically with such other environmental insults as alchol, radiation and asbestos, posing extraordinary dangers among some unfortunate groups.

Environmental pollutants also cause afflictions other than cancer. The fact air pollution helps cause and intensify chronic respiratory ailments such as emphysema, chronic bronchitis and asthma is well established. Lead from car exhausts and old paint is suspected of disturbing the nervous systems of some inner-city children. In Japan, dramatic outbreaks of heavy-metal pollution involved the destruction of the nervous systems of hundreds who ate mercury-laden fish from Minamata Bay, and the epidemic of *itai-itai* (literally 'ouch-ouch') disease among hundreds, whose bones agonizingly distintegrated because of cadmium in their food and water. These became worldwide symbols of industrially-induced diseases.

INDIGENOUS KNOWLEDGE FOR SUSTAINABLE DEVELOPMENT

There is now a growing global awareness of the role indigenous knowledge (IK) is going to play in keeping with modern scientific and technological intervention in social and economic development and cultural and political transformation. Local communities can contribute their indigenous knowledge systems to enhance the sustainability of development programmes. In view of the fact that indigenous knowledge is a powerful resource that enables local communities to improve and sustain their lives and because there is a need for a new vision to promote the indigenous knowledge systems of local communities to improve their social economic status, policy makers, planners, scientists, economists, national and international development institutions as well as the entire civil society and resource managers should be encouraged to understand and internalize the increasing value of indigenous knowledge and to promote its application as a key instrument for the empowerment of local communities. High priority and high level of support to the national IK development strategy need to be given as means of increasing opportunities to sustainable solutions for socio-economic development.

Trainer (1998) postulated the following dicta : (a) A satisfactory society cannot have a growth economy; and (b) and satisfactory society cannot have per capita rates of non-renewable resource use in rich countries. These negative principles generate the following positive corollaries: (a) Our material living standards should be simple and frugal; and (b) we should develop small, self-sufficient settlements and economies. A sustainable society emanates out of a cooperative and participatory social movement that dovetails sharing of materials and servicing the marginalized, aged, invalid and disabled. We badly need a new order of a sustainable economy-driven society that has a much lesser role for market forces and which recognizes the need for a zero-growth or steady-state economy. There is need to develop systems that will enable us to achieve this goal. Good education can influence such an objective. There exists a wide divergence between knowledge and education and between the developed and the developing countries. This conflict in understanding and translating the seriousness of issues stems from the procedures employed in (a) identification of environmental crises; (b) validation of values placed on environmental issues; and (c) empiricization of the means required to deal with such issues. A good understanding of the environmental issues in developing countries should be the first step towards effective action because it is the poor and the powerless who are generally held responsible for environmental problems. The fact that the international capital, trade relations and agreements and technology-intensive activities strongly affect the environment is seldom, if ever, acknowledged.

A global perspective on the environment needs to be integrated into today's educational system. One serious barrier in this effort is that our educational models are based on reductionism—we tend to focus more on parts rather than the whole. Consequently, we cannot recognize holism and its fundamental importance.

The human species is solely responsible for the overall environmental and ecological uncertainties owing to technology driven by greed and the hastened events of global environmental change. Those that help in sustaining human life—agriculture, manufacturing, transportation—contribute to one or many of the major environmental issues confronting the world (Raman et al., 2000). Consilient, interdisciplinary scientific activity is the quintessence of heterogeneous expertise on critical environmental problems at a global scale and contributes to a strong conceptual framework.

Sustainability is undoubtedly the bedrock for a better future of human existence. It is essential to integrate concepts and practice of environmentalism to create a sustainable society. An integrated, holistic and consilient approach is needed for linking human resources and their potentials—an approach in which human resource development and education play an increasingly important role.

POLLUTION PREVENTION

Just as pollution does not recognize geopolitical boundaries, it also has become clear that solutions to environmental problems will, in many cases, be shaped in the global arena. International cooperation is one of the central elements in the strategy for environmental protection worldwide. The obvious benefits of international cooperation are better protection of public health and the environment globally. But there are other practical benefits as well. All countries face resource constraints and a review and evaluation of existing chemicals is time-consuming and resource-intensive. International cooperation, involving many concerned nations, can help avoid duplication of effort and improve the reliability and acceptability of information-gathering, assessment and control efforts.

As far as toxic substances are concerned, we need to focus our attention on the following aspects:

1. Adopting a 'pollution prevention' perspective as we develop strategies for addressing identified risks. Pollution prevention refers to on approach to resource use that emphasizes innovative technologies to substitute nontoxic for toxic products, use resources more efficiently, generate less waste in the first place and reuse chemicals and other materials;
2. Looking at ecological as well as health effects as we characterize chemical hazards; and
3. Examining risk reduction measures that rely on information disclosure and market incentive approaches.

Prior Informed Consent

Prior Informed Consent (PIC) refers to a procedure developed by the United Nations to enable importing countries to obtain information about certain chemicals whose use is banned or greatly restricted in other countries. The importing country makes use of such information to decide whether to allow, restrict, or stop future imports of these chemicals.

PIC procedures comprise the following four components:

1. Based on information submitted by UN member countries, a list of chemicals subject to bans or other restrictive actions can be established.
2. Exporting countries can be informed of decisions by importing countries regarding future shipments of these chemicals.
3. If some country bans or restricts a chemical in some way, that country should provide this information to importing countries. Notification should occur the first time that the chemical is exported and periodically thereafter if any significant development concerning the control action occurs.
4. Exporting countries are required to implement appropriate regulatory controls to ensure that chemicals are not exported to those nations which do not want those chemicals to enter the country.

International Register of Potentially Toxic Chemicals (IRPTC)

The IRPTC is an informational network established by the UN Environment Program. Through this, countries have easy access to data on the production, distribution, exposure, release, disposal, toxicity and regulation of existing chemicals. The IRPTC stores information gathered through various OECD efforts. The IRPTC also develops chemical data profiles that contain information concerning hazard identification and risk assessment of chemicals.

Reducing Lead Exposures

The strategy for reducing lead exposures, especially to children, focuses on developing ways to identify and address the most serious existing exposures; fixing priorities for remedial action to reduce health risks from lead exposures; investigating ways to prevent pollution by limiting unnecessary uses of lead and by encouraging recycling; strengthening the partnership among national agencies, states, localities and the private sector to minimize lead exposures; and educating the public about risks from lead exposures and how to reduce them.

Lead can be a pernicious problem and every effort needs to be made to reduce risk to public health and the environment. The strategy consists of the following regulatory and nonregulatory initiatives. Efforts to reduce exposure to lead-based paint include initiatives to train workers in the safe removal of lead-based paint. Studies to identify and evaluate cost-effective abatement methods and measurement techniques have also been made.

Urban Soil and Dust

Lead contamination of urban soil results from both nonindustrial sources (paint, gasoline and used oil) and industrial sources (battery factories, mining and milling sites and smelters). Studies in some cities have been carried out to examine the effectiveness of removal of lead-contaminated soil and dust. The sources of lead-contaminated soil and dust are also being identified.

Drinking Water

Lead in drinking water results primarily from corrosion of lead-bearing materials in water supply distribution systems as well as in household plumbing.

The feasibility of further limiting the manufacture and sale of lead solder used to join water pipes is attracting interest in several countries.

To ensure further reduction of lead in drinking water, limits are needed on the amount of lead allowed to leach from brass and bronze faucets and other plumbing components.

Battery Recycling

A high proportion of all lead consumed is used to produce automotive batteries, e.g. lead-acid batteries. Increasing the rate of battery recycling can, therefore, reduce the amount of lead discarded in the environment and new lead mined.

Air Quality

Lead levels in ambient air have fallen markedly in urban areas in North America over the last two decades as a result of the elimination of lead from most gasoline and other control measures. In these areas, the National Ambient Air Quality Standard for lead has been satisfied in all places except near large stationary sources such as lead smelters, refineries and remelters.

HUMAN SETTLEMENTS

The most vulnerable human settlements are those especially exposed to natural hazards, e.g. coastal or river flooding, severe drought, landslides, severe wind storms and tropical cyclones. The most vulnerable populations are in developing countries, in the lower income groups, residents of coastal lowlands and islands, populations in semi-arid grasslands, and the urban poor in slums, especially in megacities. Important health impacts are possible, especially in large urban areas, owing to changes in availability of water and food. Health problems may increase due to heat stress which spreads infections. Changes in precipitation and temperature could radically alter the patterns of vector-borne and viral diseases by shifting them to higher latitudes, thus putting large populations at risk. These changes could initiate large migrations of people, leading over a number of years to severe disruptions of settlement patterns and social instability in some areas.

Global warming may possibly influence the availability of water resources and biomass, both of which are major sources of energy in many developing countries. Such changes in areas which lose water may jeopardize energy supply and materials essential for human habitation and energy. Climate change itself may exert different effects between regions on the availability of other forms of renewable energy such as wind and solar power. In developed countries, some of the greatest impacts on the energy, transport and industrial sectors are likely to be determined by policy responses to climate change such as fuel regulations, emission fees or policies promoting greater use of mass transit. In developing countries, climate-related changes in the availability and price of production resources such as energy, water, food and fibre may affect the competitive position of many industries (IPCC, 1990).

Health and Income Relationships

One of the best-known relations in international development is the positive correlation between health and income per capita. This correlation is generally believed to reflect a causal link from income to health. Higher income gives greater command over many of the goods and services that promote health, such as better nutrition and access to safe water, sanitation and high quality medical services. But recently, another intriguing possibility has emerged, viz., that the health-income correlation may parly be explained by a causal link in the reverse direction—from health to income. Four likely mechanisms may account for this relation.

1. **Productivity.** Healthier populations have higher labour productivity as healthy workers are physically more energetic, mentally more robust and suffer fewer lost workdays from illneses.

2. **Education.** Healthier people who live longer have stronger incentives to obtain good education. Increased schooling promotes greater productivity and, in turn, higher income.

3. **Investment in physical capital.** Longer life creates a greater need for people to save for their retirement. Increased savings lead to increased investment so workers will have access to more capital and their incomes will rise. A healthy and educated workforce is like a strong magnet for foreign investment.

4. **Demographic dividend.** There has been a strong and rapid transition from high to low rates of mortality and fertility in many developing countries. Mortality declines among children initiate the transition and trigger subsequent declines in fertility. An initial rise in the numbers of young dependents gradually gives way to an increase in the proportion of the population that is of working age. With this income per capita can rise dramatically, if the new workers are absorbed into productive employment (Birdsall et al., 2000).

All the above mechanisms can be plausible ways in which health improvements lead to income growth. Health status (as measured by life expectancy) is a significant predictor of subsequent economic growth. This strong effect goes above and beyond other influences on economic growth (see Bloom and Canning, 2000; Bloom et al., 2000).

As health improvements fortify the economy, they also relieve poverty. Economic growth can dramatically reduce poverty among the over one billion people living on less than US$1 per day. Increases in average income translate into increases in the income of the poor. In addition, health improvements are disproportionately beneficial for the poor, as they depend on their labour power more than any other segment of the population (Bloom and Canning, 2000).

Not only the direct effects of life expectancy on economic growth are important but so also are the indirect effects of improvements in health status that operate via demographic change. In East Asia, for example, the working-age population grew several times faster than the dependent population between 1965 and 1990, triggered by declining child and infant mortality, which itself happened by the development of antibiotics and antimicrobials discovered and introduced in the 1920s, 1930s, and 1940s. Other factors were the use of DDT (since the mid-1940s), and public health improvements related to safe water and sanitation. Health improvement was one of the pillars upon which East Asia's phenomenal economic achievements were based, with the demographic dividend accounting for perhaps one-third of its economic miracle (Bloom and Canning, 2000).

By contrast, poor health slows down demographic transition and inhibits growth, as e.g. in Sub-Saharan Africa.

Energy use patterns also influence the interactions between health, demography and income. The rural poor rely heavily on wood, dung and biomass. The resulting smoke and particulates impair human health and diminish people's productivity. For many countries, there exists a close association between demographic and health indicators and the traditional use of biomass.

Health improvements stimulate economic development. The development process is inherently dynamic—with health improvements promoting economic growth, which in turn promotes better health. Health improvements and economic growth can be mutually reinforcing in another way. As rising incomes reduce fertility, consequential benefits accrue to the health of mothers and children.

Sadly, the mutual association between health and income can also operate in reverse. Declines in health status in some countries can have adverse impacts on economic well-being. The AIDS epidemic in Africa is slowing economic growth and depleting the resources to deal with other diseases—such as diarrhoea, hepatitis, malaria and tuberculosis—that are also ravaging health in many countries.

Health-Led Development

Health is an instrument of self-sustaining economic growth and human progress. Poor health is more than just a consequence of low income; it is also one of its basic causes. Of course, health and demography are not the only influences on economic growth, but they certainly are among the most potent (Bloom and Canning, 2000).

PEOPLE-CENTRED DEVELOPMENT

People-centred development is based on the idea that people themselves can direct their own development process, consistent with their aspirations. It anchors development programmes in local knowledge and local skills. It involves a process by which the members of a society learn to mobilize and manage resources so as to produce sustainable and equitably distributed improvements to their quality of life. For people-centred development to succeed, individuals have to be empowered to participate in their own development process. Empowerment must be rooted in the knowledge of the people and in ways of dealing with their environment which they have successfully used in the past. Building on local knowledge and resources reduces the probability that a development intervention will 'de-skill' people or increase their dependence on external experts. On the contrary, building on local knowledge empowers people by increasing their self-reliance (Van Vlaenderen, 2000).

One central issue in development is the capacity to solve problems arising from changing socio-economic conditions. In situations of socie-oconomic change, development relies largely on the generation of problem solving strategies. Development programmes should, therefore, be based on local understanding of the notion 'problem solving' and on indigenous problem solving skills (see Figs. 12.5, 12.6, 12.7).

HUMAN DEVELOPMENT: THE CHALLENGES AHEAD

Many people will remember the twentieth century as a period of great achievement in human endeavour and of enormous economic growth and prosperity. Achievements in medical research, from eradicating several infectious diseases to laser surgery; in engineering, and from the transistor to space exploration contributed much to well-being in the world (World Bank, 1999). During the twentieth century, colonialism, Fascism, Nazism, communism and brutal dictatorship were beaten by progressive forces, which received great impetus in the past two decades by globalization and localization. These twin processes are now shaping the international system through the functional integration of economies on one level, and by decentralizing political power and decision making closer to people within countries, on another. Much of the world seems to be a safer, more prosperous place at the end of the millennium than it has been for most of the twentieth century. Unfortunately, this safety and prosperity is not enjoyed evenly throughout the world.

While the global economy has grown significantly, poverty and underdevelopment continue to blight vast communities around the world. Inequity the between the rich and the poor is growing. Forests are being

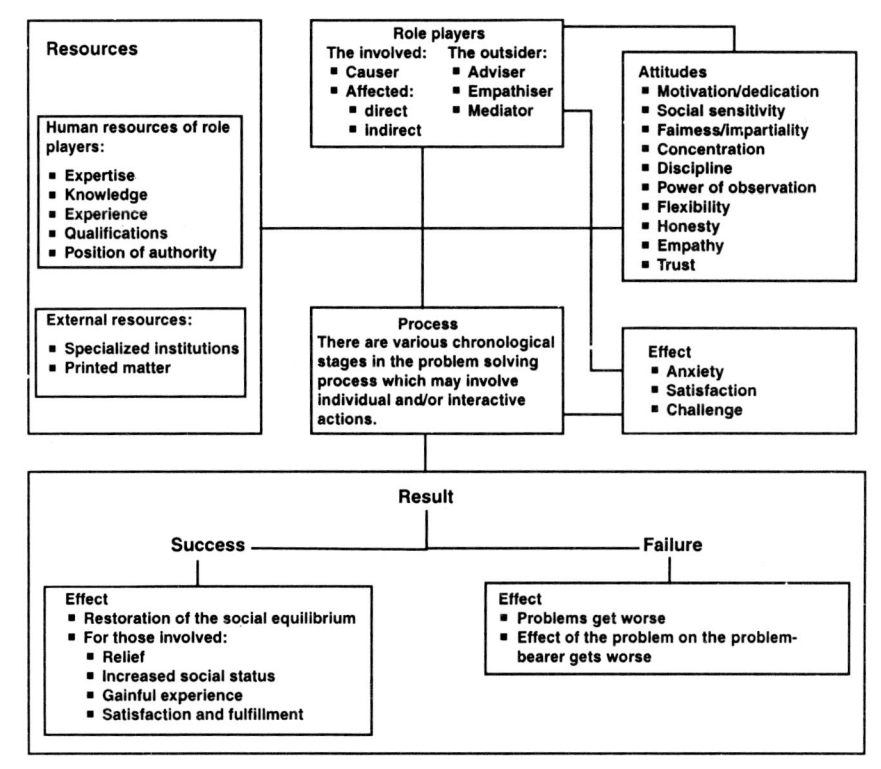

Fig. 12.5 Local knowledge of the concept problem solving (after Van Vlaenderen, 2000).

degraded at the rate of an acre a second. More than a billion people lack access to clean water and two billion lack access to sewerage (Wolfensohn, 1999). Some 80 to 90 million people are added to the population in the developing world each year. Growth, prosperity, relative peace and stability are, thus, not universal.

Globalization and Localization

Benefits of globalization can be enormous. It has brought economic prosperity to distant markets and opened the global economy to many communities around the world. At the same time, localization is forcing states to accommodate demands being made from the local level. The responses of governments to these twin forces will, by and large, determine whether low-income countries will catch up economically with those of industrialized countries and whether efforts to uproot poverty will succeed in the coming decades.

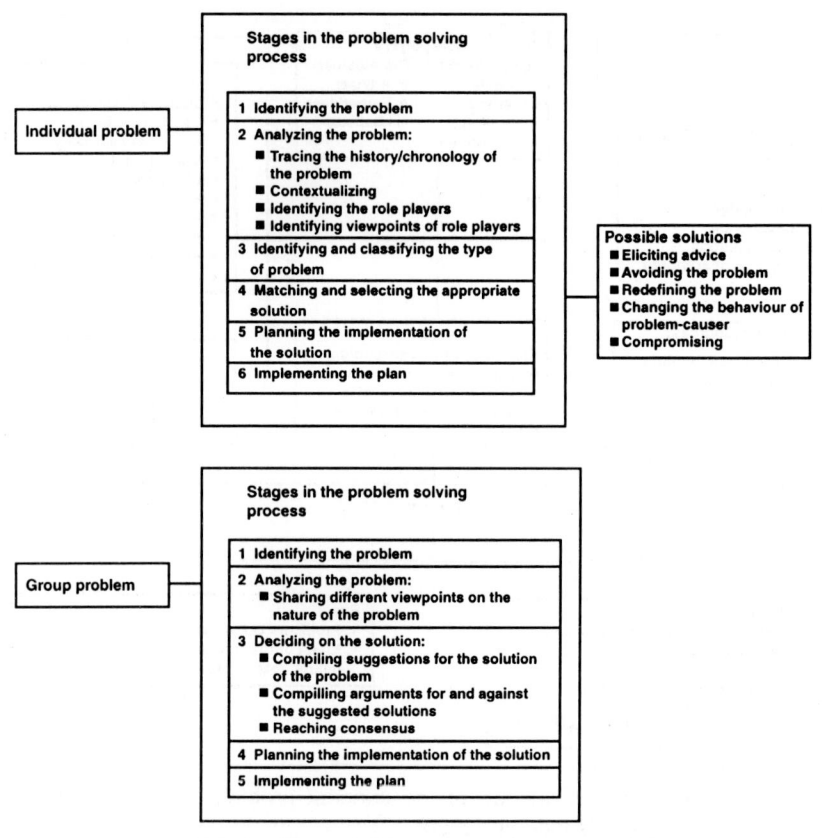

570 Environmental Technology and Biosphere Management

Individual problem

Stages in the problem solving process
1. Identifying the problem
2. Analyzing the problem:
   - Tracing the history/chronology of the problem
   - Contextualizing
   - Identifying the role players
   - Identifying viewpoints of role players
3. Identifying and classifying the type of problem
4. Matching and selecting the appropriate solution
5. Planning the implementation of the solution
6. Implementing the plan

Possible solutions
- Eliciting advice
- Avoiding the problem
- Redefining the problem
- Changing the behaviour of problem-causer
- Compromising

Group problem

Stages in the problem solving process
1. Identifying the problem
2. Analyzing the problem:
   - Sharing different viewpoints on the nature of the problem
3. Deciding on the solution:
   - Compiling suggestions for the solution of the problem
   - Compiling arguments for and against the suggested solutions
   - Reaching consensus
4. Planning the implementation of the solution
5. Implementing the plan

Fig. 12.6 The problem solving process: actions (after Van Vlaenderen, 2000).

Globalization catapulted global issues to the attention of the nation-state and also increasingly circumscribed its choices. Localization includes the demand for autonomy and political voice evinced by regions and communities. Dissatisfaction with the ability of the state to deliver on promises of development is its major cause. Another cause is the growing competition between subnational units in an open environment, compounded by a reluctance of richer companies to share resources with their less well-off neighbours (Wolfensohn, 1999).

The pull of local identity is manifested in the proliferation of nation-states from 96 in 1980 to 192 in 1998.

Urbanization has driven localization and contributed to the emerging sense of local identity. About one-half of the world's population is living in urban areas today. As recently as 1975, this share was only about one-third and by 2025, it will rise to almost two-thirds. The fastest change will

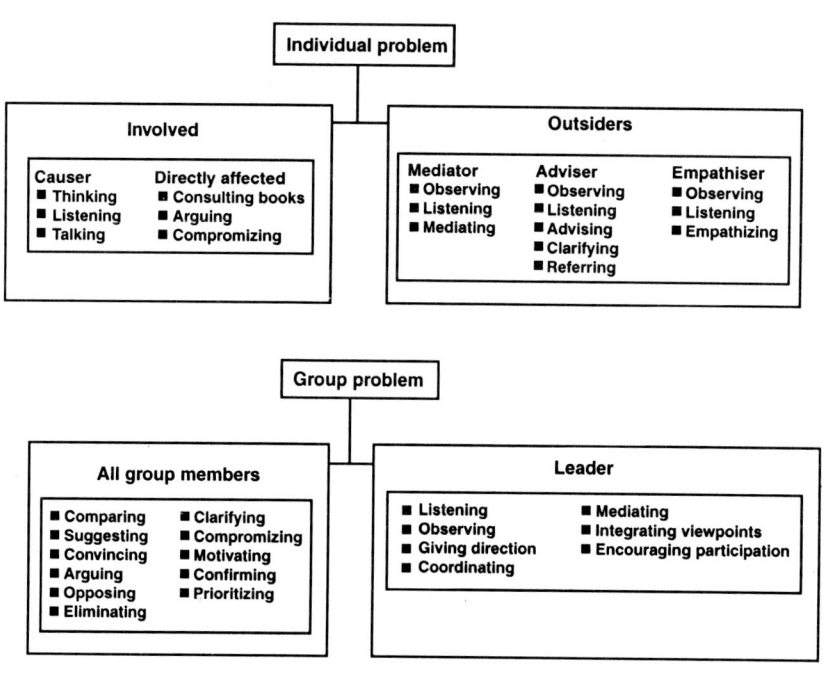

```
                    ┌──────────────────────┐
                    │ Individual problem   │
                    └──────────┬───────────┘
         ┌─────────────────────┴─────────────────────────────┐
┌────────────────────────────┐  ┌──────────────────────────────────────────┐
│         Involved           │  │                 Outsiders                │
│ Causer      Directly       │  │ Mediator     Adviser      Empathiser     │
│             affected       │  │ ■Observing   ■Observing   ■Observing     │
│ ■Thinking   ■Consulting    │  │ ■Listening   ■Listening   ■Listening     │
│ ■Listening   books         │  │ ■Mediating   ■Advising    ■Empathizing   │
│ ■Talking    ■Arguing       │  │              ■Clarifying                 │
│             ■Compromizing  │  │              ■Referring                  │
└────────────────────────────┘  └──────────────────────────────────────────┘

                    ┌──────────────────────┐
                    │    Group problem     │
                    └──────────┬───────────┘
         ┌─────────────────────┴─────────────────────────────┐
┌────────────────────────────┐  ┌──────────────────────────────────────────┐
│     All group members      │  │                 Leader                   │
│ ■Comparing   ■Clarifying   │  │ ■Listening      ■Mediating               │
│ ■Suggesting  ■Compromizing │  │ ■Observing      ■Integrating viewpoints  │
│ ■Convincing  ■Motivating   │  │ ■Giving direction ■Encouraging participation │
│ ■Arguing     ■Confirming   │  │ ■Coordinating                            │
│ ■Opposing    ■Prioritizing │  │                                          │
│ ■Eliminating               │  │                                          │
└────────────────────────────┘  └──────────────────────────────────────────┘
```

Fig. 12.7 The problem solving process: actions (after Van Vlaenderen, 2000).

occur in developing countries, where almost three-quarters of those added to the world's population in the present century will live in cities. The pace of urbanization and the sheer numbers involved are going to post a major development challenge of the twenty-first century.

Globalization and localization—the latter associated with the spread of participatory democracy and urbanization—can potentially improve the prospects for rapid and sustainable growth. This may happen through the increased availability and more efficient allocation of resources, the freer circulation of knowledge, a more open competitive environment and improved governance. But, as always, there are some risks also. Globalization entails greater exposure to external shocks, such as capital volatility, along with a number of threatening environmental issues. Decentralizing measures introduced to satisfy local demands may create macroeconomic instability in the absence of effective financial discipline.

Globalization can create novel opportunities for expanded markets and the spread of technology and management expertise which, in turn, can help to increase productivity and generate a higher standard of living. But it can also result in increased pressure on workers and communities in general, through the fear of job losses to competition from imports. In developing countries where this quality of administration and management is not high, the poor management of globalization can potentially have adverse effects (Wolfensohn, 1999).

Comprehensive Approach to Development

Against the background of uneven development and underdevelopment, the proliferation of regional and interstate conflict and rapid globalization and localization, the World Bank has put forward its Comprehensive Development Framework (CDF). This is a set of guidelines for a more holistic, integrated and realistic approach to development. The CDF suggests a holistic approach to development that is sensitive to the domestic social and economic demands and expectations within a particular country. It favours the ideal of sustainable economic growth while recognizing that the developmental needs of one country can differ from those of another and that underdevelopment and uneven development may be symptoms of deeper (and unique) social, political and economic problems (Wolfensohn, 1999).

The CDF represents a perspective that includes democratic, egalitarian and sustainable development, and higher standards of living and education. It is essentially a strategy for development in the twenty-first century that gives greater importance to the structural, social and human aspects of development. This approach goes much beyond the familiar statistics of infant and maternal mortality, unemployment, etc., to address fundamental long-term issues such as societal development. The CDF is shaped by four main interrelated ideas:

1. The ultimate goal of development is to raise people's standard of living (higher per capita income; health, education; a cleaner environment and inter-generational equity).
2. Governments have a role in development to assure investment in human capital for the protection of property rights, for establishing and maintaining the requisite regulatory environment, etc.
3. No single development policy can make much difference in an otherwise hostile policy environment. Increasing human welfare requires an integrated policy package in an institutional environment that rewards good outcomes, minimizes perverse incentives, encourages initiative and facilitates inclusiveness.
4. Processes are as important as policies. The sustainability of policy outcomes depends on the breadth of support they enjoy, the support being fostered by modes of planning and decision making that are based on consensus, participation and transparency. Institutions of good governance that embody these processes are a key to sustained development and these institutions will often require partnerships

and arrangements between government, the private sector, non-governmental organizations and other institutions of civil society. For combatting the threats of poverty and social disruption, these processes cannot be de-linked from one another.

The miraculous achievements of East Asia in the post-war period best exemplifies what is possible in development. Between 1960 and 1995, eight economies in East Asia grew about three times as fast as Latin America and South Asia, and five times faster than sub-Saharan Africa. By the early 1990s, these economies enjoyed high growth, declining inequality and dramatic increases in life expectancy, from 56 years in 1960 to 71 years in 1990. They registered dramatic improvements in living standards, education, health and income and the fruits of this growth were shared widely. The pace of poverty reduction was faster than in any other region of the developing world. Government intervention played a great role in East Asia's enormous successes (World Bank, 1993, 1999).

What is desirable now is to emulate and exploit the best practices from East Asian development experience and make them available to the other developing countries so that they can benefit from the same.

ENCLOSED ECOSYSTEM (BIOSPHERE 2)

A giant greenhouse known as Biosphere-2 has been built in 1988 just north of Tucson, Arizona (USA) to explore how humans might colonize space.

Although this 3.2-acre ecosystem consumes huge amounts of energy, it loses only about 10% of its atmosphere each year. This low rate may permit valuable studies of sustainable agriculture and community ecology as well as of how trace elements and gases move between plants and the environment.

Certainly, no other enclosed ecosystem is as large and none can boast Biosphere 2's seven different 'biomes', which include a contrived rain forest and a tiny ocean with a chemistry unlike any body of water on the earth. It has equipment that can possibly make the facility the most intensively monitored patch of vegetation in the world.

Tourists who came from far off places to see this magnificent edifice contributed about a third of the Biosphere 2's $6-million annual budget. But tourism on that scale is incompatible with research, so some wealthy Americans are now contributing funds. Unfortunately, there are serious obstacles: the Biosphere has no provision for doing multiple duplicate experiments, known as controls. Replicates are essential for any statistical treatment and for scientific rigour.

Carbon dioxide levels have fallen markedly as a result of a mammoth ventilation exercise aimed at replenishing the facility's atmosphere. The subsequent slowdown in the growth rates of some plants seems to have prompted a similar reduction in the production of CO_2 by soil microorganisms. The implication of this unconfirmed observation is that in a rising CO_2 world, planting trees to sequester the gas might not help much as it might stimulate microorganisms to produce more carbon dioxide, thereby offsetting the amount the trees absorb (Beardsley, 1995).

But the result needs to be confirmed. Without separate chambers, the only way to do so is to repeat the change of atmosphere in the entire Biosphere, a monumental undertaking that could impede other experiments. And, in any case the findings are scientifically suspect in view of the fact that the facility is unlike the real world: its glass blocks light from the sun by about 50%, and it does not have a normal insect population.

Also CO_2 levels fluctuate levels each day by far more than they do in nature.

In 1994, the levels of nitrous oxide in Biosphere-2 were high enough to prompt health concerns and water contaminated with nitrates was raining on the vegetation. Cockroaches and ants flourished, while more desirable species perished (Beardsley, 1995).

REFERENCES

Ayensu, E. et al., *Science* p. 685 (22 Oct. 1999).
Beardsley, T. Down to earth. *Sci. Amer.* 83: 23-28 (1995).
Berg, A. *The Nutrition Factor*. Brookings Institution, Washington, D.C. (1973).
Berkes, F. (Ed.) *Common Property Resources*. Belhaven, London (1989).
Birdsall, N., Kelley, A.C., Sinding, S. (eds.). *Population Does Matter: Demography, Growth, and Poverty in the Developing World*. Oxford Univ. Press, New York (2000).
Black, P.E. Biosphere management: Some tools of the trade. *Science* 287: 234-235 (2000).
Bloom, D.E., Canning, D. The health and wealth of nations. *Science* 287: 1207-1209 (2000).
Bloom, D.E., Craig, P., Malaney, P.N. *The Quality of Life in Rural Asia*. Oxford University Press, Hong Kong (2000).
Costanza, R., d'Arge, R., de Groot, R., (and 10 other authors). The value of the world's ecosystem services and natural capital. *Nature* 387: 253-260 (1997).
Curran, L.M. et al. *Science* 286: 2184 (1999).
Daily, G. (Ed.). *Nature's Services : Societal Dependence on Natural Ecosystems*. Island Press, Washington DC (1997).
Douglas, I. Hydrological investigations of forest disturbance and land cover impacts in South-East Asia: a review. *Phil. Trans. R. Soc. Lond.* B 354: 1725-1738 (1999).
Eckholm, E. *Down to Earth: Environment and Human Needs*. Affil. East West Press, New Delhi (1991).
Eliah, E. Working with the wonder weed. *Gate* 2/98: 44-47 (1998).

Gadgil, M., Berkes, F. Traditional resource management systems. *Resource Management and Optimization* 8 (3-4): 127-141 (1991).
Gazea, B., Adam, K., Kontopoulos, A. A review of passive systems for the treatment of acid mine drainage. *Minerals Engineering* 9: 23-42 (1995).
Gusek, J.J., Wildeman, T.R. New developments in passive treatment of acid rock drainage. In: *Proc. of the Engineering Foundation Conference on Technological Solutions for Pollution Prevention in the Mining and Mineral Processing Industries.* Palm Coast, Florida (Jan. 23, 1995).
Haman, W.G. Total assessment audits (TAA) in Iowa. *Resources, Conservation and Recycling* 28: 185-198 (2000).
Hartshorn, G., Bynum, N. Tropical forest synergies. *Science* 286: 2093-2094 (1999).
Hedin, R.S., Hyman, D.M. Treatment of coal mine drainage with constructed wetlands. In Scheiner, B.J., Doyle, F.M., Kawatia, S.K. (eds.) *Biotechnology in Minerals and Metal Processing.* AIME, Littleton, Colorado (1989).
Hedin, R.S., Naim, R.W. Contaminant removal capabilities of wetlands constructed to treat coal mine drainage. In Moshiri, G.N. (Ed.) *Constructed Wetlands for Water Quality Improvement.* pp. 187-195. Lewis Publishers, Boca Raton (1993).
IPCC (Intergovernmental Panel on Climate Change). Policymakers' Summary of the Potential Impact of Climate Change. Report from Working Group II to IPCC. Austr. Govt. Publishing Service, Canberra (1990).
König, G. The long road from the sea to the pharmacy. *German Research*, pp. 22-26, Wiley-VCH, Weinheim (1999).
Kreutzmann, H. Water as a development factor in semi-arid mountainous areas of settlement: Systematic approach and development potential. *Natural Resources and Development* 49/50: 99-116 (1999).
Kumar, H.D. *Biodiversity and Sustainable Conservation.* Oxford and IBH., New Delhi (1999).
Kumar, H.D., Häder, D.P. *Global Aquatic and Atmospheric Environment.* Springer, Heidelberg (1999).
Linnik, P.M., Zubenko, I.B. Role of bottom sediments in the secondary pollution of aquatic environments by heavy-metal compounds. *Lakes & Reservoirs: Research and Management* 5: 11-21 (2000).
Neate, J., Assessing, evaluating and verifying technology performance. *(UNEP) INSIGHT Newsletter*, 4-6 (Dec. 1999).
Newbery, D.M., Clutton-Brock, T.H., Prance, G.T. Preface. *Phil. Trans. R. Soc. London* B 354: 1723-1724 (1999).
Noss, R.F., Cooperider, A.Y. *Saving Nature's Legacy.* Island Press, Washington, D.C. (1994).
Omen, O.S. *Natural Resource Conservation, An Ecological Approach.* Macmillan, New York (1971).
Puffer, R.R., Serrano, C.W. *Patterns of Mortality in Childhood.* Pan American Health Organization, Washington, D.C. (1973).
Raman, A., Raghu, S., Sreenath, S. Integrating environment, education, and employment for a sustainable society: An HRD agenda for developing countries. *Current Science* 78: 241-247 (2000).

Sager, M. Chemical speciation and environmental mobility of heavy metals in sediments and soils. In: Stoeppler, M. (Ed.) *Hazardous Metals in the Environment*. pp. 133-75, Elsevier Science Publishers, Amsterdam, (1992).

Samrit, D., Srikajorn, M., Paladsongkram, S., Siva-atitkul, K. *Study on soil erosion and plant nutrient transport from various soil series and cropping practices*. Division of Soil and Water Conservation. Land Development Department, Ministry of Agriculture and Co-operatives, Bangkok (In Thai) (1997).

Smith, J.V. Geology, mineralogy, and human welfare. *Proc. Natl. Acad. Sci. USA* 96: 3348-3349 (1999).

Teck, F.H. Soil loss experimental plots. In *Annual report*. Department of Agriculture, Kuching, Sarawak (1992).

Trainer, T. *Saving the Environment*. UNSW Press, Sydney (1998).

UNRISD. Progress Report on UNRISD Activities, 1990/1991. United Nations Research Institute for Social Development, Geneva (1991).

Van Vlaenderen, H. Problem solving : a local perspective. *Indigenous Knowledge and Development Monitor* 8 (1): 3-6 (2000).

Wolfensohn, J.D. Entering the 21st century: the challenges for development. *Phil. Trans. R. Soc. Lond.* B 354: 1943-1948 (1999).

World Bank. *Entering the 21st Century*. Oxford University Press, Oxford (1999).

World Bank. *The East Asian miracle: economic growth and public policy*. pp. 27-77. Oxford University Press, Oxford (1993).

Abbreviations and Acronyms

AAC	Annual Allowable Cut
AAC	Annual Allowable Cut
ACT	Advanced Control Technology
AMD	Acid Mine Drainage
APELL	Awareness and Preparedness for Emergencies at Local Level
AR	Acetylene Reduction
ASSINSEL	Association Internationale des Séléctionneurs pour la Protection des Obtentions Végétales (International Association of Plant Breeders for the Protection of Plant Varieties)
BAT	Best Available Technology
BCT	Basic Control Technology
BIPV	Building-integrated Photovoltaics
BMPs	Best Management Practice
BNF	Biological Nitrogen Fixation
BOD	Biological Oxygen Demand
CBD	Convention on Biological Diversity
CBDC	Community Biodiversity and Development Conservation Program
CCN	Cloud Condensation Nuclei
CFCs	Chlorofluorocarbons
CGBs	Community Gene Banks
CGIAR	Consultative Group on International Agricultural Research (also called the CG System)
CGRFA	Commission on Genetic Resources for Food and Agriculture

CH_4	Methane
CIAT	International Centre for Tropical Agriculture
CIMMYT	Centro Internacional de la Mejoramiento de Maiz y Trigo (International Maize and Wheat Improvement Centre)
CIP	Centro Internacional de la Papa (International Potato Centre)
CO_2	Carbon Dioxide
CONPS	Carbon, Oxygen, Nitrogen, Phosphorus and Sulphur
COP	Conference of the Parties (to the Convention on Biological Diversity)
CSO	Civil Society Organization
DOM	Dissolved Organic Matter
DUS	Distinctness, Uniformity and Stability (requirements for seed certification)
EEA	European Environmental Agency
ELISA	Enzyme Linked Immunosorbent Assay
EMBO	European Molecular Biology Organization
ENSO	El Nino Southern Oscillation
ERBE	Earth Radiation Budget Experiment
ESI	Electricity Supply Industry
EST	Expressed Sequence Tag
ETV	Environmental Technology Verification
FAO	Food and Agriculture Organization (of the United Nations)
FCs	Fluorocarbons
GATT	General Agreement on Tariffs and Trade
GE	Genetically Engineered
GHG(s)	Green House Gases
GIS	Geographical Information System
GMO	Genetically Modified Organism
GMR	Giant Magnetoresistance
GPA	Global Plan of Action
GPP	Gross Primary Productivity
GPS	Geographic Positioning System
GRFA	Genetic Resources for Food and Agriculture
Gt	Giga tonnes
GURT	Genetic Use Restriction Technology
GUS	b-Glucurunidase
GWP	Global Warming Potential
H_2	Hydrogen
HAC	Human Artificial Chromosome
HDPE	Higher Density Polyethylene

HEMM	Heavy Earth-moving Machinery
HFCs	Hydrofluorocarbons
HM	Heavy Metals
HPR	Host Plant Resistance
HR	Hypersensitive Reaction
HRT	Hydraulic Retention Time
ICDPs	Integrated Conservation and Development Projects
ICP	Insecticidal Crystal Protein
ICPIC	International Cleaner Production Information Clearing house
IDDs	Iodine Deficiency Disorders
IDE	International Development Enterprises
IDRC	International Development Research Centre
IKS	Indigenous Knowledge System
INM	Integrated Nutrient Management
IP	Intellectual Property
IPCC	intergovernmental Panel on Climate Change
IPGRI	International Plant Genetic Resources Institute (former IBPGR)
IPM	Integrated Pest Management
IPNS	Integrated Plant Nutrient Systems
IPRs	Intellectual Property Rights
IRPTC	International Register of Potentially Toxic Chemicals
IRRI	International Rice Research Institute, Manila
ISAAA	International Service for the Acquisition of Agribiotech Applications
ISNAR	International Service for National Agricultural Research
IUCN	The World Conservation Union
LDCs	Least Developed Countries
LDPE	Low Density Polyethylene
LED	Light Emitting Diodes
LLDPE	Linear Low Density Polyethylene
LMO	Living Modified Organism
LS	Lower Stratosphere
LSCs	Life Science Companies
LWR	Long Wave Radiative (Radiation)
MIRCENs	Microbial Resources Centres
MSE	Material Science and Engineering
MSW	Municipal Solid Waste
NAFTA	North American Free Trade Agreement
NAPL	Nonaqueous Phase Liquid
NBPGR	National Bureau of Plant Genetic Resources, New Delhi

NFTs	Nitrogen Fixing Trees
NGOs	Non-governmental Organizations
NMR	Nuclear Magnetic Resonance
NPP	Net Primary Productivity
NPS	Non Point Source
NPV	Nuclear Polyhedrosis Virus
NRC	National Research Council (USA)
OECD	Organization for Economic Cooperation and Development
OH	Hydroxyl Radical
OPEC	Organization of Petroleum Exporting Countries
OTEs	Oxygen Transfer Efficiencies
PAHs	Polyaromatic Hydrocarbons
PBRs	Plant Breeders' Rights
PCBs	Polychlorinated Biphenyls
PCR	Polymerase Chain Reaction
PEM	Proton Exchange Membrane
PF	Precision Farming
Pg	Peta gram
PGRFA	Plant Genetic Resources for Food and Agriculture
PNAS	Proceedings of National Academy of Sciences (USA)
PPB	Participatory Plant Breeding
PPM (=ppm)	Parts Per Million
PV	Photovoltaic
PVP	Photovoltaic (Water) Pumps
PVS	Participatory Variety Selection
QTL	Quantitative Trait Loci
RAFI	Rural Advancement Foundation International
RAPD	Random Amplification of Polymorphic DNA
RFLP	Restriction Fragment Length Polymorphism
ROS	Reactive Oxygen Species
RS	Remote Sensing
RUE	Rational-use-of-energy
SAGE	Stratospheric Aerosol and Gas Experiment
SAR	Systemic Acquired Resistance
SAT	Semi-arid Tropics
SCM	Single Column Model
SCP	Single Cell Protein
SDV	Silica Deposition Vesicle
SHS	Solar Home System
SINGER	CGIAR's System-wide Information Network for Genetic Resources
SNP	Single Nucleotide Polymorphism

Abbreviations and Acronyms

SOAs	Sustainable Development Assessments
SWR	Short Wave Radiative (Radiation)
TAA	Total Assessment Audit
TAC	Technical Advisory Committee
TIBTECH	Trends in Biotechnology *(journal)*
TNCs	Transnational Corporations
TOA	Top of the Atmosphere
TREE	Trends in Ecology and Evolution *(journal)*
TRIPs	Trade-Related Intellectual Property Rights
UA	Urban Agriculture
UNCED	United Nations Conference on Environment and Development
UNDP	United Nations Development Programme
UNESCO	United Nations Educational, Scientific and Cultural Organization
URA	Uruguay Round (Trade) Agreement
UT	Upper Troposphere
VAD	Vitamin A Deficiency
VCU	Value for Cultivation and Use (seed certification requirement)
VFA	Volatile Fatty Acids
VOCs	Volatile Organic Compounds
WCED	World Commission on Environment and Development
WCMC	World Conservation Monitoring Centre
WIPO	World Intellectual Property Organization
WTO	World Trade Organization
WRI	World Resources Institute

Glossary

Abiotic factor. A nonorganic variable within the ecosystem, affecting the life of organisms. Examples include temperature, light and soil structure. Can be harmful to the environment, as when sulphur dioxide emissions from power stations produce acid rain.

Acid rain. Acidic deposition caused principally by the pollutant gases sulphur dioxide (SO_2) and the nitrogen oxides. It is linked with damage to and the death of forests and lake organisms and also causes damage to buildings and statues. Its main effect is to damage the chemical balance of soil, causing leaching of important minerals including magnesium and aluminium. Plants living in such soils suffer from mineral loss and become more prone to infection. The minerals from the soil pass into lakes and rivers, disturbing aquatic life, for instance, by damaging the gills of young fish. Lakes and rivers suffer more direct damage as well because they become acidified by rainfall draining directly from their drainage basin. Adding alkalis to the lakes in the form of lime neutralizes the acidity, but treatment must be frequent and in itself has some side effects.

Adaptation. A genetically-determined characteristic that enhances the ability of an organism to cope with its environment.

Adaptive Management. The process of implementing policy decisions as scientifically-driven management experiments that test predictions and assumptions in management plans and using the resulting information to improve the plans.

Adaptive zone. A particular type of environment requiring unique adaptations. Species in different adaptive zones usually differ by major morphological or physiological characteristics.

Aerosol. Particulate material, other than water or ice, in the atmosphere ranging in size from aproximately 10^{-3} to larger than 10^2 mm in radius. Aerosols are important in the atmosphere as nuclei for the condensation of water droplets and ice crystals, as participants in various chemical cycles and as absorbers and scatterers of solar radiation, thereby inflencing the radiation budget of the earth-atmosphere system which, in turn, influences the climate on the surface of the Earth.

Afforestation. Planting of trees in areas that have not previously held forests. (Reafforestation is the planting of trees in deforested areas).

Agribusiness. Commercial farming on an industrial scale, often financed by multinational corporations. Agribusiness farms are mechanized, large in size, highly structured, reliant on chemicals and are sometimes described as 'food factories'.

Agriculture. The practice of farming, including the cultivation of the soil (for raising crops) and the raising of domesticated animals. Crops are for human nourishment, animal fodder, or commodities such as cotton. Animals are raised for wool, milk, leather, dung (as fuel), or meat. The units for managing agricultural production vary from small holdings and individually owned farms to corporate-run farms and collective farms run by entire communities.

Agroecosystem. An ecological system modified by people to produce food, fibre, fuel and other products desired fro human use.

Agropastoralism. Land-use system in which arable cropping and the keeping of grazing livestock are combined.

Ammonification. Breakdown of proteins and amino acids with ammonia as an excretory by-product.

Annual. The completion of the life cycle within a year. May be a winter annual, in which case, the plant exists as a seedling during the winter months, or a spring annual, in which case the plant does not over-winter as a seedling.

Apomixis. Reproducing by seeds not resulting from fertilization; hence genetic recombination is not encountered.

Appropriate technology. Simple or samall-scale machinery and tools that, because they are cheap and easy to produce and maintain, may be of great use in the developing world; for example, hand ploughs and simple looms. This equipment may be used to supplement local crafts and traditional skills to encourage small-scale industrialization. Appropriate technology relies on easily available energy, for example, from small petrol engines, human power, wind pumps and solar panels. This avoids the need for large capital investment, a scarce resource in the developing world, but utilizes the generally large supplies of human labour.

Arid region. A region that is very dry and has little vegetation. Aridity depends on temperature, rainfall and evaporation. An arid area is usually defined as one that receives less than 250 mm (10 in) of rainfall each year.

Artificial selection. International manipulation by man of the fitnesses of individuals in a population to produce a desired evolutionary response.

Autecology. The study of organisms in relation to theeir physical environment.

Bacteroid. Mis-shapen bacterial cells of *Rhizobium* as found within leguminous root nodules.

Barren. An area with sparse vegetation owing to some physical or chemical property of the soil.

Biennial. A plant which completes its life cycle in two calendar years, and which during the first is not reproductive.

Biodiversity (biological diversity). Measure of the variety of the Earth's animal, plant and microbial species; of genetic differences within species; and of the ecosystems that support those species.

Biofuel. Any solid, liquid, or gaseous fuel produced from organic (once living) matter, either directly from plants or indirectly from industrial, commercial, domestic, or agricultural wastes.

Biogeochemical cycle—The chemical interactions among the atmosphere, biosphere, hydrosphere, and lithosphere.

Biogeography. The study of how and why plants and animals are distributed around the world, in the past as well as in the present. More specifically, a theory describing the geographical distribution of species developed by Robert MacArthur and Edward O. Wilson. The theory argues that for many species, ecological specializations mean that suitable habitats are found in a disparate pattern.

Biological weathering. Form of weathering caused by the activities of living organisms—for example, the growth of roots or the burrowing of animals. Tree roots are the most significant agents of biological weathering as they are capable of prising apart rocks by growing into cracks and joints. The action of plants in breaking up rocks in this way is an important part of the process of the development of soil.

Biomass accumulation ratio. The ratio of weight to annual production, usually applied to vegetation.

Biomass. Weight of living material, usually expressed as a dry weight, in all or part of an organism, population, or community. Commonly expressed as weight per unit area, a biomass density.

Biosphere. The narrow zone that supports life on our planet. It is limited to the waters of the Earth, a fraction of its crust and the lower regions of the atmosphere.

Biota. Fauna and flora together.

Biotic environment. Biological components of an organism's surroundings that interact with it, including competitors, predators, parasites and prey.

Calcification. Deposition of calcium and other soluble salts in soils where evaporation greatly exceeds precipitation.

Carbon cycle. All parts (reservoirs) and fluxes of carbon; usually thought of as a series of the four main reservoirs of carbon interconnected by pathways of exchange. The four reservoirs, regions of the Earth in which carbon behaves in a systematic manner, are the atmosphere, terrestrial biosphere (usually includes freshwater systems), oceans and sediments (includes fossil fuels). Each of these global reservoirs may be subdivided into smaller pools ranging in size from individual communities or ecosystems to the total of all living organisms (biota). Carbon exchanges from reservoir to reservoir by various chemical, physical, geological and biological processes.

Carbon dioxide fertilization. Enhancement of plant growth or of the net primary production by CO_2 enrichment that could occur in natural or agricultural systems as a result of an increase in the atmospheric concentration of CO_2.

Carbon flux. The rate of exchange of carbon between pools (reservoirs).

Carbon pool. The reservoir containing carbon as a principal element in the geochemical cycle.

Carrying capacity. Number of individuals that the resources of a habitat can support.

Cash crop. Crop grown solely for sale rather than for the farmer's own use, for example, coffee, cotton, or sugar beet. Many Third World countries grow cash crops to meet their debt repayments rather than grow food for their own people, thus adding to the problems of malnutrition and starvation. Cash crops are grown intensively, usually as a monoculture, so environmental damage by pesticides and fertilizers is often unavoidable.

Cation-exchange capacity. The ability of soil particles to absorb positively charged ions, such as hydrogen (H^+) and calcium (Ca^{++}).

Cereal. Any member of the Gramineae the grains of which are used mainly as food for man after being milled and ground.

Chlorofluorocarbons. A family of inert nontoxic and easily liquified chemicals used in refrigeration, air conditioning, packaging and insulation or as solvents or aerosol propellants. Becaust they are not destroyed in the lower atmosphere, they drift into the upper atmosphere where their chlorine components destroy **ozone.**

Clear cutting. A forest-management technique that involves harvesting all the trees in one area at one time.

Climate change. The long-term fluctuations in temperature, precipitation, wind and all other aspects of the Earth's climate. External processes, such as solar-irradiance variations, variations of the Earth's orbital parameters (eccentricity, precession and inclination), lithosphere motions and volcanic activity, are factors in climatic variation. Internal variations of the climate system also produce fluctuations of sufficient magnitude and variability to explain observed climate change through the feedback processes interrelating the components of the climate system.

Climate system. The five physical components (atmosphere, hydrosphere, cryosphere, lithosphere, and biosphere) that are responsible for the climate and its variations.

Climate. The statistical collection and representation of the weather conditions for a specified area during a specified time interval, usually decades, together with a description of the state of the external system or boundary conditions. The properties that characterize the climate are thermal (temperatures of the surface air, water, land and ice), kinetic (wind and ocean currents, together with associated vertical motions and the motions of **air masses**, aqueous humidity, cloudiness and cloud water content, **groundwater**, lake lands and water content of snow on land and sea ice), and static (pressure and density of the atmosphere and ocean composition of the dry air, salinity of the oceans, and the geometric boundaries and physical constants of the system). These properties are interconnected by the various physical processes such as precipitation, evaporation, infrared radiation, convection, advection and turbulence.

Climograph. A diagram on which localities are represented by the annual cycle of their temperature and rainfall.

Cline. A graded series of variants which may be correlated with other features associated with the species, e.g. ecological factors which exhibit gradual change. Usually prefixed to indicate the independent variable.

Clone. A collection of individuals all derived by vegetative reproduction from a common ancestor and therefore each of identical genetical composition.

Coadaptation. Evolution of characteristics of two or more species to their mutual advantage.

Coarse-grained. Referring to qualities of the environment that occur in large patches, with respect to the activity patterns of an organism and, therefore, among which the organism can select.

Coir. The fibrous mesocarp of coconut used in the manufacture of coarse matting and ropes.

Common land. Unenclosed wasteland, forest and pasture used in common by the community at large. Poor people have throughout history gathered fruit, nuts, wood, reeds, roots, game, and so on from common land. In dry regions of India, the landless derive about 20% of their annual income in this way, together with much of their food and fuel.

Community. An association of interacting populations, usually defined by the nature of their interaction at the place in which they live.

Competition. Use or defense of a resource by one individual that reduces the availability of that resource to other individuals.

Competitive exclusion principle. The hypothesis that two or more species cannot coexist on a single resource that is scarce relative to the demand for it.

Compost. Organic material decomposed by bacteria under controlled conditions to make a nutrient-rich natural fertilizer for use in gardening or farming. A well-made compost heap reaches a high temperature during the composting process, killing most weed seeds that might be present.

Conservation. Action taken to protect and preserve the natural world from pollution, overexploitation and other harmful features of human activity. In attempts to save particular species or habitats, a distinction is often made between preservation, that is maintaining the pristine state of nature exactly as it was or might have been, and conservation, the management of natural resources in such a way as to integrate the requirements of the local human population with those of the animals, plants, or the habitat being conserved.

Continuum index. An artificial scale of an environmental gradient based on changes in community composition.

Continuum. A gradient of environmental characteristics or of change in the composition of communities.

Crop rotation. System of regularly changing the crops grown on a piece of land. The crops are grown in a particular order to utilize and add to the nutrients in the soil and to prevent the build-up of insect and fungal pests. Including a legume crop such as peas or beans in the rotation helps build up nitrate in the soil because the roots contain bacteria capable of fixing nitrogen from the air.

Cross-inoculation. The process of attempting to establish a successful combination between the bacteria isolated from one group of host plant and the members of another group of host plants. Cross-inoculation groups are established when the limits of successful inoculation pairs are defined.

Cross-pollination. The transferring of pollen from the anthers of one flower to the stigmatic surface of the carpels of another different flower.

Cross-resistance. Resistance or immunity to one disease organism resulting from infection by another, usually closely related organism.

Cultivar. The name given to a commercially-distinct group of individuals. Considered as a taxon in its own right.

Debt-for-nature swap. Agreement under which a proportion of a country's debts are written off in exchange for a commitment by the debtor country to undertake projects for environmental protection. Debt-for-nature swaps were set up by environment groups in the 1980s in an attempt to reduce the debt problem of poor countries, while simultaneously promoting conservation. Most debt-for-nature swaps have focused on setting aside areas of tropical rain forest for protection and have involved private conservation foundations.

Decomposition. Metabolic breakdown of organic materials; the by-products are released energy and simple organic and inorganic compounds.

Denitrification. The reduction by microorganisms of nitrate and nitrite to nitrogen.

Detritivore. An organism that feeds on freshly dead or partially decomposed organic matter.

Detritus. Freshly dead or partially decomposed organic matter.

Developed world or First World or the North. The countries that have a money economy, a highly developed industrial sector, a high degree of urbanization, a complex communications network, high gross domestic product (over $2,000) per person, low birth and death rates, high energy consumption, and a large proportion of the workforce employed in manufacturing or service industries (secondary to quaternary industrial sectors). the developed world includes the USA, Canada, Europe, Japan, Australia and New Zealand.

Developing world or Third World or The South. Countries with a largely subsistence economy where the output per person and the average income are low. Typically have low life expectancy, high birth and death rates, poor communications and health facilities, low literacy levels, high national debt, and low energy consumption per person. The developing world includes much of Africa and parts of Asia and South America.

Dioecious. Possessing staminate and carpellate flowers but carried on separate plants.

Dispersal. Movement of organisms away from the place of birth or from centers of population density.

Diversity. The number of species in a community or region. Alpha diversity refers to the diversity of a particular habitat, beta diversity to the species added by pooling habitats within a region. Also, a measure of the variety of species in a community that takes into account the relative abundance of each species.

Ecocline. A geographical gradient of vegetation structure associated with one or more environmental variables.

Ecological efficiency. Percentage of energy in the biomass produced by one trophic level that is incorporated into biomass produced by the next highest trophic level.

Ecological isolation. Avoidance of competition between two species by differences in food, habitat, activity period, or geographical range.

Ecological release. Expansion of habitat and resource utilization by populations in regions of low species diversity, resulting from reduced interspecific competition.

Ecosystem Management. An approach to maintaining or restoring the composition, structure, and function of natural and modified ecosystems for the goal of long-term sustainability. It is based on a collaboratively developed vision of desired future conditions that integrates ecological, socioeconomic, and institutional perspectives, applied within a geographic framework defined primarily by natural ecological boundaries.

Ecosystem. All the interacting parts of the physical and biological worlds.

Ecotone. A habitat created by the juxtaposition of distinctly different habitats; an edge habitat; a zone of transition between habitat types.

Ecotype. A genetically-differentiated sub-population that is restricted to a specific habitat. Free gene exchange can occur between ecotypes.

Edaphic. Pertaining to, or influenced by, soil conditions.

Endemic. Confined to a certain region.

Endosperm. That tissue resulting from the development of the triple fusion nucleus after fertilization. May or may not persist until maturation of the seed. A nutritive tissue for the developing embryo.

Energy balance models. An analytical technique to study the solar radiation incident on the Earth in which explicit calculations of atmospheric motions are omitted. In the zero-dimensional models, only the incoming and outgoing radiation is considered. The outgoing **infrared radiation** is a linear function of global mean surface air temperature, and the reflected solar radiation is dependent on the surface albedo. The albedo is a step function of the global mean surface air temperatures, and equilibrium temperatures are computed for a range of values of the solar constant. The one-dimensional models have surface air temperature as a function of latitude. At each latitude, a balance between incoming and outgoing radiation and horizontal transport of heat is computed. (Abbreviated as EBM).

Environment. Surroundings of an organism, including the plants and animals with which it interacts.

Environmental gradient. A continuum of conditions ranging between extremes, as the gradation from hot to cold environments.

Erosion. The natural removal of the surface of land by the action of wind or water.

Evapotranspiration. Amount of water transferred from the soil to the atmosphere by evaporation and plant transpiration.

Factory farming. Intensive rearing of animals (e.g. chicken) for food, usually on high-protein foodstuffs in confined quarters. Antibiotics and growth hormones are commonly used to increase yield, but some countries now restrict this practice because they can persist in the flesh of the animals after slaughter. The emphasis is on productive yield rather than animal welfare so that conditions for the animals are often very poor. Many people object to factory farming on moral as well as health grounds.

Fertilizer. Substance containing some or all of a range of about 20 chemical elements necessary for healthy plant growth, used to compensate for the deficiencies of poor or depleted soil. Fertilizers may be organic, for example farmyard manure, composts, bonemeal, blood and fishmeal; or inorganic, in the form of compounds, mainly of nitrogen, phosphate and potash. Inorganic fertilizer use has increased by between 5 and 10 times during the last 50 years in a drive to increase yields. Because externally applied fertilizers tend to be in excess of plant requirements and drain away to affect lakes and rivers, attention has turned to the modification of crop plants themselves.

Plants of the legume family, including beans and clover, live in symbiosis with bacteria located in root nodules, which fix nitrogen from the atmosphere.

Field capacity. The amount of water that soil can hold against the pull of gravity.

Fine-grained. Referring to qualities of the environment that occur in small patches with respect to the activity patterns of an organism and among which the organism cannot usefully distinguish.

Fodder. The product derived from a forage crop.

Food chain. An abstract representation of the passage of energy through populations in the community.

Food technology. The commercial processing of foodstuffs in order to render them more palatable or digestible, or to preserve them from spoilage.

Food web. An abstract representation of the various paths of energy flow through populations in the community.

Forage. Any plant grown for its leaves and stems which are then consumed by domesticated animals.

Fossil fuel. Any hydrocarbon deposit that can be burned for heat or power, such as petroleum, coal and natural gas.

Gene bank. Collection of seeds or other forms of genetic material, such as tubers, spores, bacterial or yeast cultures, live animals and plants, frozen sperm and eggs, or frozen embryos. These are stored for possible future use in agriculture, plant and animal breeding, or the restocking of wild habitats where species have become extinct.

Gene flow. Exchange of genetic traits between populations by movement of individuals, gametes, or spores.

Gene mapping. Determining the relative position of genes on chromosomes.

Genetic engineering. Deliberate manipulation of genetic material by biochemical techniques. It is often achieved by the introduction of new DNA, usually by means of a virus or plasmid, to breed functionally specific plants, animals, or bacteria. These organisms with a foreign gene added are said to be transgenic.

Genetic feedback. Evolutionary response of a population to the adaptations of competitors, predators, or prey.

Genetic load. Selective deaths sustained by a population due to genotypes that deviate from the genotype with the maximum fitness.

Genome. The total set of hereditary elements in a cell or organism.

Genotype. All the genetic characteristics that determine the structure and functioning of an organism; often applied to a single-gene locus to distinguish one allele, or combination of alelles, from another.

Germplasm. The sum total of genetic variability available to a particular population of organisms.

Green accounting. The inclusion of the economic losses caused by environmental degradation in traditional profit and loss accounting systems. The idea arose in the 1980s when profitability was the main tool for judging the value of an action. By such crude measures, killing elephants for ivory, destroying tropical rainforests for hardwood and continuing whaling, all make economic sense. However, if the future values of these resources are included so that, for example, tourism, protection of biodiversity and ecosystem stability are all given a notional economic value, even in purely economic terms, it makes no sense to destroy habitats or hunt animals to extinction.

Green audit. Inspection of a company to assess the total environmental impact of its activities or of a particular product or process. For example, a green audit of a manufactured product considers the impact of production (including energy use and the extraction of raw materials used in manufacture), use (which may cause pollution and other hazards), and disposal (potential for recycling and whether waste causes pollution). Such 'cradle-to-grave' surveys allow a widening of the traditional scope of economics by ascribing costs to variables that are usually ignored such as despoilation of the countryside or air pollution.

Green belt. Area surrounding a large city, officially designated not to be built upon but preserved where possible as open space (for agricultural and recreational use).

Green computing. The gradual movement by computer companies toward incorporating energy-saving measures in the design of systems and hardware. The increasing use of energy-saving devices, so that a computer partially shuts down during periods of inactivity, but can reactivate at the touch of a key, could play a significant role in energy conservation.

Green consumerism. Catch-phrase especially popular during the 1980s when consumers became increasingly concerned about the environment. Labels such as 'eco-friendly' became a common marketing tool as companies attempted to show that their goods had no negative effect on the environment.

Green manure. Any plant grown for incorporation into soil to improve fertility; usually buried by ploughing.

Green movement. Collective term for the individuals and organizations involved in efforts to protect the environment.

Green revolution. A popular term for the change in methods of arable farming in Third World countries. The intent is to provide more and better food, albeit with a heavy reliance on chemicals and machinery. Much of the food produced is exported as cash crops, so that local diet does not always improve. The green revolution tended to benefit primarily those landowners who could afford the investment necessary for such intensive agriculture. Without a dosage of 70-90 kg/154-198 lb of expensive nitrogen fertilizers per hectare, the high-yield varieties will not grow properly. Hence, rich farmers tend to obtain bigger yields while smallholders are unable to benefit from the new methods. In terms of production, the green revolution was initially successful in SE Asia; India doubled its wheat yield in 15 years, and the rice yield in the Philippines rose by 75%. However, yields have since levelled off in many areas; some countries that cannot afford the dams, fertilizers and machinery required, have adopted intermediate technologies.

Greenhouse effect. Phenomenon of the Earth's atmosphere by which the energy of solar radiation, absorbed by the ground and re-emitted as infrared energy, is prevented from escaping by various gases in the air. This results in a rise in the Earth's temperature. The increasing levels of so-called greenhouse gases—such as carbon dioxide, methane and chlorofluorocarbons (CFCs)—are enhancing the effect to such as extent that global warming and dramatic climate change are becoming likely.

Gross production. The total energy or nutrients assimilated by an organism, a population, or an entire community.

Groundwater. The water contained in porous underground strata as a result of infiltration from the surface.

Habitat. Place where an animal or plant normally lives, often characterized by a dominant plant form or physical characteristic (i.e. the stream habitat, the forest habitat).

Halophyte. A plant which grows in soils with higher than average salt content.

Heat island effect. A 'dome' of elevated temperatures over an urban area caused by the heat absorbed by structures and pavement.

Hectare. A unit of area, equal to $10000 m^2$. Symbol: ha.

Herbicide. Any chemical used to destroy plants or check their growth.

Herbivore. An organism that consumes living plants or their parts.

Heredity. Genetic transmission of traits from parents to offspring.

Hermaphrodite. An organism that has the reproductive organs of both sexes.

Heterogeneity. The variety of qualities found in an environment (habitat patches) or a population (genotypic variation).

Heterosis. The greater (more vigorous) growth exhibited by hybrids when compared with their parents.

Heterotroph. An organism that utilizes organic materials as a source of energy and nutrients.

Heterozygous. Containing two forms (alleles) of a gene, one derived from each parent.

High-yield variety. Crop that has been specially bred or selected to produce more than the natural varieties of the same species. During the 1950s and 1960s, new strains of wheat and maize were developed to reduce the food shortages in poor countries. Later, IR8, a new variety of rice that increased yields by up to six times, was developed in the Philippines. Strains of crops resistant to drought and disease were also developed. High-yield varieties require large amounts of expensive artificial fertilizers and sometimes pesticides for best results.

Holdridge life zone. A climate category defined by three weighted climatic indexes, namely, mean annual heat, precipitation and atmospheric moisture.

Homeostasis. Maintenance of constant internal conditions in the face of a varying external environment.

Homozygous. Containing two identical alleles at a gene locus.

Humus. Organic matter present in the soil and decomposed to such an extent that its original structure cannot be determined; generally colloidal and dark brown in colour.

Hybridization. Crossing of individuals from genetically different strains, populations, or, sometimes, species.

Hydrologic cycle. The process of evaporation, vertical and horizontal transport of vapor, condensation, precipitation and the flow of water from continents to oceans. It is a major factor in determining climate through its influence on surface vegetation, the clouds, snow and ice, and soil moisture. The hydrologic cycle is responsible for 25 to 30% of the mid-latitudes' heat transport from the equatorial to polar regions.

Hydrosphere. The water component of the Earth, usually encompassing the oceans, rivers, streams, lakes, groundwater, and atmospheric water vapour.

Hydrosphere. The aqueous envelope of the Earth, including the oceans, freshwater lakes, rivers, saline lakes and inland seas, soil moisture groundwaters, and atmospheric vapor.

Inbreeding. A breeding system where mating is between close relatives. In plants with bisexual flowers the maximum degree of inbreeding arises from self-fertilization. Long term inbreeding gives rise to genetically uniform progeny.

Infiltration. The passage of water into the soil. The rate of absorption of surface water by soil (the infiltration capacity) depends on the amount of surface water, the permeability and compactness of the soil and the extent to which it is already saturated with water. Once in the soil, water may pass into the bedrock to form groundwater.

Infrared radiation. Electromagnetic radiation lying in the wavelength interval from 0.7 mm to 1000 mm. Its lower limit is bounded by visible radiation, and its upper limit by microwave radiation. Most of the energy emitted by the Earth and its atmosphere is at infrared wavelength. **Infrared radiation** is generated almost entirely by large-scale intramolecular processes. The tri-atomic gases, such as water vapor, carbon dioxide, and ozone, absorb infrared radiation and play important roles in the propagation of infrared radiation in the atmosphere. Abbreviated IR; also called **'longwave radiation'**.

Insolation. The solar radiation incident on a unit horizontal surface at the top of the atmosphere. It is sometimes referred to as solar irradiance. The latitudinal variation of insolation supplies the energy for the general circulation of the atmosphere. Insolation depends on the angle of incidence of the solar beam and on the **solar constant**.

Integrated Pest Management (IPM). The use of a coordinated array of methods to control pests, including biological control, chemical pesticides, crop rotation, and avoiding monoculture. By cutting back on the level of chemicals used the system can be both economical and beneficial to human health and the environmental.

Intermediate technology. Application of mechanics, electrical engineering, and other technologies based on designs developed in the developed world but utilizing materials, assembly and maintenance found in the developing world.

Inversion. An anomaly in the normal positive **lapse rate**; usually refers to a thermal inversion, in which temperature increases rather than decreases with height.

Ion exchange. Process whereby the ions in one compound replace the ions in another. The exchange occurs because one of the products is insoluble in water. For example, when hard water is passed over an ion-exchange resin, the dissolved calcium and magnesium ions are replaced by either sodium or hydrogen ions, so the hardness is removed. The addition of washing-soda crystals to hard water is also an example of ion exchange.

Irrigation. Artificial water supply for dry agricultural areas by means of dams and channels. Drawbacks are that it tends to concentrate salts, ultimately causing soil infertility and that rich river silt is retained at dams, to the impoverishment of the land and fisheries below them.

Karyotype. Characteristic chromosomes of a particular species.

Landfill site. Huge holes used for dumping household and commercial waste. The sites can release toxins and other leachates into the soil. Decomposing organic matter releases methane, which can be explosive, although many sites collect the gas and burn it for energy.

Laterite. A hard substance rich in oxides of iron and aluminium; frequently formed when tropical soils weather under alkaline conditions.

Laterization. Leaching of silica from soil, usually in warm, moist regions with an alkaline soil reaction.

Leaching. Process by which substances are washed out of the soil. Fertilizers leached out of the soil drain into rivers, lakes and ponds and cause water pollution. The risk of nitrates and pesticides leaching into aquifers is greatest in sandy or shallow soils. In tropical areas, leaching of the soil after the destruction of forests removes scarce nutrients and can lead to a dramatic loss of soil fertility. The leaching of soluble minerals in soils can lead to the formation of distinct soil horizons as different minerals are deposited at successively lower levels.

Life zone. A more or less distinct belt of vegetation occurring within, and characteristic of, a particular range of latitude or elevation.

Livestock. Animals may be semi-domesticated, such as reindeer, or fully domesticated but nomadic (where naturally growing or cultivated food supplies are sparse), or kept in one location. Animal farming involves accommodation (buildings, fencing, or pasture), feeding, breeding, gathering the produce (eggs, milk, or wool) slaughtering, and then further processing such as tanning.

Loam. Soil that is a mixture of coarse sand particles, fine silt, clay particles, and organic matter.

Lodging. The phenomenon seen in cereals where, as a result of disease or adverse weather, the culm is laid to the horizontal.

Lysimeter. A device used to measure the quantity or rate of water movement through or from a block of soil or other material, such as solid waste, or used to collect percolated water for quality analysis.

Manure. Primarily the excreta of animals; may contain some spilled feed or bedding.

Mesic. Referring to habitats with plentiful rainfall and well-drained soils.

Mixed farming. Farming system where both arable and pastoral farming is carried out. Mixed farming is a lower-risk strategy than monoculture. If climate, pests, or market prices are unfavourable for one crop or type of livestock, another may be more successful and the risk is shared. Animals provide manure for the fields and help to maintain soil fertility.

Monoculture. Farming system where only one crop is grown. In developing countries, this is often a cash crop, grown on plantations. Cereal crops in the indistrualized world are also frequently grown on a monoculture basis; for example, wheat in the Canadian prairies. Monoculture allows the farmer to tailor production methods to the requirements of one crop, but it is a high-risk strategy since the crop may fail (because of pests, disease, or bad weather) or world prices for the crop may crash. Monoculture without crop rotation can result in reduced soil quality despite the addition of artificial fertilizers and it contributes to soil erosion.

Mulch. A layer of organic material applied to the surface of the ground to retain moisture in it or in the roots of plants.

Mulching. The spreading of leaves, straw, or other loose material on the ground to prevent erosion, evaporation, freezing of plant roots, etc.

Multiple gene inheritance. Determination of a quantitatively varying trait by the additive effects of more than one allele.

Mutation. Any change in the genotype of an organism occurring at the gene, chromosome, or genome level; usually applied to changes in genes to new allelic forms.

Mutualism. Relationship between two species that benefits both parties.

Mycorrhizae. Close association of fungi and tree roots in the soil that facilitates the uptake of minerals by trees.

Natural selection. Change in the frequency of genetic traits in a population through differential survival and reproduction of individuals bearing those traits.

Negative feedback. Tendency of a system to counteract externally imposed change and return to a stable state.

Net aboveground productivity (NAP). Accumulation of biomass in aboveground parts of plants (trunks, branches, leaves, flowers and fruits), over a specified period; usually expressed on an annual basis.

Net production. The total energy or nutrients accumulated by an organism by growth and reproduction; gross production minus respiration.

Net reproductive rate. Number of offspring that females are expected to bear on average during their lifetimes.

Niche. All the components of the environment with which the organism or population interacts.

Nitrification. Breakdown of nitrogen-containing organic compounds by micro-organisms, yielding nitrates and nitrites.

Nitrogen cycle. A series of processes in which atmospheric nitrogen is converted into nitrates in soil and other nitrogenous compounds in plants, from which—directly, by way of animals, or through the intermediate stage of coal—ammonium compounds are formed in the soil. These ammonium compounds eventually break down to return the nitrogen to the atmosphere or to another part of the cycle.

Nitrogen fixation. Biological assimilation of atmospheric nitrogen to form organic nitrogen-containing compounds.

Nutrient cycle. The path of an element through the ecosystem including its assimilation by organisms and its release in a reusable inorganic form.

Nutrient. Any substance required by organisms for normal growth and maintenance. (Mineral nutrients usually refer to inorganic substances taken from soil or water.)

Organic agriculture. Using more sophisticated natural methods in agriculture without chemical sprays and fertilizers. These methods are desirable because nitrates have been seeping into the groundwater; insecticides are found in lethal concentrations at the top of the food chain; some herbicides are associated with human birth defects and hormones fed to animals to promote fast growth have damaging effects on humans.

Organic farming. Farming without the use of synthetic fertilizers (such as nitrates and phosphates) or pesticides (herbicides, insecticides, and fungicides) or other agrochemicals (such as hormones, growth stimulants, or fruit regulators). In place of artificial fertilizers, compost, manure, seaweed, or other substances derived from living things are used (hence the name 'organic'). Growing a crop of a nitrogen-fixing plant such as lucerne, then ploughing it back into the soil, fertilizes the ground. Some organic farmers use naturally occurring chemicals such as nicotine or pyrethrum to kill pests, but control by non-chemical methods is preferred. Those methods include removal by hand, intercropping (planting with companion plants which deter pests), mechanical barriers to infestation, crop rotation, better cultivation methods, and biological control. Weeds can be controlled by hoeing, mulching (covering with manure, straw, or black plastic), or burning off. Organic farming methods produce food with minimal pesticide residues and greatly reduce pollution of the environment. They are more labour intensive and, therefore, more expensive, but use less fossil fuel. Soil structure is improved by organic methods and a conventional farm can lose four times as much soil through erosion as an organic farm, although the loss may not be immediately obvious.

Osmoregulation. Regulation of the salt concentration in cells and body fluids.

Ozone. A molecule made up of three atoms of oxygen. In the **sratosphere**, it occurs naturally and it provides a protective layer shielding the Earth from ultraviolet radiation and subsequent harmful health effects on humans and the **environment**. In the **troposphere**, it is a chemical oxidant and major component of **photochemical smog**.

Parasite. An organism that consumes part of the blood or tissues of its host, usually without killing the host. Parasites may live entirely within the host (endoparasites) or on its surface (ectoparasites).

Parasitoid. Any of a number of so-called parasitic insects whose larvae live within and consume their host, usually another insect.

Parthenocarpy. The production of fruits in the absence of fertilization.

Pathogen. A disease-causing organism.

Pedology. The study of soil, is significant because of the relative importance of different soil types to agriculture.

Percolation. The movement of water through a permeable stratum.

Perennial. A plant which persists for more than two years.

Pericarp. The fruit covering derived from the ovary wall.

Permeability. In groundwater hydrology, the property of a material that permits the passage of water through it at an appreciable rate at normal pressures. A material that permits perceptible passage of water is said to be *permeable;* a material that does not is said to be *impermeable* (the terms 'pervious' and 'impervious', respectively, are sometimes used).

Pest. An organism (insect, mite, weed, fungus, disease, animal, etc.) that humans wish to control or eliminate for any of various reasons, including possible harm to crops, animals or structures.

Pest management. Manipulation of pest or potential-pest populations so as to diminish their injury or render them harmless.

Pesticide. Any chemical used in farming, gardening or indoors to combat pests. Pesticides are of three main types: *insecticides* (to kill insects), *fungicides* (to kill fungal diseases), and *herbicides* (to kill plants, mainly those considered weeds). Pesticides cause pollution problems through spray drift onto surrounding areas, direct contamination of users or the public, and as residues on food. The safest pesticides include those made from plants, such as the insecticides pyrethrum and derris. More potent are synthetic products, such as chlorinated hydrocarbons. These products, including DDT and dieldrin, are highly toxic to wildlife and often to human beings, so their use is now banned or restricted by law in some areas and is declining. Safer pesticides, such as malathion, are based on organic phosphorus compounds, but they still present hazards to health.

Phenotype. Physical expression in the organism of the interaction between the genotype and the environment; outward appearance of the organism.

Pheromones. Chemical substances used for communication between individuals.

Photoautotroph. An organism that utilizes sunlight as its primary energy source for the synthesis of organic compounds.

Photochemical smog. Air pollution caused by chemical reactions among various substances and pollutants in **the atmosphere.**

Photoperiodism. Seasonal response by organisms to change in the length of the daylight period. (Flowering, germination of seeds, reproduction, migration and diapause are frequently under photoperiodic control.)

Podsolization. Breakdown and removal of clay particles from the acidic soils of cold, moist regions.

Pollutant. Any undesirable solid, liquid, or gaseous matter in a gaseous or liquid medium. *Primary p.*, a pollutant emitted into the atmosphere from an identifiable source. *Secondary p.*, a pollutant formed by chemical reaction in the atmosphere.

Pollution. The introduction of pollutants into a liquid or gaseous medium, the presence of pollutants in a liquid or gaseous medium, or any undesirable modification of the composition of a liquid or gaseous medium. An 'undesirable modification' is one that has injurious or deleterious effects.

Poly(vinyl chloride). A polymer of vinyl chloride, frequently referred to by the initials PVC. The term is also applied loosely to poly (vinyl chloride) plastics, which are widely used for many different purposes.

Polycarpic. With the capacity to flower and fruit an indefinite number of times.

Polychlorinated biphenyl (PCB). Any of a series of organochlorine compounds containing two linked phenyl rings and a variable proportion of chlorine. Most commercial products contain several different isomers, and some contain polychlorinated terphenyls also. They are widely used in the electrical industry, particularly in transformers, and have also been used as lubricants, hydraulic fluids, plasticizers, flame retardants, etc. They are very stable and have become widespread in the environment, and considerable concern has been expressed over their effects on health.

Polyethylene. A polymer of ethylene (or, in systematic nomenclature, of ethene, the polymer then being termed polyethene). The term is also applied loosely to polyethylene plastics, which are widely used for electric cable insulation, laboratory equipment and many other purposes.

Polymorphism. Occurrence of more than one distinct form of individuals in a population.

Porosity. Ratio of the volume of the interstices in a given sample of a porous medium, e.g. soil, to the gross volume of the porous medium, inclusive of voids.

Potential evapotranspiration. The amount of transpiration by plants and evaporation from the soil that would occur, given the local temperature and humidity, if water were not limited.

Precipitation. The thickness of the layer of water which accumulates on a horizontal surface, as the result of one or more falls of precipitation, in the absence of infiltration or evaporation, and if any part of the precipitation falling as snow or ice were melted. It is usually measured in terms of depth divided by time, e.g. millimetres per day or per year.

Primary consumer. An herbivore, the lowermost eater on the food chain.

Primary producer. A green plant that assimilates the energy of light to synthesize organic compounds.

Primary productivity. Rate of assimilation (gross primary productivity) or accumulation (net primary productivity) of energy and nutrients by green plants and other autotrophs.

Proximate factors. Aspects of the environment that organisms use as cues for behaviour; for example, daylength. (Proximate factors are often not directly important to the organism's well-being).

Proxy climate indicators. Dateable evidence of a biological or geological phenomenon whose condition, at least in part, is attributable to climatic conditions at the time of its formation. Proxy data are any material that provides an indirect measure of climate and include documentary evidence of crop yields, harvest dates, glacier movements, tree rings, glaciers and snow lines, insect remains, pollen remains, marine microfauna, isotope measurements: ^{18}O, in ice sheets, ^{18}O, ^{2}H, and ^{13}C in tree rings; $CaCO_3$ in sediments; and speleothems. There are three main problems in using proxy data: (1) dating; (2) lag and response time; and (3) meteorological interpretation. Tree rings, pollen deposits from varved lakes, and ice cores are the most promising proxy data sources for reconstructing the climate of the last five millennia because the dating are precise on an annual basis while other proxy data sources may only be precise to ± 100 years.

Race. A group of individuals of a species probably of common genetic origin and recognizably distinct from other such groups. Could be equated with variety.

Radiatively active gases. Gases that absorb incoming solar radiation or outgoing **infrared radiation,** thus affecting the vertical temperature profile of the atmosphere. Most frequently being cited as being radiatively active gases are water vapor, CO_2, methane, nitrous oxide, chlorofluorocarbons, and ozone.

Radiosonde. A balloon-borne instrument for the simultaneous measurement and transmission of meteorological data up to a height of approximately 30,000 meters (100,000 feet). The height of each pressure level of the observation is computed from data received via radio signals.

Recharge. The process by which water is added to a **reservoir** or zone of saturation, often by **runoff** or **percolation** from the soil surface.

Recharging. The addition, by natural or artificial means, of water to an underground aquifer.

Reclamation. The improvement of land or its recovery from sea or swamp.

Recombinant DNA (rDNA). Segments of DNA from two different organisms spliced together in the laboratory into a single molecule.

Recombinant DNA technology. Methods for transferring genes or groups of genes from one organism to another.

Recombinant. The new molecule, organism or cell that is the product of rDNA methods.

Recruitment. Addition of new individuals to a population by reproduction.

Recycling. The recovery, return and reuse of scrap or waste material for manufacturing or resource purposes. *Direct r.*, the use of a recovered material for manufacture of a similar product. *Indirect r.*, the use of recovered material for manufacturing a different product or one of less critical specification. *Thermal r.*, conversion of the waste into energy, with or without by-products.

Relative abundance. Proportional representation of a species in a sample or a community.

Repressor. A protein that regulates DNA transcription by preventing RNA polymerase from attaching to a DNA promoter site.

Reproductive isolation. Prevention of successful interbreeding between individuals of opposite sex. By definition such individuals belong to different species.

Resource. A substance or object required by an organism for normal maintenance, growth, and reproduction. If the resource is scarce relative to demand, it is referred to as a limiting resource. Non-renewable resources (such as space) occur in fixed amounts and can be fully utilized; renewable resources (such as food) are produced at a rate that may be partly determined by their utilization.

Restriction enzymes. Enzymes that cleave DNA at specific sites.

Runoff. The part of precipitation that flows towards the stream on the ground surface (surface runoff) or within the soil (subsurface runoff or interflow).

Salinization. The accumulation of salt in water or soil; it is a factor in desertification.

Secondary substances. Organic compounds produced by plants, not directly involved in metabolic pathways, that are implicated as chemical defenses.

Seepage. The slow movement of water or gas through the pores and fissures in soil, rock, etc., without the formation of definite channels.

Self-incompatibility. Inability to produce zygotes by union of gametes produced by same plant, hence on self-pollination seeds are not formed.

Senescence. Gradual deterioration of function in an organism leading to an increased probability of death; aging.

Shifting cultivation. Farming system where farmers move on from one place to another. The most common form is slash-and-burn agriculture: land is cleared by burning, so that crops can be grown. After a few years, soil fertility is reduced and the land is abandoned. A new area is cleared while the old land recovers its fertility. Slash-and-burn is practised in many tropical forest areas. The system works well while population levels are low, but where there is overpopulation, the old land will be reused before soil fertility has been restored. A variation of this system is rotational bush following.

Silage. The product obtained by a controlled anaerobic fermentation of green moist plant parts. Used as an animal foodstuff.

Silviculture. Management of forest land for timber.

Slash and burn. Simple agricultural method whereby the natural vegetation is cut and burned and the clearing then farmed for a few years until the soil loses its fertility, whereupon farmers move on and leave the area to regrow. It works well with a small, widely dispersed population, but becomes unsustainable with more people and is now a form of deforestation. Slash and burn is particularly inappropriate in tropical rainforest areas because the soil is already poor and once the trees are removed, it becomes still further impoverished.

Slurry. Form of manure composed mainly of liquids. Slurry is collected and stored on many farms, especially when large numbers of animals are kept in factory units. When slurry tanks are accidentally or deliberately breached, large amounts can spill into rivers, killing fish and causing eutrophication. Some slurry is spread on fields as a fertilizer.

Soft wheat. Any wheat which gives a flour more suited to the production of biscuits and pasta, but not suited for baking bread (cf. hard wheat).

Soil cohesion. The mutual attraction between soil particles as a result of molecular forces and moisture films.

Soil depletion. Decrease in soil quality over time. Causes include loss of nutrients caused by overfarming, erosion by wind, and chemical imbalances caused by acid rain.

Soil erosion. The wearing away and redistribution of the Earth's soil layer, caused by the action of water, wind, and ice, and also by improper methods of agriculture. If unchecked, it results in desertification. If the rate of erosion exceeds the rate of soil formation (from rock and decomposing organic matter), the land declines and eventually becomes infertile. The removal of forests or other vegetation often leads to serious soil erosion, because plant roots bind soil, and without them the soil is free to wash or blow away. The effect is worse on hillsides and there has been devastating loss of soil where forests have been cleared from mountain-sides. Improved agricultural practices such as contour ploughing help combat soil erosion. Windbreaks, such as hedges or strips planted with coarse grass, are valuable and organic farming can reduce soil erosion greatly.

Soil. The solid substrate of terrestrial communities resulting from the interaction of weather and biological activities with the underlying geological formation.

Solid waste management. The purposeful, systematic control of the generation, storage, collection, transport, separation, processing, recycling, recovery and disposal of solid wastes.

Solid waste. Any refuse or waste material, including semisolid sludges, produced from domestic, commercial or industrial premises or processes including mining and agricultural operations and water treatment plants.

Species. A group of actually or potentially interbreeding populations that are reproductively isolated from all other kinds of organisms.

Stability. Inherent capacity of any system to resist change.

Stabilizing factors. Factors that tend to restore a system to its equilibrium state; specifically, the class of density-dependent factors that act to restore populations to equilibrium size.

Standing crop. The total number of individuals of a given species alive in a particular area at any moment. It is sometimes measured as the weight (or biomass) of a given species in a sample section.

Stochastic. Referring to patterns resulting from random factors.

Stratosphere. The region of the upper atmosphere extending from the tropopause (8 to 15 km altitude) to about 50 km. The thermal structure is determined by its radiation balance and is generally very stable with low humidity.

Subsidence. Settling or sinking of the land surface due to many factors such as the decomposition of organic material, consolidation, drainage, and underground failures.

Surface albedo. The fraction of solar radiation incident on the Earth's surface that is reflected by it. Reflectivity varies with ground cover, and during the winter months it varies greatly with the amount of snow cover (depth and areal extent). Roughness of terrain, moisture content, solar angle and angular and spectral distribution of ground-level irradiations are other factors affecting surface albedo.

Sustainable. Capable of being continued indefinitely. For example, the sustainable yield of a forest is equivalent to the amount that grows back.

Sustained-yield cropping. The removal of surplus individuals from a population of organisms so that the population maintains a constant size. This usually requires selective removal of animals of all ages and both sexes to ensure a balanced population structure. Taking too many individuals can result in a population decline, as in overfishing.

Symbiosis. Intimate, and often obligatory, association of two species, usually involving coevolution. Symbiotic relationships can be parasitic or mutualistic.

Synecology. The relationship of organisms and populations to biotic factors in the environment.

Third World. Term originally applied collectively to those countries of Africa, Asia, and Latin America that were not aligned with either the Western bloc (First World) or Communist bloc (Second World). The term later took on economic connotations and was applied to the countries of the developing world.

Tissue culture. In vitro culture of cells obtained from tissues.

Topsoil. The topmost layer of soil; usually refers to soil that contains humus and is capable of supporting good plant growth.

Transpiration. Evaporation of water from leaves and other plant parts.

Trophic level. Position in the food chain determined by the number of energy-transfer steps to that level.

Trophic structure. Organization of the community based on feeding relationships of populations.

Trophic. Pertaining to food or nutrition.

Tropopause. The boundary between the troposphere and the stratosphere (about 8 km in polar regions and about 15 km in tropical regions), usually characterized by an abrupt change of lapse rate. The regions above the troposphere have increased atmospheric stability than those below. The tropopause marks the vertical limit of most clouds and storms.

Troposphere—The inner layer of the atmosphere below about 15 km, within which there is normally a steady decrease of temperature with increasing altitude. Nearly all clouds form and weather conditions manifest themselves within this region, and its thermal structure is caused primarily by the heating of the Earth's surface by solar radiation, followed by heat transfer by turbulent mixing and convection.

Vector. The agent used to transfer DNA into a host cell, e.g., plasmid or bacteriophage.

Water table. The depth below which the ground is saturated with water.

Weathering. Process by which exposed rocks are broken down on the spot by the action of rain, frost, wind, and other elements of the weather. It differs from erosion in that no movement or transportation of the broken-down material takes place. Three types of weathering are physical, chemical, and biological. They usually occur together.

Wilting capacity. The minimum water content of the soil at which plants can obtain water.

Xeric. Referring to habitats in which plant production is limited by availability of water.

Index

Acanthella klethra 558
Acetivibrio cellulolyticus 273
acid mine drainage (AMD) 533
— rain 2, 3, 37, 55, 84, 107, 143, 144, 168, 177, 190, 220, 239, 244, 342, 343, 383, 406, 477, 502, 512
activated sludge treatment 296
active carbon cycle 120, 122, 163
administrative charges 26
advanced control technology (ACT) 217
aerosol and gas experiment (SAGE) 158
aerosols 43, 99, 106, 115, 116, 117, 124, 125, 150, 152, 154-158, 161, 162, 163, 164, 243, 244, 530
agricultural mineralogy 530
— residues 198, 265
Agrobacterium-mediated transformation 441
agrobiodiversity 283
agroecology 10, 11, 14, 43
agroforestry 79, 132, 138, 197, 431, 432, 465, 467, 552
air emission taxes 26
alcohol fuels 200
alternate feedstocks 275
alternative fuels 132, 167, 470
— technologies 72, 315

amyliferous 278
animal draught power 347
— wastes 59, 197, 221, 265, 275, 343
appropriate energy technology (AET) 269
aquaculture 55, 312, 336, 379
aquatic biodiversity 79
aquifers 13, 46, 53, 55, 70, 71, 305, 533, 553
Asterionella 94
Azadirachta indica 379

Bacillus sphaericus 456
— *thuringiensis* (Bt.) 455
Bacteroides 273
basic control technology (BCT) 217
bed gasifiers 199
beneficiation 209, 210
best available technology (BAT) 217
bioconversion of agricultural residues 449
Biocultural diversity 280, 281
biodegradation 292, 294, 298, 302, 304, 308, 312, 320, 321, 322, 326, 327, 329, 330, 449
biodiesel 195, 278
biodiversity 14, 15, 18, 53, 77, 79, 85, 92, 94, 95, 96, 133, 166, 195, 200, 208, 249, 264, 278-284, 288, 292,

294, 298, 302, 304, 308, 312, 320, 321, 322, 326, 327, 329, 330, 337, 338, 346, 362, 363, 364, 367, 408, 412-417, 427-432, 439, 440, 441, 449, 466, 468, 472, 473, 480, 482, 485, 490, 494, 522, 523, 526, 545, 549, 551, 558, 575
biofertilizers 446
biogas 72, 139, 178, 183, 191, 192, 198, 199, 240, 254, 272-276, 278, 279, 288, 516
biogeochemical cycles 4, 105, 106, 162, 234
bioleaching 449
biological oxygen demand (BOD) 292
— pump 102
biomass-based power generation 140
biopesticides 446, 447
bioprocess engineering 445
biosphere 4, 12, 13, 43, 100-102, 106, 107, 110, 117, 120, 121, 138, 141, 146, 163, 165, 342, 383, 474, 480, 493, 505, 525, 526, 529, 573, 574
biospheric catastrophe 111, 112
— processes 109, 110
biotechnology 20, 64, 229, 230, 231, 239, 408, 411, 413, 422, 425, 426, 429, 440-453, 456, 466, 467, 468, 520, 575
blue water 74
bovine growth hormone 446
breeders rights 414
building-integrated photovoltaics 182

Chaoborus 95, 96
charismatic megavertebrates 358
chemical coal cleaning 242
chemically Active Fluid Bed (CAFB) 242
chipko movement 2, 485
chlorofluorocarbons (CFCs) 8, 146
clean production 231, 232, 234, 235, 236, 249, 260
clearcutting 75, 330, 545, 549
climate perturbation 111, 115
Clostridium 273
— *thermocellum* 273

cloud condensation nuclei (CCN) 158
co-generation power plants 203, 204
coastal upwelling 134
— zones 77, 101, 120, 126
command and control approach 24, 25, 27
— and control principle 24
communicable diseases 386, 449
community-based conservation 415
composting 264, 265, 292, 323, 328, 340, 429, 446
— toilets 77, 316
comprehensive development framework (CDF) 572
concept of sustainable development 112, 401, 473, 479, 482, 492
conservation ethic 409
— of Biodiversity 412, 415, 467, 473
conservation tillage 306, 440, 441
constructed wetlands 306, 316, 331, 535, 575
controlled biodegradation 292, 294
conversion technologies 241, 247
crop residues 27, 198, 199, 221, 277, 335, 440
Cymbastela hooperi 558

debt-for-nature swaps 505
decarbonized energy 236, 261
deforestation 3, 13, 18, 26, 30, 42, 72, 77, 80, 99, 101, 102, 130, 138, 196-198, 287, 393, 396, 399, 400, 438, 469, 479, 493, 503, 509, 513, 516, 523, 543, 550, 551
dematerialization 509
demographic imperative 481
demography 15, 508, 567, 574
desertification 3, 30, 40, 119, 176, 197, 263, 393, 416, 477, 503
development ethic 409
— wheel 285, 286
differentiated dues 378
disease resistance 356, 448
disturbance regulation 529
drip irrigation 71, 81, 251, 253
droughts 33, 34, 42, 46, 47, 66, 67, 100, 101, 149, 393, 441, 477

earth radiation budget experiment (ERBE) 158
— system 3, 4, 29, 30, 33, 34, 69, 102, 105, 164, 493
economic ethics 113
— theory 113
— transition 28, 500
ecosystem services 482, 488, 527, 528, 529, 574
effect 530, 533, 535, 556, 566
Eichhornia crassipes 536
El Niño 33, 35, 36, 100, 134, 136, 549
— Southern Oscillation (ENSO) 100
electrokinetics 327
emerging technologies 231, 292, 447
enabling mechanisms 445, 450
enclosed ecosystem (biosphere 2) 573
end-of-pipe remedial approaches 229
end-use approach 211, 212, 213
energy 227
— and equity 248
— choices 244
— conservation 181, 183, 222, 223, 240, 249, 291, 301, 372
— intensity 23, 212, 213, 240, 248, 514
— management 100
— management and conservation 28
— myths 245, 246, 247
— plants 277, 278
— sources 168
— storage 184, 186, 206
— storage technologies 184
— tetchnology programme 240
— transition 220, 221, 227
environment-development Interface 502
environmental assessment 478
— engineering 291, 294, 331
— ethics 112
— health 308, 310, 312, 387, 389, 407, 558
— impact assessment 165, 499
— management technologies 388
— planning 64, 96, 527
— sanitation 75, 263, 330

— technology 228, 229, 531
— technology verification (ETV) 531
environmentally sustainable development 286
estuaries 77, 126, 130, 307, 308, 312
ethics 112, 113, 364, 409, 410, 411, 416, 465, 467, 514
ethnoecology 11, 44
Eubacterium cellulosolvens 273
euro-EST 376
eutrophication 77, 79, 85, 89, 90, 92, 94, 95, 96, 216, 290, 344
excavation/disposal 327
exhaust gas recirculation 219
exorbitant profits myth 246

fairway dues 377, 378
farm forestry 187
— management 341, 426
farmers rights 413, 414
feed supplements 352, 354, 355, 363
fiduciary trust 113
fire regime 39, 548
firewood 72, 221, 258, 277, 417, 503
fisheries 47, 69, 70, 79, 95, 96, 136, 258, 330, 394, 400, 434, 500
flooding 47, 54, 61, 68, 70, 72, 84, 116, 120, 126, 149, 387, 486, 565
fluvial morphology 79
— systems 77
flywheels 184
food and nutritional security 49
— security 49, 57, 68, 74, 79, 250, 346, 408, 409, 412, 416-422, 427-429, 433, 437, 440-442, 465-468, 530
food-energy nexus 373, 375
forest fires 36, 37, 39, 42
— resources management 29
fossil energy 8, 9, 169, 171, 176, 210, 211, 226, 246
— fuels 3, 13, 18, 37, 104, 106, 116, 117, 121, 122, 124, 138, 140, 144, 147, 167, 175, 177, 180-184, 189, 195, 200, 201, 204, 210, 217, 220, 236, 241, 242, 248, 257, 260, 338, 345, 399, 470, 482, 495, 509-513, 517
freshwater management 81

fuel cell 167, 185, 186, 214, 215, 226, 236, 237, 244, 260, 514
— efficient vehicles 377, 503
fuelwood 27, 41, 103, 139, 176, 180, 187, 188, 197, 198, 204, 240, 244, 399, 400, 472, 508

gas dome 317
— regulation 529
genetic engineering 245, 325, 426, 443
genetic resources conservation 409, 413
geochemistry and sustainable development 493
geographic positioning system GPS-based agriculture 453
geographical information system (GIS) 453
geosphere 163, 505
geothermal pyrolysis 293
global warming 2, 7, 43, 99, 101, 106, 115, 120, 122, 126, 130, 133, 147, 148, 162, 163, 164, 177, 180, 183, 210, 220, 222, 241, 243, 260, 341, 364, 394, 395, 505, 512, 565
— warming potential (GWP) 8
globalization 7, 19-21, 238, 279, 280, 405, 406, 429, 487, 568, 569, 570, 571, 572
GMOs 325, 440
grazing 39, 41, 75, 197, 356, 399, 416, 431, 504, 521, 540
green taxes 113
— water 74
— greenhouse effect 7, 8, 13, 18, 38, 44, 109, 111, 115, 116, 118, 142, 160, 161, 164, 201, 208, 263, 342, 343, 383, 495, 502
— gases (GHG) 8
groundwater 45, 46, 47, 53, 55-59, 63, 65, 66, 71, 72, 74, 76, 79, 80, 84, 97, 266, 298, 303, 305, 314, 319, 324, 325, 328, 330, 331, 370, 520

harvesting 40, 62, 84, 198, 200, 326, 327, 469, 498, 537, 539, 540, 542, 544

hazard maps 29, 30, 34
hazardous wastes 234, 397, 450
haze shelters 37
HDPE (high density polyethylene) 329
Helianthus tuberosus 302
herbicide-resistant crops 464
high-rate anaerobic digestion 301
high-response varieties (HRVs) 434
high-tech farming 439
hoarding myth 245
Homo sapiens 14, 473
horizontal integration 74
host plant resistance (HPR) 456
humane urban development 390
hunting tourism 361, 362, 365
hybrid fuel engines 254
hydrocarbon reserves 168
hydrogen (H_2) 28, 42, 157, 167, 184-186, 199, 214, 215, 236, 237, 272-274, 278, 317, 343, 501, 513, 514, 517
— economy 237
hydrologic 47, 48, 55, 60, 66, 513
hydrological cycle 4, 47, 63, 65, 75, 102, 103, 107, 208, 472
— forecasting 257
hydropower 65, 71, 116, 166, 177, 239, 244, 486, 493, 512
hypercar 23

ICDPs 415
improved cereals 444
in-house water conservation 61
industrial ecology 22, 488, 499, 523
— metabolism 499, 500, 509, 520
integrated conservation and development projects 415
— disease management (IDM) 464
— pest management (IPM) 463
— resource use 346
— rural development (IRD) 287
intellectual property right 12, 238, 282, 338, 414, 431, 450, 452
intensive animals 346
— fish culture 379
intergenerational equity 317, 416, 481

Index

internally sustainable development 285
irrigation farming 49

L-City 372
land cover conversion 106
— tenure patterns 498
landfills 55, 292, 313, 316, 317, 326, 327, 328
landslides 29, 384, 393, 565
latitudinal contrasts 86, 95
Lemna 276, 291
Leucaena 379
— *leucocephala* 191
life-cycle analysis 235
linear fresnel reflectors (LFR) 270
liquid fuels cleaning 242
lithosphere 29
livestock farming 334, 335, 364

macroenvironmental issues 148
malnutrition-infection complex 51
managed forests 138
mangrove forest 130, 131, 132, 162, 164, 490
— plants 131, 140
marine pharmaceuticals 557
MARKAL 222, 223, 224, 225
mechanized arable farming 347
megacities 29, 30, 84, 287, 366, 376, 406, 565
methane 8, 9, 10, 13, 14, 28, 38, 39, 42, 72, 73, 76, 102, 106, 107, 109, 111, 115, 119, 123, 124, 128, 144, 146, 157, 160, 199, 208, 211, 214, 215, 222, 243, 244, 272-278, 295, 310, 313, 316, 317, 342, 343, 352, 514, 517
Methanobacterium thermoautotrophicum 274
Methanosarcina 273, 274
microfiltration 292, 293, 331
micropollutants 77, 111, 112
mining 46, 47, 55, 59, 71, 77, 79, 166, 209, 210, 213, 271, 289, 290, 319, 493, 494, 504, 532, 533, 551, 564, 575

mitigation options 105, 139
mixed liquor suspended solids (MLSS) 297, 298
monopoly myth 246
Mt. Pinatubo 115, 158, 163
multi-purpose livestock 356
multijunction photovoltaics 237
multiple-use forests 549
municipal solid waste management (MSWM) 490
myths about urban agriculture 382
— that we may run out of oil 246

nanotechnology 229
natural disasters 31, 33, 34, 477
— conservation 361, 364, 365
— -centered recycling 316
— step progamme 488
neighbourhood sewage wall 316
neoclassical economics 147
new forestry 548
nitrogen fixation 447, 449, 494
nitrogen leakage 345
no-till farming 440
noise taxes 26
non-governmental organization (NGO) 426
noncommunicable 387
nonpoint source (NPS) pollution 304
nuclear energy 71, 72, 189, 247, 493, 495, 503
— power 71, 161, 171, 178, 190, 239, 244, 495, 501, 512
— power reactors 30
nutritionally-fortified food 444

oil shale 242, 247
oleochemical industry 517, 518
onchocerciasis (river blindness) 77
online monitoring 292
opencast mining 208, 209
optimization 9, 64, 81, 225, 294, 297, 299, 301, 302, 331, 575
organic agriculture 264, 283
— recycling 313
Oryza 414
oxygen activated sludge 301

ozone depletion 2, 99, 110, 114, 116, 118, 144
— -depleting substances 471

paradox of the plankton 53
particle bombardment transformation 441
particulates 37, 79, 210, 226, 242, 328, 398, 567
passive treatment schemes 534
pathology of natural-resource management 488
pelagic pollution 307
people-wildlife conflicts 358
peri-urban agriculture 380, 382
PET (polyethyleneterephthalate) 328
photobiological systems 185
photoelectrolysis 185
photovoltaic collectors 177
— driven pumps (PVP) 205
phytoremediation 303, 325-327, 330, 331
pitcher irrigation 70, 71
plant sewage bed 317
— viral vectors 441
plantation forests 545
plants 11, 12, 30, 67, 68, 72, 74, 76, 81, 94, 95, 100, 102, 106, 129, 130, 131, 139, 141, 143, 144, 146, 161, 178, 181, 183, 198-210, 217, 238, 240, 242, 248, 252, 275, 277-283, 295, 296, 299, 307, 313, 315, 316, 325, 330, 339, 352, 354, 370, 385, 387, 396, 398, 409, 413, 417, 433, 437, 441, 442, 445, 455-464, 470, 495, 501, 512-514, 517, 535, 538, 540, 541, 545, 553, 554, 557, 573, 574
Plasmodium falciparum 558
plastic recycling 329
pollution charges 27, 28, 230
polyaromatic hydrocarbons (PAHs) 325
Portieria hornemanni 558
precision farming 441, 453, 467
primary production 90, 91, 95, 96, 97, 107

principles of sustainability 475
prior informed consent (PIC) 563
product taxes 26
protected areas 361, 411, 415, 416, 550
proton-exchange membrane (PEM) fuel cell 237
Pseudomonas aeruginosa 304
PV-systems 205, 250
— -technology 205

rainwater harvesting 84
rational-use-of-energy (RUE) 201
recombinant-DNA technologies 414
remote sensing (RS) techniques 138, 453
renewable energy sources 9, 104, 128, 133, 169, 175, 178, 182, 187, 203, 244, 501, 511, 512
— technologies 105, 139, 183, 190, 192, 227
restraints on resource use 539
reverse osmosis 317, 318
Rhizophora 139
rhizosphere 326, 327
risk assessment 33, 112, 555, 564
river blindness 76
rotating disc filter 317
Ruminococcus 272, 273
rural cooking stoves 188
— -urban linkages 369
— -urban interfaces 403

sacchariferous 278
Salix 326
salting 69
saltwater intrusion 55
savannahs 39, 40, 41, 42, 208
schistosomiasis 77, 367
scientific ethic 409
sewage sludge 76
shifting cultivation 416, 432, 467, 485, 516, 552, 553
Shorea robusta 101, 487
short rotation forestry plantation 138, 328
sidestream dewatering 301

silting 68, 69, 210
site-specific and precision farming (PF) 453
sludge digestion 292, 295
social forestry 187, 486
soil erosion 68, 79, 102, 198, 263, 330, 393, 394, 396, 399, 416, 440, 441, 493, 503, 552, 553, 576
— flushing 327
— washing 320, 327
solar cells 182, 183, 191, 370, 513
— cooking 277
— energy 72, 183, 185, 189-191, 232, 250, 260, 316, 513
— home systems (SHS) 205
— homes 182
— pump 72
solidification 327
solubility pump 102
Sphagnum 534, 535
spinning and weaving 265
sprinkler irrigation 252
stabilization 117, 127, 133, 161, 222, 315, 326, 327, 443
sterile insect technique 449, 495
stochasticity 550
Stress tolerance 356
stripping 317
subsidies and market creation 26
superconductivity 184
— ultracapacitors, pumped hydro a 184
supply management 81
sustainability 5, 79, 230, 259, 263, 264, 327, 335, 341, 346, 349, 365, 373, 403, 409, 411, 416, 427, 440, 454, 468-476, 483, 485, 486, 488, 490, 495-501, 505, 514, 521-523, 527, 541, 544, 561, 562
sustainable agricultural management 10
— economies 23
— food production 31, 455
— food systems 10, 11
— forest management 327
— management 6, 79, 164, 367, 470, 488, 525, 526, 556

— production 6
synthetic fuels 190, 242, 244
systems analysis 193, 507, 509, 510, 522, 523
systemic/sequential concept 301

taiga 42
tailing ponds 209, 210
tapered aeration 298, 301
tax differentials 25, 26
technological change 99, 222, 228, 244, 245, 479, 506, 520
— dynamism 238
technology assessment 116, 223, 261, 505
— transition 500
termites for wood fibre degradation 292
trilogy of sustainability 476
thermophilic anaerobic digestion 301
thin-film 237
Thiobacillus ferrooxidans 533
tiny textile mills 266
total assessment audit (TAA) 554
tourism 100, 216, 271, 312, 361, 362, 365, 480, 504, 573
trade-related intellectual property rights agreement 238
— -related investment measures agreement (TRIMs) 238
tragedy of the commons 481
transgenic plants 352, 455-462
transnational corporations (TNCs) 271
transportation 3, 45, 69, 73, 103, 116, 123, 131, 167, 177, 184-186, 189, 212, 213, 216, 219, 220, 227, 231, 291, 294, 373, 376, 377, 383, 384, 386, 389, 480, 500, 526, 537, 562
treadle (pedal) pump 266
trenchless technologies 530
tropical lakes 85-97
— limnology 85, 97
— rain forests 40, 42
— soils 74
tropopause 108, 109, 114, 115, 151
two-stage anaerobic digestion 301

Typha latifolia 535

U-city 372, 373
ultra-violet 292
ultracapacitors, 184
URAs 421
urban agriculture 379, 380, 382, 405, 407
— bias 372
— ecosystem 373, 374
— transition 391
— wastes 290, 306
Uruguay Round Trade Agreements (URA) 421
user charges 25, 26
UV–B radiation 115

vaporization 317
vegetable farming 380
verification system 532
vertical integration 74
virus-free seedlings 446
volatile organic compounds (VOCs) 160
volcanic aerosols 115, 161, 243
— eruptions 29, 39, 115

waste disposal 3, 75, 93, 96, 119, 244, 259, 291, 292, 313, 328, 331, 341, 376, 387, 396, 398, 400, 474, 503
— emission taxes 26
— management 287, 290, 294, 328, 329, 332, 388, 490, 491, 507, 516
— reclamation 291
waste-to-energy (WTE) facilities 329
wastewater aeration 293
water demand management 63
— emission taxes 26
— erosion 440
— hyacinth 70, 536, 537, 538
— management 29, 59-63, 74, 76, 81, 82, 84
— resource management 59, 62, 63, 74, 82
— wastage and overuse 80
waterlogging 69, 70
weed management 464
wetlands 57, 76, 79, 105, 116, 121, 149, 281, 306, 316, 327, 331, 380, 527, 529, 534, 535, 536, 575
wildlife management 100, 333, 357, 358, 363, 364
wind pumps 268